Advances in Intelligent and Soft Computing   129

Editor-in-Chief: J. Kacprzyk

Advances in Intelligent and Soft Computing

Editor-in-Chief

Prof. Janusz Kacprzyk
Systems Research Institute
Polish Academy of Sciences
ul. Newelska 6
01-447 Warsaw
Poland
E-mail: kacprzyk@ibspan.waw.pl

Further volumes of this series can be found on our homepage: springer.com

David Jin and Sally Lin (Eds.)

Advances in Multimedia, Software Engineering and Computing Vol.2

Proceedings of the 2011 MSEC International Conference on Multimedia, Software Engineering and Computing, November 26–27, Wuhan, China

 Springer

Editors

Prof. David Jin
International Science & Education
Researcher Association
Wuhan Section
Special No.1, Jiangxia Road of Wuhan
Wuhan
China, People's Republic
E-mail: dayang1818@163.com

Dr. Sally Lin
ISER Association
Guangshan Road, Baoli Garden 1
430072 Wuhan
China, People's Republic
E-mail: 1375170731@qq.com

ISBN 978-3-642-25985-2   e-ISBN 978-3-642-25986-9

DOI 10.1007/978-3-642-25986-9

Advances in Intelligent and Soft Computing   ISSN 1867-5662

Library of Congress Control Number: 2011943102

© 2011 Springer-Verlag Berlin Heidelberg

Typeset by Scientific Publishing Services Pvt. Ltd., Chennai, India

Printed on acid-free paper

5 4 3 2 1 0

springer.com

Preface

International Science & Education Researcher Association (ISER) puts her focus on studying and exchanging academic achievements of international teaching and scientific research, and she also promotes education reform in the world. In addition, she serves herself on academic discussion and communication too, which is beneficial for education and scientific research. Thus it will stimulate the research interests of all researchers to stir up academic resonance.

MSEC2011 is an integrated conference concentrating its focus upon Multimedia, Software Engineering, Computing and Education. In the proceeding, you can learn much more knowledge about Multimedia, Software Engineering, Computing and Education of researchers all around the world. The main role of the proceeding is to be used as an exchange pillar for researchers who are working in the mentioned field. In order to meet high standard of Springer, AISC series, the organization committee has made their efforts to do the following things. Firstly, poor quality paper has been refused after reviewing course by anonymous referee experts. Secondly, periodically review meetings have been held around the reviewers about five times for exchanging reviewing suggestions. Finally, the conference organization had several preliminary sessions before the conference. Through efforts of different people and departments, the conference will be successful and fruitful.

MSEC2011 is co-sponsored by International Science & Education Researcher Association, Beijing Gireida Education Co.Ltd and Wuhan University of Science and Technology,China. The goal of the conference is to provide researchers from Multimedia, Software Engineering, Computing and Education based on modern information technology with a free exchanging forum to share the new ideas, new innovation and solutions with each other. In addition, the conference organizer will invite some famous keynote speaker to deliver their speech in the conference. All participants will have chance to discuss with the speakers face to face, which is very helpful for participants.

During the organization course, we have got help from different people, different departments, different institutions. Here, we would like to show our first sincere thanks to publishers of Springer, AISC series for their kind and enthusiastic help and best support for our conference. Secondly, the authors should be thanked too for their enthusiastic writing attitudes toward their papers. Thirdly, all members of program chairs, reviewers and program committees should also be appreciated for their hard work.

In a word, it is the different team efforts that they make our conference be successful on November 26–27, Wuhan, China. We hope that all of participants can give us good suggestions to improve our working efficiency and service in the future. And we also hope to get your supporting all the way. Next year, In 2012, we look forward to seeing all of you at MSEC2012.

September, 2011   ISER Association

Organizing Committee

Honor Chairs

Prof. Chen Bin   Beijing Normal University, China
Prof. Hu Chen   Peking University, China
Chunhua Tan   Beijing Normal University, China
Helen Zhang   University of Munich, China

Program Committee Chairs

Xiong Huang   International Science & Education Researcher
     Association, China
LiDing   International Science & Education Researcher
     Association, China
Zhihua Xu   International Science & Education Researcher
     Association, China

Organizing Chair

ZongMing Tu   Beijing Gireida Education Co.Ltd, China
Jijun Wang   Beijing Spon Technology Research Institution,
     China
Quanxiang   Beijing Prophet Science and Education Research
     Center, China

Publication Chair

Song Lin   International Science & Education Researcher
     Association, China
Xionghuang   International Science & Education Researcher
     Association, China

International Committees

Sally Wang	Beijing normal university, China
LiLi	Dongguan University of Technology, China
BingXiao	Anhui University, China
Z.L. Wang	Wuhan University, China
Moon Seho	Hoseo University, Korea
Kongel Arearak	Suranaree University of Technology, Thailand
Zhihua Xu	International Science & Education Researcher Association, China

Co-sponsored by

International Science & Education Researcher Association, China
VIP Information Conference Center, China

Reviewers of MSEC2011

Chunlin Xie	Wuhan University of Science and Technology, China
LinQi	Hubei University of Technology, China
Xiong Huang	International Science & Education Researcher Association, China
Gangshen	International Science & Education Researcher Association, China
Xiangrong Jiang	Wuhan University of Technology, China
LiHu	Linguistic and Linguidtic Education Association, China
Moon Hyan	Sungkyunkwan University, Korea
Guangwen	South China University of Technology, China
Jack H. Li	George Mason University, USA
Marry Y. Feng	University of Technology Sydney, Australia
Feng Quan	Zhongnan University of Finance and Economics, China
PengDing	Hubei University, China
XiaoLie Nan	International Science & Education Researcher Association, China
ZhiYu	International Science & Education Researcher Association, China
XueJin	International Science & Education Researcher Association, China
Zhihua Xu	International Science & Education Researcher Association, China
WuYang	International Science & Education Researcher Association, China

QinXiao   International Science & Education Researcher
                        Association, China
Weifeng Guo   International Science & Education Researcher
                        Association, China
Li Hu   Wuhan University of Science and Technology,
                        China
ZhongYan   Wuhan University of Science and Technology,
                        China
Haiquan Huang   Hubei University of Technology, China
Xiao Bing   Wuhan University, China
Brown Wu   Sun Yat-Sen University, China

Contents

A Quick Emergency Response Model for Micro-blog Public Opinion Crisis Oriented to Mobile Internet Services: Design and Implementation

Hanxiang Wu and Mingjun Xin

School of Computer Engineering and Science, Shanghai University, Shanghai 20072, China
{newwhx,xinmj}@shu.edu.cn

Abstract. On the basis of discussing the monitoring mechanism for public opinions in China, we construct a public opinions corpus on the content of micro-blog, and classify the blogs into four levels named red, orange, yellow and green by their sentiment intensities. In this paper, it firstly describes the micro-blog cases base and emergency response plans library based on web ontology language (OWL). Then, it presents a quick emergency response model (QERM) for the micro-blog public opinions crisis oriented to mobile Internet services. Furthermore, we continue to research on how to update cases under the subjects and quick response processes for micro-blog case library. Finally, we design a test experiment which shows some merits of QERM in the time cost. Thus, it will meet the quick emergency response demand on the micro-blog public opinions crisis under Mobile Internet environment.

Keywords: public opinion crisis, emergency response, CBR, Mobile Services.

1   Introduction

Micro-blog is a kind of blogging variants arising under mobile Internet services in recent years, and also becomes one of the typical applications for mobile services. It gains more and more attention and recognition for its short format and real-time characteristics, and is becoming an important platform for public opinion expression. In Wikipedia, micro-blog is described as "a broadcast medium in the form of blogging allows users to exchange small elements of content such as short sentences, individual images, or video links [1]." Because of the content of micro-blog is shorter (generally no more than 140 chars or Chinese words), the transmission speed among users is faster, and the expression means are more variant and free.

The "2010 third-quarter Assessment Analysis Report of China's Response capacity to Social public opinion", published in October 2010 by Shanghai Jiaotong University, claimed that micro-blog was becoming an important channel for enterprises and individuals to respond to public opinion. In 2010, from the event of 'Yihuang self-immolation caused by demolition' in Jiangxi province, the protagonist Zhong Rujiu registered a micro-blog account and made a live publication about the incident's development. Many blogs written by Zhong were reproduced by many net friends to be a hot topic in micro-blog network. During the Japan earthquake happened in 2011, some rumors were widely spread over the network and caused a rush of salt, which

D. Jin and S. Lin (Eds.): Advances in MSEC Vol. 2, AISC 129, pp. 1–7.
springerlink.com   © Springer-Verlag Berlin Heidelberg 2011

claimed that because of the contaminated sea water, the production of sea salt was unhealthy by nuclear radiation. Therefore, new challenges to discover public opinion trends are brought to the government monitoring department. At present, research on micro-blog for public opinion in China has just started, and sophisticated systems and applications are lacked. Especially, there are not enough experience and integrated emergency response framework on how to handle public opinions crisis quickly.

2   Related Work

Ontological knowledge representation is a kind of explicit description about the concept and the relationship in some Web service applications, which could provide a syntax or semantic standard for communication between human and computers and improve system reliability and knowledge acquisition capacity [2]. Web Ontology Language (OWL) is a part of series W3C web-related and expanding standard, which takes with strong representation and reasoning ability. OWL provides three increasingly expressive sublanguages (OWL Lite, OWL DL and OWL Full) designed for use by specific communities of implementers and users.

Emergency response is an extreme important stage during the process of dealing with emergencies [3] [8] [9]. The result of response would directly influence the quantity of casualties and the degree of property loss and environment damaging [5] [6] [7]. Wang believes that emergency response relies on the successful execution of one or more contingency plans, often managed by a command and control center [4]. A common approach is to use the key technology of case-based reasoning (CBR) to integrate experts' knowledge and emergency response cases.

3   OWL-Based Micro-blog Cases and Response Plans Description

3.1   Subject Class Description

The *Subject* Class is defined by two aspects: for one thing, according to the content of public opinion, the subjects are classified into political, economic, cultural, social and others as the first level of classification. For the other thing, it establishes different child subject categories by extracting keywords from micro-blog texts according to micro-blog text under each first-level classification. The structure of subject class is shown in Fig. 1.

As listed in Fig.1, the class *Category* is the top Class, and classes *Polity*, *Economy*, *Society*, *Culture* and *Others* are its subclasses, which represent different public opinion categories respectively. Each public opinion classification has many different subclasses except the *Other* class. In default, micro-blogs in the *Other* class are unclassified and the reasoning tool would select blogs in the *Other* class to classify into other different public opinion categories.

Each subject class has two data type attributes named 'start_time' and 'keywords' inherited from its parent class. The reasoning mechanism will decide the blogs' categories by these two attributes. Besides, there are three other attributes, which called 'ID' for a unique number in the library, 'opinion_grade' for representing the subject's opinion grade and 'panID' for connecting the response plan separately.

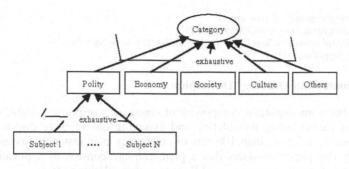

Fig. 1. The Structure of Subject Class

3.2 Micro-blog Class Description

Description of *Micro-blog* content is a knowledge representation of micro-blog information. In this paper, it defines a Micro-blog class as a blueprint of micro-blog, and treats each real micro-blog text as an instance or individual of Micro-blog class. According to the guide of OWL, the concept of attribute is defined as a binary relation, which could be specified a number of ways to restrict like the domain and range. The Micro-blog class has many attributes include data type and object type to describe the general fact of blog instances and relationship between with Category class. Part of attributes of the Micro-blog class is listed in Table 1.

Table 1. Part of Attributes of Micro-blog Class

Attribute Name	Attribute Type	Attribute Description
reference_from	object type	point at the referenced or reproduced blog
reference_at	data type	point at the referenced blog's url
blog_date	data type	published time
meta_information	data type	including provider, client type and so on
blog_keywords	data type	keywords of micro-blog

Class *Micro-blog* include two object type attributes 'reference_from' and 'belong_to' and eight data type attributes: 'blog_ID', 'author', 'date', 'meta_information' and 'content'. The object type attributes reveal relations among instances of class *Micro-blog* and *Category*. 'reference_from' is used to point at a referenced micro-blog, and 'belong_to' to point at an individual of *Category* class which the micro-blog belongs to. The data type attribute 'meta_information' includes the Micro-blog Provider like 'sina', publication client type like web or mobile and IP information. As described above, some definition program of class *Micro-blog* is shown as follow.

```
<owl:Class rdf:ID="MicroBlog"/>
<!-- object type attribute: reference_from -->
<owl:ObjectProperty rdf:ID="refence_from">
    <rdfs:domain rdf:resource="#MicroBlog"/>
    <rdfs:range rdf:resource="#MicroBlog"/>
</owl:ObjectProperty>
<!--data type attribute: date -->
```

```
<owl:DatatypeProperty rdf:ID="date">
    <rdfs:domain rdf:resource="#MicroBlog"/>
<rdfs:range rdf:resource="http://www.w3.org/2001/XMLSchema#dateTime"/>
    </owl:DatatypeProperty>
```

3.3   Response Plan Ontology Descripion

Response plan is an important component of emergency response system, which is a process template including formulating and executing one or more disposal options. After the analysis of more than 100 sets of emergency plan instances and some reference papers, this paper considers that a plan template consists of application scope, organizational structure, resource, workflow and other relative content.

As discussed above, it describes organization, resource, event and workflow as an entity class respectively in this paper. In the definition of class plan includes attributes like '*planID*' for a unique number, '*planAim*' for the plan's aim and '*planPrinciple*' for the principle of formulating the plan. Some attributes of plan ontology are detailed in table 2.

Table 2. Attributes of Plan Ontology

	Name	Type	Description
	organization	object type	organization structure
	resource	object type	resource need
Plan	event	object type	event
	workflow	object type	workflow
	tag	date type	resource name
	quantity	date type	quantity
Resource	status	date type	status, like 'ready'
	eventType	date type	event type
Event	eventSummary	date type	event summary
	eventLevel	date type	event level
...	...	...	...

The response plan ontology consists of class *Plan, Organization, Resource, Event* and *Workflow*. The organization in the plan means that those who directly response for execution of the whole plan, and the one in the *Workflow* that people who for one special task execution. The event levels in the Event class are defined according to the National Accidents Classification Standards as particularly significant (I level), major (II level), large (III level) and general (IV level).

4   The Implementation of QREM

Emergency response is an information sharing process. The QREM works on the R5 model of CBR based on case base and response plan library, which is driven by topic tracking and OWL reasoning mechanism. The response engine is composed of topic tracking, case-based reasoning and the automatically updated case base and response plan library etc. The workflow of QREM is shown in Figure 2.

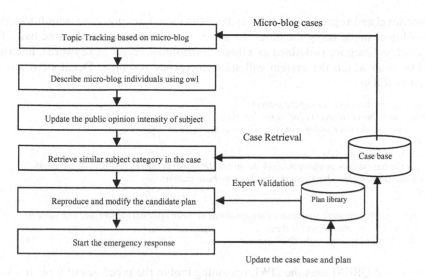

Fig. 2. The Workflow of QREM

4.1 Keywords-Based Topic Tracking

The purpose of topic tracking based on micro-blog keywords is to make the instances in the subjects' categories more affluent so as to get a more accurate response. Its main idea is described as follows:

1) *Sort the cases by their sentiment intensity from the big to small in one subject. And select N micro-blog cases as seed cases;*
2) *Extract the key attributes of the seed cases;*
3) *Track the micro-blog with the url address value of attribute 'reference_at' until the address is null;*
4) *If the publication client is mobile, then get the base station position through the mobile IP stored in the 'meta_information' attribute; And continue tracking the similar micro-blog examples around the base station;*
5) *Handle the blog examples by semantic analysis and sentiment intensity computing.*
6) *Decide the examples category using reasoner*
7) *Update the public opinion intensity of the subject category*

The micro-blog sentiment intensity is implemented by using the *public opinion classification algorithm* presented by our research team. The public opinion intensity of one subject category is calculated by the linear addition of individual sentiment intensity for each gathered micro-blog.

Case retrieval and reproduction is an important part of emergency response system based on case base. In this topic tracking model, it makes use of subject keywords and public opinion intensity to retrieve the case base.

4.2 The Reasoning Process for QERM

The quick emergency response model is based on case base and response plan library. The most important part is the rescanning engine which consists of topic tracking and case-retrieval and reproduction subsystems. The topic tracking subsystem tracks new micro-blog example using the blog case as seeds, and then updates the case base. The subject class Category is defined as a three-dimensional vector :< keywords, intensity, planID>, with which the system will start emergency response. Detail steps are described as follows:

```
//  extract the key attributes of the subject
while  the value of 'reference_at' is not null then
            tracking the micro-blog with the address at the value of 'reference_at'
end
if type is mobileType then
      get the base station position through the mobile IP stored in the 'meta_information' attribute;
track the similar micro-blog examples around the base station;
end
...
//  computing the semantic similarity of the keywords between optional subject and new subject
if similarity >  the threshold TH then
          decide the blog case as the final optional
end
```

The proposed QREM uses the OWL reasoning tool in the processes of topic tracking, case retrieval and plan reuse to implement the case base and plan library update. It provides a quick response to the micro-blog public opinion, which can assist the government monitoring department and experts to emergency incidents.

5   Experiment and Results Analysis

To test the performance of the proposed model, we design a simulation experiment using the case base as the meta data set, and then it chooses subjects 'Guo MM event', 'grab salt incident', and '7.23 high-speed rail event' separately as the test subjects. The experiment takes the time as the cost target value in which the system gets a reliable plan as the test result. The results are shown in Fig. 3.

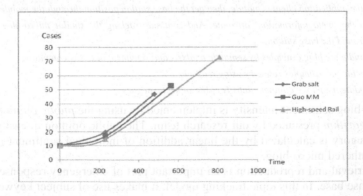

Fig. 3. Results of Experiment

It can be seen from Fig.3 that the proposed QREM has some merits in the response time (about 15 minutes). It could meet the quick response demand in the micro-blog public opinion emergency event. However, the proposal model has some disadvantages that the performance of topic tracking is not enough better. Thus, this will be the future work for our research team to be continued.

6   Conclusion

Recently, the Micro-blog as a new media network service is becoming an important channel for people to get and publish their information and ideas. With that the micro-blog public opinion events discussed in this paper, it continues to study the quick micro-blog emergency response model (QREM) by using OWL reasoning tools on the base of lab's research result. The QREM is driven by topic tracking, and approached by OWL reasoning mechanism. On the other hand, its core engine is composed of CBR-based topic tracking, the case base and response plan library. The test experiment proves the superior of the quick emergency response model, which also provides a better technical support for government monitoring department to handle the emergency public opinion incident quickly and successfully.

Acknowledgements. Our research work is supported by National Natural Science Foundation of China (Project Number. 61074135), Shanghai Creative Foundation project of Educational Development (Project Number. 09YZ14), and Shanghai Leading Academic Discipline Project (Project Number.J50103), Great thanks to all of our hard working fellows in the above projects.

References

1. Wikipedia, http://en.wikipedia.org/wiki/Micro-blog
2. Web Ontology Language Guide, http://www.w3.org/TR/2004/REC-owl-guide-20040210/
3. Zimin, Z., Ying, Z., Xi, M.: Emergency Response Information Model Based on Information Sharing (Part I): Model Definion. China Safety Science Journal 20(8), 154–160 (2010)
4. Wang, W., Meng, F.: Research on Ontology-based Emergency Response Plan Template. Computer Engineering 32(19), 170–172 (2006)
5. Jennex, M.E.: Modeling emergency response sysetems. In: 40 th Annual Hawaii International Conference on System Sciences, Hawaii, USA, pp. 22–29 (2007)
6. Mendonca, D., Beroggi, G.E.G., Wallace, W.A.: Evaluating Support for Improvisation in Simulated Emergency Scenarios. In: Proc. of the HICSS (2003)
7. Dyer, D., Cross, S.: Planning with Templates. IEEE Intelligent Systems 20(2) (2005)
8. Dokasa, I.M., Karrasb, D.A., Panagiotakopoulosc, D.C.: Fault tree analysis and fuzzy expert systems: Early warning and emergency response of landfill operations. Environmental Modelling & Software 24(1), 8–25 (2009)
9. Liao, Z., Liu, Y.: Emergency plan system for emergency pollution incident emergency response on the base of case-based reasoning. Environmental Pollution and Control 31(1), 86–89 (2009)

Effectiveness Evaluation on Material Modularity Storage and Transportation by Grey-Analytical Hierarchy Process

HuiJun Yang[1,2], Jun Meng[2], and YanXia Liu[2]

[1] School of Management, Tianjin University, Tianjin, P.R. China
[2] Automobile Transport Command Department, Military Transportation University, Tianjin 300161. P.R. China

Abstract. The grey-analytical hierarchy process(GAHP) is a method for effectiveness evaluation in decision making activity, and material modularity storage and transportation is a trend in logistics services. We may seek a optimum plan for material modularity storage and transportation based on the GAHP.

Keywords: Effectiveness evaluation, material modularity storage and transportation, grey-analytical hierarchy process.

1 Introduction

grey-analytical hierarchy process is the combination of grey system theory and analytical hierarchy process. in detail, it means that the value number of decision-making in different layer is based on grey system theory. It is useful for decision makers to evaluate the effectiveness of material modularity storage and transportation on GAHP.

2 Computation Method[2]

Firstly, to set up step-by-step hierarchy structure of evaluated object. To form step-by-step hierarchy structure shown as figure 1[1], in which the meaning of one layer' elements don't intersect, and elements relations in neighboring lower and above layer like ones between farther and son. The bottom elements are the evaluation index to seek.

Secondly, to computer the combined weight value of elements in bottom layer by experts on the weight value computation of lower layer versus above one, namely $W = (w_1, w_2, \cdots w_n)^T$.

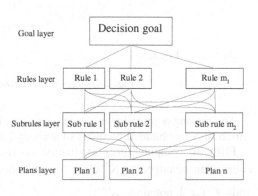

Fig. 1. Layer structure of evaluation system

D. Jin and S. Lin (Eds.): Advances in MSEC Vol. 2, AISC 129, pp. 9–15.
springerlink.com   © Springer-Verlag Berlin Heidelberg 2011

Thirdly, to seek matrix $D_{ij}^{(A)}$ of evaluated index value.

$$D_{ij}^{(A)} = \begin{pmatrix} d_{11}^{(A)} & d_{12}^{(A)} & \cdots & d_{1i}^{(A)} \\ d_{21}^{(A)} & d_{22}^{(A)} & \cdots & d_{2i}^{(A)} \\ \cdots & \cdots & \ddots & \\ d_{j1}^{(A)} & d_{j2}^{(A)} & \cdots & d_{ji}^{(A)} \end{pmatrix} \tag{1}$$

$d_{11}^{(A)}$ express A^{th} element evaluation index value matrix of evaluated J given by judger. The weight value may be gained through added value-average method on different importance among experts

Fourth, to confirm the evaluation grey-class which means to confirm the class number, the grey number and whitenization weighted function. The normal whitenization weighted functions are as follows:

(1) The first class (up), the grey number is $\oplus \in [d_1, \infty)$, and its whitenization weighted functions is shown as figure 2.

(2) The second class (middle), the grey number is $\oplus \in [0, d_1, 2d_1]$, and its whitenization weighted functions is shown as figure 3.

(3) The third class (down), the grey number is $\oplus \in [0, d_n, d_2]$, and its whitenization weighted functions is shown as figure 4.

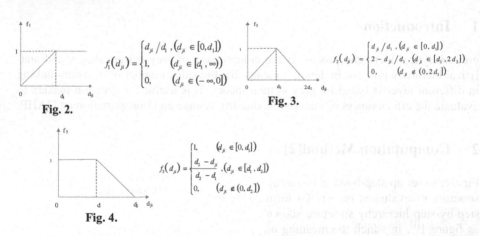

Fig. 2.

$$f_1(d_{ji}) = \begin{cases} d_{ji}/d_1, & (d_{ji} \in [0, d_1]) \\ 1, & (d_{ji} \in [d_1, \infty)) \\ 0, & (d_{ji} \in (-\infty, 0]) \end{cases}$$

Fig. 3.

$$f_2(d_{ji}) = \begin{cases} d_{ji}/d_1, & (d_{ji} \in [0, d_1]) \\ 2 - d_{ji}/d_1, & (d_{ji} \in [d_1, 2d_1]) \\ 0, & (d_{ji} \notin (0, 2d_1]) \end{cases}$$

Fig. 4.

$$f_3(d_{ji}) = \begin{cases} 1, & (d_{ji} \in [0, d_1]) \\ \dfrac{d_2 - d_{ji}}{d_2 - d_1}, & (d_{ji} \in [d_1, d_2]) \\ 0, & (d_{ji} \notin (0, d_2]) \end{cases}$$

Fig. 2-4. Class 1-3 whiteization weight fuctions

The turning point value of whitenization weighted function is called threshold value, which may be obtained by mean of analogy based on rules or experiences.

Fifth, to computer grey evaluation coefficient. to computer the grey evaluation coefficient belonged to K^{th} class for evaluated J versus evaluation index A on $d_{ji}^{(A)}$ and $f_k(d_{ji})$, noted as $n_{ji}^{(A)}$:

$$n_{JI}^{(A)} = \sum_{I=1}^{i} f_k\left(d_{JI}^{(A)}\right) \tag{2}$$

and the total grey evaluation coefficient $n_J^{(A)}$ of evaluated J belonged to each grey evaluation class for the evaluation index A.

$$n_J^{(A)} = \sum_{i=1}^{k} n_{Ji}^{(A)} \tag{3}$$

Sixth, to computer grey evaluation weight vector and weight matrix. We may computer the evaluation weight value $r_{JK}^{(A)}$ and weight value vector $r_J^{(A)}$ belonged to the kth grey class for the evaluated J^{th} versus the evaluation index A on the $n_{JK}^{(A)}$ and $n_J^{(A)}$.

$$r_{JK}^{(A)} = \frac{n_{JK}^{(A)}}{n_J^{(A)}} \tag{4}$$

Consider $K = 1, 2, \cdots k,$ then we have grey evaluation weight value row vector $r_{JK}^{(A)}$:

$$r_{JK}^{(A)} = \left[r_{j1}^{(A)}, r_{j2}^{(A)}, \cdots r_{jK}^{(A)} \right] \tag{5}$$

Consider $K = 1, 2, \cdots j,$ then we have grey evaluation weight value column vector $r_{JK}^{(A)}$:

$$r_{JK}^{(A)} = \left[r_{j1}^{(A)}, r_{j2}^{(A)}, \cdots r_{jK}^{(A)} \right] \tag{6}$$

Furthermore, we may seek grey evaluation weight value matrix $R^{(A)} = \{r_{JK}^{(A)}\}$ for all the evaluated ones versus evaluation index A, namely

$$R^{(A)} = \begin{bmatrix} r_{11}^{(A)} & r_{12}^{(A)} & \cdots & r_{1k}^{(A)} \\ r_{21}^{(A)} & r_{22}^{(A)} & \cdots & r_{2k}^{(A)} \\ \cdots & \cdots & \ddots & \cdots \\ r_{j1}^{(A)} & r_{j2}^{(A)} & \cdots & r_{jk}^{(A)} \end{bmatrix} \tag{7}$$

Seventh, to conduct the evaluation about different evaluation index, we may seek $r_J^{*(A)}$ from $R^{(A)}$.

$$r_J^{*(A)} = \max_k \{r_{JK}^{(A)}\} \tag{8}$$

then we have the evaluation index weight value vector

$$r^{*(A)} = \left\{ r_1^{*(A)}, r_2^{*(A)}, \cdots r_j^{*(A)} \right\} \tag{9}$$

We may gain the grey class for the different index versus the evaluated one based on $r^{*(A)}$, then to order grey class according to its good or bad.

Eighth, to carry out comprehensive evaluation.

(1) To ensure the grey class for the evaluated one based on all elements. to row $r^{*(A)}(A=1,2,\cdots a)$ as matrix r^*, then we may take comprehensive evaluation weight value vector based on all index $r_J^*(J=1,2,\cdots j)$.

We may gain total evaluation weight value when the evaluated one belong to different grey class according to r_J^*, then assure grey class the evaluated one belongs to based on all index.

(2) to queue the evaluated one according to all index, namely

$$r_J = \sum_{k=1}^{k} B_k R_{Jk} \qquad (10)$$

Thereinto,

$B_K(K=1,2,\cdots k)$--weight value coefficient of different grey class which may be set ahead of computation; $B_{JK}(J=1,2,\cdots j)$--the total evaluation weight value for the evaluated one belong to different grey class.

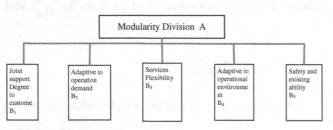

Fig. 5. Hierarchy structure about modularity storage and transportation of general material

To order the evaluated one' series of good or bad concerning all index according to the value of r_J.

3   Case Study

Firstly, Index system of grey hierarchy analytic for storage and transportation effectiveness of general material is shown as figure 5.

Secondly, to computer the combined weight value of elements in bottom layer for A-B layer index by experts, then table 1 is taken.

Table 1. Rules- modularity storage and transportation effectiveness of general material

index element / relatively importance	utmost important	Very important	relatively important	less important	no Important
class	9	7	5	3	1
B₁	√				
B₂					
B₃					
B₄		√			
B₅			√		
Note: The neighboring value is 8,6,4,2.					

$$A = \begin{bmatrix} 1 & 1.5 & 1.125 & 1.286 & 1.8 \\ 0.667 & 1 & 0.75 & 0.857 & 1.2 \\ 0.889 & 1.333 & 1 & 1.143 & 1.6 \\ 0.778 & 1.167 & 0.875 & 1 & 1.4 \\ 0.556 & 0.833 & 0.625 & 0.714 & 1 \end{bmatrix} \tag{11}$$

A's maximum eigenvalue is 5, and the correspondingly eigenvector $W = (0.257, 0.172, 0.229, 0.2, 0.143)^T$, its sect volume is the combined weight value of the element versus the goal. The consistency test CI=0, RI=1.12, CR<0.1 meet the demand[3].

Third, to give evaluation value matrix $D_{JI}^{(A)}$ of evaluation index A. Suppose there are 5 groups of evaluator, noted as I, II, III, IV, V; 3 evaluated one, noted as 1^O, 2^O, 3°; 5 evaluation index, namely B=1,2,3,4,5, which is the 5 elements in the bottom layer shown as figure 5. we gain evaluation index value matrix $D_{JI}^{(1)} \sim D_{JI}^{(5)}$ of evaluation index 1~5 according to 5 group evaluators' score table.

$$\begin{matrix} & \text{I II III IV V} & & \\ D_{JI}^{(1)} = \begin{bmatrix} 8 & 7 & 6 & 7 & 5 \\ 8 & 6 & 5 & 7 & 5 \\ 9 & 7 & 9 & 8 & 9 \end{bmatrix} \begin{matrix} 1' \\ 2' \\ 3' \end{matrix}, & D_{JI}^{(2)} = \begin{bmatrix} 8 & 7 & 8 & 9 & 8 \\ 8 & 7 & 7 & 7 & 7 \\ 9 & 3 & 9 & 5 & 9 \end{bmatrix}, & D_{JI}^{(3)} = \begin{bmatrix} 8 & 7 & 8 & 7 & 5 \\ 9 & 6 & 5 & 9 & 5 \\ 9 & 6 & 9 & 7 & 9 \end{bmatrix} \end{matrix} \tag{12}$$

$$D_{JI}^{(4)} = \begin{bmatrix} 5 & 7 & 9 & 7 & 5 \\ 9 & 6 & 5 & 7 & 5 \\ 6 & 7 & 6 & 8 & 7 \end{bmatrix}, \quad D_{JI}^{(5)} = \begin{bmatrix} 9 & 7 & 6 & 7 & 8 \\ 8 & 6 & 5 & 7 & 9 \\ 9 & 9 & 7 & 8 & 6 \end{bmatrix} \tag{13}$$

Fourth, suppose there are 4 evaluation grey class and corresponding grey number and whitenization weighted function shown as figure 6 ~ 9.

Fifth, to computer evaluation coefficient evaluated 1^O belonged to every grey class versus evaluated index 1.0020

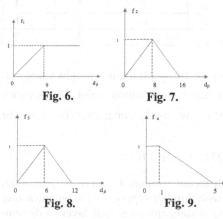

Fig. 6.   **Fig. 7.**

Fig. 8.   **Fig. 9.**

Fig. 6-9. Class 1-4 whitenization weighted fuction

$k=1 \quad n_{11}^{(1)} = 3.67, k=2 \quad n_{12}^{(1)} = 4.125,$

$k=3 \quad n_{13}^{(1)} = 4.5, \quad k=4 \quad n_{14}^{(1)} = 0$

Then the total evaluation coefficient of evaluated 1^O versus evaluation index 1 is

$$n_1^{(1)} = \sum_{i=1}^{4} n_{1i}^{(1)} = f_{11}^{(1)} + f_{12}^{(1)} + f_{13}^{(1)} + f_{14}^{(1)} = 12.295$$

Sixth, We may gain grey evaluation weight vector $r_1^{(1)}$ of evaluated 1^O, 2^O, 3° versus evaluation index 1 on $\{n_{ki}^{(1)}\}$, $\{n_k^{(1)}\}, k=1,2,3$,

$r_1^{(1)} = \{r_{11}^{(1)}, r_{12}^{(1)}, r_{13}^{(1)}, r_{14}^{(1)}\} = (0.2985, 0.3355, 0.3660, 0)$,

then to form evaluation weight value matrix $R^{(1)} \sim R^{(5)}$ of every evaluated one versus evaluation index 1.

$$R^{(1)} = \begin{bmatrix} r_{11}^{(1)} & r_{12}^{(1)} & r_{13}^{(1)} & r_{14}^{(1)} \\ r_{21}^{(1)} & r_{22}^{(1)} & r_{23}^{(1)} & r_{24}^{(1)} \\ r_{31}^{(1)} & r_{32}^{(1)} & r_{33}^{(1)} & r_{34}^{(1)} \end{bmatrix} = \begin{bmatrix} 0.2985 & 0.3355 & 0.3660 & 0 \\ 0.283 & 0.32 & 0.398 & 0 \\ 0.361 & 0.406 & 0.232 & 0 \end{bmatrix} \qquad (14)$$

$R^{(2)} \sim R^{(5)}$ is then omitted.

Seventh, we gain the maximum grey evaluation weight value $r_1^{*(1)} \sim r_3^{*(1)}$ of evaluated $1°$, $2°$, $3°$ versus evaluation index 1 on $R^{(1)} \sim R^{(3)}$:

$$r_1^{*(1)} = \max_i \{ r_{1i}^{(1)} \} = \max\{0.2985, 0.3355, 0.3660, 0\} = 0.366 \qquad (15)$$

Then the grey evaluation weight value vector of evaluated $1°$, $2°$, $3°$ versus evaluation index 1 $r^{*(1)} = \left(r_1^{*(1)}, r_2^{*(1)}, r_3^{*(1)} \right) = (0.366, 0.398, 0.406)$. on the same score, we may gain grey evaluation weight value vector $r^{*(2)} \sim r^{*(5)}$ of evaluation index 2~5, then to form evaluation weight value matrix r^*:

$$r^* = \begin{bmatrix} r_1^{*(1)} & r_2^{*(1)} & r_3^{*(1)} \\ r_1^{*(2)} & r_2^{*(2)} & r_3^{*(2)} \\ r_1^{*(3)} & r_2^{*(3)} & r_3^{*(3)} \\ r_1^{*(4)} & r_2^{*(4)} & r_3^{*(4)} \\ r_1^{*(5)} & r_2^{*(5)} & r_3^{*(5)} \end{bmatrix} = \begin{bmatrix} 0.366 & 0.398 & 0.406 \\ 0.391 & 0.36 & 0.352 \\ 0.352 & 0.35 & 0.391 \\ 0.366 & 0.382 & 0.35 \\ 0.368 & 0.352 & 0.384 \end{bmatrix} \qquad (16)$$

Then we may gain the queuing result of evaluated ones versus different evaluator from 5 row vector $r^{*(1)} \sim r^{*(5)}$ in the r^* shown as table 2.

Table. 2. The queuing order of evaluated one versus different evaluator

order of evaluation index	1	2	3	4	5
Number 1	$3°$	$1°$	$3°$	$2°$	$3°$
Number 2	$2°$	$2°$	$1°$	$1°$	$1°$
Number 3	$1°$	$3°$	$2°$	$3°$	$2°$

Eighth, the transformation vector of column vector in evaluation weight value matrix r^* is noted as r_J, then r_J is the comprehensive evaluation weight value vector based on all evaluation index, then to computer $r_J W$ as the comprehensive score for every evaluated one versus evaluation goal:

$$r_1 W = 0.3677, \quad r_2 W = 0.3711, \quad r_3 W = 0.3793$$

And its queuing order is $3°>2°>1°$, which means evaluated $3°$ has the maximum score, then the decision makers are provided with number support and avoid attracting by surface phenomenon. This paper' example is only involves A-B layer to demonstrate the computer process, and the evaluation for total effectiveness should be based on what mentioned in this paper' theory introduction part.

References

1. Wu, Y., Du, G.: Foundation of management science, p. 225. University publisher, Tianjin (2001)
2. Guo, Q.S., Zhi, Z., Yang, R.: General introduction for equipment effectiveness evaluation, pp. 100–103. National Defense Industry Press (August 2005)
3. Wu, Y., Du, G.: Foundation of management science, p. 226. University publisher, Tianjin (2001)

Study on Logistics Decision Making by Interval AHP

HuiJun Yang[1,2], Juan Wu[2], and Jian Zhang[2]

[1] School of Management, Tianjin University, Tianjin, P.R. China
[2] Automobile Transport Command Department, Military Transportation University,
Tianjin 300161, P.R. China
yanghuijun110@163.com, wjuan926@sina.com, 183334212@qq.com

Abstract. Based on the Interval Analytic Hierarchy Process to a logistics support, the paper find the different importance among various selections. It is useful for decision makers to meet the goal using limited resources efficiency and effectively.

Keywords: decision making, Interval analytic hierarchy process.

1 Introduction to the Interval Analysis

Compared with the traditional appraise and decision making, The decision making based on interval appraise has the following features: insufficient information, uncertain restraint, dynamic system, interval data, flexible model and appraise, limited rational decision making. It is better to analyze the decision making process of logistics support by Interval Analytic Hierarchy process(IAHP) than AHP. The Interval Analysis is originated from the number errors analysis, such as an observed value x with error ε, then the accurate value is [x–ε, x+ε].

Suppose interval number

$$A = [\underline{a}, \overline{a}] = \{x : \underline{a} \le x \le \overline{a}, x \in \Re\} \tag{1}$$

and its another form:

$$m(A) = \frac{1}{2}(\underline{a} + \overline{a}) \tag{2}$$

thereinto,

$$A = <m(A), w(A)>, \quad w(A) = \frac{1}{2}(\overline{a} - \underline{a}) \tag{3}$$

The definition of interval number arithmetic:

$$A * B = \{x * y \mid x \in A, \ y \in B\} \tag{4}$$

thereinto, $* \in \{+, -, \times, /\}$

$$[\underline{a}, \overline{a}] - [\underline{b}, \overline{b}] = [\underline{a} - \overline{b}, \overline{a} - \underline{b}] \tag{5}$$

D. Jin and S. Lin (Eds.): Advances in MSEC Vol. 2, AISC 129, pp. 17–23.
springerlink.com   © Springer-Verlag Berlin Heidelberg 2011

$$[\underline{a},\overline{a}]+[\underline{b},\overline{b}]=[\underline{a}+\underline{b},\overline{a}+\overline{b}] \tag{6}$$

$$[\underline{a},\overline{a}]\times[\underline{b},\overline{b}]=[\min(\underline{ab},\overline{ab},\overline{ab},\underline{ab}),\max(\underline{ab},\overline{ab},\overline{ab},\underline{ab}] \tag{7}$$

$$[\underline{a},\overline{a}]/[\underline{b},\overline{b}]=[\underline{a},\overline{a}]\times\left[\frac{1}{\overline{b}},\frac{1}{\underline{b}}\right], \quad 0\notin[\underline{b},\overline{b}] \tag{8}$$

especially,

$$\frac{1}{[\underline{a},\overline{a}]}=\left[\frac{1}{\overline{a}},\frac{1}{\underline{a}}\right], \quad 0\notin[\underline{a},\overline{a}] \tag{9}$$

interval vector $X=(X_1,X_2,\cdots,X_n)^T$, thereinto X_i is interval number; interval matrix

$$A=\begin{bmatrix} A_{11} & A_{12} & \cdots & A_{1n} \\ A_{21} & A_{22} & \cdots & A_{2n} \\ \cdots & \cdots & \cdots & \cdots \\ A_{m1} & A_{m2} & \cdots & A_{mn} \end{bmatrix} \tag{10}$$

thereinto A_{ij} is interval number.

The arithmetic of interval number and interval matrix is as same as the one of general vector and matrix.

2   The Models and Methods for Interval Appraisal

2.1   Interval AHP

IAHP method includes four steps as follows:

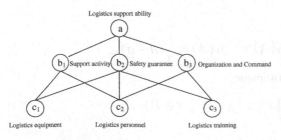

Logistics support ability

a

b₁ Support activity   b₂ Safety guarantee   b₃ Organization and Command

c₁   c₂   c₃

Logistics equipment   Logistics personnel   Logistics trainning

Fig. 1. Step-by-step hierarchy structure

1) To set up step-by-step hierarchy structure as same as AHP method shown as figure 1.

2) To structure interval judge matrix shown as formula (15)～(18) based on judge scale with normal distribution shown as figure 2 and traditional AHP weight value method shown as table 1.

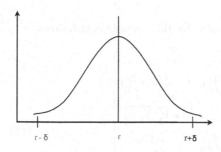

Fig. 2. The Normal Distribution of r_{ij}

Suppose the estimated value of judge scale r is r_{ij}, then we divided scale r_{ij} into three intervals to represent three situations namely certain, basically certain and possible which reflect the reliability of the scale made by decision maker.

We can structure a interval judge matrix as follows:

$$A = [a_{ij}] = [\underline{a}_{ij}, \overline{a}_{ij}] \quad (11)$$

Table 1. The Meaning of Scales in AHP Method[1]

scale	meaning
1	A_i is equally important with A_j.
3	A_i is little important than A_j.
5	A_i is a obviously important than A_j.
7	A_i is intensively important than A_j.
9	A_i is extremely important than A_j.
2、4、6、8	the middle value between five scales mentioned above.
reciprocal	if scale is r_{ij} created from comparison A_i with A_j, then the scale $r_{ji} = 1/r_{ij}$ is created from comparison A_j with A_i.

$$\underline{a}_{ij} = \begin{cases} r_{ij} & certain \\ r_{ij} - \delta & basically \quad certain \\ r_{ij} - 2\delta & possible \end{cases} \quad (12)$$

$$\overline{a}_{ij} = \begin{cases} r_{ij} & certain \\ r_{ij} + \delta & basically \quad certain, \\ r_{ij} + 2\delta & possible \end{cases} \quad (13)$$

$$a_{ij} = \frac{1}{a_{ji}}, \quad (\underline{a}_{ij}, \overline{a}_{ij}) = (\frac{1}{\overline{a}_{ji}}, \frac{1}{\underline{a}_{ji}}) \quad (14)$$

Suppose $\delta = \dfrac{1}{6}$, and the interval judge matrix for the example is as follows:

$$aB = \begin{bmatrix} (1,1) & (5-\tfrac{1}{3},5+\tfrac{1}{3}) & (3-\tfrac{1}{6},3+\tfrac{1}{6}) \\ (\dfrac{1}{5+\tfrac{1}{3}},\dfrac{1}{5-\tfrac{1}{3}}) & (1,1) & (\dfrac{1}{3+\tfrac{1}{3}},\dfrac{1}{3-\tfrac{1}{3}}) \\ (\dfrac{1}{3+\tfrac{1}{6}},\dfrac{1}{3-\tfrac{1}{6}}) & (3-\tfrac{1}{3},3+\tfrac{1}{3}) & (1,1) \end{bmatrix}$$

$$= \begin{bmatrix} (1.000,1.000) & (4.667,5.333) & (2.833,3.1667) \\ (0.187,0.214) & (1.000,1.000) & (0.300,0.370) \\ (0.315,0.353) & (2.703,3.333) & (1.000,1.000) \end{bmatrix} \qquad (15)$$

$$b_1 c = \begin{bmatrix} (1,1) & (3-\tfrac{1}{3},3+\tfrac{1}{3}) \\ (\dfrac{1}{3+\tfrac{1}{3}},\dfrac{1}{3-\tfrac{1}{3}}) & (1,1) \end{bmatrix}$$

$$= \begin{bmatrix} (1.000,1.000) & (2.667,3.333) \\ (0.300,0.375) & (1.000,1.000) \end{bmatrix} \qquad (16)$$

$$b_2 c = \begin{bmatrix} (1,1) & \left(\dfrac{1}{5+\tfrac{1}{6}},\dfrac{1}{5-\tfrac{1}{6}}\right) & \left(\dfrac{1}{3+\tfrac{1}{3}},\dfrac{1}{3-\tfrac{1}{3}}\right) \\ \left(5-\tfrac{1}{6},5+\tfrac{1}{6}\right) & (1,1) & \left(3-\tfrac{1}{3},3+\tfrac{1}{3}\right) \\ \left(3-\tfrac{1}{3},3+\tfrac{1}{3}\right) & \left(\dfrac{1}{3+\tfrac{1}{3}},\dfrac{1}{3-\tfrac{1}{3}}\right) & (1,1) \end{bmatrix}$$

$$= \begin{bmatrix} (1,1) & (0.194,0.207) & (0.300,0.375) \\ (4.833,5.167) & (1,1) & (2.700,3.300) \\ (2.670,3.330) & (0.300,0.370) & (1,1) \end{bmatrix} \qquad (17)$$

$$b_3 c = \begin{bmatrix} (1,1) & (\dfrac{1}{3+\tfrac{1}{6}},\dfrac{1}{3-\tfrac{1}{6}}) \\ (3-\tfrac{1}{6},3+\tfrac{1}{6}) & (1,1) \end{bmatrix}$$

$$= \begin{bmatrix} (1.0000,1.0000) & (0.3158,0.3529) \\ (2.8333,3.1667) & (1.0000,1.0000) \end{bmatrix} \qquad (18)$$

We may ensure which situation mentioned above is belong to decision maker' proposal after further exchanging views between analyzer and decision maker.

3) To computer the eigenvector for the maximum latent root concerning the interval judge matrix.

Suppose A is an interval matrix, λ is a real number, if there is a non-zero interval number vector x to make $Ax=\lambda x$, then λ is called an eigenvalue of A, x is an eigenvector of A to λ. There is following relation formula about maximum eigenvalue shown as formula (19):

$$AW = \lambda_{\max}W \tag{19}$$

Herein, $\lambda_{\max}$ is maximum eigenvalue, W is corresponding eigenvector.

After computing the eigenvector of the judge matrix, it is also needed to conduct coherence test with the formula in AHP shown as follows:

$$C \cdot R = \frac{C \cdot I}{R \cdot I} < 0.1^{[2]} \tag{20}$$

Based on computing, the value of C.R satisfies the demand. Then the eigenvalue of the interval judge matrix for the example is shown as formula $(21) \sim (22)$.

$$W_{ab} = \begin{bmatrix} (0.619, 0.654) \\ (0.105, 0.105) \\ (0.242, 0.276) \end{bmatrix} \qquad W_{b_1 c} = \begin{bmatrix} (0.728, 0.769) \\ (0.231, 0.272) \end{bmatrix} \tag{21}$$

$$W_{b_2 c} = \begin{bmatrix} (0.100, 0.110) \\ (0.622, 0.652) \\ (0.2387, 0.277) \end{bmatrix} \qquad W_{b_3 c} = \begin{bmatrix} (0.2226, 0.2780) \\ (0.6666, 0.8328) \end{bmatrix} \tag{22}$$

4) To computer the total weight value interval vector of each element in lower layer by means mentioned below.

Suppose we has taken the weight value interval vector by total goal compositor concerning n_{k-1} elements in $(k-1)^{th}$ layer.

$$W^{(k-1)} = (w_1^{(k-1)}, w_2^{(k-1)}, \cdots, w_n^{(k-1)})^T \tag{23}$$

the weight value interval vector of the n_k^{th} element in k^{th} layer concerning the jth element in $(k-1)^{th}$ layer is:

$$P_j^{(k)} = (p_{1j}^{(k)}, p_{2j}^{(k)}, \cdots, p_{n_{k-1}j}^{(k)}), P^{(k)} = (p_1^{(k)}, p_2^{(k)}, \cdots, p_{n_{k-1}}^{(k)}) \tag{24}$$

it is a $n_k \times n_{k-1}$ matrix, then the total compositor interval vector of elements in k^{th} layer is as formula (25)、 (26) and figure 3.

$$W^{(k)} = (w_1^{(k)}, w_2^{(k)}, \cdots, w_{n_k}^{(k)})^T = P^{(k)} \cdot W^{(k-1)} \tag{25}$$

$$W^{(k)} = P^{(k)} P^{(k-1)} \cdots W^{(2)} \tag{26}$$

According to formula (21), (22) and (24), we have formula (27).

$$P = \begin{bmatrix} (0.728, 0.769) & (0.100, 0.110) & (0.0000, 0.0000) \\ (0.231, 0.272) & (0.622, 0.652) & (0.2226, 0.2780) \\ (0.0000, 0.0000) & (0.2387, 0.277) & (0.6666, 0.8328) \end{bmatrix} \tag{27}$$

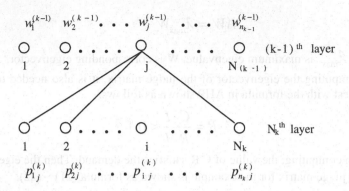

Fig. 3. The weight value relations in different layer

Then the eigenvalue of every interval matrix for the example is as follows:

$$W = \begin{bmatrix} w_1 \\ w_2 \\ w_3 \end{bmatrix} = P \cdot W_{ab} = \begin{bmatrix} (0.4611, 0.5150) \\ (0.2620, 0.3240) \\ (0.1860, 0.2590) \end{bmatrix} \tag{28}$$

Through unification, we can get

$$\bar{W} = \begin{bmatrix} \bar{w}_1 \\ \bar{w}_2 \\ \bar{w}_3 \end{bmatrix} = \begin{bmatrix} (0.4690.5070) \\ (0.2880, 0.2950) \\ (0.2050, 0.2360) \end{bmatrix} \tag{29}$$

Interval $\bar{w}_1 = (0.4690, 0.5070)$, $\bar{w}_2 = (0.2880, 0.2950)$, $\bar{w}_3 = (0.2050, 0.2360)$ represent the weight value interval of element c_1, c_2 and c_3 shown as figure 4. Obviously, the result of ordering is $w_1 > w_2 > w_3$.

Consequently, we should firstly meet C_3 demand under limited resources, then second C_2 demand if we have more resources, and finally C_1 if we have enough resources.

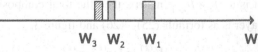

Fig. 4. Chart about c1、c2、c3

Thus we may make rational and scientific decisions and make limited resources maximum benefits.

References

1. Hu, Y., Guo, Y.: Teaching course of operations research, p. 437. Tsinghwa university publisher(2th published) (May 2003)
2. Wu, Y., Du, G.: Foundation of management science, p. 226. University publisher, Tianjin (2001)

A Kind of Privacy-Preserving Data Mining Algorithm Oriented to Data User

Li Cai[1] and JianYing Su[2]

[1] School of Information Engineering, Chongqing City Management College,
401331 Chongqing, China
cailitg@163.com
[2] Library, Chongqing City Management College,
401331 Chongqing, China
tsgsujy@163.com

Abstract. The database quality hidden after rules is always judged by data user. Only high quality to data users is real high quality. In many privacy-preserving association rule mining algorithms, there is no accurate quantitative assessment on effect of hiding on database. To address the problem, the paper presented a kind of personalized privacy-preserving association rule mining algorithm that has less damage to database from view of data user. The algorithm selects data to be deleted according to importance index expensed by each support number of each restrictive rule. In order to evaluate algorithm performance, a standard to measure damage survived from database cleaning oriented to users was also provided.

Keywords: association rule, privacy-preserving, user-oriented, data mining.

1 Introduction

In recent years, privacy-preserving data mining has been a hot point in database research. Among them, association rules hidden attracted more attention. The called rule hidden in association rule mining means hiding restrictive knowledge that data owner does not expect to be mined by data user before association rule mining. Meanwhile, other non-restrictive rule should remain to be mined. As to hiding in association rule mining, it does not involve individual privacy hidden. Knowledge hidden technologies mainly include four methods of data transformation, data blocking, data reconstruction and data sampling [1].

The manner of item deleting is mainly used in the paper to achieve restrictive association rule hidden. There have been many achievements in this area. E. Dassem *et al* presented Algola [2], Algolb[3], Algo2a[2], Algo2b[3], Algo2c[3], Naive [4], MinFIA [4], MaxFIA [4], IGA[4], RRA [5], RA[5] and SWA[6]. The [2] and [3] all assume that there is no intersect in restrictive rules to be hidden. Each time a rule been selected, the goal of rule hidden can only is achieved by decrease support or confidence of transaction. The [4-6] introduced concept of conflict and implemented rule hidden by decreasing support. It can hide restrictive rule set with intersect and only needs to scan database twice to complete rule hidden. The [7, 8] respectively

D. Jin and S. Lin (Eds.): Advances in MSEC Vol. 2, AISC 129, pp. 25–30.

proposed greedy algorithm and optimization method based on integer as well as quantitatively assess negative effect caused by rule hidden. In addition, there were related results in [9, 10]. However, previous studies did not always consider this issue from the perspective of data users. Data in the database for user data is not some equivalent place, which has commercial value. The commercial value of different data may be different. The database quality hidden after rules is always judged by data user. Only high quality to data users is real high quality. The paper aims at proposing a kind of association rule mining algorithm closer to user from the view of data user.

The paper is organized as follows: section 2 describes problem to be researched and gives related definitions; section 3 presents design idea of (Data User Oriented Privacy-preserving Data Mining) DUOPPDM algorithm and implement flow; section 4 concludes our work.

2   Problem Description and Related Definitions

2.1   Problem Description

Considering about demand instance proposed in [9], assume we are manager in a large supermarket BigMart. We have reached an agreement with Dedtrees. The company wants to provide products with a lower price, but they need us to provide them with transaction database. After they got our data, the company began to mine association rule hidden in our transaction database. They found customers who buy cold remedies always buy facial tissue afterwards, which is helpful for them purchasing products for us in advance. At the same time, they also found that customers who buy skim milk also purchase green paper, so they conducted some promotions. If customers buy Dedtrees products after purchasing skim milk, it will be lower price of 50 point, which seriously strike sales of green paper. Meanwhile, as the sales amount of green paper decrease, the price of green paper will increase. At the next time when we want to cooperate with Dedtrees, they will deny to provide us products with lower price considering about competition from competitors. We will also lose part opportunities for improper cooperation with green paper. Therefore, we should perform knowledge hidden processing before providing data to Dedtrees.

Based on the above examples, the task is to decrease support of frequent itemset {skim milk, green paper} below minimum support threshold, so that the Dedtrees can not mine frequent itemset {skim milk, green paper} in the transaction database mining. At the same time, other non-restrictive frequent itmeset should remain frequent to help Dedtrees cooperate with us.

2.2   Related Definitions

Set database D as transaction database, which consist of many transactions. Each transaction is data in once time. The transaction is made up of many items. Each item is a product in the transaction. R means all association rules contained in the database D and RR (restrictive rule) is restrictive association rule.

Importance index I of item means harm factor of database caused by deleting the item in database cleaning, which is determined by importance of the item on data user.

Restrictive matters means transaction at least support one restrictive rule.

Restrictive item means that item that contained in restrictive rules supported by the transaction.

Database damage index is the reduction importance factor of database.

Hidden efficiency of item means ratio of damage to database deleting the item to summation supported all restrictive itemsets.

3   DUOPPDM Algorithm

3.1   Design Idea

Based on above problem description, our task is to hide restrictive frequent itemsets with the manner of deleting items to clean database as to avoid vicious competition. Meanwhile, we should also ensure the cleaned database has greatest value to data user. As a result, we should consider selecting delete items in the view of data user in cleaning process. Such as frequent itemset {*skim milk*, *green paper*} in the above instance, as to data user Dedtrees, they may pay more attention to data of our corporation and regard them as accordance of some commercial decision within the corporation. So we should delete the item green paper in prior in accordance with the idea. Secondly, if some restrictive frequent itemset {A, B} is product of data user, as the price and profit of A and B are different, the company want to get high-profit goods data are more realistic, and low-cost sales volume of its total sales volume of goods is key to delete little impact. Thirdly, assume the data user cooperation is company C notable for notebook and A is the product in C, while C is the by-product screen protection film, the corporation hope data from A is more realistic. Thus the concept of item importance index was proposed, which can be divided into 10 levels from 1 to 10.

3.2   Algorithm Description

Privacy-preserving association rule oriented to data user is a algorithm to maximize satisfactory degree of user on data in the premise of ensuring hidden quality of restrictive rule. It selects data to be deleted with greedy method each lower support each restrictive to sacrifice several important indicators. The DUOPPDM algorithm introduces concept of window, making it no longer is a algorithm based on memory and can be used for large-scale database. The concept of window firstly appeared in [7].

Suppose the window size is K, the algorithm flow is as following:

Step 1: Compute transaction number $Num[1:n]$ of each restrictive itemset.

Step 2: Compute the number need to be deleted $DelNum[1:n]$ of each restrictive itemset and the proportion of each itemset in total transaction number, namely openness $p[1:n]$.

Step 3: Read K transactions each time till there is no data in the database. Obtain restrictive transaction for K transactions.

(*a*) Compute support number of each restrictive transaction in K transactions and multiply $P[m]$($1 \leq m \leq n$). After rounding, we can obtain transaction number supporting the restrictive rule that should be deleted $D[m]$.

(*b*) Compute hidden efficiency of each item in database and delete the item with highest hidden efficiency. When there are multiple itemsets with the highest hidden efficiency, randomly select an item and delete it. Then delete hidden efficiency of other items and modify corresponding $D[m]$. If the transaction is no longer restrictive to the window, delete the transaction. If D decrease to 0, delete the restrictive rule from all rules and re-compute the hidden efficiency of each restrictive transaction. Repeat (*b*) till the $D[m]$ of last restrictive rule decrease to 0.

3.3   Computation Example

Simply take the database D data shown in Table 1.

Table 1. Database D

transaction	item			
transaction1	A	B		D
transaction2	A	B	C	D
transaction3	A		C	
transaction4			C	D

Suppose importance index of item A is 5 and that of item B, C and D are all 2; the minimum support threshold is 2; the restrictive frequent itemsets are AB and AC. The window size is 4.

(*a*) Obtain restrictive itemsets AB and AC, and then we can arrive at $Num_1=1$, $Num_2=1$, and thus $P_1=1.2$, $P_2=1/2$.

(*b*) Read four transactions, and calculate transaction number $D_1=K*P_1=1$ and $D_2=K*P_2=1$ that support it to be deleted of two transactions AB and AC.

(*c*) Compute hidden efficiency of three restrictive transactions in the database, we can get Table 2.

Table 2. Hidden efficiency of transaction items

Transaction	Items		
	A	B	C
transaction1	5/1	2/1	
transaction2	5/2	2/1	2/1
transaction3	5/1		2/1

(*d*) Select the item with lowest hidden efficiency to delete, so we can delete item B in transaction 1. Then D_1 decrease 1 and become 0. Delete frequent itemset AB. Meanwhile, we should also modify hidden efficiency of restrictive transaction in the database to arrive at Table 3.

Table 3. The hidden efficiency of transaction items

Transaction	Item	
	A	C
transaction2	5/2	2/1
transaction3	5/1	2/1

(e) Delete item C, then D_2 of the last restrictive items minus 1 to 0. The operation of this window completes.

(f) Determine whether there is data in the database, if there is not, the algorithm ends; otherwise go to (a).

3.4   Setting of Important Parameters

Importance index is vital in the algorithm. For the owner of database, they can not directly know the concern of data user on data, so the setting of importance index should follow several principles:

(a) Obtain assistance from data user as possible so that they can provide more precise use of these data.

(b) Respect purpose of user statement. For data user more concerned with our corporation data, we should set higher importance index.

(c) According to characters of data user, the importance index of commodity with higher price is usually higher. But for some company has main product of themselves, the importance of main product is higher than that of others.

(d) Based on data emergency frequency, the importance index of product with lower frequency will be higher.

3.5   Algorithm Assessment Standard

In the last, the people's knowledge on effect of knowledge hidden on sanitized database often used to hide all non-restrictive rules in the measure before it can be tapped, but virtually every rule only in a particular threshold is valid. Such as minimum support of 30% is often meaningful, at this time, the study on many other thresholds of AB has no sense. The database owner can not accurately predict the threshold, so the performance result has large difference with user expected under this kind of evaluation method. The paper presents a database damage index of concern degree of user on data to evaluate the of negative impact database.

As an important index of the data items that the user data for the importance of his coefficient of database damage index was defined as the process of being removed only to hide an important index of the item and, in order to determine the databases for data users it has much impact. Smaller damage index, then the database is hidden through after higher the quality.

3.6   Performance Analysis

Firstly, from the algorithm itself we can know that it can reach the goal of decrease support of frequent itemset below minimum support threshold. Secondly, as the algorithm keeps damage coefficient of database to minimum based on greedy algorithm, the data can satisfy data user to great extent.

4 Conclusion

The paper presented a personalized privacy-preserving association rule mining algorithm in the view of data user so that requirements from data user can be more considered while hiding rules. It can maximize profit of data user in the premise of rule hidden. In the next future, we will focus on how to consider more about data users in the process of rule hidden.

References

1. Guo, Y.-H., Tong, Y.-H., Tang, S.-W., et al.: A survey on knowledge hiding in database. Journal of Software 18, 2782–2799 (2007)
2. Dasseni, E., Verykios, V.S., Elmagarmid, A., et al.: Hiding association rules by using confidence and support. In: Proc of the 4th Int'1 Information Hiding Workshop, pp. 369–383. Springer, Berlin (2001)
3. Verykios, V.S., Elmagarmid, A., Bertino, E., et al.: Association rule hiding. IEEE Trans. on Knowledge and Data Engineering 16, 434–447 (2004)
4. Oliveira, S.R.M., Zaiane, O.R.: Privacy preserving frequent itemset mining. In: Proc. of IEEE ICDM Workshop on Privacy, Security and Data Mining: Computer Society, pp. 43–54 (2002)
5. Oliveira, S.R.M., Szaiane, O.R.: Algorithms for balancing privacy and knowledge discovery in association rule mining. In: Proc. of the 7th Int'l Database Engineering and Application Symposium, pp. 54–64. IEEE Computer Society, Hong Kong (2003)
6. Oliveira, S.R.M., Szaiane, O.R.: Protecting sensitive knowledge by data sanitation. In: Proc. of the 3rd Int'1 Conf. on Data Mining, pp. 613–616. IEEE Coputer Society (2003)
7. Sun, X.-Z., Yu, P.S.: Border-based approach for hiding restrictive frequent itemsets. In: Proc. of the 5th IEEE Int'1 Conf on Data Mining, pp. 426–433. IEEE Computer Society (2005)
8. Menon, S., Sarkar, S., Mukhejee, S.: Maximizing accuracy of shared databases when concealing sensitive patterns. Information Systems Research 16, 256–270 (2005)
9. Wang, E.T., Lee, G., Lin, Y.T.: A novel method for protecting restrictive knowledge in association rules mining. In: Proc. of the 29th Annual Int'1 Conf. on Computer Software and Application, pp. 511–516. IEEE Computer Society, Edinburgh (2005)

Design of Industrial Instrument Manipulator

Xinsheng Che and Dongxue Fan

School of Information Science & Engineering,
Shenyang University of Technology, Shenyang, P.R. China
coffice@126.com

Abstract. In view of the process control system in the system operation, maintenance and repair of special circumstances, an instrument manipulator is designed for back-up operation which consists of microcontroller, A/D converter, D/A converter, graphic dot matrix liquid crystal and optical couplers etc. Analog quantity input mainly records 4~20mA current signal, analog quantity output by the D/A outputs 1~5V, then converts to 4~20mA current signal to control the valves. Switch quantity input and output achieve by the shift register and the relay. All input and output changes are storage and record.

Keywords: manipulator, analog quantity, switch valve, standard signal, storage and record.

1 Introduction

During the early 1980s to the 1990s, with the rapid development of electronic technology, electronic instrument manipulator gets rapid development which has visual display, high measurement accuracy, PID auto-tuning function and large storage capacity of measurement data. The instrument manipulator plays an important role in the industrial automation system and improves the automation level of China's petroleum, chemical, metallurgy, machinery and other industries.

In recent years, domestic instrument manipulator had great progress. Control theory and instrumentation technology development continuously as human scientific and technological progress. With the development of computers, controllers, communications and display technology, the instrument manipulator has improved to many kinds of intelligent operation devices of D type, Q type and so on [1, 2].

In the instrumentation industry, electronic and computer technology have a great development. A simple light beam display or digital display can't transmitting complex information clearly. A liquid crystal display is used for showing the analog voltage waveform curve which can display large amount of information and the operation simple and easy [3].

Instrument manipulator which is controlled by the microcontroller has 8 analog signal input channels, 8 switch input and output channels, 4 analog output channels with display, storage and communication function [4]. The system is mainly composed of A/D converter, D/A converter, graphic dot matrix liquid crystal, optical couplers and other components. It also can be able to store huge amounts of data, and be able to communicate with the host computer to complete data analysis and processing.

D. Jin and S. Lin (Eds.): Advances in MSEC Vol. 2, AISC 129, pp. 31–36.
springerlink.com   © Springer-Verlag Berlin Heidelberg 2011

2 Hardware System Design and Implementation

Instrument manipulator mainly completes many functions, such as analog quantity input, switch quantity input and output, valve opening control, analog quantity input display, storage and storage data and output data display through the button. The system centers on AT89S51 microcontroller, using multi-channel analog switch, A/D converter time division multiplexing, shift registers, D/A converter for completing the design. The overall system block diagram is shown in figure 1.

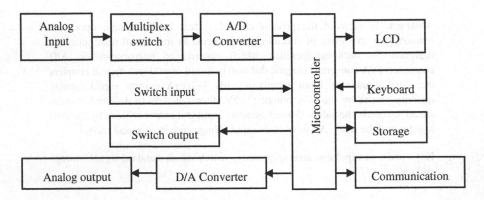

Fig. 1. Overall system block diagram

Analog quantity input 4~20mA current signal converts into 1~5V voltage through I/V conversion. Analog switch selector channel puts voltage into. And then the digital quantity which is converted by the A/D converter is inputted into the microcontroller. The microcontroller converts the data into a curve which displays by LCD. The analog quantity input and display are completed.

Switch quantity input and output are controlled by microcontroller and relay. Input the switch quantity into the microcontroller. Then the microcontroller outputs the switch quantity to the relay. The relay executes the corresponding operation that according to the received signal, in order to achieve the external device detection, identification and external implementation of components drive and control.

Electric control valve opening is controlled through the 4~20mA current signal. The microcontroller output digital value is converted into 1~5V analog signal by D/A converter. V/I conversion controls valve opening through the 4~20mA current signal. The LCD displays the current output curve.

2.1 8 Channel Analog input Circuit

4~20mA current signal is converted to 1~5V voltage signal, and then the 1~5V voltage signal is sent into A/D converter by the analog switch. The microcontroller controls signal waveform curve displaying and data recording.

The specific method is that a 250Ω resistor is connected in series in the 4~20mA current signal circuit, and then the 4~20mA current signal is converted to 1~5V voltage signal. 8-to-1 analog switches CD4051 and A/D conversion chip ADS7816 achieve analog quantity input. ADS7816 is a 12-bit, 200kHz sampling analog-to-digital converter. It features low power operation with automatic power down, a synchronous serial interface and a differential input.

Considering the industrial field, microcontroller applications and the anti-interference, the optical coupler is used for achieving signal isolation between the A/D and the microcontroller. Not only can it achieve better measurement accuracy, but also can eliminate interference components. Circuit diagram is shown in figure 2.

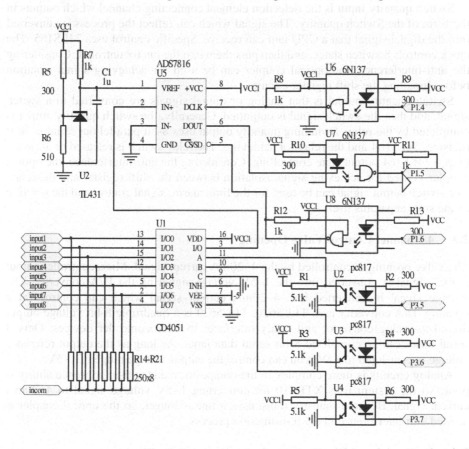

Fig. 2. 8 channel analog input circuit

As shown in the figure 2, CD4051 is a single 8-Channel multiplexer having three binary control inputs A, B and C, and an INH input. It also has low "ON" resistance and very low "OFF" leakage current. ADS7816 is +5V power supply. The input voltage is range of 0~5V. R7 and C1 purify its power source.

Voltage signal inputs into ADS7816 through the CD4051 switching. The ADS7816 output is connected with the microcontroller by means of the serial mode. DCLK provides an output clock signal. DOUT is output pin. When $\overline{CS}$ is in high level, the output is the high resistance state. When a falling $\overline{CS}$ signal comes, the clock inputs. The first 1.5 to 2.0 clock periods of the conversion cycle are used to sample the input signal. For the next 12 clock periods, DOUT will output the conversion data, and the data are transmitted on the falling edge of DCLK.

2.2   8 Switch Input and Output Control

"On" and "off" are the most basic and typical electrical functions. Switch quantity is a corresponding value that controls relay connection or disconnection.

Switch quantity input is the detection element connecting channel which outputs in the form of the Switch quantity. The signal which can reflect the process is converted into the digital signal that a CPU unit can receive. Specific control uses 74LS165. The clock controls 8 switch states, and then puts them into the microcontroller. Considering the anti-interference, the optical coupler can be used for achieving signal isolation before inputting the shift register circuit.

Switch quantity output is that analog or digital signals are converted to a switch signal, and then the switch signal is outputted. Generally, the switch quantity output is completed by the relay. Switching quantity output uses 8-bit parallel out serial in shift registers 74LS164 and the relay for achieving. The relay switch is released or attracted by the 74LS164 output state controlling. Considering the anti-interference, the optical coupler is used for achieving signal isolation between the shift register and the relay. The switch output signal can be used for the limit alarm, signal control and the reaction of the state apparatus itself.

2.3   4 Channel Control Valve Opening

The valve opening is controlled by the 4~20mA current signal. Microcontroller output digital signal is converted into the 1~5V voltage signal by the D/A converter. The voltage signal is converted into 4~20mA current signal which can control valve opening. D/A converter uses TLC5620. TLC5620 is a quadruple 8-bit voltage output digital-to-analog converter and easily interfaces to microcontroller devices. Only 4 serial buses can be completed 8-bit serial data input. As long as the output reference voltage is connected to +5V, you can control the output voltage range of 1~5V.

Analog circuits is more complex in the composition and anti-interference ability is poor, so the system uses XTR110 for converting 1~5V voltage signal to 4~20mA current signal. Because industry transmission line is longer, so the optical coupler is needed anti-interference in the transmission process.

2.4   Keyboard and Display

A 4x4 keyboards and graphic display OCMJ4X8C12864 are used. Liquid crystal display has low power consumption, small volume, light weight, thin and others incomparable advantages [5]. The Liquid crystal displays the current analog waveform, data, switch state and the valve opening. In order to save the microcontroller's I/O port, liquid crystal display uses serial connection.

2.5   Communication

Because RS485 can achieve remote, stable and accurate data transmission. Working in a multi-node system also has an extensive application [6]. This system adopts RS485 communication, and requires only 2 signal lines. RS485 interface uses unshielded twisted pair transmission.

3   System Programming

The main function completes microcontroller initialization, button, display, output control and the valve opening control. Initialization is started at the head of the main function. The A/D sampling timer interrupt, liquid crystal display initialization and A/D initialization are set so as to ensure that the system can work properly.

In the process of the analog quantity sampling, because the liquid crystal display's horizontal is 128 data points, so it needs sample 128 times. And then A/D conversion is achieved. Sampling is completed in the timer interrupt by using 16-bit mode. Setting time is 0.11ms, namely sampling time is 0.11ms. Sampling frequency is 9 kHz. When enter to the interruption, the interrupt flag adds 1, A/D sampling and conversion are completed. The result is stored in the table. Judging whether the interrupt flag is 128, if the interrupt flag is 128, interrupt flag is cleared. The table reloads the data. the Interrupt function flow chart is shown in Figure 3.

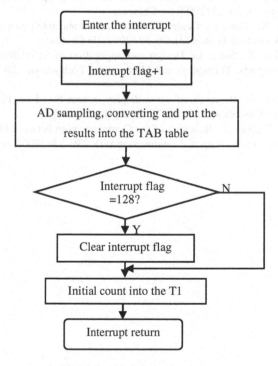

Fig. 3. Interrupt flowchart

4   Conclusion

The system is controlled by microcontroller and is achieved by the hardware circuit and software programming. Using keil C program, and the various parts are debugged. Using liquid crystal display can clearly observe voltage curve. There will be a wide range of reference value.

As time is limited, the followings are not study:

(1) For the waveform curve, it can not directly read the voltage according to coordinates;

(2) Only can it debug an analog signal, it can't achieve 8 channels' display;

(3) Only can it debug a valve opening control, it can't achieve the 4 channels control.

Acknowledgments. This project is supported by the Shenyang Science & Technology Planning Item No. F10-213-1-00.

References

1. Li, J., Zhang, H.: Improvement of D-type electric manipulator. Process Automation Instrumentation 18(2), 22–24 (1997) (in Chinese)
2. Wang, J., Yang, M., Lu, W.: Design of Multifunction Q-type Manipulator System. China Instrumentation (8), 20–23 (2003) (in Chinese)
3. Yang, M., Li, X., Zhao, L.: Design of multifunction and intelligence manipulator system. Automation & Instrumentation (3), 36–39 (2003) (in Chinese)
4. Zhang, Q., Gao, X., Song, L.: Design and application of an intelligent manipulator with communication ports. Transducer and Microsystem Technologies 25(11), 79–81 (2006) (in Chinese)
5. Pan, D., Huang, P.: Research of wave display system based on 12864 LCD. Electronic Instrumentation Customer 15(3), 28–29 (2008) (in Chinese)
6. Bi, B., Wang, C., Sun, S.: Research of the Communication between PC and SCM Based on RS-485. Science Technology and Engineering 8(1), 236–238 (2008) (in Chinese)

Facial Expression Recognition Based on MB-LGBP Feature and Multi-level Classification

Zheng Zhang[1], Chao Xu[2], Jiaxin Wang[1], and Xiangning Chen[1]

[1] College of Computer Science and Software, Tianjin Polytechnic University,
Tianjin, China, 300387
[2] College of Computer Science and Technology, Tianjin University,
Tianjin 300072, China
aaron_boy_2000@hotmail.com

Abstract. This paper presents a facial expression recognition approach based on MB-LGBP feature and multi-level classification. First, the multi-scale block local Gabor binary patterns (MB-LGBP) operator is extracted to achieve both locally and globally informative features. Then a two-level classification method is proposed. At the coarse level, two expression candidates with the first two high decision confidence are selected from 7 basic expression classes based on MB-LGBP features. At the fine level, one of the two candidate classes is verified as final expression class based on more delicate 2D MB-LGBP features. The promising result proves the superiority of our method to some other popular paradigms in expression recognition.

Keywords: MB-LGBP, Multi-Level Classification, Expression Recognition.

1 Introduction

Recently, facial expression recognition has become a very active topic in machine vision community. More and more technical papers are concerning this area, and a brief tutorial overview can be found in [1]. And due to less information for expression actions is available from the static images, expression recognition from static images [2, 3, 4] is more difficult than that from image sequences [5]. However, in many applications it is also useful to recognize expression of a single image, so we focus on subject-independent expression recognition based on static image in this paper.

In light of recent advances in image and signal processing, various features have been proposed for recognizing facial expression in the last few years. The current trend is to extract features at multiple scales and make a rational fusion [2] to utilize the features to their largest extent. So in [6], we follow this fashion and propose the Multi-scale Block Local Gabor Binary Patterns (MB-LGBP) to describe facial expression at varying levels of detail from coarse to fine.

The discrimination of different expressions needs more precise description of local textures. As 1st-order descriptor, LBP can not encode spatial structure information. And this problem can not be solved in nature only by partitioning. So we utilize Gray Level Co-occurrence Matrix (GLCM) instead of the traditional statistical histogram in MB-LGBP encoding and present the so-called 2-dimensional LGBP features in [7].

D. Jin and S. Lin (Eds.): Advances in MSEC Vol. 2, AISC 129, pp. 37–42.
springerlink.com

To further promote the performance of subject-independent expression recognition, we present a multi-level classification framework. Low-dimensional MB-LGBP features are utilized for the coarse level, in which two expression candidates are selected from seven basic expressions based on their decision confidence. In fine classification level, relatively high-dimensional 2D MB-LGBP features are exploited to verify one of the candidates as the final class label.

2   Facial Expression Database

The database we use in our experiment is the Japanese Female Facial Expression (JAFFE) Database [8]. The database contains 213 images of ten expressers posed 3 or 4 examples of each of the 7 basic expressions - happiness, sadness, surprise, anger, disgust, fear and neutral. Some examples from database are shown in the first row in Fig. 1. In preprocessing 6 fiducial points are interactively marked, and the region around the fiducial points is called Region of Expression (ROE), see the second row in Fig. 1 for illustration. In our experiment, it is only from these ROEs that our expression features are extracted.

Fig. 1. Samples from JAFFE (the first row), their fiducial points and ROEs (the second row)

3   Feature Extraction

3.1   MB-LGBP Composite Features

Gabor filters have been proved to be effective for expression recognition because of its superior capability of multi-scale representation, while MB-LBP is a powerful descriptor for encoding local-holistic textures. In [6] we combine the idea of Multi-scale Gabor representation with the concept of MB-LBP encoding to achieve both locally and globally informative MB-LGBP features. First, we adopt Gabor filters to extract multi-scale representations of the expression regions in an image. Then we utilize MB-LBP with certain block size to encode Gabor representations at the according scale, which yields our so-called MB-LGBP composite features.

3.2   2D MB-LGBP Composite Features

The discrimination of different expressions needs more precise description of local textures. As 1st-order descriptor, LBP can not encode spatial structure information. And this problem can not be solved in nature only by partitioning. So we utilize Gray Level Co-occurrence Matrix (GLCM) instead of the traditional statistical histogram in

MB-LGBP encoding and present the so-called 2-dimensional LBP features (2D MB-LGBP), and get the 2D MB-LGBP composite features for classification.

Gray Level Co-occurrence Matrix. As a 2nd-order descriptor, GLCM is a matrix or distribution that is defined over an image to be the distribution of co-occurring values at a given offset. Mathematically, a co-occurrence matrix C is defined over an n × m image I, parameterized by an offset $\delta(\Delta x, \Delta y)$, as:

$$C_{\delta(\Delta x, \Delta y)}(i, j) = \sum_{p=1}^{n} \sum_{q=1}^{m} \begin{cases} 1, & \text{if } I(p, q) = i \text{ and } I(p + \Delta x, q + \Delta y) = j \\ 0, & \text{otherwise} \end{cases} \quad (1)$$

As can be seen from Eq. (1), for a L-grayscale image, C is a LxL matrix. In our experiment, the input image is compressed to 8 grayscales. And the following four kinds of offset δ are taken into consideration.[7], as shown in Fig.2 ($D_x = D_y = 2$).

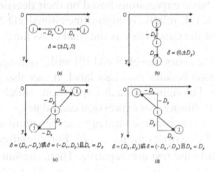

Fig. 2. Four kinds of taken into account in GLCM computing

4 Multi-level Classification

4.1 Traditional SVM Classification with Dichotomy-Dependent Weights

In [6] we introduce dichotomy-dependent weights mechanism for SVM classification and get about 75% average recognition rate for expressions of novel individuals from JAFFE Database, which is acceptable for subject-independent expression recognition based on static images.

Table 1 gives detailed information of classification results. Each row indicates how many samples from one class are classified into all seven expressions.

From Table 1 we can see that the majority of misclassified samples fall into one or two wrong class labels (italic number in table 1), which means it is much easier to recognize some expression candidates rather than a specific expression. So we can select two expression candidates first to make latter refinement of the categorization possible.

Table 1. Classification result for each expression

classified exp Real exp	AN	DI	FE	HA	NE	SA	SU
AN	23	1	1	2	3	0	0
DI	4	22	3	0	0	0	0
FE	1	4	22	0	3	0	2
HA	0	0	0	26	5	0	0
NE	0	0	0	0	28	0	2
SA	0	2	1	0	0	26	2
SU	0	0	3	2	4	0	21

4.2   Two-Level Classification

We present a two-level recognition framework in this paper. In the coarse level, low-dimensional MB-LGBP are utilized as features, in which two expression candidates are selected from seven basic expressions based on their decision confidence given by SVM classifier. In fine level, relatively high-dimensional 2D MB-LGBP features are exploited to verify one of the candidates as the final class label.

The Coarse-Level. In the coarse-level, SVM [9] with "one-against-all" max response strategy is selected because besides the class label, it can also provide the confidence information of a sample belonging to each expression, which is very useful for our first-level classification to filtrate two expression candidates.

The "one-against-all" max response strategy can be summarized as follows: In the training stage, we train 7 classifiers h_1, h_2, ..., h_7. For classifier h_i, samples belonging to $class_i$ are positive, while the rest are negative. Then, the sample $\vec{x}$ is classified by each of the 7 classifiers, suppose the responses are $f_1(\vec{x})$, $f_2(\vec{x})$, ..., $f_7(\vec{x})$. The decision is $\arg\max_i (f_i(\vec{x}))$, $1 <= I <= 7$, and response $f_i(\vec{x})$ can be regarded as confidence of $\vec{x}$ belonging to $class_i$.

By computing the confidence information of a sample belonging to all of the 7 expressions we get the confidence table for each sample. Table 2 gives the corresponding result of some sample images from JAFFE database belonging to each expression.

Table 2. Confidence table for some JAFFE samples

Expression Sample	AN	DI	FE	HA	NE	SA	SU
KA_AN2	0.62	0.28	0.21	0.36	0.04	0.44	0.46
KA_DI3	0.26	0.39	0.47	0.11	0.15	0.35	0.24
KA_SU1	0.14	0.24	0.33	0.02	0.44	0.19	0.58

We say that a sample is classified correctly if the class with the highest confidence is its correct expression, else it is misclassified. Thus we get total recognition accuracy of 74.18%. However, if two classes with the first two high confidences are filtrated as expression candidates, the probability of the correct class in them is about 87%. And if three classes with the first three high confidences are selected as candidates, the probability can be promoted to 93%. Fig.3 gives an illustration of the probability when the candidates number varies from 2 to 7.

In experiment, we choose the number of expression candidates to be 2. The reasons are: (1) The probability in Fig.3 increases most when the number of candidates varies from 1 to 2. (2) Two candidates can be easily classified by a SVM dichotomy in latter refinement.

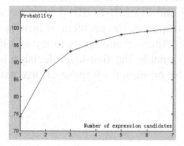

Fig. 3. Probability of the correct class in expression candidates

The Fine-Level. In the fine level, more delicate 2D MB-LGBP features are utilized to further discriminate two expression candidates. SVM with dichotomy-dependent weights mechanism [6] is selected as classifier, and the weights are assigned according to the two expressions to be dichotomized.

5   Experimental Results

In order to evaluate the multi-level expression recognition framework we present in this paper more objectively, we compare the performance of our system with several other popular recognition paradigms, such as MB-LGBP+SVM [6], GaborHistogram+SVM, and Gabor+Adaboost, the results are shown in Fig.4. In experiments, we use the leave-one-group-out cross-validation method to use the database adequately when testing on standard database. That is, we separate the whole database into 10 folds according to the people it contains. In each fold only images of one person is selected as testset and the left nine groups are trained.

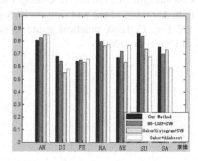

Fig. 4. Person-independent recognition rate of each expression on JAFFE images

6   Conclusions and Future Work

In this paper, a distributed expression recognition approach based on MB-LGBP feature and multi-level classification is proposed and experiments are designed to verify the validation of our method by comparing the recognition accuracy of our system with several other popular paradigms. As shown in Fig.4, our method can promote the accuracy of facial expression recognition prominently.

However, in MB-LGBP and 2D MB-LGBP feature extraction, we only make use of the amplitude of Gabor response, the problem of how to utilize the phase part still needs further consideration. Another attempt is to try low-dimensional feature in the coarse-level and selectively enable the fine-level decision routine. We hope the efficiency of our system can be promoted by these efforts without any obvious loss in recognition accuracy.

References

1. Claude, C.C., Fabrice, B.: Facial Expression Recognition: A Brief Tutorial Overview (2002)
2. Liu, W.F., Wang, Z.F.: Facial Expression Recognition Based on Fusion of Multiple Gabor Features. In: 18th International Conference on Pattern Recognition (ICPR 2006), vol. 3, pp. 536–539 (2006)
3. He, L.H., Zou, C.R., Zhao, L., Hu, D.: An enhanced LBP feature based on facial expression recognition. In: 27th Annual Conference Proceedings of the IEEE Engineering in Medicine and Biology, Shanghai,China (2005)
4. Tan, H.-C., Zhang, Y.-J., Chen, H.: Person-independent expression recognition based on person-similarity weighted expression feature. Journal of Systems Engineering and Electronics, 118–126 (April 2010)
5. Aleksic, P.S., Katsaggelos, A.K.: Automatic facial expression recognition using facial animation parameters and multistream HMMs. J. IEEE Transactions on Information Forensics and Security 1(1), 3–11 (2006)
6. Zhang, Z., Zhao, Z.: Expression Recognition based on Multi-Scale Block Local Gabor Binary Patterns with Dichotomy-Dependent Weights. In: Yu, W., He, H., Zhang, N. (eds.) ISNN 2009. LNCS, vol. 5552, pp. 895–903. Springer, Heidelberg (2009)
7. Zheng, Z., Zheng, Z., Tiantian, Y.: Expression Recognition Based on 2 Dimensional MB-LGBP Features. Journal of Computer Applications (4), 964–966 (2010)
8. Lyons, M., Akamastu, S., Kamachi, M., Gyoba, J.: Coding Facial Expressions with Gabor Wavelets. In: IEEE Conf. on Automatic Face and Gesture Recognition, pp. 200–205 (1998)
9. Hsu, C.-W., Chang, C.-C., Lin, C.-J.: A practical guide to support vector classification, pp. 1–12. National Taiwan University (2004)

Semantic Processing on Big Data

ZhenXin Qu

School of Information and Safety Engineering,
Zhongnan University of Economics and Law. 430073 Wuhan, China
{ZhenXin.Qu,quzx72}@gmail.com

Abstract. To process Big Data more efficiently and intelligently, new algorithm was proposed. Big Data was represented with RDFs; schema of data was transformed into Finite Semantic Graph. Using Map/Reduce computing model, reasoning algorithm was designed to process mass data. Query was also transformed into Finite Semantic Graph, and semantic matched with full graph. Experiment has shown that algorithm is effective.

Keywords: big data, semantic, map/reduce.

1 Introduction

Big Data [1] is not new concept, which was talked about again and again by Bill Inmon in 1990s. But it becomes hot spot in recent years. That's due to development of Internet, cloud computing and Internet of things. With these developments, mass data are produced continually. These data are too big to be processed effectively. It is more difficult to acquire useful information from boundless data. Traditional technologies cannot solve these problems. Semantic technology [2] can make data machine readable, then data can be processed intelligently. Big Data can be processed by semantic technology.

But existing ontology reasoner cannot process mass data. For example, Jena [3] and Pellet [4] can process triples not more than 10 millions in single computer; BigOWLIM can process triples more than 100 millions in single computer with bad performance.

Cloud computing [5] is a good technology to process Big Data, which has powerful computing ability.

Based on analysis of these problems, a new algorithm is proposed in this paper, semantic technology and cloud computing are integrated to reason and query Big Data. This paper is organized as follows. Section 2 talks about relative research. Section 3 describes how data are represented. Section 4 elaborates how mass ontology is reasoned. Section 5 describes the course of query. Section 6 reports experiments. Section 7 gives a summary and further work direction.

2 Relative Works

Many researchers have developed new technologies to solve the problem.

D. Jin and S. Lin (Eds.): Advances in MSEC Vol. 2, AISC 129, pp. 43–48.
springerlink.com   © Springer-Verlag Berlin Heidelberg 2011

Distributed cache, distributed database, distributed file system and many kinds of NoSQL databases are designed to process Big Data. But data still are not machine readable, data sharing between systems is difficult.

Seidenberg [6] has development ontology partitioning technology to reduce data scale, only concerned data are processed. But his work mainly focuses on schema, not instance, so Big Data cannot be processed effectively in his way.

To reason on mass ontology, Eyal [7] has developed reasoning algorithm using P2P technology, Ramakrishna [8] has developed parallel reasoning algorithm. But there is no evidence proving these algorithms can process Big Data.

3   Representations of Big Data

All data are represented in triple form, which have the structure "subject-predicate-object". With this form, data are described with all kinds of vocabularies. Furthermore, data should be expressed with normative method to be processed automatically and intelligently. Resource Description Framework (RDF) is a good choice. RDF specification is one of W3C recommendations published in 2004, which has XML grammar, formalization description method and perfect theoretical principle. It's a general model to describe resource. RDF triple has the structure "resource-property-property value".

But only resource description framework is defined by RDF, no vocabularies are defined to describe resource. RDF allow anybody creates own vocabulary set with vocabulary set description language. RDF schema (RDFs) is the language. RDFs stipulate how to describe vocabulary set with RDF. It also provides a vocabulary set to describe vocabulary set of RDF.

With RDF and RDFs, Big Data are made machine readable. So data are represented with RDF triple in our algorithm, and schema of data is described with RDFs.

Instance data are in large quantity in Big Data, not schema data. To improve efficiency of data processing, schema data are transformed into graph. Semantic reasoning and query are based on the graph.

Concept is represented by class and relationship between concepts is represented by property in RDFs. And there are some constructors to build more complex expression. Vocabularies of RDFs are as follows.

class (Thing, Nothing)	rdfs: domain
rdfs: subClassOf	rdfs: range
rdf: Property	rdfs: subPropertyOf

Having analyzed these vocabularies and their semantics, finite semantic graph (FSG) [9] is defined as follow.

Definition 1 Finite Semantic Graph. Finite semantic graph (FSG) is a quintuple.

$$G= (K, V, E, Pv, Pe)$$

K is ontology knowledge base. V is set of vertexes, which is composed of concepts and properties of ontology. E is set of directional edges. Pv is set of vertex's properties, which have value TYPE. Pe is set of edge's properties, which have value DOMAIN, RANGE, SUBCLASSOF and SUBPROPERTYOF.

It doesn't make sense to search entire FSG when search FSG. FSG should be divided into sub-graphs according to properties of edges. For example, when searching ancestor relationship, only edges having property SUBCLASSOF are reserved, FSG is degenerated into a sub-graph, searching work is carried out on the sub-graph.

4   Reasoning

RDFs can describe semantics of Big Data; we can reason new knowledge from existing knowledge.

How to reason? W3C has given 14 entailment rules [10], based on which we have designed new algorithm. Parts of these rules do not help to deduce new knowledge or deduce important new knowledge, so are ignored. Only rule 2,3,5,7,8,9,11,12 and 13 are reserved.

Some of these 9 rules can deduce schema data; the others can deduce instance data. Based on the analysis, the course of reasoning is divided into 2 parts. Inputted instance data are matched rule 2,3,7,9, inputted schema data are matched rule 5, 8, 11, 12 and 13. Matching sequence is shown as Fig. 1.

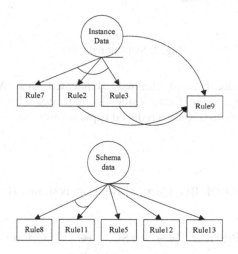

Fig. 1. Reducing rules

Rules linked by arc in Fig. 1 means input data should be matched one by one, that is AND relationship. Relationship between the other rules is OR. Rule 9 is triggered by instance like v rdf:type u, while rule 2 and 3 just reduce instance like it, so reduction results of rule 2 and 3 trigger rule 9 again.

To process mass data, our algorithm uses Map/Reduce computing model. The course of reasoning is divided into two steps: Map and Reduce, which is shown as

Fig. 2. Big Data are divided into splits automatically which are processed firstly by local Mapper. Later, intermediate result is shuffled, sorted and merged by Hadoop. Then it was passed on to Reducer.

Main work of Mapper is to realize rules shown in Fig. 1. First, Mappers outputs all original instances. Because they only know local data and can not judge replicated triples, so they outputs original triples and reduced triples to be judged by reducer latter. Second, triples are treated according to their type. If they are instances, they are matched with rule 2, 3, 7 and 9. If they are schema data, they are matched with rule 5, 8, 11, 12 and 13. During the course of reducing, relations between rules also be realized.

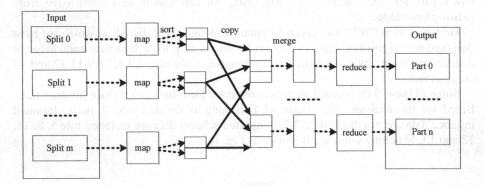

Fig. 2. Steps of reducing

Main work of Reducer is to eliminate duplicated data. All same triples are combined into a <key, value_list>. Value_list indicates who are original data or reduced data. According to it, only a reduced triple is reserved.

5   Query

Query language is SPARQL BG. Query of Big Data is shown as Fig. 3.

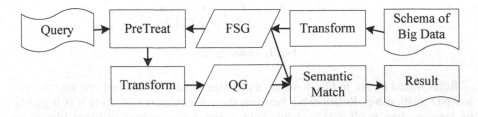

Fig. 3. Steps of querying

First, complete Big Data schema is converted into FSG and stored.

Second, query is pretreated and transformed into conjunction express.

Third, conjunction express is converted into FSG named QG. If QG is unconnected, it means that query includes meaningless condition, query should not be processed.

Fourth, search all sub-graphs from FSG matched QG. The match looks like sub-graph isomorphism problem. In fact it is not. The match is not precise match but semantic match.

At last, result triples are obtained according to matched sub-graphs. Detail algorithm has been described in literature [9].

6   Experiment

Experiment environment has 30 PCs, every PC has dual-core CPU and 2G ram. Experiment platform is based on Hadoop-0.20.2. Test data is generated with LUMB. Eight hundreds universities are generated, and there are 106 millions triples. Query listed below is executed.

```
PREFIX rdfs: <http://www.w3.org/2000/01/rdf-schema#>
PREFIX ub:
<http://www.lehigh.edu/~zhp2/2004/0401/univ-bench.owl#>
SELECT   ?teacher ?student
WHERE
{
    ?teacher   rdfs:type   LUMB:Person.
    ?student   ub:Advisor   ?teacher
}
```

The query wants to list all teacher-student relationship. Professor is subclass of Person, Associate Professor and Assistant Professor are subclass of Professor. After 263 seconds result is got. All types of teacher and their students are listed.

7   Conclusion

With Map/Reduce computing model, we have ability of processing mass data. With RDFs, we have ability of describing semantics of Big Data, and then data have the basis of being processed intelligently. Based on both, we design algorithm to reason and query on Big Data, semantics of Big Data can be expressed and processed. Big Data can be understood by people or computer more precisely and be used and shared more effectively. But semantics of Big Data are described with RDFs, RDFs has limited semantic representation ability. We will use OWL to perfect the work in the future, which has more powerful ability to describe semantics of data.

Acknowledgement. The authors would like to thank Zhongnan University of Economics and Law, who provides special fund from basic professional charge for research for central higher education institutions (2009092).

References

1. Jiangjie: Technical Review of Big Data. Programmer 8, 28–30 (2011)
2. Lu, J.: Theory and Technology of Semantic Web. Science Press, Beijing (2007)
3. Brian, M.: Jena: A semantic web toolkit. IEEE Internet Computing, IEEE 6(6), 55–59 (2002)
4. Sirin, E., Parsia, B., Grau, B.C., Kalyanpur, A., Katz, Y.: Pellet: A practical OWL-DL reasoner. Web Semantics: Science, Services and Agents on the World Wide Web 5(2), 51–53 (2007)
5. Tom, W.: Hadoop: The Definitive Guide. Tsinghua Press, Beijing (2010)
6. Julian, S., Alan, R.: Web ontology segmentation: Analysis, classification and use. In: Proceedings of the World Wide Web Conference, pp. 13–22. ACM Press, New York (2006)
7. Eyal, O., Spyros, K., George, A., Ronald, S., Annetteten, T., Van Frank, H.: http://www.larkc.eu/marvin/btc2008.pdf
8. Prasanna, R.S.: Parallel inferencing for OWL knowledge bases. In: Proceedings of the 37th International Conference on Parallel Processing, Oregon, pp. 75–82. IEEE Computer Society (2008)
9. Qu, Z.: Semantic Web and Interoperation of E-Government Information System. Hubei Science and Technology Press, Wuhan (2010)
10. Patrick Hayes, http://www.w3.org/TR/rdf-mt

Study on Land Use Structural Change Based on Information Entropy

Saiming Yang

College of Economy & Management, Taiyuan University of Science and Technology,
Taiyuan, Shanxi, 030024, China
yangsaiming@163.com

Abstract. Based on the two temporal remote sensing（RS）images in 1978 and 2007, land-use structural change of Natun coal mining area is studied over the past nearly thirty years based on GIS technology. The results show: The information entropy of land-use composition increases from 1.14 to 1.81, the balanced index from 0.52 to 0.82, and the dominance index decreases from 0.48 to 0.18. This explains that land-use system of Nantun coal mining area evolutes towards the relatively disordered state.

Keywords: information entropy, land-use structure, change.

1   Introduction

Nantun coal mining area locates in Zoucheng City Shandong Province of East China, one of the most developed area of China and the frontal area for coal transferring from north to south. It is adjacent to world major coal importing countries - Japan, Korea and enjoys convenient transportation network.

Nantun coal mining area includes Nantun coal mine and its surrounding farmland, forest land, rivers, roads, residential area and so on. Its area is 35.21km2.

Nantun coal mine was started construction in 1966 and put into commercial operation in 1973. Its original designed capacity was 1.5 Mt/a and expanded to 2.4 Mt/a in 1993 after reconstruction. The main mining coal seam is constituted by two stratums with average thickness of 5.35 meters for the upper layer and 3.21 meters for the lower layer. Up to 31, December 2007, its proven and probable reserve is totaled 125.4 million tones. Original designed capacity of Nantun Coal Mine preparation plant is of 1.80 Mt/a. The plant mainly produces No.2 clean coal/steam coal by jig machine. And most of its machines are manufactured in China.

2   Land Use Classification

According to national standards of land use classification (GB/T21010-2007), with the actual situation of land resources and information in study area,land-use types of Natun coal mining area are divided into Arable land, woodland, Garden land, Waters, House land, Storage land, Transportation land, Digging land and Collapse land.

D. Jin and S. Lin (Eds.): Advances in MSEC Vol. 2, AISC 129, pp. 49–51.

3   Land Use Structural Change Based on Information Entropy

Area of each land-use type got through remote sensing and geographic information system software is shown in table 1.

Table 1. Area of each land-use type in Nantun coal mining area from 1978 to 2007 (hm^2)

Land use types	1978	2007
Arable land	2434.85	1477.41
woodland	404.99	489.54
Garden land	164.90	204.74
Waters	168.56	188.86
House land	162.08	328.32
Storage land	85.11	303.37
Transportation land	85.24	102.08
Digging land	11.40	72.20
Collapse land	3.41	354.02
Total area	3520.54	3520.54

Land use structural change reflected intervention trends of human use of land resources and land use acts,which can use the information entropy, balanced index and dominance index to reflect and measure[1-3].

Information entropy are used to describe the extent of land use division, to a certain extent, reflect the influence of human activities on land use. The greater information entropy indexes of land use in the region are higher the degree of segmentation. The formula is shown in below. H is information entropy; Pi is the percentage of each land use types in total land area, n are land use types.

$$H = -\sum_{i=1}^{n} P_i \cdot \ln P_i \tag{1}$$

Balanced index and dominance index reflect the equilibrium level of land allocation in different land-use types. Balanced index is the ratio of the Information entropy and the maximum entropy.The value range of the balanced index is from 0 to 1.When E=0, land use is in the state of the most non-equilibrium. When E=1, land use is in the state of the most equilibrium. The formula is shown in below.E is the balanced index; Hmax=lnm shows the maximum of diversity Index; m is the quantity of land-use type; Pi is the percentage of the "i" land use types in total land area.

$$E = H/H_{max} = -\sum_{i=1}^{n} (P_i \cdot \ln P_i)/\ln m \tag{2}$$

Dominance index is opposite with the balanced index. The formula is shown in below. D is dominance index, which reflects the dominance degree of one or several land-use types to land-use types in the region.

$$D=1-E \qquad\qquad (3)$$

The information entropy, balanced index and dominance index got from the above formula and land-use data are shown in table 2.

Table 2. Land-use structural indexes and change of Nantun coal mining area

Indexes	1978	2007
Information entropy	1.14	1.81
Balanced index	0.52	0.82
Dominance index	0.48	0.18

4   Conclusion

The information entropy of land-use composition of Nantun coal mining area from 1978 to 2007increases from 1.14 to 1.81, the balanced index from 0.52 to 0.82, and the dominance index decreases from 0.48 to 0.18. This explains that land-use system of Nantun coal mining area evolutes towards the relatively disordered state.

Acknowledgement. This work is supported by the Doctor Research Foundation of Taiyuan University of Science and Technology (W20112003).

References

1. Reid, R.S., Russell, L.K., Nvawira, M., et al.: Land-use and land-cover dynamics in response to changes in climatic,biological and socio-political forces: the case of southwestern Ethiop. Landscape Ecology 15, 339–355 (2000)
2. Weber, A., Nicola, F., Detlex, M.: Long-term land use changes in a mesoscale watershed due to socio-economic factors-effects on landscape structures and functions. Ecological Modelling 140, 125–140 (2001)
3. Lepers, E., Eric, F.L.: A synthesis of information on rapid land-cover change for the period 1981-2000. Bioscience, 115–124 (2005)

Game Analysis on Adoption of Incentive Strategies in Embedded Rural Company Cluster: Case Study Based on Shangli Firework Industry

Jianlin Zhang, Ting Wu, and Xinglin Wu

School of Management, Zhejiang University
Hangzhou, Zhejiang, China
maomaojitt@163.com

Abstract. This article studies the game process when township enterprises introduce incentive strategies in Shangli fireworks industry in Jiangxi Province, where social network is based on relationship between families, friends, neighbors and other relationship. Compared with peers in the cluster, if employees feel unfair about festival welfare, bonus, pay systems, they would quit. Therefore reputation is very important for companies, and sometimes companies have to follow incentive strategies of competitors to attract employees. In addition, cost is also an important factor for them to introduce incentive strategies.

Keywords: game, incentive strategies, embeddedness, cluster, social network.

1 Introduction

In China's rural areas, there are a lot of spontaneously formed industrial clusters existed. Only in Zhejiang and Guangdong Province, hundreds of specialized town have emerged in clothing, ceramics, lighting, watches, footwear, hardware, toys, furniture and other traditional industries[1].However, for most economically backward areas, these spontaneously formed industries or enterprise clusters are under-development and irregular. According the definition by Lynn Mytelka, Fulvia Farinelli, this kind of industrial clusters are informal clusters[2],in which, relatively rational, standardized chain model have't been established, due to restrictions on knowledge and technology diffusion. In this areas, business is operated mainly by local government rules, cultural constraints, and non-contract agreement [3].This is the so called embedded cluster[4].When enterprises are social enviroment of this kind, they have to game with the particular social network,and adjust stategies to survive and develop.

Information communicates between enterprises in the cluster can classified into three kinds: technical information, market information and management experience. However, most of current research focuses on technology diffusion, organizational learning, as Tang Chang'an[5] studied technology diffusion clusters in the the growth and mature stage, Min Xue[6] studied organizational learning and knowledge diffusion. However there are limited research on tacit information like management experience diffusion. Because of restrictions on economic in Shangli town, relationships in the social network play a very important part in the firework industry there. This paper analyzes the spread of management experience based on certain social network. According to the fireworks industry in Shangli, this paper will analyze the diffusion of incentive strategies which staff concerned most. And latter, we will discuss the game process with competitors in cluster when some company want to use incentive strategy.

2   Basic Concepts of Embedded Cluster

Business is always embedded in a certain environment. Embeddedness means that because of the history root of a certain group, group members developed conventions and stable relationship in the long-term contact. And when group members want to do something, the conventions and relationship will affect their actions and behavioral tendencies. That is to say one person's action is affected by a certain social network.

In reality, embeddedness is inevitable. For economic actors, when they embedded in certain social network, they will have their decisions and subsequent actions been affected. For network participants, the networking between actors is an important channel to obtian information and resources. According to Granovetter, embeddedness can be classified into relationship embeddedness and structural embededdness [4]. Corresponding, clusters can also be assorted into relationship embedded and structure embedded cluster.

Relational embeddedness means bilateral relations. In this relationship, both sides attach importance to the needs and objectives of each other, and take action based on credits ,trust and information sharing . Structural embeddedness refers to organization in the group with bilateral relations, and also of the same relationship with a third party. So,It enables members in this social network to connect with each other. and then to form an association structure characterized by systematization. Therefore, the structural embeddedness is a function of interaction among numerous participants, which makes information in the network flow horizontally, vertically, or diagonal movement. Meanwhile, Granovertte defined the degree of embeddedness by four dimensions, times and frequency of interaction, emotional intensity, intimacy and the degree of reciprocity, and then defined the strength of the relationship as two kinds. That is strong connections and weak connection. Strong connection refers the connection time, affection and intimacy (trust each other) between organizations are strong,. The contrary is weak connection[4].

3   Case Selection

In this article, we chose Shangli town fireworks industry in Jiangxi Province as the research sample. And,we chose the Shangli fireworks industry as a typical case based on three reasons. First, the fireworks industry in Shangli town is a typical case with embedding social networks. Because the proportion of migrant workers is only about 2% in the total labor force, while the rest is the closely related local labor force[7].

Second, the Shangli fireworks industry relies on non-contract social networks. Since companies do not have regulate employment system, employees and businesses don't sign any agreement, which enables employees to replaced company and jobs more freely. Employees will consider changing companies, if some regimes or salaries are beyond the acceptable range of staff,or they feel unfairness when compared with competitive firms. And because the social environment of the cluster belongs to embeding networks, and people closely contact with each other, if a company loses popular support, the company would find it difficult to employe by word of mouth of staff. Thus in order to ensure the fairness of regime, games between enterprises, between enterprises and employees take place time to time.

Third, the researchers have a deep understanding of fireworks industry in Shangli, and collected large amounts of data about relevant companies and industries. Because researchers used to be local employees, we learn about this industry from childhood. Thus we have a deep understanding of the way information exchanges, and of games between enterprises, between managers and employees.

4 Games on Use of Incentive Strategies in Shangli Fireworks Industry

As embedded in the social networks of strong connection, rooted in specific social and cultural environment, communication between enterprises and employees will affected by appropriate social, cultural and environment. So when rural enterprises in local cluster want to update their technology, equipment, knowledge, systems, management experience enterprises, etc., they should think about influence to the stakeholders. In Shangli fireworks industry cluster, the introduce and use of incentive strategy catches the most concern, and is also the most common game between enterprises.

4.1 Types of Incentive Strategy

There are three mainly employee incentive systems in Shangli fireworks cluster, namely: the festivals welfare system, employees year-end bonuses system, and Pay rise system.

Festivals welfare system. Festivals welfare system began 7 years ago in the Shangli area, because of the celebration Birthday of fireworks founder Lee Tin. As the fireworks industry is a quite, dangerous business, when regulation of government is not strict, accidents took place almost every year in Shangli. Therefore, the memorial for Lee Tin expressed both respect and memory for the founder. People want to be blessed and peaceful, so they want souvenirs in any form in this day of celebration. Welfare payment to employees give them psychological comfort.

Bonus. Fireworks manufacturers in the region adopted a piecework system, pay the staff according to the workload. Because of constraints in economic and management, companies has not raised the bonus system, and employees have no consciousness for the award money. Until 3years ago, as sales in short supply, the current workload of the staff were insufficient to meet the demand. Because of the given number of employees, some of the pioneers decided to give bonuses to employees whose attendance and production have reached a high degree, to stimulate them to work overtime and leave a good impression in the hearts of the staff to ensure that next year's source of workers.

Pay rise. Companies develop this system because they want to attract workers, and ensure the output of year.

According to the motivation-hygiene theory of Frederick Herberg, welfare, bonus and salary are hygiene fators. However, in the case of the same level of effort, compared to other companies in the industry, when workers have a sense of unfairness about the corporate welfare, bonuses and payment., that will cause a negative impact for enterprises in the coming year.Thus the development of incentive system should game with companies in the industry.

4.2 Model and Analysis

Since companies are geographically near in Shangli fireworks industry, they understand each other very well. Meanwhile, both products and staff skills are homogenous in the areas, and comumunication between workers and companies are quite frequent. People are very sensitive to innovation and change of the incentive system.

Thus we assumed that game in this situation, belongs to complete information static game. When two homogeneous companies in this cluster get awareness of an incentive strategy at the same time, they make decision simultaneously whether to adopt this strategy. Their decisions have no order.

In the view of stable state of the total labor force and production, which we assumed that the total production of the cluster is changeless, changes in economic efficiency are cause by movements of the staff.

Assuming that under normal condition, the total income of the cluster s is R. That is to say the total income of production is R. Here R>0. Under normal production conditions, the two companies get a benefit of $R/2$. When only one companies use incentive strategy, workers would flow to the adopt company to avoid disadvantages. Assuming the cost of the only adopting company is C_1, and the extra benefits of using incentive strategies is M. Then the total income of the single using company is $R/2 + M-C_1$, while the other is $R/2-M$. Note that, although the incentive strategy will result in the movement of workers, but due to geographical factors and the existence of affinities, not all employees will choose to change companies, so $R/2> M$.

When both the two companies have adopted incentive strategies, we assume that the cost of the implementation of incentive strategies is C_2. Since there are no movements of workers, workers in each company almost the same, therefore we assume the relation between the two costs is $C_2<C_1<2C_2$.Both of the two companies get the same benefits $R/2-C_2$.

The benefit matrix of the game is shown in Fig. 1:

Company1 \ Company2	Do not use	Use
Do not use	R/2, R/2	R/2-M, R/2+M-C_1
Use	R/2+M-C_1, R/2-M	R/2-C_2, R/2-C_2

Fig. 1. Benefit matrix of using incentive strategies

Because all the games take place on the basis of equal profit R/2 , so benefit matrix in Fig. 2 can be simplified as follows:

Company1 \ Company2	Do not use	Use
Do not use	0, 0	-M, M-C_1
Use	M-C_1, -M	-C_2, -C_2

Fig. 2. Simplified benefit matrix

Case 1: When M-C_1<=0 , the use of incentive strategics will reduce the company's overall Benefits, therefore the Nash equilibrium at this time is (do not use, do not use).

Case 2: When M-C_1>=0, the use of incentive strategies will increase the company's overall Benefits, therefore the Nash equilibrium at this time is (do not use, do not use).As M-C_1>0 and C_1>C_2, M>C_2 , . Nash equilibrium at this time is (use, use).

Both cases are pure Nash equilibrium.

5   Conclusion and Analysis

Festival welfare in Shangli region, means mainly Founder's birthday and traditional festival (Dragon Boat Festival, Mid-Autumn Festival and Spring Festival). In general, festival welfare mainly are small gifts and fruit,which have a low cost. Workers would be very pleased if companies celebrate festival with the staff. If companies showed their generosity, it will be easier for them to recruit. In this situation, because of the low cost and the need for recruitment, others have to pay festival welfare as well. This kind of game meets the the second case.

Bonus. Year-end awards in the areas is similar with festival welfare. In most cases, companies pay 100 to 300 *yuan* each to those workers who have a high degree of attendance and high yield. Because of the relatively low cost and the need of a good reputation to attract workers for next year's, most enterprises have to adopt similar bonus strategies. This kind of game meets the second case.

Pay rise. In Shangli fireworks industry, companies take a salary calculation system of piecework. Salary is calculated at the end of year. So, salary increases will relate to full production of relevant products within one year. Compared to 100 to 300 *yuan* bonuses or holiday gifts, cost of salary increases will be greatly increased. Therefore, companies in the the cluster do not want to raise the payroll, that the reason why wage levels in the cluster has not been improved.

In short, characterized with frequent communication and close relations, if cluster embedded in certain social networks, they needed to consider the cost, the reputation of the company and the impact of employees when making decisions.

6 Prospects

This paper only studied games about three representative incentive strategies. Games of these three can not stand for all other incentive strategies, expecially diffusions of other management experience. Second, cases study in this reasearch with special social network, it meets most rural clusters in China, but not all township clusters. Future research would pay more attention to diffusions of other management experience in cluster. Also, future reasearch would study the operation of rural clusters in China, and to find suitable path for the development.

References

1. Qiu, B.: Smalll enterprises cluster. Fudan University Press, Shanghai (1999)
2. Lynn, M., Fulvia, F.: In: Local Clusters: Innovation Systems and Sustained Competitiveness (2000)
3. Su, J.: Study on ecological inbeing of Industrial Cluster. Fudan University (2004)
4. Granovertte: Economic action and social structure: the problem of embeddedness. The American, Journal of Sociology 91(3), 481–510 (1985)
5. Tang, C.: Game analysis on technological innovation diffusion on the early stage of industrial cluster. Science and Technology Management Research (7), 538–510 (2008)
6. Min, X.: Impact mechanism based on the absorption capacity to innovation performance. Zhejiang University (2010)
7. Development Plan for Shangli fireworks industry (2009), http://www.fireworkcn.net/ypnew_view.asp?id=3551

Research and Application on Organization Method Used in 3D Basic Geographic Data

Xiaopeng Leng[1], Fang Miao[2], Wenhui Yang[1], and Xiang Ni[1]

[1] College of Information Science and Technology, Chengdu University of Technology,
[2] College of Geophysics, Chengdu University of Technology,
610059 Chengdu, China
{lengxiaopeng,miaofang,yangwenhui,nixiang}@cdut.cn

Abstract. In recent years, along with digital earth and smart planet concepts launched and the research deepening, 3D GIS as a foundation platform has become a hotspot. The basic geographic data of 3D visualization contains image data and terrain data, how to efficient organize and manage mass geographic data is the key problem of 3D visualization. G/S model is a new type of spatial information network service model. Based on G/S model, a method that realize organization of 3D geographic data by hgml is raised. Finally, an application is used for support our design.

Keywords: G/S model, HGML, 3D visualization, data organization.

1 Overview

With the GIS technology applied in all walks of life, and the further development and mature of computer graphics and 3D technology and computer science, persons are more and more requirements to deal with GIS problems from 2D map to 3D space. 3D GIS not only broke the 2D limitations of the spatial information display,but also for the data expression and spatial analysis provides a 3D visual effect, and can provide more Intuitive show and decision support for the popularization and Industry application.The basic data for 3D geographical information system includes image data and terrain data, usually it consists of base maps or aerial photos or satellite images and DEM(Digital Elevation Model), In the largescale application of 3D GIS,how to efficient manage and organize the mass basic geographic data has been one of the emphases and hotspots at the field of 3D GIS.

3D GIS realize stereo visualization of geographical information based on traditional 2D GIS, this not only needs system has the outstanding performance of 3D display, but also needs ability of storage and organize massive geographic data. Based on the technical framework of G/S model, the paper discusses a method of organization and management of geographic data used HGML, verifies the feasibility of the 3D GIS application scheme in the geological disaster monitoring, and provides a solution for massive data organization in 3D GIS.

D. Jin and S. Lin (Eds.): Advances in MSEC Vol. 2, AISC 129, pp. 59–64.
springerlink.com   © Springer-Verlag Berlin Heidelberg 2011

2   G/S Model Framework

2.1   G/S Model Brief

G/S (Geo-Browser/ Distributed Spatial Data Servers,G/S) model, is a kind of spatial information service model, was raised by Prof. Miao Fang at Xiangshan-Science Conferences. Based on Internet environment, G/S model adopt open standard which compatible with xml, realize the organization, management and exchange of spatial information in the distributed network environment, and provide aggregation service for client by Geo-Browser. Server side mainly includes basic spatial data servers and industry data servers, basic spatial data servers group used to organize and manage such as remote sensing data, navigation satellite data, terrain data and other basic spatial information data; industry data servers group used to organize surveying and mapping data,geological data, environment data and other different industries project data. As core of data organization between GeoBrowser and Servers, HGML is a bridge of data exchange.

2.2   Hyper Geographic Markup Language(HGML)

As the core of G/S model, HGML adopt the XML Schema format, based on data type which defined by XML standards, and learn from KML and GML, can be compatible with relevant OGC standards. In the data organization, HGML can describe spatial data features such as type, file description, property, structure and levels. In the data representation, HGML can achieve an abstract description of the basic components of space entity, including point, line, area, volume object and spatiallocation relationships, and scale of raster data, data quality and spatial scope descriptions.

3   Data Organization

Visual data of 3D GIS mainly include the basic geographic data based on DOM(Digital Orthophoto Map) and DEM(Digital Elevation Model) , and the surface building 3D data which consist of the surface building model and texture mapping. Among them, the basic geographic data through the digital elevation data and the DOM which made by the remote sensing image or aerial photograph overlapped, can actually reflect the characteristics of the surface topography and ground image, with a vivid display effect, it's the core of 3D GIS data. Because usually basic geographical data in large scale and hierarchical structure relative complex, the mechanism of geographical data organization is the key research. And the surface building 3D visualization mainly in 3D model establishment and the display rendering technique, the author will discuss it elsewhere.

3.1   DEM Data

DEM is the core of the real terrain data, it's an entity ground model that express surface elevation by an array of sequential value. Using DEM data, create surface topography model and form digital terrain surface, then through overlap image data can constitute a realistic 3D effect of surficial undulation.

3D terrain visualization usually realized by DEM according to Regular Square Grid model and Triangulated Irregular Network model.

TIN(Triangulated Irregular Network), it's a method to achieve DEM surface modeling based on Triangle. The triangle represents the inclined surface, the mathematical expression of plane coordinates (x, y) and elevation(Z) data collection is:

$$Z = a_0 + a_1 x + a_2 y \ . \tag{1}$$

In this way, terrain surface consists of series of adjacent triangles, simple structure,but because of irregular network structure, data storage is more complex than regular grid approach.

RSG(Regular Surface Grid), is a kind of raster structure,realize DEM surface modeling based on the square grid, through four elevation points in square grid constitute a double linear surface, mathematical expression is:

$$Z = a_0 + a_1 x + a_2 y + a_3 xy \ . \tag{2}$$

In this way,the surface modeling based on grid, the terrain surface is composed of a series of mutual adjoining double linear surface, same with raster map, data storage and processing is very simple, especially suitable for large area terrain and continuous DEM modeling. DEM comes from direct rule grid sampling points or by irregular data points interpolation produce. Because computer deal with the matrix based on grid is very convenient, the elevation matrix of regular grid to become the most commonly DEM.

Based on the above characteristics of regular grid, the 3D GIS of large scale terrain data usually used regulay grid to form grid storage,such as the bil which NASA used, types of raster file,adopt continuous binary data storage, the bil file's data is relative sea-level altitude.

3.2 Image Data

The realistic effect of 3D GIS depend on overlaying of DEM with DOM, with the continuous improvement of the earth observation technology, high resolution remote sensing images already used more and more, high resolution image means more data volume. In order to solve the massive data organization and scheduling problems, usually the remote sensing image data are provided in the form of map tile pyramid, tile pyramid is a multi resolution and multi level model, raw data generate layers by re-sampling with different resolution, the same layer data get tiles through slice processing with specified grid size, tiles resolution more and more low from the bottom to the top,but covering the same geographical area. Building tile pyramid need slicing and layering according to the map projection degree first, often sliced into square tiles of the same size, which n-1 (n> 1) layer on the n layer is always quadtree inheritance, such as Fig.1:

Level 0

Level 1

Level 2

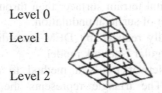

Fig. 1. Tile pyramid model

By quadtree structure map slices into square tiles for building tile pyramid model. When the original underlying data(maximum resolution base map) size is w*h, tile size is ts*ts pixel(such as 512*512), two adjacent layers of resolution change rate is m(m>1, generally an even number, often value is 2,by 2*2 pixels synthesis of 1 pixels to generate upper tile matrix), the total layers number of pyramid is:

$$l_{tot}=max\{[\log_m(w/ts)],[\log_m(h/ts)]\} .\qquad(3)$$

Hypothesis original data resolution for r_0, 0 level tile's latitude and longitude scope is d. Then when the level is l, relationship of data resolution r_l, single tile scope d_l and tiles number n_l as follows:

$$r_l= r_0/m^{\ l_{tot}-l-1} .\qquad(4)$$

$$d_l=d/2^l .\qquad(5)$$

$$n_l= (180/d)*(90/d)*4^l .\qquad(6)$$

For example, 0 level and slice layer by 36 degrees, map tiles of each level as shown in Fig.2

Level 0	Level 1	Level 2

Level	Degree	Tiles
Level 0	36	50
Level 1	18	200
Level 2	9	800

Fig. 2. Tiles level relationship

From (4),(5),(6) formula, the number of tiles increase in exponent with the increase of tile pyramid levels, the coverage range is invariable, but individual tile's geographical range is reduced and tile's resolution improved. Based on this kind of multi-resolution level structure, it is easy to realize LOD(Levels of Detail) technology. LOD is a effective method for improving display speed and performance, moreover, according to linear quadtree index the tiles could be located convenient by longitude and latitude.

Using Cartesian coordinates, the origin (x=0, y=0) in the projection coordinate lower left, namely South Pole(-90,-180). According to longitude and latitude coordinates(lat , lon) and level l and slice degree deg, we can obtain corresponding Cartesian coordinates of the image grid.

$$X =[(lon+180) / deg] * 2^l .\qquad(7)$$

$$Y =[(lat+90) / deg] * 2^l .\qquad(8)$$

3.3 Data Organization

After slice digital elevation and image raster data to tiles and distributed deploy, also need HGML to describe organizational relationship and management of data. GeoBrowser as displaying platform complete aggregation of resource needed through organization definition in HGML, achieve visualize representation of 3D data.

By HGML organization's main syntax is as follows:

(1) Raster data type description

```
<ImageFormat>elevation/bil16</ImageFormat>
```

Describe elevation data for the bil type file, and the grid files is continuous stored as 16 bit data.

```
<ImageFormat>Image/png</ImageFormat>
```

Show that image file format for png file, if need to support other file types ,such as png,dds,tif and other common image format, can increase defined as follows:

```
<AvailableImageFormats>png;dds;tif;
</AvailableImageFormats>
```

(2) description of the tile pyramid

```
<PyrLevels><num>15</num><empty>0</empty></PyrLevels>
```

The count tag is used to describe the number of pyramid levels, 15 shows that level numbers from 0 to 14. The empty tag shows display level number, 0 shows load data from minimum zoom level.

(3) description of the tile property

```
<TileOrigin>
    <units>degrees</units><latitude>-90</latitude >
    <longitude>-180<longitude></TileOrigin>
```

The tag used for describe basic property of tiles, adopt longitude and latitude to describe origin in Cartesian coordinate.

```
<LevelZeroTile>
    <units>degrees</units><latitude>36</ latitude >
    <longitude>36<longitude></LevelZeroTile>
```

Describe geographical scope that 0 level tiles coverage, the example means divide area with 36 degree.

```
<TileSize><width>512</width><height>512</height>
</TileSize>
```

For describe size of tiles , the example shows a square slice with width and length are 512 pixels.

4   Application Example

Based on the G/S model, use HGML for basic geographic data organization, the instance shows digital elevation data overlapped with image data in GeoBrowser. Some running results as follows, Fig.3 shows effect that loading elevation data and rendering surface by coordinate grid at the left, and image data overlapped at the right.

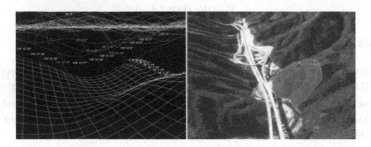

Fig. 3. Elevation model and image data loading

5   Conclusion

G/S model is a new spatial information network service model, overall framework based on distribute network structure, it has a uniform data standards HGML and realize client aggregation service full use of client computer processing ability. The paper represented a method that organize digital elevation and image raster data based on HGML, and realized gathering and displaying by GeoBrowser, achieved the expected results, through the actual application confirms the feability of this design and reveals good application prospect in 3D GIS field.

References

1. Miao, F., Ye, C.-M., Liu, R.: Discussion on digital earth platform and the technology architecture of digital China. Science of Surveying and Mapping 32(6), 157–158 (2007)
2. Dai, C.-G., Zhang, Y.-S., Deng, X.-Q.: An Organization and Management Approach of Data for Real-Time Visualization of Massive Terrain Dataset. Acta Simulata Systematica Sinica 17(2), 406–410 (2005)
3. Li, D.-R.: On Generalized and Specialized Spatial Information Grid. Journal of Remote Sensing 9(5), 513–519 (2005)
4. Ben, J., Tong, X.-C., Zhang, H.: A Pyramid Model Based on TIFF and XML. Journal of Institute of Surveying and Mapping 22(3), 184–186 (2005)
5. Yang, J.-Y., Zhang, Y.-S., Ji, S.: Image pyramid model dynamically reconstruction based on multi-resolution analysis of wavelet transform. Science of Surveying and Mapping 32(5), 50–51 (2007)

The Application of BP Neural Network in Hospitalization Expense Research

Chao Chen, Sufeng Yin, Dong Wang, Jianhui Wu, and Guoli Wang

Department of Epidemiology and Health Statistics,
School of Public Health, Hebei United University
Tang Shan 063000, China
ing_008@163.com

Abstract. Data on hospitalization expense for coronary heart disease from 2009 to 2010 in one tertiary hospital were taken as examples, so as to construct fitting models for hospitalization expense based on BP neural network. Sensitivity analysis of influence factors was executed to evaluate the effect degrees of influence factors on hospitalization expense. The purpose of this study was to evaluate application value of BP neural network used in hospitalization expense study, explore the analyze and evaluation method fitting for the construction and character of hospitalization expense, use medical costs rationally and control irrationally increase of hospitalization expense. The results showed that the main influence factors of hospitalization expense were hospitalization days, treatment outcome, times of rescuing and age, the comprehensive influences of which were 0.83101, 0.76113, 0.73227 and 0.44537 respectively.

Keywords: hospitalization expense, BP neural network, influence factors, sensitivity analysis.

1 Introduction

Artificial neural network is a type of mathematical models which simulate biological neural network and are connected by massive simple neurons. Artificial neural network is a complicated network system that can parallel process and nonlinear transfer information[1]. The application of artificial neural network has infiltrated into medical area recently. As one of the fullest developed and most used network models, BP neural network, which was most fully developed and widest applicated, was named after its special algorithm, which used error back propagation algorithm according to the adjusted regulations of network weight[2].

Nowadays, study of hospitalization expense has become the hot spot issue of our society[3]. As the distribution of data on hospitalization expense always displayed skewed and can be affected by many complicated factors which are related to each other, there exists some limits for traditional statistical methods to study hospitalization expense. As BP neural network has no requirements for type or distribution of the material, has the ability of fault tolerance, and is able to define the complex mapping relation between input variables and output variables, BP neural network has supported a brand new method for processing data with complicated relationships[4].

D. Jin and S. Lin (Eds.): Advances in MSEC Vol. 2, AISC 129, pp. 65–70.
springerlink.com   © Springer-Verlag Berlin Heidelberg 2011

Our study took coronary heart disease for instance, constructed the BP neural network for hospitalization expense of hospitalized cases to figure out the main influence factors of hospitalization expense, in order to explore the analyze and evaluation method fitting for the construction and character of hospitalization expense, and could support basis for taking target-oriented measures to use the medical resource rationally and control the growth of hospitalization expense.

2   Basic Theory of BP Neural Network

2.1   Basic Thought of BP Neural Network

BP neural network, which was first proposed by a research group led by Rumelhart and McCelland in 1986[5], is a kind of multilayer forward propagation networks and one of the most used neural networks currently. The classical network topology structure of BP neural network is composed by input layer, hidden layer and output layer. It has been theoretically proved that function approximation of any accuracy from N dimensional to M dimensional can be implemented by a BP neural network with one hidden layer for any continuous function on a closed interval. Compared with traditional statistical methods, advantages of BP neural network are as followed: has no requirements for type or distribution of the material, is able to define the complex mapping relation between input variables and output variables by self-study and self-organization, and has much better ability to deal with non-linear problems than traditional statistical methods.

2.2   Algorithm and Principle of BP Neural Network

BP neural network is based on gradient descent algorithm, the training of the whole network is composed by forward propagation and back propagation. According to forward process, sample signals are put into input layer, out put from output layer after processing by weight, threshold and transfer functions of neuron. If the error between output value and expected value is over expected error, the error back propagation stage is initiated to modify, error signals are sent back through the former connecting passage, error signals are minimized by modifying weights of neuron of each layer. The process to constantly revise weights, is the process of network training. The loop ends until the output error has been reduced to allowable value or has reached the fixed training number.

2.3   Basic Steps of BP Modeling

Simply speaking, there are three steps to construct BP neural network: 1. network initialization; 2. training; 3. simulation. Network initialization is to select and fix some network parameters, such as number of network layers, number of neurons in each layer, transmission function of each layer, speed of training, weights, threshold and algorithm of training. Make sure to avoid overfitting during training.

2.4   Select the Optimization Algorithm

As standard BP algorithm can easily be immerged in partial minimum and convergence speed is too slow, the training time might be very long[6]. Many optimization algorithms have appeared nowadays, such as LM algorithm and Newton algorithm. The convergence speeds of these new algorithms were greatly improved[7].

2.5 Sensitivity Analysis

By changing some part of network input to observe the corresponding change of network output, the significance of this part for predicting output can be determined. The concrete methods are to change the recorded variables of samples in succession, then record the maximum and minimum output during the changing process, and calculate the percentage of the difference value of maximum and minimum output in maximum output, and the sensitivity is the mean of all the recorded percentages.

3 Application

Matlab 7.1 developed by MathWorks Company was used to constructed BP neural network model in our study, and influence degrees of influence factors on hospitalization expense were evaluated by editing programs of sensitivity analysis based on MATLAB.

3.1 Data Sources

Data in our study come from front pages of medical record with coronary heart disease in one tertiary hospital in Tangshan city from 2009 to 2010, and number of total cases was 2437.

3.2 Data Initialization

All of the data were recoded by computers, cases with missing values and illogical cases were rejected. The number of valid cases was 2126, and proportion of valid cases in all cases was 87.26%.

Table 1. Quantized methods for influence factors

Factors	Code	Quantized methods
Gender	x1	Male=1, Female=2
Marital status	x2	In marriage=1, Others=2
Ages	x3	Years
Admission times	x4	Once =1, Twice and more=2
Rescue	x5	Yes=0, No=1
Payment pattern	x6	Insured=1, uninsured=2
Hospitalization days	x7	days
Treatment outcome	x8	Cured=1, Improved=2, Uncured=3, Dead=4

3.3   Database Division

Quasi-Newton Method (OSS algorithm) was used in our study. Sequence database by numbers of medical records, select number 3, 8, 13, 18, … (in equal intervals) in succession as testing set, the others as training set. Make sure database was divided into training set (80%) and testing set (20%). Training set was used to construct BP neural network model, and testing set was used to test network and its generalization ability.

3.4   Performance Results of Training Network with Different Numbers of Neurons in Hidden Layer

Networks with different numbers of neurons in hidden layer were randomly tested 100 times in our study, the results were as followed (Table 2).

Table 2. Performance outcomes of different training networks

Training times	5		15		20	
	Testing set	Training set	Testing set	Training set	Testing set	Training set
1	0.7460	0.7162	0.7009	0.7188	0.7059	0.7173
2	0.7455	0.7171	0.7068	0.7149	0.6978	0.7180
3	0.7214	0.6938	0.7003	0.7183	0.6997	0.7174
…	…	…	…	…	…	…
98	0.7451	0.7159	0.7008	0.7182	0.6981	0.7188
99	0.7483	0.7188	0.7009	0.7186	0.6999	0.7184
100	0.7465	0.7179	0.6991	0.7178	0.7015	0.7184

Table 3. Parameters of BP neural network on hospitalization expense

Network structure parameters	Network training parameters	Simulation results of testing set	Fitting results of training set
hidden layers: one	Training algorithm: OSS algorithm:	R=0.86209	R=0.84331
neurons in hidden layer: 15	A total cessation of training iterations: 20	R^2=0.74321	R^2=0.71118
neurons in input layer: 8	Leaning speed: 0.01	$R^2_{adj} = 0.41534$	$R^2_{adj} = 0.52435$
neurons in output layer:1	Performance function: SSE	SSE=5.2389e+009	SSE=2.081e+010
	Stop training SSE*=1.22463	MSE=1.6738e+007 RMSE=4091.2	MSE=1.7327e+007 RMSE=4162.6

* SEE was for normalized data and RMSE was for reverse normalized data.

3.5   Results of Hospitalization Expense Modeling

A network model with both satisfactory generalization ability and fitting ability was selected after multiple training. The BP neural network model on hospitalization expense and the influence factors of hospitalized patients with coronary heart disease was finally constructed (Table 3).

3.6   Results of Sensitivity Analysis on Influence Factors of Hospitalization Expense

Sensitivities of influence factors were analyzed in our study, the results were as followed (Table 4).

Table 4. Ranking of influence degrees of influence factors for hospitalization expense

Rank	Influence factor	Sensitivity
1	Hospitalization days	0.83101
2	Treatment outcomes	0.76113
3	Times of rescuing	0.73227
4	Age	0.44537
5	Admission times	0.41421
6	Marital status	0.40873
7	Payment pattern	0.33751
8	Gender	0.07391

As can be seen from Table 5, the factor with the largest influence degree was hospitalization days, which was followed by treatment outcomes, times of rescuing, age, admission times and so on. Factors with less influence degree were marital status, payment pattern and gender. The results were concordant with theories, actual situation and the other literature reports.

4   Conclusion

The theories of artificial neural network have been gradually improved nowadays. As one of the most used models, the medical applications of BP neural network become more and more popular. BP neural network was based on developed computer technology and has abandoned traditional statistic analysis methods which hypothesis should be put forward before verified. As no hypothesis was required for the study problems, the application prospect of BP neural network for exploring etiology was outstanding, especially when the relationships of variables are unknown. The application has shown that, the applicability and prospects of BP neural network for hospitalization expense and influence factors were satisfactory.

70   C. Chen et al.

References

1. Ge, Z., Sun, Z.: Neural network theory and MATLAB R2007 simulation, pp. 1–2. Electrics Industry, Beijing (2007)
2. Liu, B.: Research of model construction for hospitalization expense based on BP neural network. Zhejiang University (2006)
3. Jiang, Y., Yin, J.: The study of reasons and countermeasures for rapid rise of medical service costs. Cost Theory and Application (2006)
4. Wang, J., Li, M., Hu, Y., et al.: The Analysis on the Influencing Factors of Hospitalization Expenses of the Patients with Gastric Cancer by Using BP Neural Network. Chinese Journal of Health Statistics 26(5), 499–501 (2009)
5. Rumelart, D.E., Meclelland, J.L., The PDP Research Group (eds.): Parallel Distributed Processing, vol. 1, 2. The M.I.T.Press, Cambrige (1986)
6. SPSS Inc. Neural Network Algorithms.Chicago, Illinois, pp. 15-17 (2003)
7. Chen, Y.: Improved algorithms of BP neural network and application. University of electronic science and technology of China (2003)

Application of GM(1,1) Model on Predicating the Outpatient Amount

Dong Wang, Sufeng Yin, Chao Chen, Jianhui Wu, and Guoli Wang

Department of Epidemiology and Health Statistics,
School of Public Health, Hebei United University
Tangshan 063000, China
w_dong850224@163.com

Abstract. Based on the outpatient amount of one hospital in Tangshan from 2005 to 2009, we can make the forecast model of outpatient amount using GM(1,1) model in grey system theory. The result of testing the accuracy of the model shows that: the adapting accuracy of grey forecast model is high, the effect of the forecast is good. The GM (1,1) model have applicative value on predicating the outpatient amount and can be used in hospital.

Keywords: grey system, GM (1,1), forecast model, outpatient amount.

1 Introduction

In recent years, the research on forecast model of time series is carrying on. It is centralized on the election and establishment of the model. The Grey Forecast Model is one of the popular models. It has no special requirements on the distribution of the data while making model. The number of data for establishing model is little, the calculation is simple and the accuracy of short-term predication is high. When conducting forecast analysis based on partial known and partial unknown grey information, this model is better than others. We make GM (1,1) model in the grey system theory based on the outpatient amount, the purpose of which is to explore the application of grey model in forecasting the outpatient amount.

2 Principles and Methods

Grey system theory was established by Professor Julong Deng of Huazhong University of Science and Technology in 1982[1]. GM (1,1) model is the basic model of the grey system theory. It makes the seeming discrete data series form the regular one through the data accumulation, then we use the new series to set up the first-order differential equation and require the model through resolving it. We can get the fitted values of original series through the inverse accumulation of data. When compared it with the practical numerical value, if the accuracy is not high, we need to use the residual to revise the model for improving the accuracy. In all GM (n, N) models, only GM (n,1) model can be used to predicate. GM (1,1) is the most common model.

D. Jin and S. Lin (Eds.): Advances in MSEC Vol. 2, AISC 129, pp. 71–76.

The principles and calculation methods of GM (1, 1) model are as follows [2] [3]:

2.1   Modeling

If there is a data series:

$$X(t) = \{x(1), x(2), ..., x(n)\} \tag{1}$$

2.1.1   Data Accumulation and Mean Generation

Conducting an accumulation of $X(t)$ to form a new series $Y(t)$

$$Y(t) = \sum_{t=1}^{t} x(i) \qquad t = 1, 2, ..., n \tag{2}$$

The process weakens the randomness of the time series and intensifies the regularity. Then we can generate the mean series of $Y(t)$.

$$Z(t) = \frac{1}{2}[Y(t) + Y(t-1)] \qquad t = 2, 3, ..., n \tag{3}$$

2.1.2   Establishment of GM (1,1) Model

Making the first-order linear differential equation about $Y(t)$

$$\frac{dY(t)}{dt} + \alpha Y(t) = \mu \tag{4}$$

Obtaining the solution of the differential equation:

$$\hat{Y}(t) = [x(1) - \frac{\mu}{\alpha}]e^{-\alpha(t-1)} + \frac{\mu}{\alpha} \tag{5}$$

In the equation, α and μ are the parameters of the model. α is the developing coefficient, μ is the grey actuating quantity. Based on the theory of least sum of squares, we can get:

$$\alpha = \frac{1}{D}\{(n-1)[-\sum_{t=2}^{n} X(t)Z(t) + \sum_{t=2}^{n} Z(t)\sum_{t=2}^{n} X(t)]\} \tag{6}$$

$$\mu = \frac{1}{D}\{[\sum_{t=2}^{n} Z(t)][-\sum_{t=2}^{n} X(t)Z(t)] + [\sum_{t=2}^{n} X(t)]\sum_{t=2}^{n} Z^2(t)\} \tag{7}$$

therein, $$D = (n-1)\sum_{t=2}^{n} Z^2(t) - [\sum_{t=2}^{n} Z(x)]^2 \tag{8}$$

2.2   Inverse of Data Accumulation

We get the estimate $\hat{Y}(t)$ series through the formula (5), then inverse accumulation of data and get the estimate $\hat{X}(t)$ series of the original data $X(t)$.

$$\hat{X}(t) = \hat{Y}(t) - Y(t-1) \tag{9}$$

2.3   Model Testing

The accuracy of grey model is commonly tested through Back-check error, the index of which is the ratio of back-check error C and the small error probability p. The table 1 is the principles of the perception accuracy of grey model.

Table 1. The judging principles of the model accuracy

Perception accuracy	p	C
First Grad (good)	$0.95 \leq p$	$C \leq 0.35$
Second Grad (Qualified)	$0.80 \leq p < 0.95$	$0.35 < C \leq 0.5$
Third Grad (Inadequacy)	$0.70 \leq p < 0.80$	$0.5 < C \leq 0.65$
Forth Grad (Unqualified)	$p < 0.70$	$0.65 < C$

therein,

$$C = S_2 / S_1 \tag{10}$$

S_1 is the mean square deviation of the original series $X(t)$, S_2 is the mean square deviation of the residual series e, then calculate through the formulas(11)and(12).

$$S_1^2 = \frac{1}{n} \sum_{t=1}^{n} [X(t) - \frac{\sum_{t=1}^{n} X(t)}{n}]^2 \tag{11}$$

$$S_2^2 = \frac{1}{n} \sum_{t=1}^{n} \left[e(t) - \frac{\sum_{t=1}^{n} e(t)}{n} \right]^2 \tag{12}$$

$$p = P \left\{ \left| e(t) - \frac{\sum_{t=1}^{n} e(t)}{n} \right| < 0.6745 S_1 \right\} \tag{13}$$

therein, $e(t) = X(t) - \hat{X}(t) = (e(1), e(2), ..., e(n))$ (14)

3   Instance Analysis

3.1 Based on the data of one hospital in Tangshan from 2005 to 2009 we can make the grey forecasting model-GM (1,1).The specific procedures are as follows.

Table 2. The outpatient amount of one hospital in Tangshan from 2005 to 2009(Unit: ten thousand)

Years	2005	2006	2007	2008	2009
Outpatients	25.05	25.94	26.29	27.38	28.69

3.1.1 According to the formulas (2) and (3), calculating the values of the accumulative series and generated mean values, then specific values and intermediate variables are as follow, see Table 3.

Table 3. The original series X (t), the accumulative series Y (t) and the value of intermediate variables

Years	t	X(t)	Y(t)	Z(t)	$Z^2(t)$	Z(t)X(t)
2005	1	25.05	-	-	-	-
2006	2	25.94	50.99	38.02	1445.520	986.234
2007	3	26.29	77.28	64.14	4113.940	1686.241
2008	4	27.38	104.66	90.97	8275.541	2490.759
2009	5	28.69	133.35	119.105	14186.001	3417.123
Σ		108.30	-	312.235	28021.002	8580.360

3.1.2 According to formula (6) (7) (8) calculating the value of D, α and μ.

$$D = (5-1) \times 28021.002 - 312.2352 = 14593.312$$

$$\alpha = \frac{1}{14593.312} [(5-1) \times (-8580.360) + 312.235 \times 108.30] = -0.0347$$

$$\mu = \frac{1}{14593.312} [312.235 \times (-8580.360) + 108.30 \times 28021.002] = 24.3664$$

Put the value of x (1) and α, μ into the formula (5) to get the GM (1,1) Model of outpatient amount.

$$\hat{Y}(t) = (25.05 + 702.2017)e^{0.0346(t-1)} - 702.2017$$

3.1.3 Getting values of the accumulative series through the established GM (1,1) model, then get forecast series of hospital outpatient amount and the residual series according to the formula (9)(14), see Table (4).

Table 4. The comparisons between the real value and forecast value

Years	t	$\hat{Y}(t)$	$Y(t)$	$\hat{X}(t)$	$X(t)$	$e(t)$
2005	1	-	-	-	25.05	-
2006	2	50.73	50.99	25.68	25.94	0.26
2007	3	77.31	77.28	26.58	26.29	- 0.29
2008	4	104.84	104.66	27.53	27.38	- 0.15
2009	5	133.34	133.55	28.50	28.69	0.19

3.1.4 Calculate the ratio of back-check error C and the small error probability p according to formulas (10) (11) (12) (13) (14).

$$S_1^2 = \frac{1}{5}\sum_{t=1}^{5}[X(t) - \frac{\sum_{t=1}^{5} X(t)}{5}]^2 = 1.5772 \qquad S_1 = 1.2559$$

$$S_2^2 = \frac{1}{5}\sum_{t=1}^{5}\left[e(t) - \frac{\sum_{t=1}^{5} e(t)}{5}\right]^2 = 0.0424 \qquad S_2 = 0.2059$$

$$C = 0.2059/1.2559 = 0.1639$$
$$0.6745 S_1 = 0.8471$$

$$p = P\left\{\left|e(t) - \frac{\sum_{t=1}^{5} e(t)}{5}\right| \langle 0.8471 \right\} = 1$$

From table 1, the forecast accuracy of GM (1, 1) model is grad 1, and the effect of the forecast is good. So GM (1, 1) model can be extrapolated.

3.2 Extrapolation forecast. We can conduct extrapolation with GM (1,1) to forecast the next two years' outpatient amount, see Table5.

Table 5. The predictive value of outpatients 2010-2011 (unit: ten thousand)

Years	2010	2011
Outpatients	29.51	30.55

4   Conclusion

Hospital outpatient amount is affected by many factors [4], such as seasonal, family income, health awareness, social policy and so on, further more some factors are constantly changing, which makes it difficult to accurately grasp. So the impact factors of outpatient show grey feature, some of which is known and some unknown. GM (1,1) model is made through accumulating data, which makes the randomness of the data weaken and regularity strengthen, and has unique advantage of processing the grey information. And GM (1,1) has no special requirements to the number and distribution of the data and is better than other statistical methods in this. We forecast hospital outpatient amount with GM (1, 1), which reflect the general trend of outpatient amount to a certain extent. This model can be conducted extrapolation due to the high forecast accuracy. But GM (1, 1) model is suitable for short-term forecast, if carrying out long-term forecasts, we need to constantly remove the old data, and populate the new data to keep adjusting GM (1, 1) model[5].

References

1. Deng, J.: Grey Forecasting and Decision-making. Huazhong University of Science and Engineering Press, Wuhan (1988)
2. Deng, J.: Grey basic theory, pp. 218–227. Huazhong University of Science and Technology Press, Wuhan (2002)
3. Deng, J.: Basic Method of Grey System, pp. 96–108. Huazhong College of Technology Press, Wuhan (1987)
4. Yang, F., Qin, Y., Liu, L.: Application of ARMA Model in Prediction of Outpatient Headcount. Chinese Journal of Hospital Administration 25(1) (2009)
5. Liang, H., Shi, J., et al.: The Application of GM（1,1）Model on Predicating The Incidence of Incidence in Nanning. Journal of Mathematical Medicine 18(3), 273–274 (2005)

Multi-user Detection on DS-CDMA UWB System Using QDPSO Algorithm

XiangBo Song[1], Jun Shi[2], Yue Chi[1], and YaTong Zhou[1]

[1] School of Information Engineering, Hebei University of Technology,
Tianjin, 300401, China
my163xinxiang@163.com
[2] School Office, Hebei University of Technology, Tianjin, 300401, China
my163xinxiang@163.com

Abstract. To effectively restrain multiple access interference (MAI) in DS-CDMA UWB communication system and lower bit error rate, on the basis of discrete particle swarm optimization (DPSO) algorithm, combining quantum computing technology, an improved quantum DPSO (QDPSO) algorithm is got, and then the multi-user detection technology based on QDPSO algorithm is proposed. It mainly includes several key steps, such as initialization on particle indicating user information, calculation on target function for multi-user detection, particle update and optimizing etc. Simulation experiment and comparative analysis show that the bit error rate performance of QDPSO detection algorithm is best and can effectively restrain the influence of MAI, its effect is much better than DPSO and traditional detection algorithm (TA).

Keywords: Multi-user Detection, DS-CDMA UWB system, Quantum Discrete Particle Swarm Optimization (QDPSO), Multiple Access Interference (MAI), Bit Error Rate.

1 Preface

UWB [1] (Ultra-wideband) is a kind of pulse communication, which is using pulse lasting time very short to carry information. In DS-CDMA UWB multi-user communication system, different users are distinguished by different pseudo random sequence, and the main interference is that of multi-access (MAI). Multi-user detection technology [2-3] can reduce or eliminate the MAI and improve Bit Error Rate (BER) detection performance of the system. Now it has become a focus topic for scholars at home and abroad in wireless communication technology. For example, traditional detection algorithm (TA) [4-5] use the matched filter to detect directly and other intelligent detection algorithms [6-7], etc.

Discrete Particle Swarm Optimization [8-9] (DPSO) algorithm is a new type of intelligent algorithm. It is different from basic PSO algorithm for the solution of function optimization problems in continuous space, and it has the advantages of high convergent speed, easy to seek the global optimal solution for solving the function Optimization problem in discrete space, especially in solving combinatorial optimization problem in practical projects. So it can be used in multi-user detection technology on DS-CDMA UWB System.

Therefore, A kind of Quantum DPSO (QDPSO) algorithm[10-11] is proposed and researched based on DPSO algorithm, that is, combined advantage of quantum computing, introducing the concept judge factors- quantum particle q, define update function and explore theory basis of multi-user detection technology with QDPSO, which provide a new idea for multi-user detection problems.

2   Methods and Principle

The basic principle of multi-user detection is to see all users' signals as useful signals in detection processing, and other user's information is predictable in a certain extent, therefore can comprehensive use all kinds of information which include the interference user's and inherent characteristics of the user waveform information joint process the received signal and to inhibit or even eliminate MAI in the maximum possible, thus achieve the purpose of more accurate detection of the target user signal and improve the performance of the receiving system. The model of multi-user detection system can be expressed in Fig. 1.

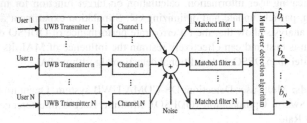

Fig. 1. Multi-user detection system model diagram

2.1   DPSO Algorithm

Kennedy and Eberhart proposed PSO algorithm in 1995, and then they proposed DPSO algorithm which can solve the problem of discrete space in 1997.The position and velocity of particles functions can be expressed as formula (1) and (2):

$$v_{id}^{t+1} = wv_{id}^{t} + c_1 r_1^{t}(P_{bestid}^{t} - x_{id}^{t}) + c_2 r_2^{t}(G_{bestid}^{t} - x_{id}^{t})$$
$$x_{id}^{t+1} = x_{id}^{t1} + v_{id}^{t+1}$$
(1)

$$x_{id}^{t} = \begin{cases} 0, rand \ge sig(v_{id}^{t+1}) \\ 1, others \end{cases}$$
(2)

$$sig(v_{id}^{t+1}) = \frac{1}{1 + \exp(-v_{id}^{t})}$$
(3)

In formula (1), v_{id}^t is the velocity of particle i after t times iteration. w is the inertia weight factor which was introduced in order to avoid DPSO getting into a local optimizing, achieved good results when $w = 1.2$. $c_j (j = 1,2)$ is the acceleration constant; r_1^t and r_2^t are the random number from 0 to 1; x_{id}^t is the current position of individual particle after t times iteration; $Pbest_{id}$ is the individual extreme value of the particles i; $Gbest_{id}$ is the global extreme value. In formula (2), rand is the random number from 0 to 1, $sig()$ is the function of which the position of particle is controlled by velocity of particle to 1 or 0.

2.2   QDPSO Algorithm

QDPSO algorithm is based on discrete thoughts of DPSO algorithm and combine quantum computing, its algorithm thoughts is roughly similar to DPSO algorithm, introduce the concept judge factors- quantum particle q, define its update function:

$$q^{t+1} = \begin{cases} p + \beta \times |mbest - q^t| \times \ln(1/u) & if \ u < 0 \\ p - \beta \times |mbest - q^t| \times \ln(1/u) & if \ u \geq 0 \end{cases} \quad (4)$$

Here, $u \in [-1,1]$. Then judgement function of QDPSO is:

$$x^{t+1} = \begin{cases} -1 & if \ \rho \geq sigmoid \ (q) \\ 1 & if \ \rho < sigmoid \ (q) \end{cases} \quad (5)$$

Multi-user detection problem in DS-CDMA UWB technology can be summed up to solve problem of maximum likelihood function in mathematics, can be seen a combinatorial optimization problem, so QDPSO algorithm is used to solve it, objective function in optimal multi-user detection is seen as the optimized fitness function. The objective function can be expressed by function (6).

$$f^b = b^T ARAb - 2y^T Ab \quad (6)$$

Here, $y = RAb + \vec{n}$, $b_i \in \{-1 1\}$, $i = 1 2 ; \cdots , N$, then solving the best vector of multi-user detection will translate into searching global optimal position of particle in particle swarm. Suppose the number of multi-user detection is N, then $b = [b_1, b_2, \cdots b_N]^T$ is the sending information sequence of the N users. $y = [y_1, y_2, \cdots y_N]^T$ is the N outputs of the matched filter. $R = [r_{ij}]_{N \times N}$ is cross-correlation matrix of PN code for each user, which r_{ij} is correlation coefficient of two code words for any two users. Because $r_{ik} = r_{ki}$, R matrix is symmetric matrices. $r_{ij} \neq 0$ when each PN code is not entirely orthogonal, that is, multiple access interference exist. Matrix A is diagonal matrix corresponding with the power of receiving signal, the diagonal elements A_{ii} is the signal energy of user i received, and the rest elements are 0.

In DS-CDMA UWB system, principle diagram of multi-user detection based on QDPSO algorithm can be expressed in Fig. 2.

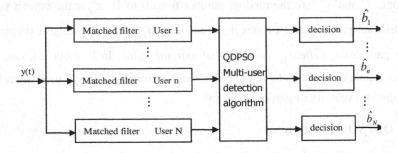

Fig. 2. Principle diagram of multi-user detection based on QDPSO

In Figure 2, the decision of discrete output results by matched filter can be seen as the initial value of algorithm, that is, i the user data bits can be seen as position x of N dimension particle.

2.3 Multi-user Detection Technology Based on QDPSO

Method steps multi-user detection based on QDPSO are as follows:

Step1: Initialize the particle swarm, that is, using the decision output y (Conventional detector output) of the matched filter to make the information of N-dimensional user as the initial particle position x, and set the parameters of the quantum particle swarm and the maximum value of iterations.

Step2: According to equation (6) to calculate objective function value (fitness value) f^{x_i} of the particle, find out $gbest$, and record the current position for each particle as $pbest$, make the iterations $T = 1$.

Step3: Adjust and update the parameters of each particle
According to equation (4) to calculate the quantum particle q, and according to equation (5) to update each particle position x_i

According to equation (6) to calculate the objective function value (fitness value) f^{x_i} of particle, update $pbest$ and $gbest$

$$\begin{cases} if \ f^{x_i} < f^{pbest_i}, & then \ pbest_i = x_i \\ if \ f^{x_i} < f^{gbest_i}, & then \ gbest_i = x_i \end{cases};$$

Step4: If the iterations $T = T_{max}$, optimizing ends, the current optimal individual is the result; otherwise, $T = T + 1$, transfer to step3.

3   Simulation Experiment and Comparative Analysis

3.1   Parameter Settings of Simulation Experiment

(1) User information sequence: user information sequence is generated by Random function rand, suppose there are N users in the system, then the user information sequence can be expressed as: $b = [b_1, b_2, \cdots, b_i, \cdots, b_N]^T$, which:

$$b_i = \begin{cases} -1 & \text{if } (temp = rand) < 0.5 \\ 1 & \text{if } (temp = rand) \geq 0.5 \end{cases}$$

(2) Spread spectrum code sequence: use Gold sequences whose length is 31.

(3) Channel: additive white gaussian noise (AWGN) channel, considering the influence of sine interference.

(4) other simulation Parameters:

Sending data:10000bit

Number of users: N =4 ;

Number of particle in the particle swarm: n=10 ;

QDPSO maximum iterations : $T_{max} = 400$;

3.2   Simulation Experiment and Comparative Analysis

In the DS-CDMA UWB system, to validate the superiority of multi-user detection technology based on QDPSO, simulation comparison experiment is performed among multi-user detection technology based on QDPSO, multi-user detection technology based on traditional TA algorithm and multi-user detection technology based on DPSO, the detection performance is analyzed among them when the average bit error rate and the number of the user increase in different SNR(Signal to Noise Ratio).

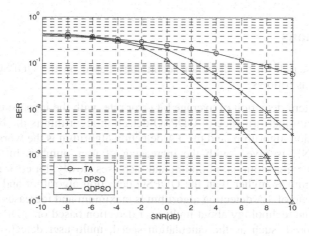

Fig. 3. Multi-user Detection Bit Error Rate Figure

Figure 3 is the bit error rate variation diagram of the three detection algorithm. As can be seen from the figure, with the increasing of the SNR, the bit error rate of TA detection algorithm changes slowly, but DPSO detection algorithm and QDPSO detection algorithm decrease rapidly, and the performance of QDPSO detection algorithm is better than the DPSO detection algorithm, this shows the result of multi–user detection based on QDPSO is remarkable.

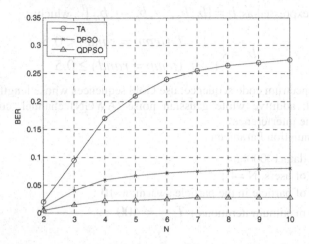

Fig. 4. BER with users number increase when SNR = 4 dB

Figure 4 is the bit error rate variation diagram of three detection algorithm with the increasing of user number. As can be seen from the figure, the bit error rate of QDPSO detection algorithm is minimum when there are more users, and its growth rate with the increasing of users change slowly. Obviously, QDPSO detection algorithm is very effective, and the detection performance is far better than the other two detection algorithms.

4   Conclusions

Through the research on the detection technology based on the QDPSO multi-user, conclusions are as follows:

(1) In the system detection of average bit error rate, with the increasing of SNR, bit error rate detection performance of QDPSO detection algorithm is the best, followed by DPSO detection algorithm, and the TA detection algorithm is the worst.

(2) When SNR is given, with the increasing of user number, bit error rate of QDPSO detection algorithm is minimum and increases slowly, bit error rate of DPSO detection algorithm is higher than the QDPSO detection algorithm and increases rapidly, bit error rate of TA detection algorithm is maximum and increases fastest.

In addition, the technology about multi-user detection based on QDPSO could still be further explored, such as the calculation speed, multi-user detection of near-far effect etc.

Acknowledgement. This work was supported by the National Natural Science Foundation of China (No.60972106), Tianjin Natural Science Foundation (No. 11JCYBJC00900) and the Science Foundation of Hebei Education Department (No.2009425). Our heartfelt thanks to Professor Xia kewen in Hebei University of Technology to review this paper.

References

1. Shayesteh, M.G., Masoumeh, N.-K.: A new TH/DS-CDMA scheme for UWB communication systems and its performance analysis. In: 2006 IEEE Radio and Wireless Symposium, San Diego, CA, United states (2006)
2. Imre, S., Balazs, F.: Quantum Multi-user Detection. In: Proc. of 1st Workshop on Wireless Services & Applications, Paris, pp. 147–154 (2001)
3. Concha, J., Poor, H.: Multi-access Quantum Channels. IEEE Trans. On Information Theory 50, 725–747 (2004)
4. Fishler, E., Poor, H.V.: Low-complexity multi-user detector for time-hopping Impulse-radio systems. IEEE Transactions on Signal Processing 52(9), 2561–2571 (2004)
5. Li, Q., Rusch, L.A.: Multiuser receivers for DS-CDMA UWB. In: Proc. IEEE Conf. Ultra Wideband Systems Technology, pp. 163–167 (2002)
6. Liu, S., Wang, J.: Blind Adaptive Multiuser Detection Using a Recurrent Neural Network. IEEE Trans. Commun. 7(4), 1071–1705 (2004)
7. Leonardo, D., Paul, J.E.: Particle Swarm and Quantum Particle Swarm Optimization Applied to DS/CDMA Multiuser Detection in Flat Rayleigh Channels. IEEE Ninth International Symposium on Spread Spectrum Techniques and Applications, 133–137 (2006)
8. Kennedy, J., Eberhart, R.: Particle swarm optimization. In: Proc. IEEE Int. Conf. on Neural Networks, Perth, pp. 1942–1948 (1995)
9. Shi, Y.H., Eberhart, R.: Parameter selection in particle swarm optimization. In: Proc. of the 7th Annual Conf. on Evolutionary Programming, Washington D.C, pp. 591–600 (1998)
10. Pan, G., Xia, K., Dong, Y., Shi, J.: An Improved LS-SVM Based on Quantum PSO Algorithm and Its Application. In: The 3rd International Conference on Natural Computation, Haikou, China (2007)
11. Imre, S., Balazs, F.: Performance Evaluation of Quantum Based Multi-user Detector. In: IEEE International Symposium on Spread Spectrum Techniques and Applications (ISSSTA), Prague, pp. 722–725 (2002)

The Characters of Two-Directional Vector-Valued Poly-Scale Small-Wave Wraps with Finite Support

Honglin Guo[*] and Rui Tian

Department of Fundamentals, Henan Polytechnic Institute, Nanyang 473009, China
sxxa66xauat@126.com

Abstract. The rise of wavelet analysis in applied mathematics is due to its applications and the flexibility. In this paper, the notion of two-direction vector-valued multiresolution analysis and the two-direction biorthogonal vector-valued wavelets are introduced. Their properties is investigated by algebra theory, means of time-frequency analysis method and, operator theory. The direct decomposition relationship is provided. Moreover, it is shown how to draw new Riesz bases of space $L^2(R, C^v)$ from these wavelet wraps.

Keywords: two-directional, vector-valued small-wave wraps, Riesz bases, iteration method, sampling theorem, Bessel sequence, functional analysis.

1 Introduction

Wavelets, firstly succeed in the application and then form the theory, has been and will be widely used in signal processing, image processing, quantum field theory, seismic exploration, voice recognition and synthesis, music, radar, CT imaging, color copies, fluid turbulence, object recognition, machine vision, machine fault diagnosis and monitoring, fractal and digital television, economicanalysis and forecasting and so on [1-3]. Multiwavelets, which can offer properties like symmetry, orthogonality, short support at the same time, has been an active research field of wavelets analysis.

To obtain some beautiful feature, Yang and Li [4] introduce the concept of two-direction scaling function and two-direction wavelets, give the definition of orthogonal two-direction scaling function and construct orthogonal two-direction wavelets, moreover, they investigate the approximation order At present, image interpolation algorithms based on wavelet transform are mainly based on multiresolution analysis of the wavelet. Raditionally, short-time Fourier Transform and Gabor Transform were used for harmonic studies of nonstationary power system waveforms which are basically Fourier Transform-based methods. To overcome the limitations in these existing methods, wavelet transform based algorithm has been developed to estimate the limitations the frequency and time information of a harmonic signal. Multiwavelets can simultaneously possess many desired properties such as short support, orthogonality, symmetry, and vanishing moments, which a single wavelet cannot possess simultaneously. Already they have led to exciting

---

[*] Corresponding author.

D. Jin and S. Lin (Eds.): Advances in MSEC Vol. 2, AISC 129, pp. 85–91.
springerlink.com   © Springer-Verlag Berlin Heidelberg 2011

applications in signal analysis [1], fractals [2] and image processing [3], and so on. Vector-valued wavelets are a sort of special multiwavelets Chen [4] introduced the notion of orthogonal vector-valued wavelets.However, vector-valued wavelets and multiwavelets are different in the following sense. Pre-filtering is usually required for discrete multiwavelet transforms [5] but not necessary for discrete vector-valued transforms. Wavelet wraps, owing to their nice characteristics, have been widely applied to signal processing [6], code theory, image compression, solving integral equation and so on. Coifman and Meyer firstly introduced the notion of univariate orthogonal wavelet wraps. Yang [7] constructed a-scale orthogonal multiwavelet wraps that were more flexible in applications. It is known that the majority of information is multidimensional information. Shen [8] introduced multivariate orthogonal wavelets which may be used in a wider field. Thus, it is necessary to generalize the concept of mu-ltivariate wavelet wraps to the case of multivariate vector-valued wavelets. The goal of this paper is to give the definition and the construction of biorthogonal two-direct-ion vector-vavalued wavelet wraps and construct new Riesz bases of $L^2(R,C^v)$.

2   The Preliminaries on Vector-Valued Function Space

Let Z and Z_+ stand for all integers and nonnegative integers, respectively. The multi-resolution analysis is one of the main approaches in the construction of wavelets. Let us introduce two-direction vector-valued multiresolution analysis and two-direction vector-valued wavelets. By $L^2(R,C^v)$, we denote the set of all vector-valued functions

$$L^2(R,C^v) := \{ \hbar(x) = (h_1(x), h_2(x), \cdots, h_v(x))^T : h_l(x) \in L^2(R), l = 1, 2, \cdots, v \} \ , \quad \text{where } T$$

means the transpose of a vector. For any $\hbar \in L^2(R,C^v)$ its integration is defined as $\int_R \hbar(x) dx = (\int_R h_1(x) dx, \int_R h_2(x) dx, \cdots, \int_R h_v(x) dx)$, and the Fourier transform of $\hbar(x)$ is defined by

$$\hat{\hbar}(\omega) := \int_R \hbar(x) \cdot \exp\{-i\langle x, \omega \rangle\} dx, \tag{1}$$

where $\langle x, \omega \rangle$ denotes the inner product of x and ω. For two vector-valued functions $\hbar(x), \lambda(x) \in L^2(R,C^v)$, by $<\hbar, \lambda>$, we denotes their symbol inner product, i.e.,

$$\langle \hbar, \lambda \rangle := \int_R \hbar(x) \lambda(x)^* dx, \tag{2}$$

where $*$ means the transpose and the complex conjugate.

Definition 1. We say that $F(x) \in L^2(R,C^v)$ is a two-direction vector-valued refinable function if $f(x)$ satisfies the following two-direction refinable equation:

$$F(x) = \sum_{n \in Z} B_n^+ F(5x - n) + \sum_{n \in Z} B_n^- F(n - 5x), \tag{3}$$

where the matrix sequences $\{B_v^+\}_{v \in Z} \in l^2(Z^{v \times v})$ and $\{B_v^-\}_{v \in Z} \in l^2(Z^{v \times v})$ are also called positive-direction mask and negative-direction mask, respectively. If all negative-direction mask are equal to 0, then two-direction refinable equation (2) become two-scale refinable equation (1). Taking the Fourier transform gives

$$\widehat{F}(\omega) = B^+(e^{-i\omega/5})\widehat{F}(\omega/5) + B^-(e^{-i\omega/5})\overline{\widehat{F}(\omega/5)}, \qquad (4)$$

where $b^+(z) = (1/5)\sum_{v \in Z} b_v^+ z^u$, $z = e^{-i\omega/5}$ is called positive-direction mask symbol, and $5b^-(z) = \sum_{v \in Z} b_v^- z^u$ is called negative-direction mask symbol. In order to investigate the existence of solutions of the two-direction refinable equation (2), we rewrite the two-direction refinable equation (2) as

$$F(-x) = \sum_{v \in Z} B_v^+ F(-5x - v) + \sum_{v \in Z} B_v^- F(v + 5x), \qquad (5)$$

By taking the Fourier transform for the both sides of (4), we have

$$\overline{F(\omega)} = B^+(e^{-i\omega/5})\overline{F(\omega/5)} + B^-(e^{-i\omega/5})F(\omega/5), \qquad (6)$$

From the refinement equation (3) and the refinement equation (5), we get that

$$\widehat{\underline{F}}(\omega) = \begin{bmatrix} \overline{\widehat{F}(\omega)} \\ \widehat{F}(\omega) \end{bmatrix} = \begin{bmatrix} B^+(e^{-i\omega/5}) & B^-(e^{-i\omega/5}) \\ B^+(e^{-i\omega/5}) & B^-(e^{-i\omega/5}) \end{bmatrix} \begin{bmatrix} \overline{\widehat{F}(e^{-i\omega/5})} \\ \widehat{F}(e^{-i\omega/5}) \end{bmatrix} \qquad (7)$$

By virtue of the positive-direction mask $\{B_v^+\}_{v \in Z}$ and the negative-direction mask $\{B_v^-\}_{v \in Z}$, we construct the following matrix equation:

$$\underline{F}(x) = \begin{bmatrix} F(-x) \\ F(x) \end{bmatrix} = \sum_{u \in Z} \begin{bmatrix} B_{-u}^- & B_{-u}^+ \\ B_u^+ & B_u^- \end{bmatrix} \underline{F}(5x - u) \qquad (8)$$

Definition 2. A pair of two-directional vector-valued function $F(x)$, $\widetilde{F}(x) \in L^2(R)$ are biorthogonal ones, if their translates satisfy

$$< F(x), \ \widetilde{F}(x - k) > = \delta_{0,k} I_v, k \in Z, \qquad (9)$$

$$< F(x), \ \tilde{F}(n-x) > = 0, \quad n \in Z, \qquad (10)$$

where $\delta_{0,k}$ is the Kronecker symbol. Define a sequence $S_j \in L^2(R, C^v)$ by

$$S_j = clos_{L^2(R,C^v)} \langle 5^{j/2} F(5^j \cdot - k), 5^{j/2} F(l - 5^j \cdot): k, l \in Z \rangle, \quad j \in Z.$$

where "clos" denote the closure of a space by a function $F(x)$.

A two-direction multiresolution analysis is a nested sequence $\{S_j\}_{j \in Z}$ genera-ted by $F(x)$, if it satisfies: (i) $\cdots \subset S_{-1} \subset S_0 \subset S_1 \subset \cdots$; (ii) $\bigcap_{j \in Z} S_j$

$= \{0\}$, $\bigcup_{j\in Z} S_j$ is dense in $L^2(R,C^v)$, where 0 is the zero vector of $L^2(R,C^v)$; (*iii*) $H(x)\in S_0$ if and only if $H(5^j x)\in V_j$, $j\in Z$; (*iv*)There exists $F(x)\in S_0$ Such that the sequence $\{F(x-k),F(n-x): k,n\in Z\}$ is a Riesz basis of S_0.

Four two-directional functions $\Psi_l(x)(l\in \Lambda \triangleq \{1,2,3,4\})$ are called two-directional wavelets with scale factor 5 associated with $F(x)$, if the family $\{\Psi_l(x-k),\ \Psi_l(n-x):\ k,n\in Z^2,l=1,2,3,4\}$ forms a Riesz basis of W_j, where $V_{j+1}=\ S_j\oplus W_j$, $j\in Z$, where $\oplus$ denotes the direct sum. Then $\Psi_l(x)$ satisfies the following equation:

$$\Psi_l(x)=\sum_{v\in Z}Q^+_{v,l}F(5x-v)+\sum_{v\in Z}Q^-_{v,l}F(v-5x),\ \ l\in\Lambda \tag{9}$$

Implementing the Fourier transform for the both sides of (10) gives

$$\widehat{\Psi}_l(\omega)=Q^+_l\left(e^{-i\omega/5}\right)\widehat{F}(\omega/5)+Q^-_l\left(e^{-i\omega/5}\right)\overline{\widehat{F}(\omega/5)},\ l\in\Lambda. \tag{10}$$

Similarly, there exist two seq.s $\{\widetilde{B}^+_u\}_{u\in Z},\{\widetilde{B}^-_u\}_{u\in Z}\in l^2(Z)$, such that

$$\widetilde{F}(x)=\sum_{u\in Z}\widetilde{B}^+_u\widetilde{F}(5x-u)+\sum_{u\in Z}\widetilde{B}^-_u\widetilde{F}(u-5x). \tag{11}$$

The Fourier transforms of refinable equation (11) becomes

$$\widehat{\widetilde{F}}(\omega)=\ \widetilde{B}^+_k(e^{-i\omega/5})\widehat{\widetilde{F}}(\omega/5)+\ \widetilde{B}^-_k(e^{-i\omega/5})\widehat{\widetilde{F}}(\omega/5). \tag{12}$$

Similarly, there also exist two sequences $\{\widetilde{q}^+_{u,l}\},\{\widetilde{q}^-_{u,l}\}\in l^2(Z)$ so that

$$\widetilde{\Psi}_l(x)=\sum_{u\in Z}\widetilde{Q}^+_{u,l}\widetilde{F}(5x-u)+\sum_{u\in Z}\widetilde{Q}^-_{u,l}\widetilde{F}(u-5x). \tag{13}$$

By taking the Fourier transforms for (14), for $l\in\Lambda$, we have

$$\widehat{\widetilde{\Psi}}_l(5\omega)=\widetilde{Q}^+_l(e^{-i\omega})\widehat{\widetilde{F}}(\omega)+\widetilde{Q}^-_l(e^{-i\omega})\overline{\widehat{\widetilde{F}}(\omega)}. \tag{14}$$

We say that $\psi_l(t),\widetilde{\psi}_l(t)\in L^2(R)$ are pairs of biorthogonal two-direction wavelets associated with a pair of biorthogonal two-direction scaling functions $F(x)$, $\widetilde{F}(x)\in L^2(R)$, if the family $\{\psi_l(x-n),\psi_l(n-x):n\in Z^2\}$ is a Riesz basis of subspace W_0, and they satisfy the following equations:

$$\left\langle F(x),\widetilde{\Psi}_l(x-u)\right\rangle=\left\langle F(x),\widetilde{\Psi}_l(u-x)\right\rangle=0,\ l\in\Lambda,u\in Z, \tag{15}$$

$$\langle \overline{F}(x), \Psi_\iota(x-u) \rangle = \langle \overline{F}(x), \Psi_\iota(u-x) \rangle = 0, \quad \iota \in \Lambda, \; u \in Z, \tag{16}$$

$$\langle \Psi_\iota(x), \Psi_\iota(x-u) \rangle = \delta_{u,0}\delta_{\iota,\iota}I_v, \quad u \in Z, \; \iota, \iota \in \Lambda, \tag{17}$$

$$\langle \Psi_\iota(x), \overline{\Psi}_\iota(u-x) \rangle = \delta_{u,0}\delta_{\iota,\iota}I_v, \quad u \in Z^2, \; \iota, \iota \in \Lambda. \tag{18}$$

$$\Psi_\iota(-x) = \sum_{v \in Z} Q_{v,\iota}^+ F(-5x - u) + \sum_{v \in Z} Q_{v,\iota}^- F(v + 5x). \tag{19}$$

The refinement equ. (10) and (20) lead to the following relation formula

$$\Psi_\iota(x) = \begin{bmatrix} \psi_\iota(-x) \\ \psi_\iota(x) \end{bmatrix} = \sum_{k \in Z^2} \begin{bmatrix} q_{-k,\iota}^- & q_{-k,\iota}^+ \\ q_{k,\iota}^+ & q_{k,\iota}^- \end{bmatrix} F(5x - k), \quad \iota \in \Lambda. \tag{20}$$

$$\widehat{\Psi}_\iota(\omega) = \begin{bmatrix} \widehat{\psi}_\iota(\omega) \\ \widehat{\psi}_\iota(\omega) \end{bmatrix} = \begin{bmatrix} q_\iota^-(e^{-i\omega/5}) & q_\iota^+(e^{-i\omega/5}) \\ q_\iota^+(e^{-i\omega/5}) & q_\iota^-(e^{-i\omega/5}) \end{bmatrix} \begin{bmatrix} \widehat{f}(\omega/5) \\ \widehat{f}(\omega/5) \end{bmatrix}, \quad \iota \in \Lambda. \tag{21}$$

$$\widetilde{\Psi}_\iota(x) = \begin{bmatrix} \widetilde{\psi}_\iota(-x) \\ \widetilde{\psi}_\iota(x) \end{bmatrix} = \sum_{v \in Z^2} \begin{bmatrix} \widetilde{q}_{-v,\iota}^- & \widetilde{q}_{-v,\iota}^+ \\ \widetilde{q}_{v,\iota}^+ & \widetilde{q}_{v,\iota}^- \end{bmatrix} F(5x - v), \quad \iota \in \Lambda. \tag{22}$$

$$\widehat{\widetilde{\Psi}}_\iota(\omega) = \begin{bmatrix} \widehat{\widetilde{\psi}}_\iota(\omega) \\ \widehat{\widetilde{\psi}}_\iota(\omega) \end{bmatrix} = \begin{bmatrix} \widetilde{q}_\iota^-(e^{-i\omega/5}) & \widetilde{q}_\iota^+(e^{-i\omega/5}) \\ \widetilde{q}_\iota^+(e^{-i\omega/5}) & \widetilde{q}_\iota^-(e^{-i\omega/5}) \end{bmatrix} \begin{bmatrix} \widehat{\widetilde{f}}(\omega/5) \\ \widehat{\widetilde{f}}(\omega/5) \end{bmatrix}, \quad \iota \in \Lambda. \tag{23}$$

Since $F(x) \in S_0 \subset S_1$, by Definition 3 and (4) there exists a finitely supported sequence of constant $v \times v$ matrice $\{\Omega_n\}_{n \in Z^2} \in \ell^2(Z)^{v \times v}$ such that

$$F(x) = \sum_{n \in Z} \Omega_n F(5x - n). \tag{24}$$

Equation (6) is called a refinement equation. Define

$$5 \cdot \Omega(\gamma) = \sum_{n \in Z} \Omega_n \cdot \exp\{-in \cdot \gamma\}, \quad \gamma \in R. \tag{25}$$

where $\Omega(\gamma)$, which is $2\pi Z$ fun., is called a symbol of $F(x)$. Thus, (26) becomes

$$\widehat{F}(5\gamma) = \Omega(\gamma)\widehat{F}(\gamma), \quad \gamma \in R. \tag{27}$$

Let $X_j, j \in Z$ be the direct complementary subspace of Y_j in Y_{j+1}. Assume that there exist 63 vector-valued functions $\psi_\mu(x) \in L^2(R, C^v), \mu \in \Gamma$ such that their translations and dilations form a Riesz basis of X_j, i.e.,

$$X_j = \overline{(span\{\Psi_\mu(5^j \cdot -n) : n \in Z, \mu \in \Gamma\})}, \quad j \in Z. \tag{28}$$

Since $\Psi_\mu(x) \in X_0 \subset Y_1, \mu \in \Gamma$, there exist 63 finitely supported sequences of constant $v \times v$ matrice $\{B_n^{(\mu)}\}_{n \in Z^4}$ such that

$$\Psi_\mu(x) = \sum_{n \in Z^2} B_n^{(\mu)} F(Mx - n), \quad \mu \in \Gamma. \tag{29}$$

We say that $\Psi_\mu(x), \tilde{\Psi}_\mu(x) \in L^2(R, C^\nu), \mu \in \Gamma$ are pairs of biorthogonal vector wavelets associated with a pair of biorthogonal vector scaling functions $F(x)$ and $\tilde{F}(x)$, if the family $\{\Psi_\mu(x-n), n \in Z, \mu \in \Gamma\}$ is a Riesz basis of subspace X_0, and

$$[F(\cdot), \tilde{\Psi}_\mu(\cdot - n)] = [\tilde{F}(\cdot), \Psi_\mu(\cdot - n)] = O, \quad \mu \in \Gamma, \quad n \in Z, \quad (30)$$

$$[\tilde{\Psi}_\lambda(\cdot), \Psi_\mu(\cdot - n)] = \delta_{\lambda, \mu} \delta_{0, n}, \quad \lambda, \mu \in \Gamma, \quad n \in Z. \quad (31)$$

$$X_j^{(\mu)} = \overline{Span\{\Psi_\mu(M^j \cdot - n) : n \in Z^2\}}, \mu \in \Gamma, j \in Z. \quad (32)$$

Similar to (5) and (9), there exist 5 finitely supported sequences of $\nu \times \nu$ complex constant matrice $\{\tilde{\Omega}_k\}_{k \in Z^2}$ and $\{\tilde{B}_k^{(\mu)}\}_{k \in Z^2}$, $\mu \in \Gamma$ such that $\tilde{F}(x)$ and $\tilde{\Psi}_\mu(x)$ satisfy the refinement equations:

$$\tilde{F}(x) = \sum_{k \in Z} \tilde{\Omega}_k \tilde{F}(5x - k) \quad (33)$$

$$\tilde{\Psi}_\mu(x) = \sum_{n \in Z} \tilde{B}_n^{(\mu)} \tilde{F}(5x - n), \quad \mu \in \Gamma. \quad (34)$$

3   The Biorthogonality Features of a Sort of Wavelet Wraps

Denoting by $G_0(x) = F(x), G_\mu(x) = \tilde{\Psi}_\mu(x), \tilde{G}_0(x) = F(x), \tilde{G}_\mu(x) = \tilde{\Psi}_\mu(x), Q_k^{(0)} = \Omega_k,$ $Q_k^{(\mu)} = B_k^{(\mu)}, \tilde{Q}_k^{(0)} = \tilde{\Omega}_k, \tilde{Q}_k^{(\mu)} = \tilde{B}_k^{(\mu)}, \mu \in \Gamma, k \in Z,$. For any $\alpha \in Z_+$ and the given vector biorthogonal scaling functions $G_0(x)$ and $\tilde{G}_0(x)$, iteratitively define,

$$G_\alpha(x) = G_{4\sigma + \mu}(x) = \sum_{k \in Z} Q_k^{(\mu)} G_\sigma(4x - k), \quad \mu \in \Gamma_0, \quad (35)$$

$$\tilde{G}_\alpha(x) = \tilde{G}_{4\sigma + \mu}(x) = \sum_{k \in Z} \tilde{Q}_k^{(\mu)} \tilde{G}_\sigma(4x - k), \quad \mu \in \Gamma_0. \quad (36)$$

where $\sigma \in Z_+^2$ is the unique element such that $\alpha = 4\sigma + \mu$, $\mu \in \Gamma_0$ follows.

Definition 3. We say that two families of vector-valued functions $\{G_{4\sigma + \mu}(x), \sigma \in Z_+,$ $\mu \in \Gamma_0\}$ and $\{\tilde{G}_{4\sigma + \mu}(x), \sigma \in Z_+, \mu \in \Gamma_0\}$ are vector-valued wavelet packets with respect to a pair of biorthogonal vector-valued scaling functions $G_0(x)$ and $\tilde{G}_0(x)$, resp., where $G_{4\sigma + \mu}(x)$ and $\tilde{G}_{4\sigma + \mu}(x)$ are given by (35) and (36), respectively.

Theorem 1[8]. Assume that $\{G_\beta(x), \beta \in Z_+\}$ and $\{\tilde{G}_\beta(x), \beta \in Z_+\}$ are vector-valued wavelet packets with respect to a pair of biorthogonal vector-valued functions $G_0(x)$ and $\tilde{G}_0(x)$, respectively. Then, for $\beta \in Z_+, \mu, \nu \in \Gamma_0$, we have

$$\left\langle G_{4\beta+\mu}(\cdot), \tilde{G}_{4\beta+\nu}(\cdot - k)\right\rangle = \delta_{\underline{0},k}\delta_{\mu,\nu}I_{\nu}, \ k \in Z. \tag{37}$$

Theorem 2[8]. If $\{G_{\beta}(x), \beta \in Z_{+}\}$ and $\{\tilde{G}_{\beta}(x), \beta \in Z_{+}\}$ are vector-valued wavelet wraps with respect to a pair of biorthogonal vector scaling functions $G_{0}(x)$ and $\tilde{G}_{0}(x)$, then for any $\alpha, \sigma \in Z_{+}$, we have

$$\left\langle G_{\alpha}(\cdot), \tilde{G}_{\sigma}(\cdot - k)\right\rangle = \delta_{\alpha,\sigma}\delta_{\underline{0},k}I_{\nu}, \ k \in Z_{+}. \tag{38}$$

References

1. Telesca, L., et al.: Multiresolution wavelet analysis of earthquakes. Chaos, Solitons & Fractals 22(3), 741–748 (2004)
2. Iovane, G., Giordano, P.: Wavelet and multiresolution analysis:Nature of ε^{∞} Cantorian space-time. Chaos, Solitons & Fractals 32(4), 896–910 (2007)
3. Zhang, N., Wu, X.: Lossless Compression of Color Mosaic Images. IEEE Trans. Image Processing 15(16), 1379–1388 (2006)
4. Chen, Q., Qu, G.: Characteristics of a class of vector-valued nonseparable higher-dimensional wavelet packet bases. Chaos, Solitons & Fractals 41(4), 1676–1683 (2009)
5. Chen, Q., Huo, A.: The research of a class of biorthogonal compactly supported vector valued wavelets. Chaos, Solitons & Fractals 41(2), 951–961 (2009)
6. Chen, Q., Shi, Z.: Biorthogonal multiple vector-valued multivariate wavelet packets associated with a dilation matrix. Chaos, Solitons & Fractals 35(2), 323–332 (2008)
7. Chen, Q., Wei, Z.: The characteristics of orthogonal trivariate wavelet packets. Information Technology Journal 8(8), 1275–1280 (2009)
8. Chen, Q., Shi, Z.: Construction and properties of orthogonal matrix-valued wavelets and wavelet packets. Chaos, Solitons & Fractals 37(1), 75–86 (2008)

The Traits of Quarternary Minimum-Energy Frames in Sobolev Space

Zhihao Tang[*]

Department of Fundamentals, Henan Polytechnic Institute, Nanyang 473009, China
sxxa11xauat@126.com

Abstract. The concept of a generalized quarternary multiresolution struccture (GQMS) of space is formulated. A class of multiple affine quarternary pseudo frames for subspaces of $L^2(R^4)$ are introduced. The construction of a GQMS of Paley-Wiener subspaces of $L^2(R^4)$ is studied. The sufficient condition for the existence of pyramid decomposition scheme is presented based on such a GQMS. A sort of affine quarternary frames are constructed by virtue of the pyramid decomsition scheme and Fourier transform. We show how to draw new orthonormal bases for space $L^2(R^4)$ from these wavelet wraps.

Keywords: Sobolev space, minimum-energy frame projection operator, quarter- nary, wavelet frames, frame multiresolution analysis, Besov space.

1 Introduction and notations

The frame theory plays an important role in the modern time-frequency analysis. It has been developed very fast over the last twenty years, especially in the context of wavelets and Gabor systems. In her celebrated paper[1], Daubechies constructed a family of compactly supported univariate orthogonal scaling functions and their corresponding orthogonal wavelets with the dilation factor 2. Since then wavelets with compact support have been widely and successfully used in various applications such as image compression and signal processing. he frame theory has been one of powerful tools for researching into wavelets. To study some deep problems in nonharmonic Fourier series, Duffin and Schaeffer[2] introduce the notion of frames for a separable Hilbert space in 1952. Basically, Duffin and Schaeffer abstracted the fundamental notion of Gabor for studying signal processing [3]. These ideas did not seem to generate much general interest outside of nonharmonic Fourier series how-ever (see Young's [4]) until the landmark paper of Daubechies, Grossmann, and Meyer [5] in 1986. After this ground breaking work, the theory of frames began to be more widely studied both in theory and in applications [6-8], such as signal process-ing, image processing, data compression and sampling theory. The notion of Frame Multiresolution Analysis (FMRA) as described by [6] generalizes the notion of MRA by allowing non-exact affine frames. However, subspaccs at different resolu-tions in a FMRA are still generated by a frame formed by translares and dilates of a single

---

[*] Corresponding author.

D. Jin and S. Lin (Eds.): Advances in MSEC Vol. 2, AISC 129, pp. 93–99.

94   Z. Tang

biv-ariate function. Inspired by [5] and [7], we introduce the norion of a Generalized Qua-rternary Multiresolution Structure (GQMS) of $L^2(R^4)$ generated by several functions of integer translates $L^2(R^4)$. We demonstrate that the GQMS has a pyramid decomposit-ion scheme and obiain a frame-like decomposition based on such a GQMS.It also lead to new constructions of affine of $L^2(R^4)$. Since the majority of information is multidimensional information, many researchers interest themselves in the investig-ation into multi-variate wavelet theory. The classical method for constructing multi-variate wavelets is that separable multivariate wavelets may be obtained by means of the tensor product of some univariate wavelet frames. It is significant to investigate nonseparable multivariate wavelet frames and pseudoframes. Let Ω be a separable Hilbert space .We recall that a sequence $\{\eta_v\}_{v\in Z^4} \subseteq \Omega$ is a frame for Ω, if there exist positive real numbers L, M such that

$$\forall \wp \in \Omega, \quad L\|\wp\|^2 \le \sum_{v\in\Lambda} |\langle \wp,\eta_v\rangle|^2 \le M\|\wp\|^2 , \tag{1}$$

A sequence $\{\eta_v\}_{v\in Z^4} \subseteq \Omega$ is called a Bessel sequence if only the upper inequality of (1) follows. If only for all element $g \in Q \subseteq \Omega$, the upper inequality of (1) holds, the sequence $\{\eta_v\}_{v\in Z^4} \subseteq \Omega$ is a Bessel sequence with respect to (w.r.t.)Q .If $\{\eta_v\}$ is a frame, there exist a dual frame $\{\eta_v^*\}$ such that

$$\forall h\in \Omega, \quad h = \sum_v \langle h,\eta_v\rangle\eta_v^* = \sum_v \langle h,\eta_v^*\rangle\eta_v. \tag{2}$$

The Fourier transform of an integrable function $h(x)\in L^2(R^4)$ is defined by

$$Fh(\omega) = \hat{h}(\omega) = \int_{R^4} h(x)e^{-2\pi i x\omega}dx, \quad \omega\in R^4. \tag{3}$$

For a sequence $c = \{c(k)\}\in \ell^2(Z^4)$, we define the discrete-time Fourier tramsform as the function in $L^2(0,1)^4$ given by

$$Fk(\omega) = K(\omega) = \sum_{v\in Z^4} k(v)\cdot e^{-2\pi i v\omega}. \tag{4}$$

For $s > 0$, we denote by $H^s(R^4)$ the Sobolev space of all quarternary functions $h(x)\in H^s(R^4)$ such that

$$\int_{R^n} |\hat{h}(\gamma)|(1+\|\gamma\|^{2s})d\gamma < +\infty.$$

The space $H^\lambda(R^n)$ is a Hilbert space equipped with the inner product given by

$$\langle h,g\rangle_{H^\lambda(R^4)} := \frac{1}{(2\pi)^4}\int_{R^4} \hat{h}(\gamma)\overline{\hat{g}(\gamma)}(1+|\gamma|^{2\lambda})\,d\gamma, \quad h,g\in L^2(R^4). \tag{5}$$

We are interested in wavelet bases for the Sobolev space $H^\lambda(R^n)$, where λ is a positive integer. In this case, we obtain

$$\langle h, g \rangle_{H^{\lambda}(R^4)} = \langle h, g \rangle + \langle h^{(\lambda)}, g^{(\lambda)} \rangle, \quad h, g \in L^2(R^4). \tag{6}$$

Suppose that $h(x)$ is a function in the Sbolev space $H^{\lambda}(R^n)$. For $j \in Z, k \in Z^n$, setting $h_{j,u}(x) = 4^j h(2^j x - u)$, we have

$$\| h_{j,k} \|_{H^{\lambda}(R^4)} = \| h_{j,k}^{(s)} \|_{L^2(R^4)} = 4^{j\lambda} \| h^{(s)} \|_{L^2(R^4)}.$$

Note that the discrete-time Fourier transform is Z^4-periodic. Let r be a fixed positive integer, and J be a finite index set, i.e., $J = \{1, 2, \cdots, r\}$. We consider the case of multiple generators, which yield multiple pseudoframes for subspaces.

Definition 1. Let $\{T_k \hbar_j\}$ and $\{T_k \tilde{\hbar}_j\}$ ($j \in J, k \in Z^4$) be two sequences in subspace $M \subset L^2(R^4)$, We say that $\{T_k \hbar_j\}$ forms a multiple pseudoframe for V_0 with respect to (w.r.t.) ($j \in J, k \in Z^4$) if

$$\forall \Upsilon \in M, \quad \Upsilon = \sum_{j \in J} \sum_{v \in Z^4} \langle \Upsilon, T_v \hbar_j \rangle T_v \tilde{\hbar}_j \tag{7}$$

where we define a translate operator, $(T_v \phi)(x) = \phi(x - v), v \in Z^4$, for a function $\phi(x) \in L^2(R^4)$. It is important to note that $\{T_k \hbar_j\}$ and $\{T_k \tilde{\hbar}_j\}$ ($j \in J, k \in Z^2$) need not be contained in M. The above example is such case. Consequently, the position of $\{T_k \hbar_j\}$ and $\{T_k \tilde{\hbar}_j\}$ are not generally commutable, i.e., there exists $\Upsilon \in M$ such that

$$\sum_{j \in J} \sum_{v \in Z^2} \langle \Upsilon, T_v \tilde{\hbar}_j \rangle T_v \hbar_j \neq \sum_{j \in J} \sum_{v \in Z^2} \langle \Upsilon, T_v \hbar_j \rangle T_v \tilde{\hbar}_j = \Upsilon.$$

Definition 2. A generalized multiresolution structure (GQMS) $\{V_j, \hbar_j, \tilde{\hbar}_j\}$ is a sequence of closed linear subspaces $\{V_j\}_{j \in Z}$ of $L^2(R^4)$ and $2r$ elements $h_j, \tilde{\hbar}_j \in L^2(R^4)$, $j \in J$ such that (a) $V_j \subset V_{j+1} \ \forall \ j \in Z$; (b) $\bigcap_{j \in Z} V_j = \{0\}; \bigcup_{j \in Z} V_j$ is dense in $L^2(R^4)$; (c) $h(x) \in V_j$ if and only if $Dh(x) \in V_{j+1}, \ \forall j \in Z^4$, where $D\phi(x) = \phi(2x)$, for $\forall \phi(x) \in L^2(R^4)$; (d) $g(x) \in V_0$ implies $T_v g(x) \in V_0$, for all $v \in Z^4$; (e) $\{T_v h_j, j \in J, v \in Z^4\}$ forms a multiple pseudoframe for V_0 with respect to $T_v \tilde{\hbar}_j, j \in J, v \in Z^4$.

2 Construction of GQMS of Paley-Wiener Subspace

A necessary and sufficient condition for the construction of multiple pseudoframe for Paley-Wiener subspaces of $L^2(R^4)$ is presented as follow.

Theorem 1. Let $h_j \in L^2(R^4)(j \in J)$ be such that $|\hat{h}_j| > 0$ a.e. on a connected neighbourhood of 0 in $[-\frac{1}{2}, \frac{1}{2})^4$, and $|\hat{h}_j| = 0$ a.e. otherwise. Define

$$\Lambda \equiv \bigcap_{j \in J} \{\gamma \in R :$$

$$|\hat{h}_j| \geq c > 0, \ j \in J\} \text{ and}$$

$$V_0 = PW_\Lambda = \{f \in L^2(R^4): \text{ supp}(\hat{f}) \subseteq \Lambda\}. \tag{8}$$

$$V_n \equiv \{\Upsilon(x) \in L^2(R^4): \Upsilon(x/2^n) \in V\}, \quad n \in Z, \tag{9}$$

Then, for $\tilde{h}_j \in L^2(R^4)$, $\{T_v h_j : j \in J, \ k \in Z\}$ is a multiple pseudoframe for V_0 with respect to $\{T_v \tilde{h}_j, j \in J, v \in Z^2\}$ if and only if

$$\sum_{j \in J} \overline{\tilde{h}_j(\gamma)} \hat{h}(\gamma) \cdot \chi_\Lambda(\gamma) = \chi_\Lambda(\gamma) \quad a.e., \tag{10}$$

where χ_Λ is the characteristic function on Λ.

Proof. For all $\Upsilon \in PW_\Lambda$ consider

$$F(\sum_{j \in J} \sum_{v \in Z^2} \langle \Upsilon, T_v h_j \rangle T_v \tilde{h}_j = (\sum_{j \in J} \sum_{v \in Z^2} \langle \Upsilon, T_v \tilde{h}_j \rangle F(T_v h_j)$$

$$= (\sum_{j \in J} \sum_{v \in Z^4} \int_{R^4} \hat{\Upsilon}(\mu) \overline{\hat{\tilde{h}}_j(\mu)} \cdot e^{2\pi i v \mu} d\mu \hat{h}_j(\gamma) e^{-2\pi j v \gamma}$$

$$= \sum_{j \in J} \sum_{v \in Z^4} \int_0^1 \sum_{n \in Z^4} \hat{\Upsilon}(\mu + n) \overline{\hat{\tilde{h}}_j(\mu + n)} \cdot e^{2\pi i v \mu} d\mu \hat{h}_j(\gamma) e^{-2\pi j v \gamma}$$

$$= \sum_{j \in J} \hat{h}_j(\gamma) \sum_{n \in Z^4} \hat{\Upsilon}(\gamma + n) \overline{\hat{\tilde{h}}_j(\gamma + n)} = \sum_{j \in J} \hat{\Upsilon}(\gamma) \hat{h}_j(\gamma) \overline{\hat{\tilde{h}}_j(\gamma)}.$$

where we have used the fact that $|\hat{h}| \neq 0$ only on $[-\frac{1}{2}, \frac{1}{2})^4$ and that $\sum_{n \in Z^2} \hat{\Upsilon}(\omega + n) \overline{\hat{\tilde{h}}(\omega + n)}$, $j \in J$ is 1-periodic function. Therefore

$$\sum_{j \in J} \hat{h}_j \overline{\hat{\tilde{h}}_j} \cdot \chi_\Lambda = \chi_\Lambda, \quad a.e.,$$

is a neccssary and sufficient condition for $\{T_k \tilde{h}_j, j \in J, \ k \in Z\}$ to be a multiple pseudoframe for V_0 with respect to $\{T_k \tilde{h}_j, j \in J, k \in Z^4\}$.

Theorem 2. Let $h_j, \tilde{h}_j \in L^2(R^4)$ have the properties specified in Theorem 1 such that (6) is satisfied. Assume V_ℓ is defined by (9). Then $\{V_\ell, h_j, \tilde{h}_j\}$ forms a GQMS.

Proof. We need to prove four axioms in Definition 2. The inclusion $V_e \subseteq V_{e+1}$ follows from the fact that V_e defined by (9) is equivalent to $PW_{2\Lambda}$ .Condition (b) is satisficd bccause the set of all band- limited signals is dense in $L^2(R^4)$ .On the other hand ,the intersection of all band-limited signals is the trivial function. Condition (c) is an immediate consequence of (9). For condition (d) to be prooved, if $f \in V_0$, then

$$f = \sum_{j \in J} \sum_{k \in Z} \langle f, T_k \tilde{\hbar}_j \rangle T_k \hbar_j \cdot \text{ By taking variable substitution, for } \forall n \in Z^4 ,$$

$$f(t-n) = \sum_{j \in J} \sum_{v \in Z^4} \langle f(\cdot), \tilde{\hbar}_j(\cdot - v) \rangle \hbar_j(t - v - n) = \sum_{j \in J} \sum_{v \in Z^4} \langle f(\cdot - n), \tilde{\hbar}_j(\cdot - v) \rangle \hbar_j(x - v)$$

That is, $T_n f = \sum_{j \in J} \sum_{v \in Z^4} \langle T_n f \cdot T_v \tilde{\hbar}_j \rangle T_v \hbar_j \cdot$ Or, it is a fact $f(T_v \hbar_j)$ has support in Ω for arbitrary $k \in Z$. Therefore, $T_n f \in V_0$.

Example 1. Let $\hbar_j \in L^2(R^4)$ be such that

$$\hbar_j(\gamma) = \begin{cases} 1/r & a.e., \quad \|\omega\| \leq \frac{1}{8}, \\ (3 - 16\|\gamma\|)1/r, & a.c., \quad \frac{1}{8} < \|\gamma\| < \frac{3}{16}, \\ 0, & otherwise. \end{cases} \qquad \overline{\hbar}(\omega) = \begin{cases} 1, a.e., & \|\gamma\| \leq \frac{1}{8}, \\ 5 - 16\|\omega\|, a.e., & \frac{1}{8} < \|\gamma\| < \frac{5}{16}, \\ 0. & otherwise. \end{cases}$$

$j \in J$. Choose $\Lambda = \{\omega \in R : |\hbar(\omega)| \geq \frac{1}{r}\} = [-\frac{1}{4}, \frac{1}{4}]$, and define $V_0 = PW_\Lambda$, select $\tilde{\hbar}_j \in L^2(R^4)$ such thatThen, since $\sum_{j \in J} \tilde{\hbar}_j(\gamma) \overline{\hbar_j(\gamma)} = 1$ a.e. on Λ, by Theorem 1, $\{T_k \hbar_j\}$ and $\{T_k \tilde{\hbar}_j\}$ form a pair of pseudoframes for $V_0 = PW_\Lambda$.

3   The Characters of Nonseparable Quarternary Wavelet Wraps

To construct wavelet packs, we introduce the following notation: $a = 2, \Lambda_0(x) = h(x)$, $\Lambda_v(x) = \psi_v(x), b^{(0)}(k) = b(k), b^{(v)}(k) = q^{(v)}(k)$, where $v \in \Delta = \{0,1,2,\cdots,15\}$ We are now in a position of introducing orthogonal trivariate nonseparable wavelet packs.

Definition 3. A family of functions $\{\Lambda_{16n+v}(x): n = 0,1,2, 3,\cdots, v \in \Delta\}$ is called a nonseparable quarternary wavelet packs with respect to the orthogonal quarternary scaling function $\Lambda_0(x)$, where

$$\Lambda_{16n+v}(x) = \sum_{k \in Z^4} b^{(v)}(k) \Lambda_n(2x - k), \tag{11}$$

where $v \in \Delta$. By taaking the Fourier transform for the both sides of (13), we have

Theorem 3[6]. If $\{\Lambda_{16n+v}(x): n = 0,1,2, 3,\cdots, v \in \Delta\}$ is called a nonseparable quarternary wavelet packs with respect to the orthogonal scaling function $\Lambda_0(x)$, we have

$$\langle \Lambda_m(\cdot), \Lambda_n(\cdot - k) \rangle = \delta_{m,n} \delta_{0,k}, \quad m, n \in Z_+, \quad k \in Z^4. \tag{12}$$

4   The Multiple Affine Fuzzy Quarternary Frames

We begin with introducing the concept of pseudoframes of translates.

Definition 4. Let $\{T_v f, \ v \in Z^4\}$ and $\{T_v \tilde{f}, \ v \in Z^4\}$ be two sequences in $L^2(R^4)$. Let U be a closed subspace of $L^2(R^4)$. We say $\{T_v f, \ v \in Z^4\}$ forms an affine pseudoframe for U with respect to $\{T_v \tilde{f}, \ v \in Z^4\}$ if

$$\forall \Upsilon(x) \in U, \quad \Upsilon(x) = \sum_{v \in Z^4} \langle \Upsilon, T_v \tilde{f} \rangle T_v f(x) \tag{13}$$

Define an operator $K : U \to \ell^2(Z^4)$ by

$$\forall \Upsilon(x) \in U, \quad K\Upsilon = \{\langle \Upsilon, T_v f \rangle\}, \tag{14}$$

and define another operator $S : \ell^2(Z^4) \to W$ such that

$$\forall c = \{c(k)\} \in \ell^2(Z^4), \quad S\{c(k)\} = \sum_{v \in Z^4} c(v) T_v \tilde{f}. \tag{15}$$

Theorem 4. Let $\{T_v f\}_{v \in Z^4} \subset L^2(R^4)$ be a Bessel sequence with respect to the subspace $U \subset L^2(R^4)$, and $\{T_v \tilde{f}\}_{v \in Z^4}$ is a Bessel sequence in $L^2(R^4)$. Assume that K be defined by (14), and S be defined by (15). Assume P is a projection from $L^2(R^4)$ onto U. Then $\{T_v \Upsilon\}_{v \in Z^4}$ is pseudoframes of translates for U with respect to $\{T_v \tilde{\Upsilon}\}_{v \in Z^4}$ if and only if

$$KSP = P. \tag{16}$$

Proof. The convergence of all summations of (7) and (8) follows from the assumptions that the family $\{T_v \Upsilon\}_{v \in Z^4}$ is a Bessel sequence with respect to the subspace Ω, and he family $\{T_v \tilde{\Upsilon}\}_{v \in Z^4}$ is a Bessel sequence in $L^2(R^4)$ with which the proof of the theorem is direct forward.

Theorem 5[8]. Let $\phi(x), \tilde{\phi}(x), \psi_\iota(x)$ and $\tilde{\psi}_\iota(x), \iota \in \Lambda$ be functions in $L^2(R^4)$. Assume conditions in Theorem 1 are satisfied. Then, for any $\hbar \in L^2(R^4)$, and $n \in Z$,

$$\sum_{k \in Z^4} \langle \hbar, \tilde{\phi}_{n,k} \rangle \phi_{n,k}(x) = \sum_{\iota=1}^{15} \sum_{v=-\infty}^{n-1} \sum_{k \in Z^4} \langle \hbar, \tilde{\psi}_{\iota:v,k} \rangle \psi_{\iota:v,k}(x). \tag{17}$$

$$\hbar(x) = \sum_{\iota=1}^{15} \sum_{v=-\infty}^{\infty} \sum_{k \in Z^4} \langle \hbar, \tilde{\psi}_{\iota:v,k} \rangle \psi_{\iota:v,k}(x). \quad \forall \hbar(s) \in L^2(R^4) \tag{18}$$

Consequently, if $\{\psi_{\iota:v,k}\}$ and $\{\tilde{\psi}_{\iota:v,k}\}$, $(\iota \in \Lambda, v \in Z, \ k \in Z^4)$ are also Bessel sequences, they are a pair of affine frames for $L^2(R^4)$.

References

1. Daubechies, I.: The wavelet transform, time-frequency localization and signal analysis. IEEE Trans. Inform. Theory 39, 961–1005 (1990)
2. Iovane, G., Giordano, P.: Wavelet and multiresolution analysis:Nature of ε^{∞}. Cantorian space-time. Chaos. Solitons & Fractals 32(4), 896–910 (2007)
3. Zhang, N., Wu, X.: Lossless Compression of Color Mosaic Images. IEEE Trans Image Processing 15(16), 1379–1388 (2006)
4. Chen, Q., Wei, Z.: The characteristics of orthogonal trivariate wavelet packets. Information Technology Journal 8(8), 1275–1280 (2009)
5. Chen, Q., Cao, H.: Construction and decomposition of biorthogonal vector-valued wavelets with compact support. Chaos, Solitons & Fractals 41(4), 2765–2778 (2009)
6. Chen, Q., Qu, X.: Characteristics of a class of vector-valued nonseparable higher-dimensional wavelet packet bases. Chaos, Solitons & Fractals 41(4), 1676–1683 (2009)
7. Chen, Q., Huo, A.: The research of a class of biorthogonal compactly supported vector-valued wavelets. Chaos, Solitons & Fractals 41(2), 951–961 (2009)
8. Chen, Q., Shi, Z., Cao, H.: The characterization of a class of subspace pseudoframes with arbit-rary real number translations. Chaos, Solitons & Fractals 42(5), 2696–2706 (2009)

Beautification of Chinese Character Stroke-Segment-Mesh Glyph Stroke Curve

MaiKu Zhang, Min Lin, and HanQuan Huang

Computer & Information Engineering College, Inner Mongolia Normal University,
Hohhot 010022, China
zhangmaiku@163.com

Abstract. Stroke-Segment-Mesh Chinese character glyph describe method can describe all kinds of Chinese character skeleton similarities and differences, and support glyph comparative computation. Since the method describe glyph with straight line-segment, so that the described glyph losses of the curve characteistics, influence the display of the glyph. This paper designs a stroke curve algorithm based on Chinese character glyph which described on Stroke-Segment-Mesh Model. This algorithm adopted cubic B-spline curve to restore the missing stroke curve characteristics, and to make the glyph display more consistent with human cognitive habits, And the user can adjust the adjustment points of corresponding strokes so as to make glyph more beautiful; experiments show that the method can achieve better results, broaden the application fields, in teaching wrongly written character, variant forms of characters in ancient literatures display and so on.

Keywords: Stroke-Segment-Mesh, cubic B-Spline, adjust point, stroke, stroke curve.

1 Introduction

Chinese is an open set of language, Chinese characters in large quantities, complex structure, description of Chinese character and characteristics calculation is a basic research of Chinese character information processing. Previous did a lot of work in description of the Chinese character, major achievements have A DICTIONARY OF CHINESE CHARACTER INFORMATION, Chinese Characters Components Specification (GF3001-1997), IDS (Ideographic Description Character Sequence), CPL (Chinese Document Processing Language), Mathematical Expression of Chinese Characters, CDL (Character Description Language)[1-7], generally the glyph of the Chinese character constructor method is classified according to the human cognitive, and described by the use of human cognitive components, strokes. These descriptions of the characters are quite effective, but there are a lot of ambiguities and the lack of description, which can not support the automatic extraction of glyph features, analysis and calculations.

Stroke-Segment-Mesh Chinese character glyph describe method[8] adopts the suitable granular degree, normalized and unambiguous base stroke segment unit to depict any possible character glyph(including right character, wrongly written

D. Jin and S. Lin (Eds.): Advances in MSEC Vol. 2, AISC 129, pp. 101–110.

character, variant forms of characters in ancient literatures and combined-characters) .The 16×16 equal base small rectangles(unit grid) constitute a square matrix, we call it Stroke-Segment-Mesh plane(stroke grid for short), in every base unit stroke segment element and its direction type is stipulated. as shown in Fig. 1 (a), every vertex link with the midpoint of each edge and two diagonal vertexes linked each other, we can obtain a line segment(stroke segment for short),the top side and the left vertical side of the small rectangle on the edge there were two strokes; every vertex to the midpoint of the edge total of $4×2$ strokes, the between vertexes total of 2 strokes, the strokes grid plane, according to the small rectangular side of the "under-closed and top-open, the open left-closed and right-open" principle, in a base small rectangle has 14 strokes, 8 direction, as shown in Figure. 1 (a) as A, B, C, D, E, F, G, H shows; by painted these strokes on the grid plane to describe grid glyph, If the stroke segment was described, it is valid, or else it is invalid. All valid stroke segments common describes a Chinese Character's skeleton, then get a Stroke-Segment-Mesh Chinese character glyph, combined characters "ZhaoCaiJinBao" as shown in Fig. 1 (b).

In the Grid Model, glyph is described with only straight line-segments, and strokes of the glyph are described with one or more straight line-segments, which result in redundant sharp angle at the connected line-segments, missing stroke curve characteristics, poor display . This paper is to design an algorithm to make up the missing curve characteristics of grid glyph, Chinese characters that processed display more consistent with human cognitive habits and better meet the needs of special-shaped display output, such as the teaching of Chinese characters typo, ancient variant of the display output, and thus broaden the applications of Stroke-Segment-Mesh Chinese character glyph model

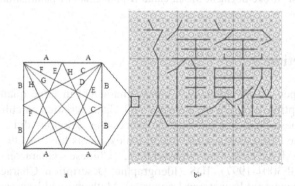

Fig. 1. Stroke segment type direction of base small rectangle

2   Algorithm of Stroke-Segment-Mesh Glyph Stroke Curve

The basic idea of the algorithm: First, extract the each stroke's [9] starting point, turning point, end point and other key point's coordinates information of the glyph; then follow the human cognitive rules of stroke curve to determine both straight line-segments lie in both sides of each turning point whether need curve process, if necessary, on both sides of the turning point on each line-segment to add a adjustment point, and add mark on adjustment points and turning points; Finally, process the marked points as follows: (1) obtain four key points which are ordered as follows: two key points before the turning point, the turning point, a key point after the turning point; (2) obtain four key point which are ordered as follows: a key points before the turning point, the turning point, two key points after the turning point; respectively, as the cubic B-spline control points to draw two curves, two curves can be achieved second-order geometric continuous at the splicing, to complete drawing curve strokes effect.

2.1 Key Points' Extraction of Stroke

Classification of Strokes Based on Stroke-Segment-Mesh Chinese Character Glyph

(1)Basic stroke: In stroke grid character glyph, the same type and the longest valid stroke segment series connected by tail with head. According to the difference of contained stroke segment type in basic stroke, it can be divided into 8 different types which are Horizontal basic stroke a, Vertical basic stroke b, Left oblique basic stroke (c,d,e), Right oblique basic stroke(f,g,h).

(2) Simple stroke: In stroke grid character glyph, the specific type and valid basic stroke series connected by tail and head.Simple stroke Heng: basic stroke a.Simple stroke Shu: basic stroke b.Simple stroke Pie: the b,e,d,c type basic stroke series connected by tail with head from top to bottom in vertical direction, where these four kinds of basic stroke types can exist in different times but one of e, d and c must exist at least.

Simple stroke Na: the b,h,g,f type basic stroke series connected by tail with head from top to bottom in vertical direction, where these four kinds of basic stroke types can exist in different times but one of h, g and f must exist at least.

Simple stroke Pie is divided into four subtypes as, PingPie, XiePie, LiPie and ShuPie. Simple stroke Na is divided into four subtypes as PingNa,XieNa,LiNa and ShuNa. Simple stroke by tail and head, head and tail, head and head, tail and tail connection can obtain more complex types of stroke. The key point coordinate of stroke are obtained on the basis of basic strokes, simple strokes and complex strokes. Due to space constraints, stroke classification details, see [8].

The Extraction of Stroke Key Points

(1) The Definition of Storage Structure of stroke segment grid Character glyph's Key Points:

Struct skeleton_point{int x;int y;}skeleton_keyPoint[][];x, y store respectively coordinate values of key point x and y; Skeleton_keyPoint[i][j] means the jth key points of the ith stroke.

(2)The Steps of Getting Stroke Key Points:

(a)Check every valid stroke segment of stroke segment grid character glyph, and combine the conjoint valid stroke segment into basic strokes.(b)Combine the basic strokes into simple strokes.(c)Combine the linked simple strokes into complex strokes. If the connection point exists multiple simple strokes, firstly combine the two strokes which have the largest angle between them, until the completion of all strokes combined.(d)Extract every independent stroke's coordinate information of start point, turning point, end point orderly, and then store them in skeleton_keyPoint.

These key points record the coordinate information of starting point, turning point and terminal point of stroke segment character glyph's corresponding stroke, Based on the original data used for beautifying character glyph.

2.2   Stroke Curve Process Judge

Determine connected two straight line-segment in a stroke whether need to curve process on the basis of having obtained the key points of Stroke-Segment-Mesh Chinese character glyph. As shown in Fig. 2, if it need curve process, add a key point D in front of turning point B and a key point E after it, else stay as it is, and make all the key points corresponding marks, preparing for the following curve drawing.

Curve Process Judge Rules of Connected Two Straight Line-Segment of Stroke

Whether connected two straight line-segment of stroke need curve process is determined by the type, angle, slope of stroke and people's perceptions of character glyph. The specific circumstances described in Fig. 3, Fig. 4, Fig. 5.

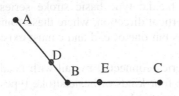

Fig. 2. Two straight segments that need curvingFig

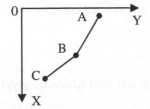

Fig. 3. The situation 1 of stroke curve

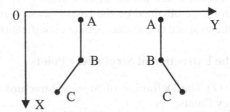

Fig. 4. The situation 2 of stroke curve   **Fig. 5.** The situation 3 of stroke curve

In the three situations of Fig. 3,in (b),the straight segment AB belongs Na type,BC belongs Na type or Heng type, AB's slope is bigger than CD's and the angel between them is bigger than $90°$.in (a),(c) the straight segments AB all belong to Na type, and the straight segments BC respectively belong to Shu and Pie type, and the angel between AB and BC is bigger than $90°$ and smaller than $180°$.

In the situation shown in Fig. 4, the straight segment AB belongs to Pie, Heng or Na type, AB's slope is smaller than BC's and the angel between them is bigger than $90°$.

In the situation shown in Fig. 5, the straight segments AB in (a),(b) belong to Shu type and the straight segments BC respectively belong to Pie and Na type. The angel between them is bigger than $90°$ and smaller than $180°$.

In algorithm, use the start point, end point's horizontal direction coordinates in the grid glyph strokes to determine the difference between the types of Pie, Shu, Na. The difference between the horizontal coordinates of line segment AB is :

$$D_{AB} = X_A - X_B ,$$ (1)

x_A, x_B respectively represent the line segment AB's start point and end point horizontal coordinate value. When value of D_{AB} are respectively greater than 0, equal to 0, less than 0 indicates the type of line segment AB belong to stroke Pie, Shu, Na. The method can accurately articulate the type of a segment of the stroke and provide gist for the judgement of stroke curve process.

Judgement and Process Steps of Stroke Curve Process

(1) The Definition of Storage Structure of stroke Key Points after the judgement of curve process: struct prettify_point{int x, y;bool add_point; bool curve_point; } prettity_keyPoint[][];x, y respectively store corresponding key points(include newly added key points) coordinate value. If this key point is new added, we assign true to add_point, or else we assign false to add_point. If this key point is the turning point of conjoint two straight segments that need curving, we assign true to curve_point, or else we assign false to curve_point. prettity_keyPoint[i][j] represents the ith stroke's jth key point.

(2) The steps of adding the new key point of stroke,i is the stroke number subscript. The initial value i=0.

Step1 Take the ith undisposed stroke element from the skeleton_keyPoint[][]. k is the subscript of ith stroke's key point. The initial value k=0. The processed results stored in prettity_keyPoint. j is the subscript of ith stroke's key point been processed.

Step2 Take out this stroke's kth key point and set it as B point. Dispose of it as follows:

2.1) If this key point is the stroke's starting or ending point, then store it in prettify_keyPiont[i][j], j++, and mark the point non curve process point,that is setting the value of add_Point and curve_Point false.

2.2) Or else, dispose of it like this:

a) Assume X axis is horizontally right, Y axis is vertically down. D_{AB}, D_{BC} are respectively the difference values of line-segment AB, BC's horizons, For the situation in Fig. 3(b), $D_{AB}<0$ and $D_{BC}<0$, the situation in Fig. 4, $D_{AB}>0$ and $D_{BC}>0$, calculate the K_{AB},K_{BC} by (2),which respectively is slope of line-segment AB, BC.

$$K_{fe} = \left(Y_f - Y_e\right)/\left(X_f - X_e\right) . \tag{2}$$

b) Judge whether the conjoint two straight segments is belong to any situations in Fig. 3, Fig. 4 and Fig. 5:

Case1: $D_{AB}<0$ and $D_{bc}=0$, as the situation in   Fig. 3 (a).

Case2: $D_{AB}<0$, $D_{bc}<0$ and $K_{AB}>K_{BC}$, as the situation in Fig. 3(b).

Case3: $D_{AB}<0$ and $D_{bc}>0$, as the situation in Fig. 3(c).

Case4: $D_{AB}>0$, $D_{bc}>0$ and $K_{AB}<K_{BC}$, as the situation in Fig. 4.

Case5: $D_{AB}=0$ and $D_{bc}>0$, as the situation in Fig. 5(a).

Case6: $D_{AB}=0$ and $D_{bc}<0$, as the situation in Fig. 5(b).

If the condition is true, then do the process of b1 to b4.

b1) Calculate the square of straight line-segment AB,BC,AC respectively by (3).

$$D_{fe} = \left(X_f - X_e\right)\times\left(X_f - X_e\right) + \left(Y_f - Y_e\right)\times\left(Y_f - Y_e\right) , \tag{3}$$

D_{ab}, D_{bc}, D_{ac} are the square value of line-segment AB, BC, AC.

b2) calculate the cosine value of the angel between AB and BC:

$$Cosin\_value = \left(D_{ab} + D_{bc} - D_{ac}\right)/\left(2\sqrt{D_{ab}}\sqrt{D_{bc}}\right) . \tag{4}$$

b3) If Cosin_value<0,respectively add a adjustment point D, E to the place which is two-fifths length of AB, BC away from point B, as Fig. 2. When exixt both straight line-segment of both sides of continuous turning point need curve process, add adjustment point which far from the turning point 2/5 on the straight line-segment, to avoid position disorder of adjustment point and gain better curve drawing effect.Store key points D, B, E in order, and mark new added points and turning point, As follows:

b3.1) Adjustment point D stored in prettify_keyPoint, j++, x,y respectively take $X_B-(X_B-X_A)\times0.4$, $Y_B-(Y_B-Y_A)\times0.4$, and mark it as new added point, that is setting add_Point,curve_Point respectively true,false.

b3.2) Turning point B stored in prettify_keyPiont[i][j], j++, x,y respectively are X_B, Y_B , and mark it as a point need curve process, that is setting add_Point、 curve_Point respectively false, true.

3.3) Adjustment point E stored in prettify_keyPiont[i][j], j++, x,y respectively are $X_B-(X_B-X_C)\times0.4, Y_B-(Y_B-Y_C)\times0.4$, and mark it as a point need curve process, that is setting add_Point,curve_Point respectively true, false.

b4) If   Cosin_value>=0, the point stored in prettify_keyPiont[i][j], j++,and marked it as a non curve process point, that is setting add_Point,curve_Point false.

c) If b) is not true.Do as what the b4) do.

Step3 k++. If skeleton_keyPoint[i][k] is not visited, jump to Step2. Until all key points of this stroke are accessed and processed one time.

Step4 i++. If the stroke element i is not disposed, jump to Step1. Until all strokes are accessed and processed one time.

2.3   Stroke Curve Drawing Method

After curve processe being judged towards stroke key point, the two straight line-segment need curve process show Fig. 2. Respectively use key points A,D,B,E,D,B,E,C as the cubic B-spline[10] control points to draw two curves, two curves can be achieved second-order geometric continuous at the splicing, D, E are new added adjustment point. User can dynamically adjust stroke curve by moving adjustment point, to make glyph display effect more consistent with human cognitive habits.

The Introduction of B-Spline Curve

B-spline curve can be obtained by basis functions weighted sum given a set of control points , the general shape of the curve to be controlled by these points.

m+n+1 plane or space vertexes $P_i(i=0,1,\cdots,m+n)$ are given, we call n times parameter curve segment:

$$P_{k,n}(t) = \sum_{i=0}^{n} P_{i+k} G_{i,n}(t) \quad t \in [0,1] \tag{5}$$

Which is n times B-Spline curve segment of the kth section(k=0,1, ... ,m). If these m+n+1 control vertexes don't superpose with each other, then generate m+1 curve sections, these curve sections joining by order generate a curve and the whole curve have n-1 order geometry continuity (Gn-1). The polygon consisting of vertexes $P_i(i=0,1,\cdots,m+n)$ is called characteristic polygon of B-Spline curve.

In above formula, the primary function is:

$$G_{i,n}(t) = \frac{1}{n!} \sum_{j=0}^{n-1} (-1)^j C_{n+1}^j (t+n-i-j)^n \quad (t \in [0,1]; i = 0,1,\cdots,n) \ . \tag{6}$$

Cubic B-spline curve segment parameter expression is:

$$P_t = \sum_{i=0}^{3} P_i F_{i,3}(t) \ . \tag{7}$$

$F_{i,3}(t)$ can get by (6), P_i are Polygon corresponding to control points.

Stroke-Segment-Mesh Glyph Drawing

Step1 Extract an unprocessed stroke from prettify_keyPoint which store stroke Key Points after the judgment of curve process.

Step2 Take out this stroke's key point in order, and set it as D. If point D is this stroke's starting point, then painter move to this point to be the starting point to draw line. Or else, dispose of it as follows:

2.1) If D.add_Point is true and subsequent point B of point D, B.curve_Point is true, then take a precursor key point A of point D, take subsequent continuous two key points B,E of point D as Fig. 2. Take these four key points as the control points of cubic B-Spline curve to weight and summate to depict a curve.

2.2) If D.add_Point is true, and let precursor of Point D be point B and B.curve_P-oint is true, then take precursor continuous two key point E,B of point D, subsequent key point C of point D as Fig. 6. Take these four key points as the control points of c-ubic B-Spline curve to weight and summate to depict a curve. The starting point of the curve joints with end point of last curve, and the two curves have two-order geometry continuity in the joint place. Reduce the sharp angle of two linked line-segment, so the linked line-segment DB,BE become a smooth curve as shown in Fig. 7. User can move to any direction to adjust stroke curve effect.

2.3) If D.add_Point and D.curve_Point are false, draw line to point (D.x,D.y).

2.4) If D.curve_Point is true, jump to Step2 to execute.

Step3 If the key point is not last point of the shoke then jump to Step2. Until all key points of this stroke are accessed and processed one time.

Step4 If the stroke is not the last stroke then jump to Step1. Until all strokes are accessed and processed one time.

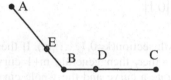   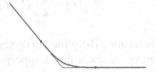

Fig. 6. Two straight segments hat need curv process   **Fig. 7.** Effect of two straight segments after curv process

3   Comparison and Experiment of before and after of Stroke Curve Process

The grid character glyph of the combined character "ZhangCaiJinBao" shown as Fig. 8.The effect of beautifying treatment algorithm shown in Fig. 9, the small dot in the figure is movable adjustment point, the user can dynamically adjust the curve segment effect by moving the dot in all directions, the effect by adjusted shown in Fig. 10; final results shown in Fig. 11, through interactive adjustment can get better display effect.

Using development tool of Stroke-Segment-Mesh described method has completed all the 20902 Chinese characters of character set GBK and more than 1,000 foreign students of Chinese writing typo and variant forms of characters description in accordance with the unique structure, upper and lower structure, left and right structure, inclusive structure, nested grid structure of DICTIONARY OF CHINESE CHARACTER INFORMATION [1]structer type, Respectively extract 100 chinese characters from Simplified and Traditional font of Stroke-Segment-Mesh glyph adopt this method curve process, reached satisfactory results.we have don a test to 200 Chinese character of font typo of foreign student achived a better effect.

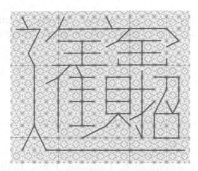

Fig. 8. The grid character glyph of the combied character "ZhangCaiJinBao"

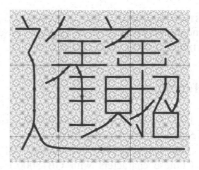

Fig. 9. The effect of grid character glyph of "ZhangCaiJinBao" after process

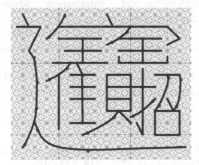

Fig. 10. The effect of grid character glyph of "ZhangCaiJinBao" after interactive adjust

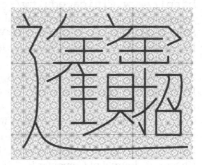

Fig. 11. The effect of grid character glyph of "ZhangCaiJinBao" finally effect

4 Conclusions

This paper designs a stroke curve algorithm based on Chinese character glyph which is described on Stroke-Segment-Mesh Model. And use cubic B-spline curve to restore the missing curve characteristics of Stroke-Segment-Mesh glyph, which displays more consistent with human cognitive habits. Stroke curve processing precision is high and speed is fast, and can move the adjustment point for further dynamic adjustment stroke curve effect, and ultimately achieve better effect of glyph display, which broaden the application fields, which Stroke-Segment-Mesh glyph method in teaching wrongly written character, variant forms of characters in ancient literatures display and so on. The next step: (1) reduce human intervention to further improve the computer automatically strokes curve beautifying effect. (2) the starting point of the grid-glyph, turning point to do more modifications to display similar to the standard font and store it according to the TrueType format, so that special characters (right character, wrongly written character, variant forms of characters in ancient literatures and combined-characters) can use as ease as set inner of characters in the office software.

110   M. Zhang, M. Lin, and H. Huang

Acknowledgment. The preferred Project Supported by National Natural Science Foundation(60863007), Post-graduate Research Innovation Foundation of Inner Mongolia Normal University(CXJJs10054).

References

1. Jiao, S.: Tong University Chinese character encoding set. Chinese charactersdictionary. Science Press, Beijing (1988)
2. State Language Work Committee. GF3001-1997 information-processing components of Chinese characters with GB13000.1 character specification. Language Press, Beijing (1997) (release December 1, 1997) (implementation May 1,1998)
3. Ideographic Descripion, http://www.unicode.org/versions/ Unicode4.0.0/ch11[S].pdf:307-309
4. http://www.sinica.edu.tw/~cdp/Institute of Information Science, Academia Sinica, Taipei, literature dealing with the laboratory site
5. Sun, X.-M., Yin, J.-P., Cheng, H.-W., Wu, Q.-Y., Jing, X.-H.: On Mathematical Expression of A Chinese Character. Journal of Computer Research And Development 39(6), 707–711 (2002)
6. Zhang, W.-Y., Sun, X.-M., Zeng, Z.-B., Wu, J.-Z.: Automatic Gener-ation of Mathmatical Expression of Chinese Characters. Journal of Computer Research And Development 41(5), 848–852 (2004)
7. Cook, R.: A Specification for CDL (Character Description Language):an extract of [Ph-D Dissertation].UC Berkeley. Dept.of Linguistics (2003)
8. Lin, M., Song, R.: A Stroke-Segment-Mesh (SSM) Glyph Description Method of Chinese Characters. Journal of Computer Research and Development 47(2), 318–327 (2010)
9. Lin, M.: The Study of Formal Description of Chinese Character glyph and Application. Beijing University of Technology, Beijing (2009)
10. Luo, X.-N., Wang, Y.-M.: Computer Graphics (in Chinese), pp. 188–196. Zhongshan University Press (2003)

Technology of GPS Data Acquisition
Based on Linux Operating System

Miaolei Zhou[1] and Yimu Guo[2]

[1] Department of Control Science and Engineering, Jilin University,
130022 Changchun, China
zml@jlu.edu.cn
[2] Department of Electrical Engineering, Changchun University of Science and Technology,
130022 Changchun, China
mooyyii@gmail.com

Abstract. A data communication method between the GPS receiver and the USB port in the Linux operating system is proposed in the paper. The data format of GPS receiver is analyzed. The extraction method is achieved about the GPS positioning data acquisition and robot positioning information using C++ programming language. In order to verify the proposed method, data collection experiment is researched. The experimental result shows that this method is effective and stable.

Keywords: GPS, Data acquisition, USB, Positioning.

1 Introduction

Linux is a real multi-user and multi-task operating system, it has the characteristics of high stability and could be customized based on need, and has the good features of high reliability, supporting multi-hardware, open source code, etc. It has been widely applied in industrial measurement and control and embedded. Appling the Linux in the industrial control, especially the military and national defense, etc, can greatly improve the system security.

GPS is the short for Navigation Satellite Timing and Ranging/Global Position System, it is composed by a 24-satellite system which could cover the whole world. The system can assure that we could observe four satellites in any time and any point at the same time, to be sure that the satellites could collect the longitude and height of the point, so that achieving the functions of navigation, position, time service, etc. The technology can guide airplanes, shippings, vehicles and individual, get the destination timely through the chosen route safely and exactly. As GPS technology owns the characteristics of all-weather, high precision and automatic measurement, it has been integrated into various fields of national economy development, national defense development and social development as the advanced measurements and new production.

Considering the Linux operating system characteristics of high reliability and GPS navigation and location, the text adopts Linux as the operating system, GPS conducts

D. Jin and S. Lin (Eds.): Advances in MSEC Vol. 2, AISC 129, pp. 111–116.

the outdoor autonomic mobile robot absolute position detection sensor, researching a method based on the Linux operating system that can be data communication with the GPS receiver through USB port.

2   Linux System USB Port Communication

The text adopts GPS receivers as the USB port, so it needs self-produce USB port communication system based on the Linux operating system to collect the GPS data. In the Linux system, files are abstract, Linux has considered all things as the files, so we can open, close, read and write the different terminal equipments as operating common files. File system Hierarchy Standard (FHS) is name and architecture standards of files contents that Linux follows, in the file system hierarchy, all the Linux equipments are stored in the contents/dev as the way of equipment file name, the equipment file that USB reaches serial ports adapter is ttyUSB. If we use n USB ports, the corresponding equipment file names are respectively ttyUSB0, ttyUSB1, ..., ttyUSBn. Therefore, when reading the data from GPS receivers, we can achieve it by visiting the equipments files that USB reaches serial ports adapter.

Operating the USB ports through the following steps:

(1) *Open the USB ports*: Opening USB port can adopt standard file Open function. Open function has two parameters, the first one is to be opened equipment file name (as: /dev/tty/USB0), the second one is open pattern of equipment files.
(2) *Set USB port communication pattern*: The main task is setting members value of struct termios. The most fundamental settings comprise baud rate setting, efficacy bit and stop bit setting, etc.

```
struct termio{
unsigned short c_iflag;     /*input mode mark*/
unsigned short c_oflag;     /*output mode mark*/
unsigned short c_cflag;     /*control mode mark*/
unsigned short c_lflag;     /*local mode*/
unsigned char c_line;       /*line control*/
unsigned char c_cc[NCC];    /*control character*/
}
```

(3) *Read USB ports data*: Reading USB ports adopt standard Read function. Reading function has three parameters, they are respectively file descriptor, the byte size of reading.
(4) *Close USB ports*: Using Close function to close the opened USB ports. The unique parameter of the function is opening serial ports file descriptor.

Through the operations above can receive GPS data. Here showing how to collect the effective latitude and longitude information from the received GPS data.

3   Collect GPS Position Information

The collected data that are read and written from USB ports comprise position time, position condition, latitude and longitude, speed and direction, etc. Therefore, it is necessary to collecting latitude and longitude information that is needed by robots position from the data. The data forms are GPRMC (suggesting using minimum GPS data form), GPGGA (GPS fixed data output statements) and GPGSV (shown satellite form). The text adopts GPRMC form.

GPRMC form as follows:

$GPRMC, <1>, <2>, <3>, <4>, <5>, <6>, <7>, <8>, <9>, <10>, <11><CR><LF>
1) Standard position time form: hhmmss.sss.
2) Communication condition, A=data valid, V=data invalid.
3) Latitude form: ddmm.mmmm.
4) Latitude distinguish, north latitude (N) or south latitude (S).
5) Longitude form: ddmm.mmmm
6) Longitude distinguish, east longitude (E) or western longitude (W).
7) Relative displacement speed, 0.0 to 1851.8 knots.
8) Relative displacement direction, 000.0 to 359.9°.
9) Date form: ddmmyy.
10) Magnetic pole variant, 000.0 to 180.0°.
11) Degree.

The information that robot position system needs is position information, therefore we can just judge whether the keywords beginning of line are "$GPRMC" when receiving GPS data, if so, receive the data and collect part5 and part6 information on that line.

4   Specific Realization

Through the Linux operating system, C++ program realizes GPS data collection. Following codes are specific realization process.

```
    int OpenDev(char* dev)        //open serial function
{
        int fd = open(dev,  O_RDWR);
        if(fd == -1){
                perror("error to open!\n");
                return -1;
        }
        else
                return fd;
}
void GPSInit()//USB ports initialization function
{
        char* port = "/dev/ttyUSB0";  //specified equipment
function
        fd = OpenDev(port);       //open USBports
        printf("Now GPS port is %s, fd = %d\n", port, fd);
        struct termios opt;       //define the structure
        tcgetattr(fd, &opt);      //receive source ports
default setting
        cfsetispeed(&opt, B4800);//input baud rate 4800bps
```

```
        cfsetospeed(&opt, B4800);//output baud rate 4800bps
        opt.c_cflag |= (CLOCAL | CREAD);//permit to receive
        opt.c_cflag &= ~CSIZE;          //shield other mark
        opt.c_cflag |= CS8;             //8 data bits
        opt.c_cflag &= ~CSTOPB;         //1 stop bit
        opt.c_cflag &= ~PARENB;         //no odd-even check
        opt.c_lflag &= (ICANON | ECHO | ECHOE | ISIG);
        //original input form, not dealing with the input
data in the buffer
        tcsetattr(fd, TCSANOW, &opt); //setting
characteristic become effective immediately
}
int ReadGPSData()        //read GPS data
 {
        int i;
        char buff[256];
        char tmpchar1[10], tmpchar2[11];
        if ((nread = read(fd,  buff,  256)) > 0)
        {
              buff[nread + 1] = '\0';
              if ( (buff[18]=='A') && ( memcmp(buff,
"$GPRMC",  4) == 0 )) {
                    for (i=0;  i<=8;  i++){
                          tmpchar1[i]=buff[20+i];
                          tmpchar2[i]=buff[32+i];
                    }     tmpchar2[9]=buff[41];
                    tmpchar1[9]='\0';
                    tmpchar2[10]='\0';
                    lati=atof(tmpchar1)/100;
//collect latitude information, and save as dd.mmmmmm
form
                    long=atof(tmpchar2)/100;  //collect
longitude information, and save as dd.mmmmmm form
                    return 0;
              }
              else
                    return -1;
        }
        else
              return -1;
}
int main()        //main program
{
        ofstream outfile;
        outfile.open("Gpsdata.txt", ios::out);
        if (!outfile) {
              cerr << "Cannot open file\n";
              exit(1);
        }
        outfile << setprecision(6) <<
setiosflags(ios::fixed);        //set saving data form
```

```
    GPSInit();   // USB ports initialization
    while (1){
         if (ReadGPSData() == 0){
              outfile << ' ' << longi << ' ' << lati
<< endl;   //output the data to files and save it
         }
    }
    close(fd);          //close USB ports
    exit(0);
}
```

Through the program above, it can realize the latitude and longitude data collection that robot position needs. Of course, if the data information of relative displacement speed, relative displacement direction and etc are needed, it also can be realized by the method.

5   Experimental Research

In order to verify the proposed method, the paper conducts the field data collection experiment research (see in Fig. 1).

Fig. 1. GPS data experimental curve

In the Fig. 1, green line is data curve that the GPS receiver obtains when autonomous mobile robot is walking on this road. The experimental curve confirms the effectiveness and stability of collecting GPS data by the method.

6   Conclusion

A collection and processing method of the GPS data under the Linux system is proposed in this paper, the USB ports data communication method is described. The method used in this paper is not only for robots position, but also applied to navigation system of vehicle, shipping, etc.

The paper's innovation:

1. Realizing the data communication method to USB ports based on the Linux operating system.
2. The proposed method is not only for GPS data collection based on the Linux operating system, but also applied to other data collection of USB equipments, such as magnetic compass, etc.

Acknowledgments. This research is supported by Program of Science and Technology Development Plan of Jilin province of China (Grant No.: 20090122), Scientific Frontiers and Interdisciplinary Innovation Project of Jilin University of China (Grant No.: 200903309).

References

1. Liu, Y.J., Yue, H.: Linux Operating System Tutorials. China Machine Press, Beijing (2005) (in Chinese)
2. Leick, A.: GPS Satellite Surveying. Wiley, Alexandria (2004)
3. Hu, W.S., Gao, C.F.: GPS Measuring Principle and Application. China Communications Press, Beijing (2004) (in Chinese)
4. Gu, B.L., Wang, L.M., Han, Y.: GPS Data Acquisition and Processing Based on VC++. Microcomputer Information 24, 203–205 (2008) (in Chinese)

Key Technology and Statistical Mapping Method of Active Emergency Service

Deguo Su[1,2], Ximin Cui[1], and Lijian Sun[2]

[1] College of Geoscience and Surveying Engineering,
China University of Mining & Technology, Beijing, China
[2] Chinese Academy of Surveying and Mapping, Beijing, China
sdg@casm.ac.cn, cxm@cumtb.edu.cn, slj@casm.ac.cn

Abstract. This paper studies the key techniques and methods of active emergency mapping service, which focuses on describing the statistical classification of map data management, statistical graphics templates design and knowledge management, etc., which were for detailed mapping of knowledge and analysis, general knowledge and professional knowledge in mapping the role of emergency mapping service.

Keywords: Active, statistical mapping services, Knowledge, emergency.

1 Introduction

With the social progress and development, all kinds of unexpected events occur, there is now a variety of platforms and a variety of emergency response management policies and preventive measures. In the field of geographic information systems, space-based information technology, emergency services, also appeared on the application of geographical information platform. From the Component GIS sent to the grid GIS, GIS development from two-dimensional virtual geographical environment, the software architecture is increasing at the same time, the application software complexity and difficulty are also rising; simultaneous service resources are not controlled, share the problem of low level, applications for the growing number of service resources, less and less control over the hardware resources, system expansion bottleneck, WEB Service technology and the emergence of grid services provide a powerful shared management control service model, but followed by a more complex system, how to choose grid services over the network has become a problem, especially in the emergency services platform, how to more efficient use of network services is currently the hot issue of the one. Therefore, this paper presents an active emergency key technologies of geographic information services and methods of research, hoping to take the initiative through the emerging field of computer technology to further promote the emergency services, geographic information service model for efficient application.

2 Active Mapping Methods

Active Emergency cartographic services research mainly includes basic emergency event definitions, mapping data classification, mapping and cartographic services to build templates that look so active.

D. Jin and S. Lin (Eds.): Advances in MSEC Vol. 2, AISC 129, pp. 117–121.

2.1   Definition of Basic Event

First, the active mapping, we must first define a variety of drawing events, according to event type or conditions of the initiative to enter the process of mapping software, such as in a large city when the accident occurred, the alarm, take the initiative to take the initiative to create mapping services the details of the accident to the city traffic maps, the command chain of command for dealing with the accident, not by the chain of command to find the type of traffic maps. Therefore, the mapping of events, including broad and narrow mapping event. Generalized mapping drive mapping event is any event mapping service system, and a narrow mapping event is driven by some non-interrupting events associated with the mapping events, such as certain intelligent devices may be associated with the alarm, while some graphics issue messages and those messages will automatically create the required mapping system to request a map.

2.2   Mapping Data Classification

For graphics, the automated processing of data is an important content, including data specification checks, and according to data visualization features of the initiative to provide solutions.

Spatial statistical data classification according to different classification criteria have different classification results. This project is the study of statistical cartography for active service data classification model, the data dimension (structure), the distribution of characteristics (demographic characteristics) is mapping the data model to express an important aspect of visualization, so the need to follow the data itself statistical characteristics and dimensions as a standard for data classification. Demographic characteristics of the main considerations: central tendency, dispersion, skewness and kurtosis; data structure with information should form a five-dimensional set of properties.

Mapping model with the application of spatial statistical properties of the main features of the data (such as central tendency, dispersion, skewness, etc.), Naive Bayes algorithm to maintain the property independent of each other under the premise of their properties by setting different weights to be the probability that the sample belongs to each category. In this way the weight of each attribute value is equal to 1 and the Rega. Let each data sample with an n-dimensional feature vector to describe the n value of the property, that is: $X = \{x1, x2, ..., xn\}$, the corresponding attribute weights, respectively: $Q = \{Qx1, Qx2, ..., Qxn\}$, both set up the following formula:

$$Qx1+Qx2+...+Qxn=1. \tag{1}$$

Thus, given an unknown data sample X (that is, no class label), if the following conditions is established, the Bayesian classification and the unknown sample X assigned to the class Ci class.

$$P(Ci|X)* Qx1>P(Cj|X)* Qxj \quad 1{\leq}j{\leq}m, \quad j{\neq}i \tag{2}$$

Statistics for spatial data classification and relationship with the mapping model classification was adaptive, with the weight of each attribute among attributes instead of as the probability distribution associated with the impact brought about, while maintaining the simple Bayesian classification features, but also reduces the between attributes are not independent of the error caused, can improve the classification accuracy.

2.3 Mapping Template to Build

The basic configuration of statistical map is divided into partitions according to the type of statistical cartography configuration, grade configuration and statistical cartography legend configuration.

Although statistics show graphics with a variety of different styles, but the statistical map for the district to set the default color and style, in the management of the configuration template with a graphic way, drawing on a variety of statistical classification of design management configuration. For different types of graphics have their own custom template parameters. Some type of custom design template interface as shown as Fig.1:

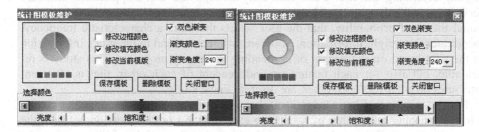

Fig. 1. Template examples of statistical mapping for cartography users

Partition statistical graphics templates to modify the main parameters of the type including border color, fill color, if gradient, gradient color, gradient angle, and color brightness and saturation, to achieve a new functional template, modify the current template and delete the current template.

Saved template used in the user mapping service application process, if the user set a different style or color, it will record the user selection information for the next time the user has selected the active drawing sent to the graphics user.

3   Knowledge Management Mapping Services Statistics

Statistics Cartography learning and knowledge management is based on the statistical classification of cartographic knowledge, knowledge, learning strategies and knowledge management research and application methods, are described below:

3.1   Statistical Classification of Cartographic Knowledge

Knowledge management and learning is to achieve active statistical basis for cartography, mapping knowledge, including knowledge of cartography and mapping the user's own information and knowledge. Mapping for their expertise with the data characteristics determine the type of mapping symbols, data classification, mapping and other color schemes, mapping the user's information, including user data and user interest information and color preferences. In the mapping process, and often use mapping knowledge to determine the various parameters and expression mapping methods. Among them, the knowledge management includes the construction of a priori knowledge of mapping and cartographic knowledge update.

3.2   Mapping Knowledge Learning Strategies

Knowledge and learning, including selection by the user drawing the user's preference for active learning, and for updating the user's knowledge correct, but also for mapping the original a priori knowledge of the update, for example, depending on the data distribution model to get the appropriate type of mapping may be change. Therefore, knowledge of the learning strategy also includes two aspects, namely the user to update their knowledge strategy and mapping to update their knowledge strategy. For users, the software services in addition to the user's understanding of the basic information describing the user other than only through the user's application software services to make the choice of learning, so the user's learning should meet the conventional theory of probability, that the user selects the more a characteristic frequency, the greater the weight information. Software will recommend to the user's information to the user based on the weight of descending to priority programs. This study is repeatable, but also for each feature are independent of each other, that is, different attributes do not affect each other, are discrete single study.

For the mapping of knowledge itself, because different parameters together constitute the mapping expression of results, so each drawing or coordination between the properties have influence. It can be used to increase the weight of the ways to improve Bayesian model to establish basic rules of statistical cartographic knowledge learning mechanism. Naive Bayesian classification model with applications to expand and mapping changes in conditions, statistics and mapping models the relationship between the adaptation may be adjusted. Naive Bayes' theorem assuming a property value impact of a given class, independent of other attribute values, while in reality this assumption often is not established, so the need to design a weight-based Naive Bayesian classifier, compensate property independent of assumptions the loss caused by the classification accuracy.

3.3   Knowledge Management and Application

With the learning strategy, the object-oriented statistical thinking to build the knowledge rule base map. Rely on a priori statistical cartographic mapping knowledge and rules to build the original knowledge base, using the current relational database technology to build knowledge of the rules of statistical map learning system, through the knowledge of statistical cartography (mapping model selection, classification, color schemes, etc.) continuous learning, self-learning knowledge maps improve the statistical system. The use of ECA (Event-Condition-Action) method to build the basic library of events and basic events have the ability to be combined into complex events, the event may be derived from the mapping engine, mapping a variety of environmental or external mapping information. The following graphics engine as an example to the application of knowledge mapping process. Among them, the cartographic analysis of the engine's primary role is key information to user needs and current major source of statistical data format, then the source of information classification, which classified design information on the different data sources under the resolution process, the final design of a uniform system accepted norms of data structure, the various data sources into a unified format for data format. As shown as Fig.2.

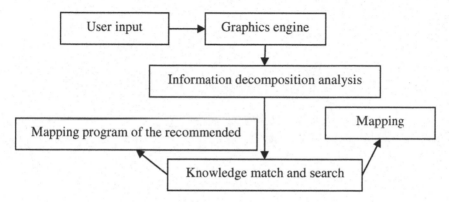

Fig. 2. The programs of mapping use some certain knowledge by matching methods

Acknowledgment. The study is financially supported by the National Natural Science Foundation of China under Grant No. 41071328 and basic Research and Development Operations under grant No. 7771114.

References

1. Aygn, E.: Modelling of Expert Knowledge in Geographic Information Systems-based Planning of the Tuz Lake Special Environmental Protection Area, Turkey. Planning Practice and Research 24(4), 435–454 (2009)
2. Buja, A., Cook, D., Swayne, D.F.: Interactive high-dimensional data visualization. Journal of Computational and Graphical Statistics 5, 78–99 (1996)
3. Frye, C., Eicher, C.L.: Modeling Active Database-Driven Cartography Within Gis Databases. In: Proceedings of the 21st International Cartographic Conference (ICC), pp. 1872–1878 (2008)
4. Andrienko, G.L., Andrienko, N.V.: Interactive maps for visual data exploration. International Journal of Geographical Information Science 13(4), 355–374 (1999)
5. Andrienko, G., Andrienko, N.: Knowledge-Based Visualization to Support Spatial Data Mining. In: Hand, D.J., Kok, J.N., Berthold, M. (eds.) IDA 1999. LNCS, vol. 1642, pp. 149–160. Springer, Heidelberg (1999)
6. Kraak, M.-J.: The cartographic visualization process: from presentation to exploration. The Cartographic Journal 35, 11–15 (1998)
7. Luan, S., Dai, G.: Fast algorithms for revision of some special propositional knowledge bases. Comput. Sci. Technol. 18(3), 388–392 (2003)
8. Michael Friendly, Milestones in the history of thematic cartography, statistical graphics, and data visualization (October 16, 2008)

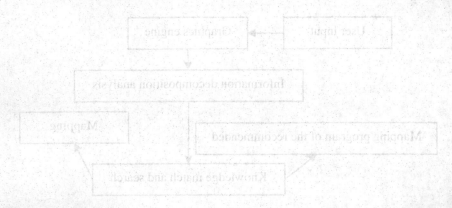

Pattern Synthesis of Array Antennas with a Kind of Quantum Particle Swarm Optimization Algorithm

Ting Wang[1], Shou-da Wang[2], Wen-mei Zhang[3], and Zhi-wei Zhang[4]

[1] School of Information Engineering, Hebei University of Technology & Chinese People's Liberation Army Air Force 93756 Forces, Tianjin, China
wangting031@126.com
[2] School of Computer Science, Hebei University of Technology, Tianjin, China
johnnywsd@gmail.com
[3] Chinese People's Liberation Army Air Force 93756 Forces, Tianjin, China
zhang.wenmeilook@163.com
[4] School of Information Engineering, Hebei University of Technology, Tianjin, China
zhangzhiwei@hebut.edu.cn

Abstract. Pattern synthesis is a key technology in smart antennas. As a result of some short comings, such as the premature convergence and bad local optimal searching capability in traditional intelligence methods for pattern synthesis, a novel array antenna pattern synthesis technique is proposed based on quantum particle swarm optimization(QPSO) with probability amplitude coding of quantum bits. In this method, the positions of particles are encoded by quantum bits and the movements of particles are performed by quantum rotation gates, which achieve particles' optimal location. The mutations of particles are performed by quantum non-gate to avoid premature. The typical application shows the novel way for pattern synthesis with multi-null and low sidelobe restrictions is feasible with high precision, and fast running speed, and has very good generalization capability.

Keywords: Array antennas, Patterns synthesis, Probability Amplitude Coding of Quantum Bits, Particle Swarm Optimization.

1 Introduction

With the rapid development of wireless communication technology, the available spectrum resources is becoming more and more limited, how to optimize the allocation of resources and improve the communication system resource utilization rate have become the key problem of wireless communication for the further development. Smart antenna technology is the one key technology in communication areas that emerged in this context. Smart antennas transmit and receive patterns by using combinations of multiple antenna array elements to automatically adjust, and achieve the optimal system parameters for different signal environment. The smart antennas are playing an increasingly important role in modern communication systems, and pattern synthesis technology is a core technology for smart antennas [1].

D. Jin and S. Lin (Eds.): Advances in MSEC Vol.2, AISC 129, pp. 123–130.

Pattern synthesis is a complex nonlinear optimization problem. The more antenna element is, the more parameters are needed to adjust. Using the classic Woodward-Law-son sampling method for such issues, the last pattern is usually with high side-lobe level, and low gain[2].When we use the fine-tuning methods to optimize, the workload is big, what's more, we need rich debugging experience and theoretical analysis. In order to improve efficiency in recent years, many experts and scholars have used a variety of intelligent optimization algorithms, with good optimization ability, such as genetic algorithms(GA)[2],particle swarm optimization (PSO)[3], immune clone selection algorithm(ICSA)[4]in the optimization of antenna pattern and all achieved good results.

These algorithms although are simple and quick, but there are still some problems to solve, for instance, not high optimal accuracy and slow convergence speed.

Quantum evolutionary algorithm is a probabilistic search algorithm developed in recent years, which is based on some theories and concepts of quantum computing such as qubits and quantum superposition state. Compared with the classical evolutionary algorithm, it has a better population diversity and global optimization ability. Li Shiyong and Li Panchi [5] fused the improved quantum evolution algorithm to PSO algorithm, and proposed a novel quantum particle swarm optimization (QPSO). The global optimization capability of the new algorithm has been greatly improved in multi-dimensional optimization problems, compared with the PSO algorithm.

We applied the algorithm (QPSO) to the antenna pattern comprehensive study, then proposed a new pattern synthesis technique. The typical application shows the novel way for pattern synthesis with multi-null and low sidelobe restrictions behaves well.

2   Algorithm Theory

The quantum particle swarm algorithm in the article came from the QEA and the PSO. In this method, the positions of particles are encoded by quantum bits and the movements of particles are performed by quantum rotation gates, which achieve particles' optimal location. The mutations of particles are performed by quantum non-gate to avoid premature. Since each quantum bit has two probability amplitudes, every particle occupies two positions in the search space, and the chance of finding the optimal position of the particles increases significantly. The QPSO algorithm has a strong search capabilities and optimization efficiency, especially for complex optimization functions and high-dimensional optimization problems.

The QPSO algorithm implementation steps are as follows.

Step1: initialize particles

In the QPSO algorithm, we use the probability of quantum bits to code as the current location of the particles directly. Considering the randomness of population initialization, use the following programs

$$P_i = \left[\begin{matrix} \cos(\theta_{i1}) & \cos(\theta_{i2}) & \cdots & \cos(\theta_{in}) \\ \sin(\theta_{i1}) & \sin(\theta_{i2}) & \cdots & \sin(\theta_{in}) \end{matrix} \right] \tag{1}$$

$\theta_{ij} = 2\pi \times Random$, The range of $Random$ is $(0\ 1)$, $i = 1, 2, ..., m$, $j = 1, 2, ..., n$, m represents the population size, and n represents the dimension of the space.

Step2: transform the solution space, and calculate the fitness.

In QPSO, each dimension of searching space for the particles is [-1,1], in order to calculate the merits of the current location for the particles, we transform the searching space. Two positions of each particle are mapped to the solution space from the unit space. Transforming formula is as follows.

$$X_{ic}^{j} = \frac{1}{2}[b_i(1 + \alpha_i^{j}) + a_i(1 - \alpha_i^{j})]$$

$$X_{is}^{j} = \frac{1}{2}[b_i(1 + \beta_i^{j}) + a_i(1 - \beta_i^{j})]$$

(2)

Thus, each particle corresponds to two solutions of optimization problems. $i=1,2,\cdots,n$; $j=1,2,\cdots,m$. Calculate the fitness of each particle and update the particle's own best position and the global best position of population.

Step3: update particles state

In QPSO, movements of particles are realized by quantum revolving door.

Assuming the best current location for the particle (p_i) is as follows.

$$p_{il} = (\cos(\theta_{il1}), \cos(\theta_{il2}), \cdots, \cos(\theta_{i\ln}))$$

(3)

And the best position of the entire population is P_g.

$$p_g = (\cos(\theta_{g1}), \cos(\theta_{g2}), \cdots, \cos(\theta_{gn}))$$

(4)

Based on the above assumptions, updates for particles can be described as follows:

(1) the rotation updates of quantum bit for the particle- p_i

$$\Delta\theta(t+1) = \omega\Delta\theta_{ij}(t) + c_1 r_1(\Delta\theta_l) + c_2 r_2(\Delta\theta_g)$$

(5)

And

$$\Delta\theta_l = \begin{cases} 2\pi + \theta_{ilj} - \theta_{ij} & (\theta_{ilj} - \theta_{ij} > -\pi) \\ \theta_{ilj} - \theta_{ij} & (-\pi \le \theta_{ilj} - \theta_{ij} \le \pi) \\ \theta_{ilj} - \theta_{ij} - 2\pi & (\theta_{ilj} - \theta_{ij} > \pi) \end{cases}$$

(6)

$$\Delta\theta_g = \begin{cases} 2\pi + \theta_{gj} - \theta_{ij} & (\theta_{gj} - \theta_{ij} > -\pi) \\ \theta_{gj} - \theta_{ij} & (-\pi \le \theta_{gj} - \theta_{ij} \le \pi) \\ \theta_{gj} - \theta_{ij} - 2\pi & (\theta_{gj} - \theta_{ij} > \pi) \end{cases}$$

(7)

(2) the probability amplitude updates of quantum-bit for the particle- p_i

$$\begin{bmatrix} \cos(\theta_{ij}(t+1)) \\ \sin(\theta_{ij}(t+1)) \end{bmatrix} = \begin{bmatrix} \cos(\Delta\theta_{ij}(t+1)) & -\sin(\Delta\theta_{ij}(t+1)) \\ \sin(\theta_{ij}(t+1)) & \cos(\Delta\theta_{ij}(t+1)) \end{bmatrix}$$

$$\times \begin{bmatrix} \cos(\theta_{ij}(t)) \\ \sin(\theta_{ij}(t)) \end{bmatrix} = \begin{bmatrix} \cos(\theta_{ij}(t)) + \Delta\theta_{ij}(t+1) \\ \sin(\theta_{ij}(t) + \Delta\theta_{ij}(t+1) \end{bmatrix}$$

(8)

Two new positions for the particle

$$\tilde{p}_{ic} = (\cos(\theta_{ij}(t) + \Delta\theta_{ij}(t+1)), \cdots, \cos(\theta_{ij}(t) + \Delta\theta_{ij}(t+1)))$$
$$\tilde{p}_{is} = (\sin(\theta_{ij}(t) + \Delta\theta_{ij}(t+1)), \cdots, \sin(\theta_{ij}(t) + \Delta\theta_{ij}(t+1)))$$
(9)

Step4: mutate particles

First, give a mutation probability p_m, and give each particle a random number $rand_i$ between 0 and 1. If $rand_i < p_m$, $\lceil n/2 \rceil$ quantum bits of the particle is randomly selected, and two probability amplitude are exchanged by quantum non-gate. The best position for the particle on memory and the angle of their vectors remain unchanged. Mutation operation is as follows:

$$\begin{bmatrix} 0 & 1 \\ 1 & 0 \end{bmatrix}\begin{bmatrix} \cos(\theta_{ij}) \\ \sin(\theta_{ij}) \end{bmatrix} = \begin{bmatrix} \sin(\theta_{ij}) \\ \cos(\theta_{ij}) \end{bmatrix} = \begin{bmatrix} \cos\left(\theta_{ij} + \dfrac{\pi}{2}\right) \\ \sin\left(\theta_{ij} + \dfrac{\pi}{2}\right) \end{bmatrix}$$
(10)

Step5: return step 2, until meet the convergence criteria or reach the maximum steps.

A typical test function (Ten-dimensional Rosenbrock) has been used to optimize to verify the effectiveness and flexibility of the QPSO algorithm. In order to demonstrate the superiority of QPSO in the simulation, the performance of QPSO is compared with LDW-PSO and PSO. To find extreme value in this function is a classic complex optimization problem, the global optimal point is located in a smooth, long and narrow parabolic valley. As provided only a small amount of information, it is difficult for algorithms to identify the direction of the searching and the chance to find the global minimum is little, it is often used to evaluate the efficiency of optimization algorithms.

Rosenbrock function expression as follows:

$$f(x) = \sum_{I=1}^{9}\left[(1 - X_i)^2 + 100(x_{i+1} - x_i^2)^2\right], \quad x_i \in [-2.048, 2.048]$$

Spatial characteristics of the function are shown in Figure 2.

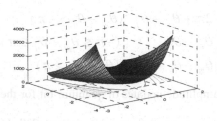

Fig. 2. Spatial characteristics of the Rastrigin

Use the function value directly as fitness function, the smaller function value is, the greater fitness is.

Experimental setup: population size $N = 20$, space dimension $D = 10$, total 30 runs are made, maximum iterant times is 300, $c_1 = c_2 = 2$, inertia weight $w = 0.5$, $P_m = 0.01$. When the fitness value is less than 5×10^{-6}, the algorithm stop, and is considered convergent. The results are listed below. It can be found from Table 1 that QPSO procedure possesses the dominant speed and precision in the optimization compared with other algorithms.

Table 1. Performance of three algorithms to optimize Rosenbrock function

Method	Convergence rate	Best fitness value	Average number of iterations
QPSO	1	2.2245e-013	122
LDW-PSO	1	1.3273e-010	176
PSO	0.98	1.3546e-009	210

3   QPSO FOR Pattern Synthesis

First, consider an N element array of a straight-line, the array factor

$$A_F(\theta) = \sum_{i=1}^{N} \omega_i \ell^{j\frac{2\pi}{\lambda} x_i \sin\theta} \tag{11}$$

Where: ω_i is the array element incentives; x_i is the antenna array element location; λ is the wavelength. For the uniform linear array, the array spacing d, that can be written as:

$$A_F(\theta) = \sum_{n=0}^{N-1} \omega_n \ell^{j2\pi kdn \sin\theta} \tag{12}$$

Consider the composition of the 20 element uniform linear array, set the initial current phase are zero (edge radio array), while the current amplitude is symmetric.

The objective function defined as follows:

$$f = \alpha|MSLL - SLVL| + \beta|NULL - PAT - NLVL| \\ + \gamma NULL - STD \tag{13}$$

$$MSLL = \max\{F(\phi)\}, \phi \in S \tag{14}$$

Where: $F(\phi)$ is the antenna pattern function, S is sidelobe region of pattern, MSLL is the highest sidelobe level, SLVL design sidelobe level, NULL-PAT is the average depth of zero settlement, NLVL is designed depth of zero trap , NULL-STD is the variance of zero trap depth, α, β, γ are the rights.

Example 1

A uniform linear array is formed by $2N = 20$ elements with fixed spacing of $d = \frac{\lambda}{2}$ between each two elements. The initial current phase is zero, and the current amplitude of elements is $I_n \in (0,1)$. In the design, the main lobe width is $2\theta_0 = 20^0$, side lobe (SLVL) level less than -40 dB.

Set of 50 particles, the maximum number of iteration step 500. Strike a fitness function according to (13), use QPSO algorithm to optimize the solution. Pattern simulation and iterative curve are shown in Figure 3,4, the optimized amplitude of the current source are shown in Table 2. As shown in figure 3,the maximum sidelobe level converged to -39.9876dB with the global population evolving approximately 180 generations. Xiao Longshuai and Huang Hua have used the NPSO algorithm to do the same optimization with the same requirement [6]. However, the last maximum sidelobe level was -39.5996dB with the global population evolving approximately 280 generations. It illustrates better performance for QPSO both in optimization speed and accuracy

Table 2. Computed element current for examples

Current amplitudes	$I_{1,20}$	$I_{2,19}$	$I_{3,18}$	$I_{4,17}$	$I_{5,16}$
Example 1	0.9986	0.9647	0.8730	0.7510	0.6342
Example 2	0.1221	0.1513	0.0307	0.2292	0.6805
Current amplitudes	$I_{6,15}$	$I_{7,14}$	$I_{8,13}$	$I_{9,12}$	$I_{10\,11}$
Example 1	0.4665	0.3597	0.2324	0.1537	0.0769
Example 2	0.8586	0.8509	0.8927	0.9656	0.9986

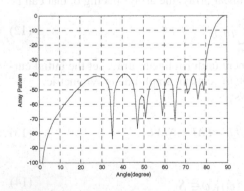

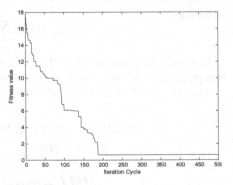

Fig. 3. Synthesized array patterns for example 1, SLVL=-40dB

Fig. 4. Convergence curve for example 1

Example 2

$2N = 20$, $d = \frac{\lambda}{2}$, The initial current phase are zero, and the current amplitude of elements is $I_n \in (0,1)$. Design specifications : the main lobe width is $2\theta_0 = 20^0$, side lobe (SLVL)level less than -20dB, six nulls imposed at 30,40,50,60,70,80,and the depth all less than -90dB.Compared with the similar example 2 in the article[4],it increases a low sidelobe requirement, and 4 locations of deep nulling angle. Use QPSO algorithm to optimize the problem. Fig.5 shows optimized field patterns, which have met the design requirements both in sidelobe level and null depth level. The side lobe level of the QPSO is -19.9950dB. Fig.6 shows the fitness function convergence curve, we can see that particles converge to about 130 steps.

As a whole, in each of these cases, the QPSO algorithm easily achieved the optimization goal.

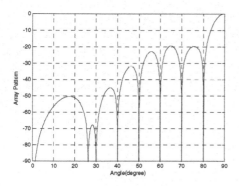
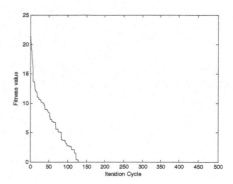

Fig. 5. Synthesized array patterns for example 2   **Fig. 6.** Convergence curve for example 2

4   Conclusions

In this paper, the QPSO algorithm with probability amplitude coding of quantum bits was compared with PSO and LDW-PSO, and we found that its convergence accuracy and speed are all the best. Then a novel array antenna pattern synthesis technique is proposed based on the QPSO algorithm. The typical applications show that the novel way for pattern synthesis with multi-null and low sidelobe restrictions is feasible with high precision, fast running speed and good stability.

Acknowledgments. This work supported by National Nature Science Foundation of China(NO.60972106) and Tianjin Natural Science Foundation (NO. 11JCYBJC00900). The corresponding authors of this article are Ting Wang and Shou-da Wang. Email: wangting031@126.com. johnnywsd@gmail.com

References

1. Jin, R.-H., Geng, J.-P., Fan, Y.: Smart antenna in wireless communication. Beijing university of posts and telecommunications press, Beijing (2006)
2. Calvete, H., Gale, C., Mateo, P.: A new approach for solving linear bilevel problems using genetic algorithm. European Journal of Operational Researchs (1), 14–18 (2008)

3. Khodier, M.M., Christodoulou, C.G.: Linear array geometry synthesis with minimum side lobe level and null control using particle swarm optimization. IEEE Transactions on Antennas and Propagation 53(8) (August 2005)
4. Chi, Y., Zhang, P.-L.: Pattern Synthesis Based on Immune Clone Selection Algorithm. Communication Technology 42(5), 71–74 (2009)
5. Li, S.-Y., Li, P.-C.: Quantum computing and quantum optimization algorithm, pp. 113–117. Harbin industrial university press, Harbin (2009)
6. Xiao, L.-S., Huang, H.: Array Antennas Beam Pattern Synthesis Based on Neighborhood Particle Swarm Optimization. Communication Technology 42(9), 52–53 (2009)

An Approach to the Algorithm for Adaptive Routing Strategy Oriented to Location-Based Services

Mingjun Xin, Guobin Song, and Qiongqiong Wang

School of Computer Engineering and Science, Shanghai University, Shanghai 20072, China
{xinmj,billsong,wqq818}@shu.edu.cn

Abstract. With the increasing popularity of location-based services (LBS), the issues caused by users' location privacy are suffering more and more complicated problems during these years. In this paper, it firstly analyzes the attacks on LBS and the existing protection methods, and then presents a new adaptive routing policy based on mobile multi-agents under LBS environment. For the unique characteristics of multi-agents, the advantages of mobile agents not only ensure the safety of the agent itself, but also protect the privacy of users' personal information. Then, it presents a multi-agent based adaptive routing policy to improve the speed of selecting the mobile path by using two-level routing tables discussed in this paper. Furthermore, the updating algorithm of improved routing table makes the mobile agents at a lower level of total cost. Finally, we design a simulation experiment to show that the routing policy is feasible, efficient and safe to the LBS platform.

Keywords: Privacy protection, mobile multi-agents, LBS, routing policy.

1 Introduction

With the rapid development of mobile Internet technology and the increasing popularity of mobile terminals, more and more location-based users can freely query any service information at any time and any place. In the past five years, although the increasing applications of LBS have brought us more and more convenience, the privacy of mobile user's information is experiencing more serious threat under the mobile service environment.

Just as the agent has its unique characteristics of autonomy, responsiveness, initiative, reasoning and social interaction, mobile agent has been applied up to LBS as a key technical tool in many ways. Firstly, it can help mobile users to complete their query, payment and other events for LBS provider. Furthermore, it will ensure the privacy of user information and the quality of mobile service. Because of the vulnerability of a single agent, multi-agents system (MAS) is needed to deal with the bottle-neck problems caused by LBS platform. The multiple individuals can communicate with each other to form several large cooperative organizations. They can compensate for the limited capacity of the individual. However, the practicality and enforceability of LBS system depends on how agents choose their routing path to the destination node. The multi-agents based routing policy is one of the current main research areas oriented to mobile Internet services, thus a good routing policy can directly determine a good performance for LBS platform. Thus, we propose a

D. Jin and S. Lin (Eds.): Advances in MSEC Vol. 2, AISC 129, pp. 131–138.

two-level policy to solve the adaptive routing strategy in this paper, so as to provide better technical support to improve the transmission efficiency for LBS.

2   Related Works

Location privacy information consists of personal identify information and dynamic location information, which will be used to describe a particular user's location track or position coordinates. Location privacy protection prevents unauthorized users from knowing a particular user's current location or historical dynamic location. According to the main idea of k-anonymous, Gruteser[1] proposes quadtree-based structure of algorithm for the k-anonymous location information blurred. Xue[2] proposes the hidden ring and hidden seed ,which fuzzy user location information to effectively protect the location privacy. Duckham[3] firstly proposes the obfuscation as a mechanism for location privacy protection, and then designs an algorithm to make use of fuzzy location information. But it does not reduce the quality of mobile service, while effectively balances the location privacy protection and the quality of mobile service. In general, these studies of location privacy protection only focus on few users to fuzzy their location information. But in the case of continuous queries of enormous concurrent users, their private information will be exposed to LBS server. Therefore, Mokbel[4] proposes a method that can handle the location privacy problem for enormous concurrent users. Although the proposed method in large part solves the location privacy problems, but higher processing power is urgently demanded on the server side. Therefore, the server must improve the efficiency of processing user queries to support the increasing LBS applications.

In the point of view for mobile Internet, mobile agent has several advantages of multiple characteristics to support LBS platform. It can be moved from one location server to another independently, and on behalf of the user to complete the assigned tasks. But mobile agent also has many security problems. Shen[5] proposes a RBAC based security access policy for mobile agent. It shows that role-based authorization achieves the privilege of access control to the resource of proxy server. Furthermore, it adopts data encryption to achieve secure transmission of information. Cui[6] combines mobility, execution and autonomy to make the appropriate safety analysis. As the characteristics of MAS oriented to LBS, the efficient and safe operation of mobile agent lies in the efficient mobile routing strategy of multi-agents. In the implementation of routing strategy for mobile multi-agents, choosing dynamic routing algorithm meets the changing and unstable network environments. Liu[7] uses a one-way function that meets combination of law but does not meet the exchange of law to construct a new mobile agent routing solution.

According to the characteristics of mobile agent and the elliptic curve algorithm of its own advantages, it adopts elliptic curve threshold signature scheme to ensure the mobile agent security and protect users' privacy in this paper. In this way, malicious attackers only invade all of the relevant privacy of mobile agents in order to obtain a meaningful information chain. Meanwhile it presents a multi-goal oriented routing policy, which improves the speed of selecting the optimized mobile agent path using two-level routing tables. Furthermore, the updating algorithm of the improved routing table makes the mobile agent cost at a lower level under LBS environment.

3   The Mobile Agent Based LBS Routing Strategy

3.1   The Distribution of Mobile Agent to Support LBS

Under the mobile multi-agents supported LBS platform, the proxy server generates several mobile agents with the total number of n, it then distributes the multi-agents to LBS platform. When mobile agents get into the LBS network, they will be arranged to next node until they reach the final LBS server. In this way, each mobile agent stops at any network node to query the routing table, thus the mobile agents will select the optimal path to LBS server, which is relied on the routing algorithm in the complex mobile Internet environment. It is shown in figure1.

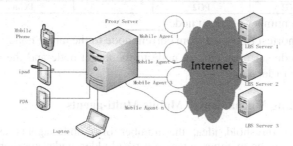

Fig. 1. Multi-agents Based LBS Platform

In the working process discussed above, the structure of routing table and the algorithm of optimal path selecting are directly related to the query efficiency and privacy safety. In this paper, it shows that LBS network node will automatically generate dynamic routing, and then construct more efficient dual routing tables.

3.2   The Design of Routing Table

When mobile users send commands to the proxy servers under LBS environment, proxy servers then create several mobile agents. Mobile agents move to their target node through the LBS network, and select a most optimal path. Then the mobile agent routing policy will decide the next node to be visited according to the adjacent node (local) routing table and the target node (overall) routing table.

Definition 1. The adjacent node(local) routing table: local routing table records the cost required for the current node accessing to adjacent nodes. Local routing table is figured by a two-dimension array $Pat[n][3]$, and its data is shown in Table 1.

Definition 2. The target node(overall) routing table: the overall routing table records the cost required   for the current node accessing to target nodes, it is also figured by a two-dimensional array $Fat[][]$, and the data is shown in Table 2.

Table 1. The Adjacent Node (local) Routing Table

1	2	3	...	n
PNo1	PNo2	PNo3	...	PNon
Pt1	Pt2	Pt3	...	Ptn
Pm1	Pm2	Pm3	...	Pmn
α1	α2	α3	...	αn

PNoi: The network node's number.

Pti: time required from current node to the adjacent node.

Pmi: times of the node selected as middle routing.

αi: the probability of the node selected as middle routing.

Table 2. The Target Node (overall) Routing Table

1	2	....	M
FPNo1	FPNo2	....	FPNom
FPt1	FPt2	....	FPtm
PG1	PG2	....	PGm

FPNoi : The number of target node *i*.

FPti : The shortest time cost from current node to the target node.

PGi : The node set from current node to the target node and the set arranged by the agent visiting order.

3.3   The Working Mechanism of Mobile Multi-agents

According to the threshold idea, the number of mobile agents and the privacy information listed in the member agents must be hidden under appropriate conditions. Only in this way, mobile agents will generate an efficient proxy signature.

The specific working process is listed as the following steps:

1)   Mobile users send their mobile service request to the mobile agent engine (MAE).
2)   MAE then assigns the information to n mobile agents. Each of them carries some requested information.
3)   Agents leave the proxy server into the LBS network environment. The related multi-agents will be stayed on each local LBS host, and then make full use of its autonomous intelligence into the routing algorithm for the shortest path selection.
4)   The mobile agents choose the best path to LBS provider.
5)   LBS providers return the appropriate service information to mobile users.

In this way, users send requests to MAE, the mobile agents then handle the related LBS request to the server host. Thus it greatly reduces the dependency of LBS network environment. Meanwhile user information is carried by k mobile agents, according to the threshold idea, even if the attacker obtains the information of some agents, they are still not fully aware of the user's personal information. However, under actual LBS platform, the mobile agents have to face with much more complexity. The quality of service (QoS), the host operating speed and the dynamic network conditions should be considered directly.

4   The Implementation Routing Path Selection Algorithm

As mobile agent moves on to the destination, some nodes will be selected as the middle node frequently in the routing path. In the case of heavy load, it is easy to form congestion. In order to avoid this phenomenon, it defines the probability of the node selected factor α. It describes the metric of intermediate node which is selected as the intermediate routing node. The bigger α is, the more times the node will be selected as the intermediate routing nodes. Each node is assigned initially probability factor α_{init}. Each time, if node is selected as the intermediate node of the path, α will reduce α_{Δ} . So the probability of the node selected factor defines as follows:

$$\alpha = \alpha_{init} - n \times \alpha_{\Delta} \tag{1}$$

Where n is the times of the node to be accessed.

4.1   Main Parameters of the Routing Policy

According to the characteristics of LBS, the routing table updating algorithm should be more efficient and optimized to serve for routing selection. Some important parameters in updating the algorithm are defined as follows:

Definition 3. Mobile agent: The data structure of mobile agent is defined as $Agent\_i\_stru\{ID,FID,HID, Pro,Cinfo,Pa\}$, ID is the identifier of the agent, FID is LBS server ID which the agents search for, HID is the host's identifier which creates mobile agents, Pro is the relevant program algorithm of mobile agent, $Cinfo$ represents the user's information, Pa is a set of cooperative nodes which the agents traverse through$\{...Pi...Pj...\}$.

Definition 4. Time spent to get adjacent node: Tij is the time required for two adjacent nodes i and j reaching to each other in network. If the coordinate of node i is (x_i, y_i) and the coordinate of node j is (x_i, y_i) , then the agent moves from node i to node j costs Tij:

$$Tij = \frac{-(ab+cd) + \sqrt{(a^2+c^2)r^2 - (ad-cb)^2}}{a^2+c^2} \tag{2}$$

In which $a = v_{fast}\cos\partial_i - v_{fast}\cos\partial_j$; $b = x_i - x_j$; $c = v_{fast}\sin\partial_i - v_{fast}\sin\partial_j$; $d = y_i - y_j$.

Definition 5. The refresh rate of the routing table: Pi is the access frequency of one node, if Pi is higher, the refresh rate is higher. The time access state decides the refresh rate, so each node derives table update interval Δt according to $F(Pi)$, every time when $NowTime = LastTime + \Delta t$ then update the current routing table.

4.2   Implementation of the Adaptive Routing Algorithm

Mobile agent $Agent\_i$ reaches a node $Post\_i$, the program Pro in mobile agent does the following operations :

>   When Agent_i reaches Post_i
>   Search overall routing table
>   If there is FPNoi = target node AND FPtl <= the reasonable time Tv
>   Routing path selection is PGi,AND Pa=Pa+PGi
>   Else Search Pti in local routing table

Select the biggest value of α corresponding to the smallest value of Pti
Calculate α=α-n×αΔ , insert into local routing table
      Records PNoi,AND insert Pa
Move to another Post_j
Routing table update algorithm: Update(local routing table):
    When Agent_i reaches CurrentPost
Read HID,Pa,$T_{currentpost}$ ∈ Pa in Agent_i_stru FPNoi(1<i<m) in current table
  If HID >< FPNoi(1<i<m)
    Insert FPNoj=HID, FPtj=$T_{currentpost}$, PGj=Pa to current routing table as j
  Else if HID=FPNoi(1<i<m)
  Compare $T_{currentpost}$ with FPti
  If $T_{currentpost}$ < FPti then FPti=$T_{currentpost}$

The algorithm will be executed constantly when mobile agents move from proxy host to LBS host. When it selects the next node at any time, a most optimized path will be constructed finally.

Assuming the local routing table size is m, the overall routing table size n, and the routing path length k. Routing optimization algorithm computational complexity mainly attributes to searching the overall routing table. In the worst case, routing optimization algorithm gets back into exhaustive traversal algorithm, and the algorithm complexity comes up to $m * n$; While in the best case, the algorithm complexity is 1. Routing updating algorithm computational complexity is decided by the routing path length, and its complexity is k. The routing table is dynamically updated and the mobile agents are always searching for optimized network path, so the overall performance of the routing optimization algorithm will be better than other algorithms.

4.3 Simulation Results

The commercial mobile payment experiment validates the performance of routing policy by using the simulation software oriented to LBS platform. It implements a simulative ten-node LBS network topology by using VC6.0, in which the network node can send information to each other. The network topology is shown in Figure 2.

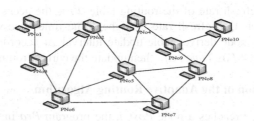

Fig. 2. The mobile payment experiment network topology

Each PC in the above figure represents a routing node, and has its routing table. The test simulates 100 clients which send payment information. Proxy host creates 5 agents every time from *PNo1* to target node *PNo10*. The back process records the reaction of each node when the program is running. This experiment can be divided into two parts, one is designed in accordance with this routing policy, and the other is designed with only local routing table. The experiment results are shown in Table 3:

Table 3. Experiment Statistics for the Adaptive Routing Algorithm

	PNo2	PNo3	PNo4	PNo5	PNo6	PNo7
Tc%	97.00	88.00	89.00	92.00	90.00	89.00
Rc%	98.57	98.89	98.33	98.57	98.42	98.33
T%	97.00	96.00	94.00	98.00	97.00	95.00
R%	99.33	99.67	99.13	99.33	99.54	99.27
AT	310	190	260	50	190	200

In table 3, T% is the percentage of one node's processing time to complete the whole algorithms below the theoretical ideal time, R% is the optimize rate of node's routing table, AT is the number of visits, and Tc% and Rc% are the experimental data using only local routing table.

The experimental data shows that the success rate is in the ideal range. As is compared to the second part experiment, the efficiency of agents that arrive at target merchant is higher. It also shows that the first one has obvious advantages. This simulation experiment proves that all of the mobile agents in mobile payment system can find the final node quickly and effectively. This routing policy discussed above is fully applicable to the mobile payment system oriented to LBS platform.

5   Conclusion

According to the characters of multi-agents under LBS environment, it proposes a routing optimization strategy in this paper. The dual routing table makes mobile agent find target node's routing path quickly. The routing optimization strategy effectively reduces the time cost of mobile agents on the route, thus shortens the waiting time of location based services for mobile users. On the other hand, the introduction of multi-agent technology meets the privacy requirements of sensitive information and also improves the quality of LBS. However, because of the volatility and unpredictability of the mobile services, the communication speed and running time among LBS network nodes needs to be further improved for its accuracy in the future.

Acknowledgements. Our research is supported by National Natural Science Foundation of China (Project Number. 61074135), Shanghai Creative Foundation project of Educational Development (Project Number. 09YZ14), and Shanghai Leading Academic Discipline Project (Project Number.J50103), Great thanks to all of our hard working fellows in the above projects.

References

1. Gruteser, M., Grunwald, D.: Anonymous Usage of Location-based Services Through Spatial and Temporal Cloaking. In: Processing of the International Conference on Mobile Systems, Applications, and Services, MobiSys 2003, pp. 163–168 (2003)

2. Xue, J., Liu, X.: Road network-oriented method of location privacy protection. Computer Science 34(5), 865–877 (2011)
3. Duckham, M.: A formal model of obfuscation and negotiation for location privacy. Pervasive (2005)
4. Mokbel, M.F.: Towards Privacy-aware Location-based Database Services. In: Proceedings of the 22nd International Conference on Data Engineering Workshops (2006)
5. Shen, X., Zhuang, C.: RBAC-based security policies for mobile agent. Computer Engineering and Design 26(2), 329–331 (2005)
6. Cui, Y.: Mobile Agents and Security Analysis. Anhui University 26(4), 21–25 (2002)
7. Liu, Y.: A new mobile agent routing solution. Computer Science 36(11), 113–115 (2009)
8. Yi, Z., Li, F.: Multi-constrained Routing Algorithm Based on Mobile Agent for Mobile Ad Hoc Networks. Journal of Southwest Jiao Tong University 45(1) (2010)
9. Peng, C., Li, X.: An Identity-Based Threshold Signcryption Scheme With Semantic Security. In: Hao, Y., Liu, J., Wang, Y.-P., Cheung, Y.-m., Yin, H., Jiao, L., Ma, J., Jiao, Y.-C. (eds.) CIS 2005. LNCS (LNAI), vol. 3802, pp. 173–179. Springer, Heidelberg (2005)

An Improved Shuffled Frog Leaping Algorithm for TSP

Zhoufang Li and Yuhua Wang

School of Information Science and Engineering
Henan University of Technology
ZhengZhou, China
lzhf1978@126.com

Abstract. With the optimization theory of Shuffled Frog Leaping Algorithm (SFLA), the paper extended the traditional model of SFLA for solution by using the encoding basing on urban-based sequence and using the new method for individual. This paper also improved the SFLA for the traveling salesman problem. The simulation results show the effectiveness of the proposed method and strategy.

Keywords: SFLA Artificial Intelligence Optimization.

1   Introduction

Traveling Salesman Problem can be dated back to 1759, Knight Travel issue proposed by Euler, currently, it has become a classic NP-hard in field of combinatorial optimization[1]. Shuffled frog leaping algorithm (SFLA) is a group intelligent algorithm resulting from simulating the characteristics of information sharing and exchange in the process of frog feeding. It was first proposed by Eusuff and Lansey in 2003, and used to solve the minimization problem of pipe size in the expansion of pipeline network successfully[5]. While SFLA is a very effective optimization algorithm, but in the algorithm, most of the relevant parameters are belonging to continuous real domain, the algorithm model only uses elementary operations, therefore, the standard SFLA is mainly applied to solve the optimization problems of continuous space domain, and not suitable for direct soluting the discrete combinatorial optimization problems such as TSP. Therefore, this article extended the solving model of thaditonal SFLA, designed the new update operator, proposed a new solving algorithm for TSP basing SFLA and improved the algorithm. The simulation in the standard example results show that: the algorithm proposed by this paper has faster convergence speed and better solution quality.

2   Description of TSP

TSP problem can be described as the following: Giving N cities and the distance between any two cities, finding the shortest route which through each city once and only once. It's description in graph theory is: Given graph G=(V,A),where V is the vertex set, A is the arc set composed by interconnected vertex, knowing connection distance between the vertex, required to identify a Hamilton circuit which has the shortest

D. Jin and S. Lin (Eds.): Advances in MSEC Vol. 2, AISC 129, pp. 139–144.
springerlink.com   © Springer-Verlag Berlin Heidelberg 2011

length, that is the shortest circuit that through all the vertices once and only once. To the city set V= {s_1, s_2, ..., s_N },supposing an access order of which is T={t_1,t_2,...,t_N},where $t_i \in$ V(i=1,2,...,N),agreeing t_{N+1}= t_1,then the mathematical model for the traveling salesman problem is:

$$\min \sum_{i=1}^{N} d_{t_i, t_{i+1}}$$

Where, $d_{t_i, t_{i+1}}$ is the distance between the city t_i and the city t_{i+1}.

3   The Basic Principle of SFLA

SFLA is a post-heuristic algorithm with a global information exchange and basing on population search. It is essentially a combination of two heuristics algorithms of both SCE and PSO of their respective advantages, and expands it into discrete engineering.SCE algorithm is a heuristic algorithm for continuous problems [6], it is very suitable for solving the optimization problem with a number of local minimum.

SFLA uses SCE's shuffle strategy in the global and uses the search mechanism similar to the discrete particle swarm algorithm in the local, paper[7] has proven its effectiveness to solve discrete problems.

The global search algorithm flow of SFLA is as following:

Initialization: Choosing m and n, m is the number of species, n is the frog's (feasible solution) number of each species, supposing the dimensions of each solution is d, therefore, the total number of initial solution is F = mn.

Step(1) Generating F initial solution: U(1),U(2),...,U(F),the i-frogs is behalf of the i-solution: U(i)=(Ui,Ui,...,Ui),Calculating the fitness value f(i) of each initial solution.

Step(2) Sorting Frog. Sorting the F initial solution according to the fitness value of the order from good to bad, recording the best frog PX which is in the initial solution.

Step(3) Aassigning all the frogs to the m species Y1 , Y2 , ..., Ym according to equation(1),there are n frogs in each species.

YK=[U(j)k,f(j)k|U(j)k=U(k+m(j-1)),f(j)k = f(k+m(j-1)),j=1,...,n],k=1,...,m   (1)

Step(4) Giving the local search in each species Yk, k=1,...,m

Step(5) Exchanging the information of species(Shuffle),updating all the frogs, and sorting them, updating PX after the completion of each species's local evolution

Step(6) Checking if it meets the convergence criteria. If the convergence criterion is satisfied, stop, otherwise go to step(3).

The local search algorithm flow of SFLA is as following:

Let iM= 0,iN=0,NM is the maximum number of iterations of local evolution.
Step(1) iM= iM+1
Step(2) iN= iN+1
Step(3) Improving the worst frog's position, sorting the frogs within the species, recording the best frog's location PB and the worst frog's location PW within a species. Calculating step according equation(2) and calculating new location according equation(3).

Positive step:$S=\min\{int[rand \cdot (P_B-P_W)]\}$

$$\text{Negative step:}S=\max\{int[rand \cdot (P_B-P_W)]\} \tag{2}$$

$$U(n) = P_W + S \tag{3}$$

rand is a random number in $[0,1]$,if the new solution $U(n)$ is in the feasible region, calculate its fitness value $f(n)$, otherwise go to step(4).If the new $f(n)$ is better than the old $f(n)$,then replacing the old with the new $U(n)$,go to step(6),otherwise go to step(4).

Step(4) If step(3) can not produce a better value, or the new value is not in the feasible region, then the step and the new location are decided by the formula(4).

$$S=\min\{int[rand*(P_X - P_W)]\}$$
$$S=\max\{int [rand*(P_X - P_W)]\} \tag{4}$$

If the new solution $U(n)$ is in the feasible region, calculating its fitness value $f(n)$, otherwise going to step(5).If the new $f(n)$ is better than the old, then replacing the old $U(n)$ with the new, going to step(6),otherwise going to step(5).

Step(5) If the new solution is neither in the feasible domain nor better than the original, then generating a solution randomly in feasible domain.

Step(6) Updating species Yim, sorting them according to the fitness value from good to bad.

Step(7) If iN < NM, then going to step(2)

Step(8) If im< m, then going to step(1), otherwise going to the global search.

4   An Improved Discrete Shuffled Frog Leaping Algorithm for TSP

Because of the continuous nature of SFLA, it is hard to directly deal with the combinatorial optimization problems similar to TSP, therefore, this article used encoding basing on city sequence and new individual updating strategy, proposed a class of discrete shuffled frog leaping algorithm for TSP.

4.1   Individual Vector's Update

In the standard SLFA, location's update of frog Pw is only an abstraction, that model is essentially only suitable for solving the continuous problems, for the discrete combinatorial optimization problems, it needs to design specific update operator. According to the SFLA, the nature of the location's update of frog Pw is the vector operations in the continuous solution space tracking its local extremum or global extremum of solution vectors which is expressed by frog.where,rand() is the frog Pw's inheritance degree from the local extremum P_b, which reflects the confidence index of P_b. In other words, equation(2),(3)express the process that P_w's learning and approaching to P_b. where, rand( ) express the degree of learning, combinign formula(2), (3),showing P_n= P_w+ rand( ) * (P_b- P_w).

Then, when rand( )=1,P_n= P_b, when rand( )=0,P_n=P_w. Defining $f(P_w,P_b)$ is the learning process from P_w to P_b, the process can be achieved through crossover. Crossing as follows:

(1)Selecting a cross-region randomly in P_b, where, rand() determines the size of cross-region.

(2)Adding the cross-region of P_b to the front or behind of the P_w, and deleting the digitals from P_w that having appeared in the cross-region of P_b. Such as:

Current position P_w= 2 6 4 3 1 7 8 10 5 9   rand( ) = 0. 4

local extremum P_b= 2 8 û 6 3 10 7 û 1 9 4 5

Assuming cross-region selected randomly is: 6 3 10 7

After crossing becomes: 6 3 10 7 2 4 1 8 5 9 or 2 4 3 1 8 5 9 6 3 10 7

Obviously, by using this update strategy, it's easy in implemention, simple in opera-tion, and the substring can inherit the effective models from the parent string, then achieving the purpose that obtaining more new information from the local extremum p_b.

4.2   Improved DSFLA

Basic SFLA for function optimization problems is poor[5],mainly the convergence speed and optimization capability is poor, the performance of which is algorithm is stagnant when it evolved into local minima. The reason why the basic SFLA will be the case, is there are some flaws that the evolutionary approach used in the producing better new individuals.

The reason of which is: Frog first jumps in the generation of new individuals, only affected by the best individual of the local(that rand.(P_B-P_W)),and the frog jumps again to generate new individuals, only associated with the global optimum individuals(that rand.(P_X-P_W)). Therefore, in the two jumps of frog, only using the individual informa-tion of local optimum and the global best, while ignoring information exchanging between individuals and between sub-groups, losing a lot of valuable information. And both using the local worst individual in the two jumps, which is often not conducive to the generation of better individuals. Second, when the two jumps of frog don't generate outstanding individuals, the basic SFLA, without restriction, generated a new indi-vidual randomly and replaced the original individual(the third jumping of frog),completely abandoned the original information of this frog, leading to the loss of some individual information having advantages in the group, to a certain extent, slowed algorithm's trend of climbing and optimization capabilities.

Obviously, the key that overcoming the shortcomings of basic SFLA in evolution is: it is necessary to keep the impact of local and global best information on the frog jump, but also pay attention to the exchange of information between individual frogs. In this paper, first two jumping methods in basic SFLA are improved as follows:

$$Pn= PX + r1*(Pg-Xp1 (t)) +r2*(PW-Xp2 (t)) \tag{5}$$

$$Pn= Pb + r3*(Pg-Xp3 (t)) \tag{6}$$

Where,Xp1(t),Xp2(t),Xp3(t) are any three different individuals which are different from X. Meanwhile, removing the sorting operation according to the fitness value of frog individual from basic SFLA, and appropriatly limiting the third frog jump. Thus, we get an an efficient modified SFLA basing on the improvements of above. In the modified algorithm, the frog individual in the subgroup generates a new individual(the first jump)by using formula(5),if the new individual is better than its parent entity then replacing the parent individual. otherwise re-generating a new individual (the frog jump again)by using(6).If better than the parent ,then replacing it. or when r4 ≤ FS(the pre-vector, its components are $0.2 \leq FSi \leq 0.4$),generating a new individual (the third frog jump) randomly and replacing parent entity.

5 Simulation Experiment

This experimental platform is Intel P4 2G CPU, 1G memory, operation system is Windows XP. The size of species F is 100, the number of subgroup N is 20,the number of frogs in subgroup m is 20.To verify the effectiveness of the algorithm in the paper, We use the standard problems of TSP about 14 cities, 30 cities. From the two tables can be seen, the average of the solution, the proportion of getting the optimal solution, he average number of iterations for the optimal solution by using DSFLA proposed in the paper is much better than GA, similar to PSO. When solving TSP by using the improved DSFLA, the number of iterations to reach the best solution is far less than DSFLA, the proportion of achieving the best solution is greater than DSFLA, it shows that the improvement strategy makes the convergence speed and effectiveness has been greatly enhanced. Each algorithm run 10 times randomly, the results that the average of the optimal solution of 10 times, the best and the worst solution in the 10 times are shown in the two tables.

Table 1. The comparison of algorithms in 14 cities

Algorithm	Average	The best	The worst	The proportion of Achieving the best	The Average number of Iterations for the best
GA	33.995	33.352	34.569	0	——
PSO	30.879	30.879	30.879	100%	81.8
DSFLA	30.879	30.879	30.879	100%	79.6
Improved DSFLA	30.879	30.879	30.879	100%	9.8

Table 2. The comparison of algorithms in 30 cities

Algorithm	Average	The best	The worst	The proportion of Achieving the best	The Average number of Iterations for the best
GA	483.46	467.684	502.574	0	——
PSO	426.45	423.741	442.571	50%	10048
DSFLA	425.52	423.741	448.525	42%	11910.2
Improved DSFLA	431.5	423.741	453.591	50%	7820.6

6   Conclusion

SFLA can not be directly used to solve problems such as TSP for the reason that the search space is discrete, the paper proposed a new SFLA for the discrete search space(Named DSSLFA). Experiments in the data set of TSPLIB show that DSSLFA has better performance in the superiority and convergence of solution. It is a new attempt that getting the approximate solution by using SFLA. Therefore, the paper's significance lies not only in solving TSP by using SFLA, but expecting researchers' attention to this new algorithm. The next step is to solve these problems such as non-linear function optimization, material distribution problems, gear problems and so on by using SFLA or the improved SFLA.

References

1. Kang, Z., Qiang, X.-L., Tong, X.-J.: Algorithm for TSP. Computer Engineering and Applications 43(29), 43–37 (2007)
2. Ye, L., Ni, Z.-W., Liu, H.-T.: An improved genetic algorithm for TSP. Computer Engineering and Applications 43(6), 65–68 (2007)
3. Zhong, Y., Yuan, C.-W.: Ant colony optimization algorithm for Chinese traveling salesman problem. Circuits and Systems Journal 9(3), 122–126 (2004)
4. Yang, L.-Y.: Simulated Annealing Algorithm for TSP. Microelectronics and Computer 24(5), 193–196 (2007)
5. Eusuff, M., Lansey, K.: Optimization of water distribution network design using the shuffled frog leaping algorithm. Water Resour Plan Manage 129(3), 210–225 (2003)
6. Thyer, M., Kuczera, G.: Probabilistic optimization for conceptual rainfall run off models: A comparison of the shuffled complex evolution and simulated annealing algorithms. Water Resources 35(3), 767–773
7. Xun, Z.-H., Gu, X.-S.: Immune scheduling algorithm for Flow Shop problems under uncertainty. Systems Engineering Journal 20(4), 374–375 (2005)

A Detection Method of Iron Ore Pipeline Transportation Leak Point Positioning

Jiande Wu[1,2], Xuyi Yuan[1], Guangyue Pu[3], Chunlei Pan[3], Yugang Fan[1]

[1] Faculty of Information Engineering and Automation, Kunming University of Science and Technology, Kunming, 650500, China
[2] Engineering Research Center for Mineral Pipeline Transportation YN, Kunming, 650500, China
[3] Yunnan Da Hong Shan Pipeline Co., Ltd., Kunming, 650302, China
wjiande@foxmail.com

Abstract. The paper introduces a fix-point detection method of iron ore transportation pipeline leak point. Once leakage occurs during the transportation of mineral, it is great harmful to the ecological environment and the pipeline running. As a result, when the pipeline is running, it is so significant to detect the leakage precisely in time that enterprise can shorten the response time and reduce the damage and loss caused by leakage. The paper depicts a fix-point method of transportation pipeline leak point mainly based on stress variation rate method and mass balance method. It is proved in practice that this method can exactly judge the leak point positioning within 10m precision. This method is feasible and available, with the application value of widely popularizing.

Keywords: iron ore, pipeline transportation, leak point positioning, detection method.

1 Introduction

The iron ore transportation pipeline of Dahongshan is 171 km, it is the longest pipeline in the domestic field; As the special terrain of yunnan-guizhou plateau, and forming multiple large U ups and downs, ore pulp is sent to the elevation difference of 1520 m, the difficulty is unprecedented in international; The delivery pressure of ore pulp also reaches the highest 24.44 Mpa in international. All these difficulties requirement stringently made on security, stability, economic and environmental protection of the pulp pipeline[1,2].

There are so many factors accelerate the slurry pipeline decayed aging with the high-pressure running, such as the solid ingredients of Slurry beating and abrasiving on the wall, the slurry of soil airoxidating and across on both inside and outside wall, continuous vibration of pipe making it fatigue due to pressure pulsation, the defects of pipe materials and pipe weld, the vandalism along the way, and so on. Before the end of the design life, the pipeline will come about problems possibly in local areas——appearing cracks or gaps, even if only a small loophole initially, it will soon be evolved into a serious spill because of the huge internal pressure. High-pressure

D. Jin and S. Lin (Eds.): Advances in MSEC Vol. 2, AISC 129, pp. 145–149.
springerlink.com   © Springer-Verlag Berlin Heidelberg 2011

pipeline leakage will not only damage the surrounding natural environment, threaten the life and property of residents along the way and other creatures, but also bring huge direct and indirect economic losses. Therefore, it is very important that the ability of timely detection of leakage occur and accurate positioning in the process of pipeline operation. It determines the emergency response time of business to reduce, and we can minimize the harm and losses caused by leakage.

In this paper, the fix-point detection method of iron ore transportation pipeline leak point mainly based on the rules, which is the pressure rate of change and mass balance. Through 3 years of operation, which based on study of the iron ore pipeline characteristics, this high positioning accuracy is dozens of meters or less.

2   A Fix-Point Detection Method of Iron Ore Transportation Pipeline Leak Point

Pipeline leak detection must meet two conditions at least:

1) The detected pipe is independent hydraulics:"independent hydraulics" refers to that pressure and flow of the pipe is independent of other pipeline section, the pressure wave can't go through the boundary of the pipe (because the border have positive displacement piston diaphragm pump, the pressures wave can't through such pump). Based on the requirement, Dahongshan pipeline of 171 km is divided into the following three separate section, which needed for each independent leak detection (may even use different leak detection methods and strategies according to the practical situation).

2) The detection of pipe in a steady state

Only in steady state of the pipe section can leak detection, otherwise we cannot tell what factors caused the changes of pressure. The pressure changes in unsteady state may come from the valve's opening and closing, pump's speed adjustment, slurry's head and tail across the pipe, rather than the leakage. Entering the "steady state" means: a pipe is working rather than stopping (jumping from the flow information), with more than 5 m3/hr flow volume and flow rate (ROC) less than 2 m3 / hr/s; there is no valve position changing in the pipeline; the main pump pumping speed at both ends of the pipe is stability; the rate of change (ROC) on main pump pumping speed less than 3%/s; the slurry's head and tail does not exist in pipe; Rate of change of concentration (ROC) less than 0.05 EU/s.

3   Leakage Detection Methods

The leak detection methods commonly used in basic: stress variation rate method and mass balance method. For a given pipeline, if the pipeline leakage and the quantity of leak detection method more than 1 at the same time, then the pipe leak exists [3].

1) Stress variation rate method

Principle: if the pipe found in a significant pressure drop, which is not caused by normal operation (including closing the valve, adjustment of the pump's speed, starting and stopping the pipeline, across pulp head / tail), then there is a leak within the pipe.

Requirements: only a pipeline section must have 2 or more pressure gauges, can use the rate of pressure change method. Applicability: it is used in the three pipes in Dahongshan pipeline. Pressure gauges on all the pipe include:

Section 1: PIT-1161 (start), PIT-2002 (end)
Section 2: PIT-2161 (start), PIT-3002 (end)
Section 3: PIT-3161 (start), PIT-5002 (end)

Take the pipeline 1 for example let us show you the texting process of the pressure change rate:

Step1. Makes gauge counter of pressure change rate to zero.

Step2. The pressure gauge at the starting position of the pipe should be sampled in the sampling frequency of 2Hz. When the sample quantity up to four, it will start to calculate the pressure's moving average (MA) and moving average rate of change(MA-ROC) from The table using the simple moving average (SMA):

$$M_t = [X_t + X_{t-1} + ... + X_{-M +1}] / M \qquad (M=2)$$

Step3. In addition to calculating the pressure MA, it is also need to calculate the rate of change of pressure MA constantly at the same time.

$$MA\_ROC_t = (MA_t - MA_{t-1})/MA_{t-1}/\Delta t \qquad (unit: \% / s)$$

In which Δt is the corresponding time difference between MA_t and MA_{t-1}.

Step4. Compare MA_ROC_t with pre-set threshold value (19% /s), if MA-ROCt not lager than threshold value, then the calculation and comparison process should be continue to repeated; Otherwise, the pressure change rate increase 1, and repeat the calculation and comparison process. Executing the circulation process constantly, if the pipe is still in steady state (It will soon become a unsteady state, because the pump speed will make up for the leakage caused by flow changes, it will enter the unsteady state) when gauge counter of the pressure change rate lager than 2, this pipeline leakage will be reported. Stop the detection of pipeline leakage, until the leak detection program manually restarted. The leak detection method of pipeline 2 and 3 is similar to pipeline 1.

2) Mass Balance Method

Principle: according to the theory of mass conservation, in a long enough time (and then we can ignore local disturbance and noise), the fluid volume through any two section of the pipe is the same. But fluid volume is equal to product of flow and time, so it is the same that the flow are equal on any two section of the pipe. If the flow on upper reaches section is greater than the lower section of the pipeline, then the leakage occurs.

Requirements: only a pipeline section must have 2 or more flowmeters (to calculate the thevirtual flowmeter can based on the main pump stroke number), can we use the quality balance method.

Applicability: it is used in the three pipes in Dahongshan pipeline.

Pressure gauges on all the pipes include:

Section 1: 100-FIT-MLP (start),   200-FIT-MLP (end)
Section 2: 200-FIT-MLP (start),   300-FIT-MLP (end)
Section 3: 300-FIT-MLP (start),   500-FIT-5011 (end)

Take the pipeline 1 for example let us show you the texting process of mass balance method.

Step1. Make the deviation overrun mark in and out of flow to zero, make the flowmeter at the starting point of the pipeline to "the main flowmeter ".

Step2. The flowmeters 100-FIT-MLP and 200-FIT-MLP at the starting and the ending position of the pipe should be sampled in the sampling frequency of 2Hz. When the sample quantity up to twenty, it will start to calculate the pressure's moving average (MA) using the simple moving average (SMA):

Mt = [Xt + Xt-1 +... + Xt-M + 1] / M ,   (M = 18)

Step3. In addition to calculating the two flows MA, it is also need to calculate the deviation of the two flows MA constantly at the same time.

Setp4. Compare flow deviation with pre-set threshold value. If the flow deviation bigger than threshold, then the deviation overrun mark in and out of flow will be set to 1, If the pipe is still in steady state (it will soon become a unsteady state, because the pump speed will make up for the leakage caused by flow changes, it will enter the unsteady state), this pipeline leakage will be reported. Stop the detection of pipeline leakage, until the leak detection program manually restarted.

The leak detection method of pipeline 2 and 3 is similar to pipeline 1.

The two kinds of method above can determine whether there is a pipeline leakage.

4   The Fix-Point Accurate Detection Method of Pipeline Leak Point

According to time difference which comes from the pressure wave transmission on bilateral pressure gauge, we use the stress variation rate method and mass balance method to get the fix-point method mathematical model on pipeline leakage and jam, and calculate the leakage position of the jam point. The leakage position calculation schemes were shown in figure 1[4].

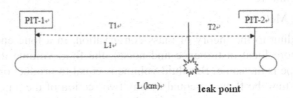

Fig. 1. The leakage position calculation schemes

Assuming the transmission speed of the pressure wave in the water and pulp is V_w and V_s respectively.

If leakage occurs in the pipeline with full of water:

$$\begin{cases} \Delta T = T_1 - T_2 \\ L_1 = V_w * T_1 \\ L = V_w * (T_1 + T_2) \end{cases} \tag{1}$$

Among them L, V_w, and ΔT are known, the leak point L1 can be derived:

$$L_1 = L/2 + V_w * \Delta T / 2 \tag{2}$$

Special case: when $\Delta T = 0$, $L_1 = L/2$, namely the leakage point is in the midpoint pipeline section. If leakage occurs in the pipeline with full of tube pulp:

$$L_1 = L/2 + V_s * \Delta T / 2 \tag{3}$$

If leakage occurs in the pipeline with water and plasma, then equations need to modified according to the dynamic position of the slurry's head and tail, and calculate L1.

5   Conclusion

It is a hot topic on leakage and jam at home and abroad. But most of the focus on the liquid or gas researching and project implementation. The iron ore transportation pipeline of Dahongshan realized the leakage and jam point positioning on solid materials after overcoming elevation difference, high concentration and many large U ups and downs problems. The technical level belongs to front row in the world. This method given in the paper is feasible and available, and being operation for 3 years, it is proved in practice that this method can exactly judge the leak point positioning within 10m precision with the application value of widely popularizing in the field of solid material pipe.

Acknowledgments. Project supported by Natural Science Foundation of China (Grant No. 51169007), Foundation of Yunnan Education Department (No. 09Y0094 & No.2010J063), Foundation of Kunming University of Science and Technology(No. KKZ1200903011).

References

1. Pu, G., Wu, J., An, J., et al.: The Design and Application of the Ore Pulp Water Treatment in Pipeline Transport of Refined Iron Ore. In: The 3rd International Symposium on Knowledge Acquisition and Modeling, pp. 192–195 (2010)
2. An, J., Wu, J., Pu, G., et al.: A Kind of Pressure Detection Method in Long Distance Ore Slurry Pipeline. In: The 3rd International Symposium on Knowledge Acquisition and Modeling, pp. 196–198 (2010)
3. Jian, F.: Researches on Intelligent Diagnosis and Localization Approach of Leak for Fluid Transporting Pipeline. Dissertation for the Doctorial Degree, Northeastern University, Shenyang (2005)
4. Guo, X., Yang, K.: Hydraulic Transient Model in Frequency Domain for Leak Detection in Pipeline Systems. Journal of Hydraulic Engineering 39(10), 1264–1271 (2008)

Design of Data Acquisition and Monitoring System Used in Complex Terrain Long Distance Pipeline Transportation of Iron Ore

Weidong Geng[1], Yu Gong[1], Li Zhang[2], Chunlei Pan[3], and Xiaodong Wang[4]

[1] Productivity Promotion Center of Kunming, Kunming, 650000, China
[2] Yunnan Academy of Scientific and Technical information, Kunming, 650051, China
[3] Yunnan Da Hong Shan Pipeline Co., Ltd., Kunming, 650302, China
[4] Faculty of Information Engineering and Automation, Kunming University of Science and Technology, Kunming, 650500, China
gwd@kmtm.org.cn

Abstract. This paper introduces the data acquisition and monitoring system used in complex terrain long distance pipeline transportation of iron ore. The system realizes that Pump Station No. 1-5 accessed OA network, SCADA networks and Internet networks, established "triple play" safely and reliably. The Pump Station No. 1-5 accessed VoIP, achieved IP video production monitoring, audio and conferencing of production scheduling accessed data acquisition and monitoring system. The systems improved productivity and reliability; achieve the scientific planning and scheduling enterprise resource.

Keywords: data acquisition, monitoring system, pipeline transportation.

1 Overview

In order to control operations of pipeline accurately, mastering slurry flow position, blocking, lets-out and acceleration flow inside the pipeline exactly is needed. Master operational person not only need to judge all the meters on the spot and information of facilities, but also needs to judge information beyond those. Hence, a mighty data acquisition and monitoring system is needed to provide real-time information needed [1-3].

Because of the demand of stable, traditional managing mode of production units can not meet the demands of monitoring and managing of the field apparatus. Engineers need to judge and solve the problems according to running condition of the pipelines and facilities in spite of not staying on the spot. Hence, a data acquisition and monitoring system is needed to reflect real-time conditions of field apparatus [4]..

In terms of these demands and background to complex terrain long distance pipeline transportation of iron ore in this paper, data acquisition and monitoring system needs to achieve: Pump Station No. 1-5 accessed OA network, SCADA networks and Internet networks, established "triple play" safely and reliably. The Pump Station No. 1-5 accessed VoIP, achieved IP video production monitoring, audio and conferencing of production scheduling accessed data acquisition and monitoring

D. Jin and S. Lin (Eds.): Advances in MSEC Vol. 2, AISC 129, pp. 151–155.
springerlink.com   © Springer-Verlag Berlin Heidelberg 2011

system; program and debug with PLC; according to the project using configuration to manage and monitor software, monitoring and data acquisition system is used for factory testing and the server application integration. Data acquisition and monitoring system includes: network communication system, data acquisition system and monitoring system.

2   Integration of Network Communication System

1000Mbps Ethernet system is used here. Based on SCADA ring network, automatic switching double-loop backup is established. At the same time switch public network backup lines automatically using digital transmission circuit provided by ISP.

2.1   Technical Requirements for Network Facilities

Network switchboard is industrial product, and supports fully redundant power and multiple power input (9.6-60VDC, 18-60VAC or 110-230VAC); the site switches use a fan-less design; work ambient temperature is -40-70°C; a variety of installation is supported (rail or cabinet); pass the general industrial product certification: cUL 508; Cul1604 Close 1 Div.2. Switches have long-term long-distance transmission capacity and transmission distance can reach over 100km without relay. Ethernet adopts modular-design, port number and types can be configured flexibly.

Network equipments have capabilities of anti-electromagnetic interference in order to meet the demand of increasing products' reliability in electromagnetic interference environment, switches reach or exceed EMC EN61000-4-2, 3, 4, 5, 6; while it has reached or exceeded the IEC 60068-2-27 shock and IEC 60068-2-6 vibration standard in the mechanical vibration resistance; it can work normally when environment humidity is less than 95%; the average trouble-free running time is more than 20 years.

2.2   Technical Requirements for Ethernet

Multiple network topology and redundancy: Ethernet switches support arbitrary network topologies (bus, star, ring) and ensure that in the events of failure it can be switched quickly. When coming to link redundancy, it supports HIPER-Ring super redundant ring protocol, realizes ring network redundancy. The switching time of 100M ring network(100 switches) is less than 200 ms, the switching time of 1000M ring network(50 switches) is less than 50 ms. Fast redundancy protocol HIPER-Ring 2 which switching time is less than 10 ms is supported to those networks that need higher real-time demands. Network Coupling is supported in order to achieve rapid recovery performance between regional branch networks and achieve fast switch less than 500 ms. Link Aggregation is supported, which provides redundant connections on the one hand, and on the other hand increases the communication bandwidth between switches.

In network management, not only the universal serial interface and Web interface network management mode but also the dedicated network management software is supported to achieve unified management of large networks. Network topology generation software is based on LLDP (IEEE802.1 AB) standard protocol, which can detected network equipments automatically, generate network topology automatically, monitor equipment state, link connection state and the state of power switch and fan, event record is took by rotate examine and SNMP alarm, the events can be notified through short message triggered, e-mail triggered and information window, all the connected history data can be viewed.

3   Integration of Data Acquisition and Monitoring System

Data acquisition and monitoring system has a lot of constituent tasks, and each task completes specific functions.   Servers situated on one or more machines are responsible for data acquisition, data processing (such as range switching, alarm inspection, events record, historical memory, execute user scripts). The server can communicate with each other. Some systems will further separate servers into several specialized servers, such as an alarm server, log server, historical server, login server. The server is a unified unit logically, but physically it can be placed on different machines. Typical hardware configuration diagram, fig.1, is listed as follow [5]:

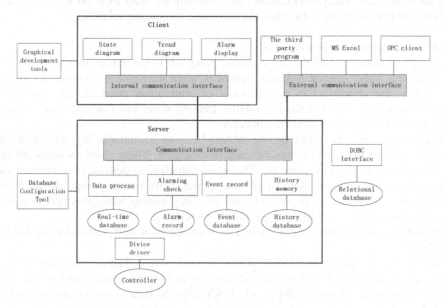

Fig. 1. Typical hardware configuration diagram of data acquisition and monitoring system

3.1   Data Acquisition System – PLC Control System

Control system processor monitors I/O via the EtherNet / IP, ControlNet, DeviceNet and Universal Remote I / O. When the PLC frame has several processor modules, or even in the control network there are more than one processor modules, all processors can read the input value from all the input modular. Any processor can control any particular output module. That each output modular is controlled by which processor is designated by system configuration. Control system fully utilizes the principles of fluid mechanics, judges the fluid in the pipeline automatically, switches mode to avoid the generation of accelerated flow automatically. During operation, when the head and tail of the slurry pass through the pump station, the pump will not be stopped and can be switched with no disturbance.

Unperturbed switchover of equipments: according to the needs it should achieve no disturbance switching between work equipments and standby equipments, and avoid "water hammer" phenomenon and large-scale fluctuations of process parameters to ensure the stable work. Switchover during pipeline transportation: according to the process requirements, when the Pump Station No.4, No. 3 and No. 5 stop working, the control system can cut out the broken -down pump station rapidly and stably and operate under the corresponding technical requirements.

3.2   Data Monitoring System – The Application and Development of Host Computer Software

Human–Machine Interaction (HMI) computer shows all the information needed to support the operating sequence. Intuitive interaction between the operator and control system is allowed. Each workstation can communicate with PLC on the spot. Each station will communicate with other stations using remote tag reference.

3.2.1   Screen and Screen Display

The graphics display of all the input points. Pump off is red, pump on is green. Valve closed is red, valve opened is green, yellow is when the valve is opening or closing. All values are displayed in the process of simulation graphics, and displayed engineering units aside. For slurry pipeline, treating water pipeline and water process pipeline are distinguished by different colors.

Each station and the entire piping system have an overall display, and show the dynamic process of split screens of each workstation. These displays show all the working state of major equipment. The whole pipeline system display showed the main process equipments' state of all feed pump, main pump, pumps entrance and outlet valve. It also shows the liquid level of slurry mixing tank, pressure, flow and density values of the main process parameter monitoring.

If there is any alarm of equipment and instrument, it will scintillate on display interface.

There is a comprehensive communication state screen and an electrical system status screen. Electrical system state screen shows voltage, current and power read from power supply monitor. HMI reads 10KV relevant data of power supply through the Ethernet.

Setting value and limit alarm. All the settings of simulations, alarm and linkage can be adjusted by operators. A unified adjustment interface is given.

Tendency. All the input analog quantity, output analog quantity has a function of data recording and trend display. The system can record the pipeline control instructions. System records all the state for all pumps (on/ off), valve (on / off). Archiving data can be stored for 90 days. Each analog quantity has its own trend interface. When clicking on the analog value it will pop up corresponding trend interface.

3.2.2   Alarm

All alarms will be generated in the PLC. All of the outputs of the controller have an associated fault alarm. All analog inputs can provide high and low alarm. To configures specific alarm screen and provide to the operator. Provide alarm history. An operator can access the last 90-days' alarm. Unacknowledged alarms are listed in an alarming title, and ready to display on a monitor.

3.2.3   Report

A detailed record of operating parameters(such as pressure, current, concentration, flow rate) of each workstation , a duty report of all the operation time of the pumps and summative scale of slurry flow and water flow accumulation ,maximum value and minimum value at the end of work, can be generated automatically. When the operators have requirements, the report can be printed on a printer, it can also be accessed the previously generated reports.

4   System Analysis and Conclusion

The system realizes that Pump Station No. 1-5 accessed OA network, SCADA networks and Internet networks, established "triple play" safely and reliably. The Pump Station No. 1-5 accessed VoIP, achieved IP video production monitoring, audio and conferencing of production scheduling accessed data acquisition and monitoring system. Actual operation proves: this system guarantees the transfer of information on the real-time, a diversified data transmission platform is built on technologies such as converged network, mobile communication, software programming. It ensures the complex terrain long distance pipeline transportation of iron ore transported and operated safety, stably, efficiently, and it is in the leading position at home and abroad in the same industry.

Acknowledgments. Project supported by Natural Science Foundation of China (Grant No. 51169007), Foundation of Yunnan Education Department (No. 09Y0094 & No.2010J063), Foundation of Kunming University of Science and Technology(No. KKZ1200903011).

References

1. Pu, G., Wu, J., An, J., et al.: The Design and Application of the Ore Pulp Water Treatment in Pipeline Transport of Refined Iron Ore. In: The 3rd International Symposium on Knowledge Acquisition and Modeling, pp. 192–195 (2010)
2. An, J., Wu, J., Pu, G., et al.: A Kind of Pressure Detection Method in Long Distance Ore Slurry Pipeline. In: The 3rd International Symposium on Knowledge Acquisition and Modeling, pp. 196–198 (2010)
3. Bai, X., Hu, S., Zhang, D., et al.: Advances and applications of solid-liquid two-phase flow in pipeline hydro-transport. Journal of hydrodynamics 16(3), 304–310 (2001)
4. Hu, S., Qin, H., Bai, X., et al.: Resistance characteristics of particulate materials in pipeline hydro-transport. Chinese Journal of Mechanical Engineering 38(10), 13–16 (2002)
5. http://article.cechina.cn/2007-04/200748075506.htm

Design of the Intelligent Simple Electrocardiograph

Jie Sun[1,2] and Xiao-li Wang[2]

[1] School of Control Science and Engineering, Shandong University, Jinan 250061, China
[2] School of Mechanical, Electrical & Information Engineering, Shandong University at Weihai,
Weihai 264209, China
sunjie_zidonghua@163.com

Abstract. This paper proposes a low-cost simple intelligent electrocardiograph (ECG) which is easy to popularize. With AT89S52 singlechip as its core controller, and based on software system developed by RTX51, the ECG is able to achieve real-time display of heart waveform and heart rateon its LCD. It adopts INA128- a low-power-consuming and highly-precise instrument amplifier- to be the pre-amplifier, and thus has a high CMRR. It has high-pass, low-pass and band-stop filter circuit to restrain various disturbances. The system is connected to the PC with serial communicate interface to display, analyze and process heart waveforms. The simple intelligent electrocardiograph is easy to use and easy to carry in practical application.

Keywords: ECG signal acquisition, INA128, Active filter circuit, AT89S5, Heart rate calculation.

1 Introduction

ECG(electrocardiograph) collects a patient's biological signal and records his ECG waveform, which is an important basis for analyzing his heart pathology activities. Heart rate is a significant parameter in evaluating the physical condition of human body. Measuring one's heart rate can reflect the heart's performance. It is quite necessary to minimize and popularize the common ECG, so that people with heart disease can observe their ECG waveforms and measure their heart rates. This article designs an intelligent simple ECG, which collects ECG signals of the human body, and display the real-time ECG waveforms and heart rates using the SCM (single-chip microcomputer) controlled LCD. The system is connected to the PC with serial communicate interface and programs with high-level languages, thus achieving the waveform's display, analysis and processing.

2 Hardware Design

The hardware circuit is mainly consists of the analog signal acquisition system, the digital control system and the PC data processing system. The analog signal acquisition system includes the signal collecting module, the amplifying and filtering module. The digital control system includes the A/D converter module, SCM controlling module, and the LCD module. The PC data processing system includes the serial communication module, the data analyzing and processing module. The system's overall hardware design is illustrated in Fig. 1..

D. Jin and S. Lin (Eds.): Advances in MSEC Vol.2, AISC 129, pp. 157–162.
springerlink.com   © Springer-Verlag Berlin Heidelberg 2011

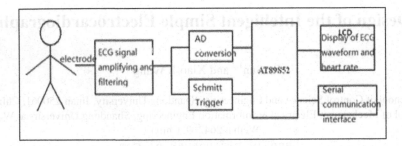

Fig. 1. Whole design plan of the system hardware circuit

2.1 Analog signal Acquisition System

The ECG observes the heart's activities by measuring the potential of living tissue surface., and we can make use of biological potential and electrode to collect ECG signals. Besides electrical signals generated by the heart, signals that collected on the body surface contains interference caused by EMG(electromyography) signals, breathing and power-line interference (50Hz). Meanwhile, ECG signals are millivolt-level small signals of low frequency, which is difficult to measure. The system amplifies the input ECG signals and have a high and low pass filter to filter out the high or low bioelectrical signals as well as their Interference. The band-stop filter circuit has the working frequency filtered. Generally the amplitude of people's ECG signal is 0.05-4mV, with the frequency of 0.05-100HZ. So we design the voltage gain of the amplifier to be 800-1000, and the bandwidth of the filter to be 0.035HZ-110HZ[1].

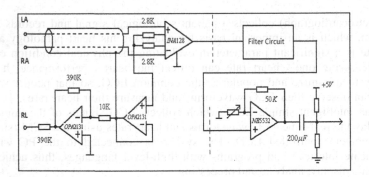

Fig. 2. Analog signal acquisition system diagram

(1)ECG signal's collecting

Here we use a 30mm-diameter round thin copper as the simple electrode to contact with the skin, and shield lines are used as the lead. The electrode is connected in accordance with lead standard I: RA(Right Arm) is connected to the inverting input of an amplifier, LA(Left Arm)is connected to the non-inverting input, RL(Right Leg)acts as the reference electrode and connects to the reference point of the ECG amplifier(Fig. 2.). The system uses RL to drive the shielded circuit, and amplifies the common-mode signals reverse. Then it returns to the body and superimposes them in order to reduce the common-mode signals that interfere[2].

(2)ECG signal's amplifying

The amplifying and filtering module adopts secondary amplifying structure: put the ECG signal through the integrated operational amplifier where the first-stage amplification happens. Filter it by the high and low pass filter and through the second-stage amplification finally adjusts the signal's amplitude with controllable range of SCM using the LV-to-HV level shifter.

Preamplifier: As ECG signals can be easily submerged by interference, the ECG requires the preamplifier has high Common Mode Rejection Ratio (CMRR). CMRR of General Operational Amplifier (Op-amp) is difficult to meet the requirements, so it cannot control the common mode interference well. This design adopts the general instrumentation amplifier with low-power consumption and high accuracy, INA128,. It has an internal design of three Op-amps, and they are integrated components whose volumes are small. INA128 has a rather low bias voltage ($50\mu V$), temperature drift ($0.5\mu V / °C$) and a high CMRR (When G=100, 120dB), so it is an ideal choice for this design. Between pin 1 and pin 8 we put an external resistor and can set the gain. The relation of amplifier gain:

$$G = 1 + \frac{50K\Omega}{R_G}$$

The ECG signal is amplified in INA128 through the shield line with a differential input(Fig. 2).

Secondary side Amplification : Adopt the same-phase proportion amplification circuit, Amplifier choose NE5532, magnification is adjustable (Fig. 2).

(3)ECG signal's filtering[3]

The design of filter circuit has a fourth-order low pass electrical filter, a second-order high pass filter and a band stop filter that is used to filter out the power-line interference (50Hz). As the active filter has good properties of high input impedance and low output impedance, its filter performance is better than passive filter, so all the formations adopted here are active filters.

Component parameters of filter are selected below: (LPF) : cutoff frequency 110HZ, By formula, choose C=33nF,R=43 。 (HPF) : cutoff frequency 0.035HZ, By formula, choose C=47nF, R=100 。 (BEF) : center frequency: 50HZ, choose C=133nF, R=24 。

quality factor $Q = \dfrac{1}{2\left|2 - \overset{\cdot}{A}_{up}\right|} = 20.5$, stopband $BW = \dfrac{f_o}{Q} = 2.44HZ$

2.2   Digital Control System

The system adopts the low-power, high-performance SCM AT89S52 as core control module. ECG signals that have been amplified are changed to digital signals through AD converter and then sent to SCM, which controls LCD to display the real-time

ECG waveforms. Simultaneously, amplified ECG signals are sent to SCM after waveform conversion with Schmitt trigger. We measure heart rates using the timer and external interrupt of SCM and show them on LCD. The schematic diagram of Digital Control System is shown in Fig. 3:

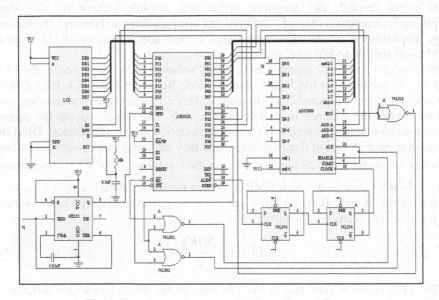

Fig. 3. The schematic diagram of Digital Control System

(1)A/D converter module

ECG signals need to be A/D converted to digital signals and sent to SCM for processing after filter and amplification. The system adopts the 8-bit A/D converter ADC0809, which is easy and convenient to use and has enough accuracy. The SCM oscillator selects 12MHZ, and ALE(2MHZ) is input to CLK of ADC0809 as its clock signal after fourth frequency division with D-flip-flop. SCM P0 port connects to data bus of ADC0809, while P0.0-P0.2 id directly connected to ADDA-ADDC. As there is address decoding latch in ADC0809, 74LS373 is not needed to latch signals[4].

(2)Schmitt trigger

Use Schmitt trigger to shape the ECG signals to rectangular pulses after filter and amplification and connect them to SCM pin INT1 as the external interrupt, then calculate the frequency of signals with the help of SCM timer. Finally transform the frequency to heart rate and display it on LCD. The Schmitt trigger is connected with 555 timers[5].

(3)LCD module

In order to simplify the design of SCM peripheral circuits, the system adopts the formation of dot matrix graphic LCD module HS12864 with internal integrated driver and controller, to which SCM can be directly connected and achieves display control. HS12864 dot matrix LCD module is an 128(columns)×64(rows) array composed of 128×64 liquid crystal display points. RAM that stores the lattice information is called display data memory. There is a definite sequential relationship between positions (rows and columns) of points in the display panel and their addresses in the memory. Displaying the graphic is that writing the lattice information into corresponding memory units.

2.3   PC Data Processing System

The logic level of SCM is TTL, while the computer's serial's is RS232 and they are not compatible. Between them there must be a level converting circuit so that the two can communicate. This design uses MAX232 produced by MAXIM to achieve logic level conversion, and the connection of circuit is shown in Fig. 4. MAX232 chip can convert the input +5V voltage to ±10V voltage that RS-232 interface needs with exterior capacitors [6].

SCM sends the ECG data that have been A/D converted to the PC through serial interface RS-232. The PC acquires real-time ECG signal data from interface RS-232, programs with advanced language VB, shows ECG waveforms on display, and then have the ECG signals analyzed, stored and printed out. The PC sends the ECG data information to the Internet and remote terminals can get them real-time through some protocol and process the data, thus achieving the doctor's remote diagnosis[7].

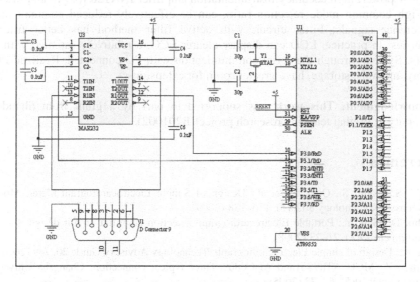

Fig. 4. The schematic diagram of MCU and PC Serial communication

3   Soft Ware Design

Software designe is based on RTX51 embedded system[8]. RTX51 is used in the micro controller (MCU), a multi-task real-time operating system,which can significantly improve system efficiency and real-time implementation[9]. The MCU selection of system is AT89S52. use Keil development environment. By writing procedures to control the LCD in real time display of ECG waveform and heart rate.

The ECG is analog and continuous signal. The LCD display of ECG curves corresponding to dot matrix display.Concrete measures are: From left to right single row scanning. After the completion of a screen display and full-screen cleared and then display the next screen. Accordingly enabling real-time dynamic display of ECG waveform.

Microcontroller INT0 pin receiving external interrupt applications,and calculated with the SCM timer with the frequency of signal, Eventually translate into heart rate and displayed on the LCD. Heart rate calculation algorithm is as follows:MCU timer set time 1/60s, MCU external interrupt pin used to receive ECG after plastic surgery,that is rectangular pulse. Between the two rising edge of external interrupt count n 1/60 s, The heart rate is calculated as 3600/n. Single chip real-time calculated heart rate and then output to the LCD display can.

4   Conclusion

An intelligent simple electrocardiograph which is low-cost and easy to popularize is designed. Due to ECG signal is weak, Vulnerable to being submerged in a variety of interference, So requires a system with a high anti-jamming capability. The system uses low power, high accuracy instrumentation amplifier INA128 for pre-amplifier, It has high common mode rejection ratio, can be effectively reduced common mode interference signals; Filter circuit with active filter method has better filtering properties. In practice, LCD can display clearer ECG, a relatively satisfied with the results. System through the serial communication with PC-connected, Realized ECG display, analysis, storage, have greater room for extensions.

Acknowledgment. This work was supported in part by a grant from Shandong University at Weihai teaching research project(B201002).

References

1. Wu, S., Yanli, S., Qian, L., et al.: Design of Simple Electrocardiogram Meter. Modern Electronic Technology 302(15), 136–138 (2009)
2. Zhu, D., Guo, Y.: Portable Electrocardiograph detection amplifier circuit design. Medical Equipment 29(5), 21–23 (2008)
3. Li, C.: Design of simple Electrocardiograph. Technology Advisory Guide 20, 7–9 (2006)
4. Yuan, Z.: MCU AT89S51and ADC0809 Three typical connection. Changsha University Technology 19(5), 69–72 (2005)
5. Yan, S.: Fundamentals of digital electronics, 5th edn. Higher Education Press, Beijing (2006)
6. Yang, S.: Principle and Application of Micro Computer Systems, 2nd edn. Tsinghua University Press, Beijing (2004)
7. Zhang, H., Qiao, X.: Family electrocardiography and remote diagnostic system. Automation Technology and Application of Heilongjiang (6), 44–46 (1999)
8. Microsoft Corporation Complete control Visual C++6.0 MFC Application program development. Tsinghua University Press, Beijing (2002)
9. Keil Cx51 V7.0 MCU High Level Language Programming And Vision2 Application. Electronic Industry Press, Beijing (2004)

Best Path Analysis in Military Highway Transportation Based on MapX

Qingge Zhang and Xuri Yin

Dept. of Transportation Command
Automobile Management Institute of PLA
Bengbu, 233011, China
{Zhangqingge,Yinxuri}@163.com

Abstract. In recent years, the development of the computer and GIS technology brings many new methods for the military highway transportation. In this paper, several key technologies to establish the road topology structure based on MapX are discussed, and an algorithm for finding the shortest path in the damaged road network based on MapX is presented. Finally, some example is used to illustrate the procedure of finding the best path on digital map, and the results show that this algorithm can obviously improve the efficiency of military command.

Keywords: MapX, best path, military transportation, geography information system, topological structure.

1   Introduction

Military highway transportation is the lifeline of military operations, it plays a key role in deploying troops, expanding the operations, meeting operational needs and wining the war. During the entire transportation process, the motorcade in transit routes is attacked at any time by enemy aircraft, missiles and artillery, and the transport routes will be interrupted. Especially in the high-tech war, the battlefield situations will request the army commander to fast work out concrete and scientific transportation plan in such a short time [1]

MapInfo MapX is an Active X Component enabling developers to embed mapping functionality to any application in the most cost-effective manner using object-oriented languages such as Visual Basic, Visual C++ or Delphi. It simplifies application development, providing a highly visual way to display and analyze location-based data so that we can make better decisions, serve customers and manage operations more effectively.

To achieve the shortest path calculation, we must have topology information about the arc and node in road network. Because MapX do not support the topology of spatial data, the shortest path analysis has become a difficult question for MapX users to develop application systems. Therefore the road network topology generation is the key to the best path analysis.

In recent years, the establishments of road topology and the shortest path algorithm based on MapX also have been proposed successively [2-5]. In this paper, several key

D. Jin and S. Lin (Eds.): Advances in MSEC Vol. 2, AISC 129, pp. 163–168.
springerlink.com   © Springer-Verlag Berlin Heidelberg 2011

technologies to establish the road topology structure are discussed, and a novel algorithm for finding the shortest path for military highway transportation based on MapX is presented. Then, an example is used to illustrate the proposed algorithm. Finally, a summary to the entire research is made.

2   Several Key Techniques

Several key techniques in constructing road topology are described as follows.

2.1   Calculate Intersections of Roads

CMapXFeatureFactory class of MapX is used to judge spatial relationships between any primitives on the map. This method CMapXFeatureFactory intersects a Feature or Features object with another Feature or Features object, and returns the resulting object as a stand-alone feature.

We can intersect roads with other roads. If two roads cross, the method IntersectFeatures will return a polyline feature with one point. The following c++ code demonstrates how to calculate the intersections of roads.

```
CMapXPoint pt;
tagVARIANT vt_ftr;
vt_ftr.vt = VT_DISPATCH;
vt_ftr.pdispVal = ftj.m_lpDispatch;
pointObj=IntersectFeatures(fti.m_lpDispatch,vt_ftr);
if(pointObj.GetParts().GetCount()!=0)//intersect
{
for(int i=1;i<=pointObj.GetParts().GetCount();i++)
for(int j=1; j<= pointObj.GetParts().Item(i).GetCount();j++)
{
pt.Set(pointObj.GetParts().Item(i).Item(j).GetX(),pointObj.GetParts().Ite
m(i).Item(j).GetY());
pts.Add(pt);
}
}
```

2.2   Calculate the Distance between Two Points on a Road

We can find the distance between two points $A(x1,y1)$ and $B(x2,y2)$ on a straight line on the map. In fact, a line object can be regarded as a polyline that consists of a number of straight-line. Therefore, the length of a line object is simply the sum of the lengths of each straight-line segment.

The following c++ code segment shows how to calculate the distance between any two points on a road.

```
for(int i=1; i <= LineObj.GetParts().GetCount(); i++)
for(int j=1; j <= LineObj.GetParts().Item(i).GetCount();j++)
{
   x=LineObj.GetParts().Item(i).Item(j).GetX();
  y=LineObj.GetParts().Item(i).Item(j).GetY();
}
```

3   Building the Road Topology

The process for best path analysis in MapX environment has two major steps. First, we must establish the road topology relation of the road network, and then calculate the shortest path using the Dijkstra algorithm. Among them, the first step is the most important in finding best path.

It is obvious that the size of road topology map directly influences the complexity of algorithm for finding the shortest path. This can be a very time-consuming process, if we would build the road topology structure for National map or war zone map. Even if the road topology structure has been established, finding for the shortest path on it is also in reality not feasible, as the computing time is too long and cannot meet the needs of the wartime transportation commands.

Therefore, reasonable choice of the scale of the network region is very important for the system to efficiently run. A feasible method is that the scale of selected road network is dynamically based on the users' interactions, that is to say, selecting road net's partial regions relays on the beginning point and end point of the transport tasks [5].

In our experiment, we find that in most cases the shortest path between two points is included in the rectangle used the two points as the diagonal. In order to avoid a few cases that the shortest path is outside the boundaries of the rectangle, we introduce a threshold value α to adjust the size of the rectangle, namely to generate a new incremented rectangle denoted by R_α, where the top, bottom, left, and right edges of the original rectangle expand by α units. Therefore, we only need to build the road topology within this rectangle.

When the road network is interrupted, the procedure of the algorithm for find the shortest path in the road network is given below:

Step 1   Create a new layer LocalMapLayer and temporary layer TmpLayer ;
Step 2   Gain the coordinates of the user-specified beginning point P_1 and end point P_2.Then get the coordinates of the user-specified interruption point P_0;
Step 3   Display P_1 ,P_2, and P_0 on TmpLayer.
Step 4   Set threshold value α ,then generate the region feature R_α
Step 5   Calculate the intersection (denoted as I) of R_α and the all road features, then add them to the layer LocalMapLayer.
Step 6   Build the road topology (denoted as G) from I.
Step 7   Apply the Dijkstra algorithm on G and compute the shortest path between P_1 and P_2 .
Step 8   Show the shortest path and length between P_1 and P_2 on the TmpLayer .

In practice, if the longitude or latitude difference between two points is relatively small, the value α may be appropriately larger; if not, the value α may be appropriately smaller. Obviously, this method can effectively reduce the scale of the road topology, thus raises the most short-path computing speed.

4   Simulation Studies

We use Visual C + + visual programming environment and realize the algorithm proposed in this paper for wartime military transport simulation analysis.

4.1   Threshold Value Analysis

We use an example to analyze the importance of α value, which is used to find best path between Bengbu and Zhengzhou.

As α value increases, the number of nodes and edges of the selected road network will also increase quickly, as shown in Table1.

Table 1. Effect of α value on network scale

α	Bengbu—Zhengzhou	
	Nodes	Edges
0.1	495	301
0.2	589	368
0.3	696	437
0.4	794	503
0.5	863	538

The scale of the road network is more and more big, the topological structure of network is more and more complicated, and so the computing time to find the best path is several times incremented too.

4.2   Best Path Analysis

We select of local area road network nearby Bengbu on the electronic map, as shown in Figure 1. It is assumed that a military mission must be carried out, transporting goods from the "Caolaoji Zhen" to "Linhuai Zhen". In addition, because there is a river way between the two places, the bridges across this river are strategic pass.In the experiment, threshold α is set to 0.3.

Applying the algorithm proposed above to the selected road network, we can calculate and display the shortest path from "Caolaoji Zhen" to "Linhuai Zhen", as shown in Fig.2. The shortest distance between two locations is about 38.82Km.

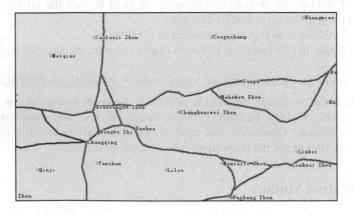

Fig. 1. The road network

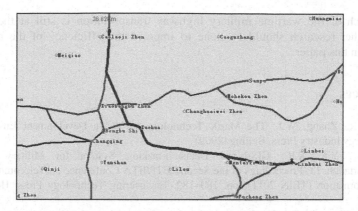

Fig. 2. The shortest path under the normal situation

Now suppose that a bridge across the river is destroyed by enemy, and military vehicles can not cross the bridge. In order to distinguish the interrupted road, we represent the interrupted road by the symbol "×". Using our algorithm above, we can obtain the shortest path from "Caolaoji Zhen" to "Linhuai Zhen", as shown in solid thick line in Fig.3. The analysis result shows that the shortest distance is increased to 40.98Km.The reason for this result is that the disruption of the bridge cause a change in the road network structure, thus making the military vehicles detour the destroyed bridge.

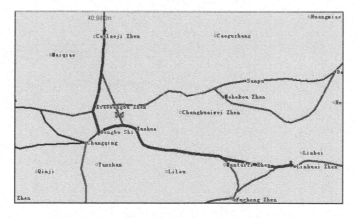

Fig. 3. The shortest path when road is destroyed

5   Conclusions

In wartime, many uncertain factors influence highway transportation, such as the enemy attack, natural disaster, and so on. Therefore the simulation analysis of the wartime military transport gradually becomes a research hotspot. In this paper, an algorithm for finding the best path in the damaged road network is presented, and several examples are also shown to prove that the algorithm is fast and effective. But the research of the wartime military highway transportation is still at the starting stage, further research should be done to improve the efficiency of the algorithm proposed in this paper.

References

1. Yin, X.R., Zhang, W.J.: The MapX Technology in VC++ Development Environment. Metallurgy Industry Press, Beijing (2009)
2. Yin, Y., Yin, X.R.: GIS-based Decision-making System for Military Highway Transportation. In: Proceedings of the Second ETP/IITA Conference on Telecommunication and Information (TEIN 2011), pp. 180–182. Engineering Technology Press, Hong Kong (2011)
3. Nie, Y., Zhao, J., Fei, X.: Discussion and implementation of building path topological relation and analyzing the shortest path based on MapX. In: Science of Surveying and Mapping, vol. 31(3), pp. 96–98. Science Press, Beijing (2006)
4. Yang, H., Zhao, C., Liu, Y.: GIS-based Inner Mongolia Grassland Fire Spread Simulation System. In: Proceedings of International Conference on Computer Science and Software Engineering, pp. 923–925. IEEE Press, New York (2008)
5. Yin, X.R.: GIS-based Simulation System for Wartime Military Highway Transportation. In: Proceedings of the 5th International Conference on Computer Science & Education, pp. 1810–1812. IEEE Press, New York (2010)

Design and Optimization of Data Process Server of ITS Base on Shared Memory

Shaochun Zhang, Weiming Wu, and Yonghao Gu

Computer Science Department, Beijing University of Posts and Telecommunications,
100876, China

Abstract. When the intelligent transportation system (ITS) operates, it has the characteristics of exchanging information frequently and data flowing instability. In order to improve efficiency, stability and intelligence of the data processing server, this paper proposes that the data processing server should be optimized by adding detecting thread and utilizing shared memory file. The detect thread can dynamically control the number of the processes, so that the number of threads and the amount of the tasks can match automatically, and the system resources can be efficiently and rationally used. In addition, we optimize the data structure in the memory file by creating physical and logical link to improve memory access speed.

Keywords: ITS Shared-Memory Multi-Thread.

1 Introduction

With the economic developing rapidly and the urbanization accelerating, large number of people pour into the city. At the same time, the buses in the large cities have begun to dramatically increase. In order to promote the sustainable development of the transport, it is necessary to design the intelligent public transport system.

Intelligent transportation system collects and forwards the data through the terminal, and interacts with the communication server through the GPRS network. The communication servers make use of the multi-processes concurrently to deal with the related data, and interact with others through Linux IPC. The operating system stores the real-time information in shared memory, and then optimizes the data structure of the memory to greatly improve the speed of processing data.

2 Smart Devices in the Intelligent Transportation System

In this paper, the intelligent transportation system includes intelligent vehicle equipment, terminal scheduling screen, smart bus schedule, intelligent scheduling system, communications server system, the monitoring command center. The distributed architecture is shown as below.

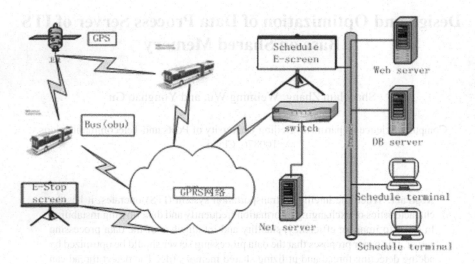

Fig. 1. Smart devices in the ITS

3   The Overall Flow of Information in Intelligent Transportation System

Because the data size and the frequency differ greatly while the intelligent terminals interact with the data communication server, the communication server adapts the method of concurrent multi-process and storing the real-time information into the shared memory to operate. Each module start a separate process, the process access the shared memory through the RW lock and they transport the data through the MSQ. System data flow diagram as shown in figure 2.

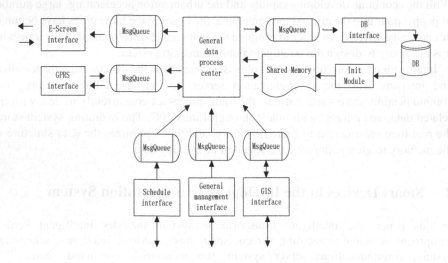

Fig. 2. Data Flow Diagram

The data transporting through the GPRS networks are also exchanged by the GPRS interface process. The peer of data transported by the GPRS network is intelligent bus terminal. The data from the web server and the communication server interact at the web server interface. And the other processes are as familiar, they separately handle with the corresponding data from the peer and complete the information interaction. Each process will deals with the receiving data packets and then delivers to the next message queue which will be further processing by the next layer,

4   The Optimization of the Data Processing

The whole data process system adapts the multi-process to operate. In order to maintain the number of the processes dynamically and improve data processing rate, the communication server starts a monitoring thread and use shared memory file to meet the demand.

4.1   Monitoring Thread

This monitoring process detects the amount of data in each message queue when the system is running. If the data in the message queue overflows the warning number, the system will automatically increase the corresponding "consumer" process. In the contrary, if the data in the message queue has been empty, while the data in the upstream message queue generates backlog, the system will automatically increase the "producer" process.

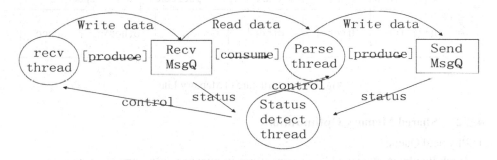

Fig. 3. Mulit-Threads System

Figure 4 illustrates the relationship that with the amount of received information increases, the parse processes dynamically grow.

Parse thread number	1	2	3	4
Receive message per second	2121	3913	5832	7786
Send thread num	1	1	2	3

Fig. 4. Test Data

4.2   Shared Memory File

Each process in the data process system delivers the message through the system V message queue. And the system records the real-time information in the shared memory so that all processes can access the current real-time intelligent transportation system data.

4.2.1   Data Structure of Memory File

The information in the shared memory can be divided into the following categories:

Shared Memory Properties: Recode the properties of the whole shared memory. It includes the size of the shared memory file, the offset of other information.

Terminal Attributes: Recode the terminal information, including the offset of the queue, the terminal number, the occupied memory size and so on.

Site Properties: Recode the information of the site, including the number of the site, the starting position and the ending position for the site in the memory and so on.

Other Properties: In addition to the information listed above, there are more information, such as the fixed message, the driver information and so on.

	Term infor	term1	term2		...		termN1	space	
	Station info	station1	station2			...	stationN2	space
	Escreen info	screen1	screen2	...		screenN3		space	
	Other info	Other 1	other2		...		otherNn	space

Fig. 5. Data Structure of Memory File

4.2.2   Shared Memory Optimization

1. Physical Queue

Each Intelligent device i (1<= i <=n) stores its own properties and status information in the shared memory .What's more, the other information such as roads and site information can also be recorded in the shared memory. We regard the information unit recorded in the memory as the information node. The information node can be divided by type, and each node can form a physical queue. At the beginning of the each information node queue, there are properties for the queue.

2 Logical Queue (virtual queue)

Vehicles belonging to the same line belong to the same logical queue. They safeguard the line list together.

The method is that we should maintain the index of belonging road information node and its predecessor and successor vehicle's information node for each vehicle's information node

3. Hash Function

Each information node's index position and the number of the node attributes such as (vehicle ID, site ID) can establish one mapping. So based on the number of the node attributes, we can quickly use harsh function to find the information in the corresponding memory.

5 Conclusions

The data processing system in the Intelligent Transportation Systems is the core of the ITS. Designing a stable and efficient data processing system is of great significance for the ITS's development. Because the LINUX IPC server has a number of advantages in data processing, it should be widely used and promoted.

In this paper, we optimize the data structures in the shared memory and establish the physical and virtual tables list, which greatly improves the query speed. In addition, the dynamic multi-process architecture and dynamic thread pool design make the system more stability, efficiency, flexibility and adaptability.

References

1. Zhao, Z., Zhou, W., Wang, N.: Shipping Monitoring System Based on GPS and GPRS Information Engineering. In: WASE International Conference ICIE 2009, vol. 07, pp. 346–349 (2009)
2. Navet, N., Song, Y., Simonot-Lion, F., Wilwert, C.: Trends in automotive communication systems. Proceedings of the IEEE 93, 1204–1223 (2005)
3. Bertoluzzol, M., Buja, G., Zuccollo, A.: Development and testing of a communication network for a drive-bywire industrial vehicle. In: 3rd IEEE International Conference on Industrial Informatics (INDIN 2005), pp. 529–534 (2005)
4. Cui, W., Pang, Y.-B.: Applying scheduling communications in Controller Area Network. Instrument Technique and Sensor 9, 27–29 (2005)
5. Lisner, J.C.: A flexible slotting scheme for TDMA-based protocols. In: Workshop Proceedings, ARCS 2004, Augsburg, pp.54–65 (2004)
6. Stevens, W.R., Rago, S.A.: Advanced Programming in the UNIX Environment. Pearson Education,Inc. (2005)
7. Bruneton, E., Lenglet, R., Coupaye, T.: ASM: a code manipulation tool to implement adaptable systems. Adaptable and Extensible Component Systems (2002)
8. Research and Innovative Technology Administration (RITA),
   http://www.its.dot.gov

A Fast Method for Large Aperture Optical Elements Surface Defects Detection

Cheng Yang, Rong Lu, and Niannian Chen

Department of Computer Science and Techenology,
Southwest University of Science and Techenology, Mian Yang 621010, China

Abstract. In such as inertial confinement fusion (ICF) and other large optical systems, there are a lot of large diameter optical elements. However, if the surface of these elements has defects, it will cause serious influence or damage to the optical system. Therefore, the effective defects detection of the optical element surface becomes the key guarantee to the quality of the optical system. Aimed at the shortages of the defects detection system, which can't handle the target with a single-pixel wide and the extraction speed of the defects feature parameters is too slow, this paper designed a rapid feature extraction algorithm based on the combination of vertex chain code and discrete green. And the method improved the extraction speed up to about three times than the traditional methods, and could handle the closed defects with a single-pixel wide. Moreover, under its better generalization ability, the recognition rate of defects classification using machine learning methods reached more than 90%.

Keywords: Defects, Vertex Chain Code, Discrete Green, Machine Learning.

1 Introduction

In inertial confinement fusion (ICF) system, the optical element surface defects (scratches, pitting, etc.) will result in different degrees of laser optical scattering, energy absorption, harmful show and diffraction patterns, which will cause damage to components' polished layer and laser which will directly affect the quality of the optical system. In order to make the focusing spot of the ICF system meet the performance requirements, the processing quality of optical components must be accurately evaluated and strictly controlled [1]. But the current detecting defects methods based on machine vision are mostly implemented by microscopic method of stitching. Although this approach is feasible, the speed of detection is very slow [3, 4] It is a key of defect detection to get the information on the characteristics of defects extraction. Defect detection systems of the past are based on Freeman chain code feature extraction methods, but the method can not characterize single pixel wide defects and slow[8, 9]. The length and width of the Defects are an important basis of defect classification and the minimum area exterior rectangle can characterize the length and width of defects well. The conventional approach is to use rotation method [3, 4, 10] to calculate the minimum area exterior rectangle, but it relies on the size of each rotation angle, the smaller the θ value is, the bigger it will compute.

D. Jin and S. Lin (Eds.): Advances in MSEC Vol. 2, AISC 129, pp. 175–184.
springerlink.com   © Springer-Verlag Berlin Heidelberg 2011

This paper presents an algorithm based on Vertex Chain Code and Discrete Green for image geometry feature extraction and extracting the characteristics of the defect information. We have designed the spindle method under both the characteristics of the discrete Green's and the Vertex Chain Code, by which we can calculate the minimum area exterior rectangle more quickly and accurately, thereby system can use the defect characteristics to classify automatically and rapidly. Experiments shown that the algorithm can better meet the requirements of line detection, what's more, the results of classification are satisfying.

2   The Principle of Defects Detection

In this paper, using the defects optical scattering theory to obtain the defects bright image that under the aperture and suitable for digital image processing on a dark background solves the tardiness of the sub-aperture stitching method. As the single pixel of existing camera cannot reach the μm level of magnitude, we obtain uniform illumination of magnified images in the dark background with light around the side by the LED [5].

Fig. 1 shows the workflow of optical surface defect digital detection system.

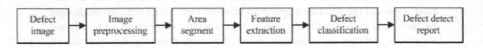

Fig. 1. Flowchart of defect detection

3   Rapid Extraction of Defect Feature

3.1   Unified Computational Framework Based on the Combination of the Discrete Green's Theory and the Vertex Chain Code

Vertex Chain Code has a feature of mapping the target area to a closed and connected region, but the Discrete Green's theory can only deal with closed and connected region, this paper combines Vertex Chain Code with the Discrete Green, transforming the geometric information on the calculation of defect method from the curve surface area integration to curve line integration, so that the calculation is greatly reduced.

As Fig.2 (a) shows, if walking along the image boundary, the current point to the next vertex of the walk direction is the direction code of the current vertex, defined respectively as the 0,1,2,3. There are three vertexes of image's pixel that have different nature, as Fig.2 (b) shows, respectively, are marked with the code 1, 2, 3, walking counterclockwise along the image boundary vertex a circle, recording the code in the image boundary pixel vertexes to get the image boundary Vertex Chain Code coding sequences. Image boundary can be uniquely determined by vertex chain coding sequence, expressed as $\{(x0, y0)/ \overline{Direction} /c0c1c2...cn\}$, coordinates of the starting vertex P is (x0, y0), $\overline{Direction} \in \{0, 1, 2, 3\}$ as the initial direction of walking, Ci $\in \{1, 2, 3\}$ as the Vertex Chain Code value of boundary pixel vertex.

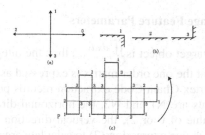

Fig. 2. Four direction of the code and the image boundary vertex chain code

According to the definition of the Vertex Chain Code, Fig.2 (a) shows the law of migration of boundary point coordinates are as follows:

$$
\begin{aligned}
&D_i = 0: x_{i+1} = x_i + t, y_{i+1} = y_i, dx = dt, dy = 0 \\
&D_i = 1: x_{i+1} = x_i, y_{i+1} = y_i - t, dx = 0, dy = -dt \\
&D_i = 2: x_{i+1} = x_i - t, y_{i+1} = y_i, dx = -dt, dy = 0 \\
&D_i = 3: x_{i+1} = x_i, y_{i+1} = y_i + t, dx = 0, dy = dt
\end{aligned} \tag{1}
$$

Denote $f(x, y) = \dfrac{\partial Q}{\partial x} - \dfrac{\partial P}{\partial y}$, Assume $f(x, y) = x^m y^n$, M, n are non-negative integers. Di=1 or Di=3 in the direction of vertical, then

$$
P = 0, f_1(x, y) = \int^x f(t, y)dt
$$

$$
Q_1 = \int_0^1 f_1(x, y - t)dt = \frac{x^{m+1}}{m+1} \sum_{v=0}^{n} \frac{(-1)^{n-v} y^v C_n^v}{n - v + 1}
$$

$$
Q_3 = -\int_0^1 f_1(x, y + t)dt = -\frac{x^{m+1}}{m+1} \sum_{v=0}^{n} \frac{y^v C_n^v}{n - v + 1} \tag{2}
$$

$$
\iint_D f(x, y)dxdy = \oint_L (Q_1 + Q_3)dy
$$

Di=0 or Di=2 in the direction of horizontal, then:

$$
Q = 0, f_1(x, y) = \int^y f(x, t)dt
$$

$$
P_0 = -\int_0^1 f_1(x + t, y)dt = -\frac{y^{n+1}}{n+1} \sum_{v=0}^{m} \frac{x^v C_m^v}{m - v + 1}
$$

$$
P_2 = \int_0^1 f_1(x - t, y)dt = \frac{y^{n+1}}{n+1} \sum_{v=0}^{m} \frac{(-1)^{m-v} x^v C_m^v}{m - v + 1} \tag{3}
$$

$$
\iint_D f(x, y)dxdy = \oint_L (P_0 + P_2)dy
$$

3.2   Calculation of Image Feature Parameters

1) **Area:** The area of the target object is $\iint_D dxdy$, then the integrand is $f(x, y) = 1$, that m = 0, n = 0, with that the one order matrix is expressed as M_{00}, means the area of the target in physical. Vertex Chain Code extraction records points number of the horizontal and vertical points are N1 and N2, the horizontal direction in Fig. 3 has 12 points (direction code value of 0 or 2), the vertical direction has 4 points (direction code value of 1 or 3), so choice expression (2) to calculate area, the area can be calculated by only 4 points in the X coordinate, namely: calculate the differ between x coordinates of its direction upward and direction points downward.

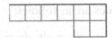

Fig. 3. Discrete Green

$$Q_1 = \frac{x^{m+1}}{m+1} \sum_{v=0}^{n} \frac{(-1)^{n-v} y^v C_n^v}{n-v+1} = x \qquad Q_3 = -\frac{x^{m+1}}{m+1} \sum_{v=0}^{n} \frac{y^v C_n^v}{n-v+1} = -x$$

$$Area = \sum Q_1 + \sum Q_3 = 0 - 4 + 6 + 6 = 8 \qquad (4)$$

2) The center of mass:

$$\bar{x} = \frac{\iint_D x dxdy}{\iint_D dxdy} = \frac{M_{10}}{M_{00}} \qquad \bar{y} = \frac{\iint_D y dxdy}{\iint_D dxdy} = \frac{M_{01}}{M_{00}} \qquad (5)$$

M_{10} means that m=1, n=0, then

$$Q_1 = \frac{x^{m+1}}{m+1} \sum_{v=0}^{n} \frac{(-1)^{n-v} y^v C_n^v}{n-v+1} = \frac{x^2}{2} \qquad Q_3 = -\frac{x^{m+1}}{m+1} \sum_{v=0}^{n} \frac{y^v C_n^v}{n-v+1} = -x$$

$$M_{10} = \sum Q_1 + \sum Q_3 = 32$$

M_{01} means that m=0, n=1, then

$$Q_1 = \frac{x^{m+1}}{m+1} \sum_{v=0}^{n} \frac{(-1)^{n-v} y^v C_n^v}{n-v+1} = x(y - \frac{1}{2}) \quad Q_3 = -\frac{x^{m+1}}{m+1} \sum_{v=0}^{n} \frac{y^v C_n^v}{n-v+1} = -x(y + \frac{1}{2})$$

$$M_{01} = \sum Q_1 + \sum Q_3 = 18$$

then $\bar{x} = \frac{M_{10}}{M_{00}} = 4$, $\bar{y} = \frac{M_{01}}{M_{00}} = \frac{18}{8}$

Calculate the coordinates of the center of mass just need only four points' coordinates.

3) Circumference the circumference of the defect is the length of Vertex Chain Code.
Fig.3 shows the target area perimeter l=16.

3.3   Fast Calculation of Minimum Area Exterior Rectangle

When calculate area of the minimum exterior rectangle, we can regard the segmented image with defect as a uniform density distribution of the plate, determine the spindle of the inertia according to the rotary inertia of thin target image, so that the spindle of inertia for any target image is the only one and accurate without any deviation, in addition, we need to pay attention that the minimum exterior rectangle is associated with the salient points of the target image only, and points have the code value of 1 in the Vertex Chain Code are just salient points, so when calculating the exterior rectangle just need to calculate the coordinates of the points which have value of 1 in Vertex Chain Code, Assume the angle of spindle X' and X axis is α.

$$\tan 2\alpha = \frac{2U_{11}}{U_{20} - U_{02}} \qquad (6)$$

Which

$$U_{11} = M_{11} - yM_{10}, \quad U_{20} = M_{20} - xM_{10}, \quad U_{02} = M_{02} - yM_{01}$$

According to the formula (2) or (3), (6), we can quickly calculate the target's spindle. Regard the centroid of the target area as the origin of coordinates, so the spindle is x' axis, the angle of the X-axis rotation direction is α, the conversion between old and new coordinate system is:

$$x' = p + x\cos\alpha + y\sin\alpha \quad y' = q - x\sin\alpha + y\cos\alpha \qquad (7)$$

Through the above steps, we get the minimum exterior rectangle of the target, assume the length as L, width as W.

4   Category Defects

By analyzing the features of scratches and pits, we use the extension, rectangular, roundness of the defect as the feature vector in [7] described. In order to select the characteristic parameters effectively, we select the feature of the extracted feature parameters of the defect by using weka, then by comparing experimental results; at last, we use the length, width, and the following parameters as the feature parameters of defects.

1) Elongation level
$$Elongation = L/W,$$

the longer and slender the target area is, the greater Elongation level is;

2) Rectangular level

$$Re\,ctDegree = M_{00}\,/(LW)\,,$$

the deviation between target image and the rectangle;

3) Circularity level

$$CircleDegree = 4\pi M_{00}\,/(Perimeter)^2\,,$$

The deviation between target image and the circular.

Literature [2] uses the neural network to approach classification, the method prone to over-fitting, and the generalization cannot be guaranteed. In order to find suitable classifier for defects, we use neural networks, supporting vector machines and strong classifier. It is found that the results of kernel function of supporting vector machine reorganization can reach about 90% by comparing, while support vector machine is based on small sample theory, with the higher the credibility of its generalization.

5   Results and Discussion

In summary, defect detection in this paper in the processing as shown in Fig.1, the working principle, first of all to eliminate dark current noise, denoising the target by using salt and pepper noise adaptive denoising filter, then split defects by using the adaptive segmentation of multi-regional segmentation algorithm, then extract defect feature by using the rapid extraction of geometric feature extraction algorithm which combines Vertex Chain Code with Discrete Green, and minimum area exterior rectangles of defects are obtained by using the spindle algorithm, select the defects classifier according to comparative experiments. In VC + +6.0 to achieve the defect detection system in this paper, operating environment is Pentium (R) 4 CPU 3.00GHZ, 1GB RAM, Windows XP operating system, high resolution CCD has a resolution of 2.1×10^7 pixel.

5.1   Feature Parameters Extraction of Defects

1) The effectiveness of minimum exterior rectangle simulation

Fig. 4. Minimum exterior rectangle

It can be seen from the above figures that the spindle method can detect the minimum size exterior rectangle in all directions.

2) The minimum exterior rectangle speed test

Table 1. Time Comparison Table ($\theta = 1$)

Image size [pixel]	1254x567	411x330	489x355
Sum of target	12	90	43
literature[2]	2250ms	3020ms	2600ms
literature[9]	1806ms	1740ms	1450ms
Spindle method	493ms	281ms	234ms

3) The effectiveness of defect's feature extraction test

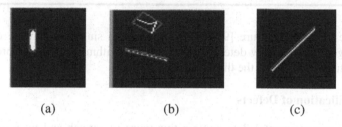

(a) (b) (c)

Fig. 5. Defects Image

Table 2. Defects feature extraction results

Picture	Fig 5(a)	Fig 5(b)		Fig 5(c)
		up	down	
Perimeter	54	68	76	128
Area	104	25	28	32
Elongation	2.84966	1.17096	13.5897	16
Rectangularity	0.740739	0.136921	0.430802	0.25026
Circularity	0.448183	0.067941	0.060917	0.024543
MER Length	20.0023	14.622	29.7197	45.2313
MER Width	7.01919	12.4872	2.18694	2.82696

4) The defect feature extraction speed test

Table 3. Defects feature extraction rate of contrast

Image Size (pixel)	Detection time of literature [9] (ms)	Detection Time of literature [2](ms)	Detection time of improved Algorithm[ms]	Area of debris[pixel]	Sum of defects
1280×104	7085	1791	342	2634	320
1280×104	9381	1362	456	4199	566
1280×104	9722	1354	654	4210	590
1280×104	22885	1880	667	10959	868
1280×104	15827	1493	735	6911	1204
1280×104	28037	6672	1345	60791	2057
1280×104	44811	1161	1678	15918	4260

Literature [2] and literature [9] cannot recognize a single pixel wide object; the proposed algorithm can fully detect (Fig. 5). The algorithm to solve this problem, extract the features in Table 2, the third and four.

5.2   Classification of Defects

Defects in the experimental data contain two types of scratch and pit a total of 140 samples, 70 samples of each type. We use 10-fold cross validation algorithm validation recognition rate.

1) Support Vector Machine test
Support vector machines mainly have the following core functions: linear kernel function $u' \cdot v$; Polynomial kernel function $(gamma \cdot u' \cdot v + coffe)^{degree}$; Radial basis function $\exp(-gamma \cdot |u - v|^2)$; Sigmoid kernel function $\tanh(gamma \cdot u' \cdot v + coffe)$.

Table 4. Performance comparison of SVM

Kernel function	Linear	Polynomial		Radial basis			Sigmoid	
gamma	\	1	0.5	1	0.5	0.5	1	
coffe	\	1	13	1	1	1.5	1	
degree	\	2	2	\	\	\	\	
Recognition rate	87.5	89.6	89.6	91.7	68.9	92.4	68.9	

As can be seen from the table, radial basis function recognition rate is higher. Finally, the above experiments using radial basis support vector machine as classifier.

2) The strong classifier test

Table 5. Performance Comparison of strong classifier

Algorithm	Boost	RTrees
Training samples	140	140
Training recognition rate	82.6111	95.6522
Test rate	78.2778	84.8276

Finally, we use radial basis support vector machine as classifier according to the above experiment.

6   Conclusions

This paper according to the optical system quality assurance for large aperture optical component surface defect monitoring requirements, realized based on the machine vision of the large optical element surface flaw detection and classification. Firstly, improve the image feature extraction algorithm, design of a image feature extraction algorithm based on the combination of vertex chain code and discrete Green. The algorithm can handle non-closed single-pixel wide features, to achieve rapid and effective parameters of flaws accurate extraction; and then making the flaws classification recognition rate about 90%. The results show that: the algorithm used in this paper can be much faster, more efficient, more accurate on extracting defect feature information and more robust to the classification for the defects.

References

1. Wang, F.-Q., Yang, Y.-Y., Sun, D.-D.: Research of digital inspection system of precise surface defect. Optical Instruments 28(3), 146–150 (2006)
2. Fan, Y., Chen, N.N., Gao, L.L., et al.: Digtal detection system of surface defects for large aperture optical elements. High Power Laser and Particle Beams 6(21), 835–840 (2009)
3. Li, A., Yang, T., Zhang, Y.: Preliminary research of surface defect recognition based on machine vision. Jouranl of Chongqing University of Post and Telecommunications: Natural Science Edition 19(4), 442–445 (2008)
4. Yang, Y., Lu, C., Liang, J., et al.: Microscopic dark-field scattering imaging and digitalization evaluation system of defects on optical devices precision surface. Acta Optical Sinica 27(6), 1031–1038 (2007)
5. Chen, X., Xu, X., Zhang, L., et al.: Defect testing of large aperture optics based on high resolution CCD camera. High Power Laser and Particle Beams 11(21), 1677–1680 (2009)
6. Xie, J.-B., Liu, T., Ren, Y., et al.: An adaptive fast filtering algorithm for removal of salt and pepper noise. Journal of Image and Graphics 14(5), 843–847 (2009)

184   C. Yang, R. Lu, and N. Chen

7. Brlek, S., Labelle, G., Lacasse, A.: The discrete Green Theorem and some applications in discrete geometry. Theoretical Computer Science 346, 200–225 (2005)
8. Chen, Y.-G., Zhang, W., Huang, S.: Algorithm for extracting contour and generating chain code tree. Journal of East China Nomal University (Natural Science) 3, 77–85 (2006)
9. Yuan, Z., Niu, X., Liu, C.: A study on inspection of rice kernel ratio with minimum enclosing rectangle method. Cereal & Feed Industry 9(1), 7–8 (2006)

Study and Optimization Based on MySQL Storage Engine

Xiaolong Pan, Weiming Wu, and Yonghao Gu

Computer Science Department, Beijing University of Posts and Telecommunications,
100876, China

Abstract. MySQL is an open database of the most common. One of the important characteristics of MySQL is that it provides rich storage engines. In the structure of Web applications, it is very important for the performance of database that the system must choose different storage engines based on different needs. But the default settings performance of MySQL is very not good, so it is necessary to optimize when the system uses MySQL. This paper discusses the most two important storage engines in MySQL, and raises the performance of MySQL by optimize MySQL based on different storage engines.

Keywords: MySQL, Storage Engines, Performance Optimization.

1   Introduction

With the increasing development of the information technology, information of the world has expanded rapidly and database has been widely used. Database has become the base and core of modern computer systems and computer applications.

As the most widely used open source database, MySQL provides different kinds of storage engine (table type). Different storage engine uses different storage mechanism, indexing techniques and lock level so they can provide different features and capabilities. The most basic task of database is storing and managing data, and the only feature that end-user can feel is the performance of the database: what is the speed of the database to process an inquiry action and return the result to the user's application. So it will be much important to improve the efficiency of dynamic growing database query.

This paper uses the MySQL application experience to compare some current important storage engines, and then elaborates MySQL database optimization depending on different storage engines.

2   MySQL Storage Engines

The data in MySQL is stored in files (or memory) by different techniques. By choosing different techniques, the users can get extra speed or functions and then improve the overall function of the application.

These different techniques and their supporting related functions are called storage engine (or table type) in MySQL as shown in figure 1. MySQL has the default configuration of different kinds of storage engines.

D. Jin and S. Lin (Eds.): Advances in MSEC Vol. 2, AISC 129, pp. 185–189.

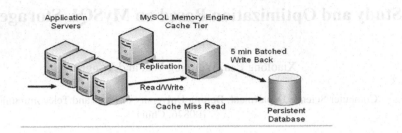

Fig. 1. The Working principle of Memory Engine

2.1   MyISAM Engine

MyISAM is the default storage engine in MySQL 5.1, and it is first provided by MySQL. MyISAM doesn't support transaction and foreign key. Its superiority is that it has a high access speed. The applications that don't have a requirement of transactional integrity or SELECT & INSERT-based applications basically use this engine to create tables.

MyISAM is divided into static MyISAM, dynamic MyISAM and compressed MyISAM:

Static MyISAM: if the length of data columns in the table is pre-fixed, the server will use this table type automatically. Because the space of every record in the table is the same, the storage and updating efficiency of this table is very high.

Dynamic MyISAM: Relative to static MyISAM, this king of table has a small occupied storage space. But because the length of every record is different, after several revisions of data, the data of the table will be stored in the memory discretely, then the efficiency will drop.

2.2   InnoDB Engine

InnoDB provides MySQL transaction-safe tables which have the transaction, rollback and crash recovery capabilities, multi-versioned concurrency control functions. These features increase multi-user and deployment functions.

The design goal of InnoDB is to maximize performance as handling large volumes of data. It has the best CPU utilization of all other disk-based relational database. Technically, InnoDB storage engines are fully integrated with the MySQL server. InnoDB storage engines maintain their own buffer pool to cache data and indexes in the memory.

InnoDB put data and indexes in table space. Which is different from others is that the table space may include multiple files. For example, in MyISAM, tables are stored in a single file. The table size of InnoDB is only limited by the file size of the operating system. Each table can also have its own table space just by staring "innodb_file_per_table".

3   Comparative study of MyISAM and InnoDB

InnoDB and MyISAM are the most common storage engines when people use MySQL. Each has advantages and disadvantages depend on different applications.

MyISAM is the extension of earl ISAM. ISAM is designs to suit for the situation that frequencies are much greater than the write rate. ISAM also occupies less space in the memory. MyISAM has most tools with check and repair functions compared with other storage engines. MyISAM tables can be compressed, and they support full text retrieval. They are not transactional-safe, and they don't support foreign key.

InnoDB is designed for the situation of high concurrent read and write. It uses MVCC (Multi-Version Concurrency Control) and row-level locking to provide transaction support complied with ACID. InnoDB supports foreign key referential integrity and it has the capability of fault recovery. It also support foreign key. InnoDB table is very fast and has richer features than BDB. So if a transactional-safe storage engine is needed, InnoDB is a good choice.

To sum up, it is clear that:

MyISAM is more suitable for the case that there is less insert and more query: It doesn't support transaction, foreign key and WAL(write ahead logging) and it can only lock the full table.

InnoDB is suitable for the case of large concurrent write and query: it supports transaction (ACID compatible), row-level locking, and foreign key. And it has its own memory buffer pool and independent table space (without large limitation).

4   Performance Comparisons and Optimization of the Two Storage Engines

4.1   Performance Comparisons

To further understand the specific differences in performance of the tow storage engines, this paper specifically give a simple test:

The tested database version: MySQL Ver 14.14 Distrib 5.1.55, for Win32 (ia32).

[Inserted data-1] (innodb_flush_log_at_trx_commit=1)

MyISAM 10W : 22/s
InnoDB 10W : 2010/s

[Inserted data-3] (innodb_buffer_pool_size=1024M)
InnoDB 1W : 3/s
InnoDB 10W : 33/s
InnoDB 100W : 607/s

[Inserteddata-4](innodb_buffer_pool_size=256M,innodb_flush_log_at_trx_commit =1, set auto commit=0)

InnoDB 1W : 3/s

InnoDB 10W : 26/s

InnoDB 100W : 379/s

It is clear that in MySQL 5.0, the performance differences between MyISAM and InnoDB is not big.

4.2   Performance Optimization

Based on the above test data, different optimization strategies are needed for different storage engines.

For MyISAM, the main optimization aspects include:

1) key_buffer_size: Number of buffers which are distributed to MyISAM index cache
2) query_cache_size: The number of caches distributed to query cache.
3) long_query_time: Set slow query time
4) external-locking: Prohibit the use of external lock and prevent deadlock
5) back_log: The number of requests can be stacked before temporary stops responding new requests. If the users need to allow a large number of collections in a short time, this valued can be raised.
6) table_cache: The number of cache data table to avoid the spending to open the table repeatedly.
7) thread_cache_size: The number of threads the cache can use to reduce the spending of creating new thread.

For InnoDB, the main optimization aspects include:

1) Put INSERT, UPDATE and DELETE into the same transaction. But the resulting efficiency should also be paid attention.
2) innodb_flush_log_at_trx_commi: Set is as 0(refresh every second), 1(real-time refresh), 2(only write log file and not refresh to the disk).
3) innodb_buffer_pool_size: Control the size of buffer pools which are distributed to cluster data and secondary index page. The default is 16MB.
4) innodb_additional_mem_pool_size: Control the size of pools which are distributed to InnoDB internal data dictionary to sort. The default is 1MB.
5) innodb_log_buffer_size: Control the size of buffers which are distributed to InnoDB memory to write log file in advance. The default is 1MB.
6) innodb_log_files_in_group: The number of log files in the log group. InnoDB write files circularly. The default is 2.
7) innodb_log_file_size: The default is 5MB. The recommended value is from 1MB to one Nth of the buffer pool. N is the number of log files in the group.

5   Conclude

Database storage engine is a important part of database. To assure the conditions of each storage engine is an important precondition of optimizing the database storage engines. This paper sums up the features of the two engines and introduces the optimization strategies by studying the two main engines. Through research and optimization, the advantages of MySQL can be played better.

References

1. Dan, A., Yu, P.S., Chung, J.-Y.: Characterization of database access pattern for analytic prediction of buffer hit probability. The VLDB Journal 4, 21–24 (1995)
2. Martin, P., Powley, W., Li, H., et al.: Managing Database Server Performance to Meet QoS Requirements in Electronic Commerce Systems. International Journal on Digital Libraries 3, 316–324 (2002)
3. Ramakrishnan, R., Gehrkc, J.: Database management systems. McGraw-Hill (2000)
4. Silberschatz, A., Kedem, Z.: Consistency in hierarchical database systems. J. ACM 27, 72–80 (1980)
5. Valentin, G., Zuliani, M., Zilio, D.C., et al.: DB2 advisor: an optimizer smart enough to recommend its own indexes. In: Proc. of 16th Intl. Conf. on Ta Engineering (2000)
6. Scbwartz, B.: High Performance MySQL. The O'Reilly Press, Sebastopol (2008)
7. Database Test Suit, http://osdldbt.sourceforge.net
8. Micro architecture Codename Nehalem, http://www.intel.com

An Analysis of Statistical Properties on Some Urban Subway Networks

Yimin Ding[1] and Zhuo Ding[2]

[1] Faculty of Physics and Electronic Engineering,
Hubei University, Wuhan 430062, China
dymhubu@sina.com
[2] School of Business Administration,
South China University of Technology, Guangzhou 510640, China

Abstract. In this paper, we present an empirical investigation on 12 urban subway networks in Asia, which include main urban subway networks in China, Japan, Korea, India, Thailand and Singapore. The size of these networks ranges from $N=47$ to 455. The clustering coefficient, the character path length and the degree distribution have been analyzed. The empirical results show that these networks have high clustering coefficient ($C>0.80$) and small character path length ($L<3.0$), which exhibit a small-world behavior (in space p).The empirical studies from different times show that the clustering coefficient C gradually decreases with the increase of the total number of subway stops N, but the characteristic path length L increases, both following a power law. In addition, we also have studied the fractal scaling of these networks and find that these subway networks exhibit some properties of fractal scaling networks.

Keywords: Complex networks, Urban subway network, Small-world, Fractal scaling.

1 Introduction

In the past decade, A lot of complex networks have drawn great researching enthusiasm from many scholars[1-3] since the seminal works on small-world networks by Watts and Strogatz [4] and on scale-free networks by Barabási and Albert [5]. These investigations are on social networks, technological networks, or biological networks [2,3]. In study on complex networks, the degree distribution, the character path length and the clustering coefficient are studied. The previous study found that some networks exhibit a power law degree distribution behavior (Internet, and film actor networks), which is the property of the scale-free networks. Barabási and Albert have proposed the BA model to explain the evolutionary mechanism of power law degree distribution. Most complex networks exhibit the small-world behavior, Watts and Strogatz have proposed the WS small-world model to explain the mechanism of small-world, Subsequent studies have found that some networks (metabolic networks) have modularity [6,7], further more, Ravasz E *et al.* [8] have proposed a hierarchical network model that explains the simultaneous emergence of

D. Jin and S. Lin (Eds.): Advances in MSEC Vol.2, AISC 129, pp. 191–196.
springerlink.com   © Springer-Verlag Berlin Heidelberg 2011

the observed hierarchical and scale-free topology of the metabolism. Song *et al.* have studied self-similarity of complex networks and investigated the network properties with the box-counting method [9].

Recently, several public transport systems (PTS) have been investigated using various concepts of statistical physics of complex networks. These investigations are on railway networks [10,11], subway networks [12,13] and urban bus transport networks [14-16]. All the railway, subway and urban bus transport networks appear to share well known small-world properties, and the degree distribution can follow a power law or can be described by an exponential function[10-16].

In the present paper, we shall report an empirical investigation on the urban subway networks of 12 major cities in Asia, The size of these networks ranges from N=47 to 455. The clustering coefficient C, the character path length L, the diameter of the network D, the average degree $<k>$ are studied. We shall also study the variation of these parameter from difference times. In addition, we shall study the fractal scaling of these networks from different times. As far as we know, our results are the first comparative survey of several subway networks in Asia.

2   Empirical Investigation

2.1   The Small-World Properties of Subway Networks

There are two common features which make apparently very different networks all "small-world" [1,7]. The first feature, the characteristic path length (as known as the average path length L) between nodes is short, compared to what might be expected based on the total number of nodes. Considering a network of N nodes, define the distance d_{ij} between two nodes i, j to be the length of the shortest path between them, then, the characteristic path length L is the average of the distances d_{ij} between all pairs of nodes [7].

$$L = \frac{1}{N(N-1)} \sum_{i,j \in N, i \neq j} d_{ij} \tag{1}$$

The second feature, the one we are noting when we agree with a new acquaintance that the world is indeed small, is that two nodes both linked to a third node are likely to have a direct link also. This feature has been called clustering (clustering coefficient C), and defined as follows [1,7]. A quantity c_i (the local clustering coefficient of node i) is first introduced, expressing how likely $a_{jm} = 1$ for two neighbors j and m of node i. Its value is obtained by counting the actual number of edges (denoted by e_i) in G_i (the subgraph of neighbors of i). Notice that G_i can be, in some cases, unconnected. The local clustering coefficient is defined as the ratio between e_i and $k_i (k_i$ $1)/2$, the maximum possible number of edges in Gi:

$$c_i = \frac{2e_i}{k_i(k_i-1)} = \frac{\sum_{j,m} a_{ij} a_{jm} a_{mi}}{k_i(k_i-1)} \tag{2}$$

The clustering coefficient of the graph is then given by the average of c_i over all the nodes in G:

$$C = \langle c \rangle = \frac{1}{N} \sum_{i \in N} c_i \qquad (3)$$

The random graphs have the first feature, they do not have the second. But many real-world networks have both the two features, in other words, they have high clustering coefficient and small character path length, and are small-worlds. Now, we shall present the empirical investigation for the urban subway networks in Asia [17], the network properties of some cities such as Beijing, Shanghai, Guangzhou , Nanjing, Tokyo, Osaka, Seoul, Delhi, Bangkok, Singapore, Taipei and Hong Kong are calculated (in the space P) and listed in Table 1.

Table 1. Network properties for each subway network (2010)

	I	M	N	D	C	L	$<k>$
Beijing	12	9	124	4	0.9303	2.364	22.15
Shanghai	15	12	234	4	0.9213	2.604	29.32
Guangzhou	7	4	62	3	0.9582	1.944	14.64
Nanjing	5.5	3	47	2	0.9687	1.631	17.06
Tokyo	13	13	202	4	0.8448	3.007	32.04
Osaka	3	9	117	3	0.9028	2.531	20.77
Seoul	10	14	445	4	0.9262	2.404	54.99
Delhi	13	4	114	3	0.9838	1.949	34.60
Bangkok	5.5	9	204	3	0.9042	2.308	35.58
Singapore	4	4	90	3	0.9148	2.349	29.01
Taipei	2.7	6	80	4	0.8853	2.589	29.15
Hong Kong	6.9	9	81	5	0.9194	2.760	14.89

I: population (million); M: total number of subway routes; N: total number of subway stops; D: diameter of the network; C: clustering coefficient; L: average shortest path length; $<k>$: average degree.

From Table 1, the subway network of Beijing in 2010 consists of nine routes, serving 124 stops. The network is decentralized, in that no stop lies on all nine routes, thus having every other stop as a direct neighbor with our definition of adjacency. In fact, the highest value of k_i is 55, and the average degree $<k>$=22.15, most nodes have high degree. The clustering coefficient C=0.9303, being so close to 1 because of the proportion of stops lie on a single train route..The characteristic path length L=2.364, and the diameter of the network is D = 4, is small when compared to the number of nodes, so the subway network of Beijing satisfies the two basic condition for being small-world. The subway networks in the other cities (see to Tab.1) have the similar properties, and they are all small-worlds. The empirical investigation results are in good agreement with Latora and Katherine's results on the subway networks of Boston (C=0.928, L=1.81) and Vienna (C=0.945, L=1.86) [12, 13].

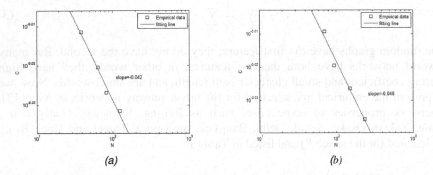

(a)   *(b)*

Fig. 1. Variation of C on N for: (a) Beijing subway networks between 2000 and 2010, and (b) Shanghai subway networks between 2000 and 2010. (squares, data; lines, fitting lines).

We have obtained the variation of network properties from different time. Fig. 1a and Fig. 1b shows the variation of C on N for Beijing and Shanghai subway networks between 2000 and 2010. The horizontal coordinate is N, N represents the total number of subway stops; the vertical coordinate is C, C represents the clustering coefficient. The linear lines are obtained by least square fitting method from the data.

The variation of C on N may be expressed as

$$C \propto N^{-\gamma_1} \tag{4}$$

Where, $\gamma_1=0.042\pm0.004$ (Beijing).$\gamma_1=0.048\pm0.004$ (Shanghai).

From then, we can find that the clustering coefficient C gradually decreases with the increase of the total number of subway stops N, but the variation of C is very small. If the networks grow up to $N=1000$, the clustering coefficient C can be calculated by equation (4) as $C=0.846$ (Beijing), $C=0.854$ (Shanghai), so the subway networks in Beijing and Shanghai from different times are all high clustering networks.

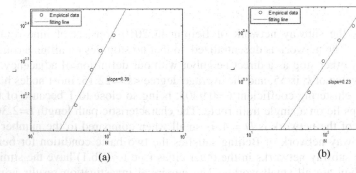

(a)   *(b)*

Fig. 2. Variation of L on N for: (a) Beijing subway networks between 2000 and 2010, and (b) Shanghai subway networks between 2000 and 2010. (circles, data; lines, fitting lines).

Fig. 2 shows the variation of L on N for Beijing and Shanghai subway networks between 2000 and 2010. The horizontal coordinate is N; the vertical coordinate is L, L represents the characteristic path length. The linear lines are obtained by least square fitting method from the data. The variation of L on N may be expressed as

$$L \propto N^{\gamma_2} \tag{5}$$

Where, $\gamma_2 = 0.39 \pm 0.01$ (Beijing).$\gamma_2 = 0.23 \pm 0.01$ (Shanghai).

It appears that the characteristic path length L gradually increases with the increase of the total number of subway stops N, but the variation of L is very small. If the network grow up to $N=1000$, the clustering coefficient L can be calculated by equation (5) as $L=5.16$(Shanghai), so the subway networks of Beijing and Shanghai from different times are all small-world networks.

2.2   Fractal Scaling of Subway Networks

We have also obtained the fractal scaling of some subway networks by node-counting method. The node-counting method [18,19] is based on the computation of $N_v(k)$ and k values. $N_v(k)$ is taken as the accumulated number of stops connected by a starting stop in k links from transportation maps. Fig. 3 shows plots of $N_v(k)$ vs.k for some subway networks (2010).

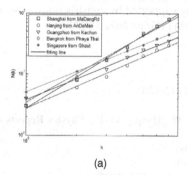
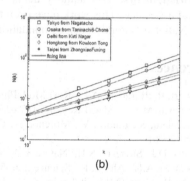

(a)   (b)

Fig. 3. (Color online). N(k) vs k of subway networks for: (a) Shanghai, Nanjing, Guangzhou, Bangkok and Singapore in 2010, and (b) Tokyo, Osaka, Delhi, Hong Kong and Taipei in 2010. Nv(k) is taken as the accumulated number of stops connected by a starting stop in k links from transportation maps.

Fig.3 shows logarithmic plots of $N_v(k)$ vs. k the 10 cities, all the data are fitted by linear lines, The slopes of the fitting lines are The slopes of the fitting lines are listed in Table 2, and $N(k)$ can be approximately described by a power law,

$$N_v(k) \propto k^D \tag{6}$$

Observational data can satisfactorily be fitted to fractal scaling described by Csányi and Szendröi[20].

Table 2. The slopes of the fitting lines

City	Shang-hai	Bang-kok	Tok-yo	Osa-ka	Del-hi	Singa-pore	Tai-pei	Hong-Kong	Guang-zhou	Nan-jing
D	1.72	1.57	1.48	1.55	1.20	1.12	1.15	1.19	1.15	1.21

3   Conclusions

We have presented an empirical investigation on 12 urban subways in Asia, and try to explain the empirical results and show a possible subway networks evolution scenario. The empirical results show that these urban subway networks have high clustering coefficient and small character path length, which exhibit a small-world behavior, further study shows that the clustering coefficient C gradually decreases with the increase of the total number of subway stops N, however, the characteristic path length L increases, both of them follow a power law. In addition, we also have studied the fractal scaling of these networks. The empirical results show that both the variation of $N(k)$ vs k and L vs N are take a power law. So the subway networks exhibit some properties of fractal scaling networks. Whether do the subway networks of other cities take the same law? We shall present the discussion elsewhere soon.

Acknowledgments. The authors thank Pro. Changping Yang for his the valuable technical assistance. This work have been partially supported by construction foundation for the National Natural Science Foundation of China (Grant Nos. 11074067).

References

1. Albert, R., Barabasi, A.L.: Rev. Mod. Phys. 74, 47 (2002)
2. Newman, M.E.J.: SIAM Rev. 45, 167 (2003)
3. Boccaletti, S., Latora, V., Moreno, Y., Chavez, M., Hwang, D.-U.: Physics Reports 424, 175 (2006)
4. Watts, D.J., Strogatz, S.H.: Nature 393, 440 (1998)
5. Barabási, A.L., Albert, R.: Science 286, 509 (1999)
6. Milo, R., Shen, O.S.S., Itzkovitz, S., et al.: Science 298, 824 (2002)
7. Yeger Lotem, E., Sattath, S., Nadav, K., et al.: Proc. Nat. Acad. Sci., USA, vol. 101, p. 5934 (2004)
8. Ravasz, E., Somera, A.L., Mongru, D.A., et al.: Science 297, 1551 (2002)
9. Chaoming, S., Shlomo, H., Hernan, A.M.: Nature 433, 392 (2005)
10. Sen, P., Dasgupta, S., Chatterjee, A., et al.: Phys. Rev. E 67, 036106 (2003)
11. Kurant, M., Thiran, P.: Phys. Rev. Lett. 96, 138701 (2006)
12. Latora, V., Marchiori, M.: Physica A 314, 109 (2002)
13. Seaton, K.A., Hackett, L.M.: Physica A 339, 635 (2004)
14. Sienkiewicz, J., Holyst, J.A.: Phy. Rev. E 72, 046127 (2005)
15. Zhang, P.-P., Chen, K., et al.: Physica A 360, 599 (2006)
16. Chen, Y.-Z., Li, N., He, D.-R.: Physica A 376, 747 (2007)
17. http://urbanrail.net/index.html
18. Antonio, D.: Physica A 388, 4658 (2009)
19. Kurant, M., Thiran, P.: Phyical Review E 74, 036114 (2006)
20. Csányi, G., Szendröï, B.: Physical Review E 70, 016122 (2004)

The Research of Applying Chaos Theory to Speech Communicating Encryption System

YunPeng Zhang[*], Fan Duan, and Xi Liu

College of Software and Microelectronics Northwestern Polytechnical University
710072 Xi'an, China
poweryp@163.com

Abstract. This article reviews the chaos voice encryption based on the existing results of chaos theory. This article concentrates on the chaos encryption of speech, putting forward the idea which is applied to speech data that combine chaos type library with data segmentation. In this paper, it analyzes characteristics of the algorism. It proves that this algorism has characteristics such as large key space, good diffusibility, and strong randomness of cipher text. And, this algorism has the ability to defend existing plaintext attack and reconstruction attack based on chaotic model

Keywords: cryptography, chaotic type library, speech, data sub block.

1 Study Background and Significance

Voice communication is one of the most common and convenient methods. it is not suitable to encrypt voice data with traditional cryptography's symmetrical and asymmetric code [1].

At present, the encryption algorithm can be divided into two categories [2]: complete encryption algorithm and select encryption algorithm. Although the encryption strength has been improved, the algorithm cannot ensure the encryption speed. So, many researchers try to use chaos[3] method to construct a fast encryption algorithm [4].

The existing voice encryption based on chaos theory usually uses single chaotic sequence, single parameter and initial conditions to produce a chaotic sequence to encrypt plaintext data. This encryption can be cracked easily and cannot meet the strength requirement of voice encryption system, and may cause some data losing.

To cope with this problem, the newly constructed algorithm adopts the method to encrypt the data which is divided into parts. Even if some data was lost during transporting, the receiver can still decrypt most of the information. Meanwhile, the new algorithm has characteristics namely good diffusibility, large key space, strong randomness of cipher text, with high security.

---

[*] Corresponding author.

D. Jin and S. Lin (Eds.): Advances in MSEC Vol. 2, AISC 129, pp. 197–202.
springerlink.com   © Springer-Verlag Berlin Heidelberg 2011

2   Algorithm Description

2.1   Algorithm Principle

First, we separate the voice information into blocks according to fixed size, which is called data sub block. Every fixed number of data sub blocks makes a data group. Encrypt different data group separated from data sub block with different chaotic sequences.

To get different chaotic sequences match different data groups, we utilize different chaotic types to generate these sequences. This algorithm constructs an extensible chaotic type library and uses different chaotic type to generate diversified chaotic sequences, thus we could enlarges the key space. The chaotic types in the extensible library are Logistic mapping, Henon mapping, Chebychev mapping, Lorenz mapping, Cat mapping and Baker mapping. For further improvement of security, this algorithm introduces a random number generator to generate random array, using the uncertainty of random number to choose the chaotic type and parameter in the library.

Use the information of encryption algorithm applied to the second data group as the seed key of the first data group, and save it in the fist data group in head-file form. Analogize the entire encrypted data group like this. The encryption information of the first data group is stored in password file. And, only the data sender and receiver have the password file. Thus, the decryption of the latter data group is depended on the seed key in the head-file of the previous data group; meanwhile, the decryption of the previous data group is depended on the decryption of the more previous data group. At last, all the encryption of the data groups will be uniquely depended on the password of the encryption of the first data group.

2.2   Algorithm Flow

The specific flow(Fig 1) is:

(1) Number the chaotic type in chaotic type library:

0001 representing Logistic mapping,
0010 representing Henon mapping,
0011 representing Chebychev mapping,
……

(2) The random number generator generates the random number between 0001 and 1000 (Supposing there are 8 chaotic types in the library), which means it has selected Logistic mapping. The Logistic mapping generates the parameter region in chaotic state, that is the next random number region is [3.5699457, 4], and random number generator gives 0101011011110111110000011 (represents $\mu=3.5699459$). Then, generate the corresponding pseudo chaotic sequence 2 to encrypt data group 2 according to formula $x_{n+1} = 3.5699459 x_n(1-x_n)$ $n \in \{1, 2, ...\}$ (Referring to the code rules in the process schematic diagram)

(3) Form a random array with the two random numbers (0001, 0101011011110111110000011, 0) (As some chaotic types have three parameters, so the array has three values. The first is used to ensure chaotic type, the second and third are parameters of the chaotic type, which is used as "head file 2" of "data group 1"). Analogize like this to get "data group 3" and encryption sequence "chaotic sequence 3", "parameter 3" and "head file 3"……

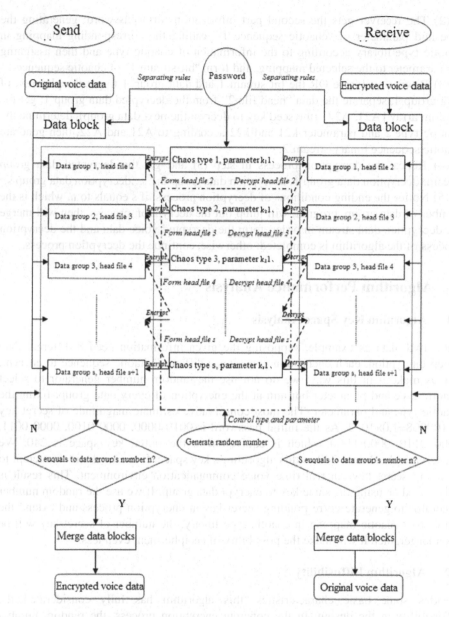

Fig. 1. Encrypt and decrypt process schematic diagram

The decryption process is similar to the one of encryption. The basic steps are as follows(See fig 1):

(1) The receiver receives password and separates information: the data dividing rules. Separate the voice information in "digital voice database" into sub blocks in binary form according to fixed size. Every fixed number of data sub blocks makes a data group.

(2) The receiver gets the second part information form password: generating the type and parameter of "chaotic sequence 1", calling the corresponding mapping in chaotic type library according to the information of chaotic type and then assigning the parameter to the selected mapping, and form "bit-stream 1" of chaotic sequence.

(3) Bitwise exclusive OR the bit stream 1 and data group 1 to get the plaintext of data group 1, separate the data "head file 2" from the decrypted data group 1; get the random array {A21, A22} (the seed key to decrypt the next data group), determine the chaotic type 2 and parameter k21 and k22 according to A21 and A22, then generate chaotic sequence binary-stream 2.

(4) Repeat the decryption process in step (3) and get the decrypted data group stream: decryption data group 1, decryption data group 2.......decryption data group s.

(5) Notice the ending conditions of decryption process: if s equals to n, which is the number of data group, program will jump out the decryption process, then will merge the decryption data stream in order, form the decrypted voice data and the decryption process of the algorithm is completed; otherwise, continue the decryption process.

3   Algorithm Performance Analysis

3.1   Algorithm Key Space Analysis

Take 1Mb data as example, supposing the given information need 8 different data group encryption, each encryption generates a different chaotic sequence to encrypt. Let us make it in this way: we do not use the random number generator to select chaotic type and parameter but aim at the encryption of every data group, fixing the chaotic type and parameter. The most conservative estimate magnitude of secret key is $10000^8 = 1.0 \times 10^{32}$. As the initial password is 0010 0000, 0000 0100, 0000 0011, 0000 0110, 0000 0000, which is equivalent to the initial key space is 2^{40}. We conservatively choose 2^{40} as this algorithm's key space. It is unrealistic to attempt to so many secret keys in real time voice communication environment. This result is only based on using the same key to encrypt data group. If we use the random number generator to generate corresponding secret key in encryption process and extend the number of chaotic mapping in chaotic type library, the number of secret key will be even larger, then it will make the possibility of decipherment even less.

3.2   Algorithm Diffusibility

Besides some basic characteristics, this algorithm has fully considered the diffusibility in the design. In the concrete encryption process, the random number generator creates different random arrays. Different chaotic sequences are generated by these arrays. Then we use these chaotic sequences to encrypt their respective data group. Therefore different data group encryption has different diffusibility, thus the diffusibility obtains the maximum enhancement:

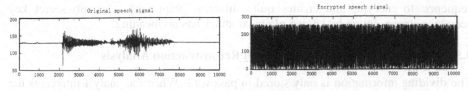

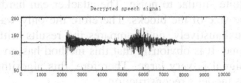

Fig. 2. Original speech signal   **Fig. 3.** Encrypted speech signal

Fig. 4. Decrypted speech signal

From Fig 2 to Fig 4 we can observe that, the encrypted speech is similar to noise. The recovered speech signal is basically similar to original signal.

3.3   Cipher Text Diffusibility Analysis

If the cipher text is output randomly, it should have good 01 balance quality. By experiment, test the pressed and encrypted data, the result is as Table 1:

Table 1. The result of 01 balance

	0 (%)	1 (%)
1	45	55
2	48	52
3	44	56
4	40	60
5	51	49
Averrage	45.6	54.4

According to the testing result, the cipher text generated by this algorithm has good 01 balance quality, that is, good randomness.

3.4   Anti Known Plaintext Attack Analysis

The chaotic sequences are completely got from password, including the determination of chaotic type. The chaotic type and sequence can be both different if the password changes. From chaotic sequence 2, all the chaotic sequences' secret keys are no longer depended on password. We draw into the "random number generator" to create chaotic sequence. As the process of generating random number is irreversible, the algorithm immensely increases the nondeterminacy of chaotic sequence. Therefore, repeating functioning this algorithm's encryption process will get quite other chaotic sequence to encrypt, and creates quite different plain text. The sub secret key sequence stream gotten by known-plaintext attack has no meaning.

3.5   Anti Attack Base on Chaos Model Reconstruction Analysis

The dividing information is only stored in password. When the analyst intercepts the cipher text and wants to use the chaotic model reconstruction attack, he must know how the data divided. However, through the infusibility analysis we know that the encrypted voice is quite similar to noise. The attacker can hardly know whether the data is divided or the size of the blocks. Therefore the only way to get the dividing rules is to analysis exhaustively, and compares the results with the obtained chaos type library one by one. It is obviously that this method has no practical value when the data of speech signal is very large. Therefore, this algorithm can overcome the attack based on chaos model's reconstruction.

4   Conclusions

The algorithm analysis and testing results show that this algorithm has large key space, good infusibility, perfect randomness of cipher text and high security. Besides, this algorithm can defend the secret analysis such as known-plaintext attacking, attack based on chaos model reconstruction, and is suitable for the encryption of voice communication.

Acknowledgements. This work is supported by Aero-Science Fund of China (2009ZD53045), Science and Technology Development   Project of Shaanxi Province Project(2010K06-22g), Basic research fund of Northwestern Polytechnical University (GAKY100101), and R Fund of   College of Software and Microelectronics of NPU (2010R001).

References

1. Zhao, L.: Speech Signal Processing. Machinery Industry Press (2006)
2. Li, K.H., Wang, D.Z., Dong, X.M.: Practical Cryptography and Computer Data Security. Northeastern University Press (2001)
3. Wang, G.Y., Yu, X.L., Chen, S.G.: The Control of Chaos, Synchronization and Utilization. National Defense Industry Press (2001)
4. Zhang, H., Wang, X.F., Li, C.H.: A Fast Image Encryption Algorithm Based on Chaos System and Henon Mapping. Computer Research and Development 42(12), 2137–2142 (2005)

The Research and Improvement of Path Optimization in Vehicle Navigation

Bo Huang, Weiming Wu, and Yuning Zheng

Computer Science Department, Beijing University of Posts and Telecommunications, China

Abstract. The shortest path finding problem is the core of route optimization algorithm. And the shortest path finding is one of the most important modules of the traffic navigation system. There are many kinds of route optimization algorithms, including some famous algorithms like: Dijkstra, Bellman-Ford, Floyd-Warshall, A* algorithm and so on. Even the same algorithm has a variety of different implementations. This paper provides an improved route optimization algorithm by improving and researching on the memory occupation and searching time consumption of Dijkstra algorithm. Finally, simulation results show that the improved algorithm has a higher efficiency than the original.

Keywords: Dynamic Route Optimization, Shortest Path Finding, Vehicle Navigation, Dijkstra Algorithm.

1   Introduction

In recent years, the rapid development of Computer, Communications, GIS (Geographic Information System) and GPS (Global Positioning System) has been a strong technological support for vehicle navigation, which also makes the vehicle navigation system an accelerated development.

Currently, the path optimization in vehicle navigation system is more complex than simple shortest path problem. Many scholars have proposed many types of path optimization algorithm. But these algorithms are mostly used in static road networks. However, in the actual road network, road weights are not fixed, but changes over time. And in practice, many algorithms cost a lot of memory and computing time. When the number of users reaches a certain level, there will be performance sharp reduce, even system crashes or other issues.

2   Classic Algorithms of Route Optimization

The prototype of path optimization algorithm is the shortest path problem. The shortest path problem has always been a hot research subject in computer science, operations research, transportation engineering and geographic information system. Classic graph theory and constant development of computer data structure and algorithm make new effective shortest path algorithm emerge.

D. Jin and S. Lin (Eds.): Advances in MSEC Vol. 2, AISC 129, pp. 203–207.
springerlink.com   © Springer-Verlag Berlin Heidelberg 2011

Among current path optimization algorithms, three of which are comparatively better: TQQ, DKA, and DKD. TQQ algorithm is based on graph theory, with two FIFO queues to achieve a double-ended queue structure to support the search process. The last two algorithms are based on Dijkstra, using bucket structure to increase the search speed of permanent markers. However, these algorithms exchange space for time, which may lead to the collapse of memory with the increase of consumers and concurrent access.

The classic Dijkstra's Shortest Path Algorithm is achieved by using adjacent matrix, storing the path information from any node to another, namely to store n^2 path information for the road network with n nodes. A large number of insignificant and repeated information lies between when n is comparatively large, which would lead to serious memory waste, and the large time cost on path optimization could not satisfy people's need for path optimization gradually.

3   Improved Algorithm

During path optimization, the moving direction is always from source node to target node. If connecting the source node and target node with a line, and then the angle between the final path and the line would mostly between -90 and 90. Therefore, a property: path angle is added to every node's path information. The path angle of each path refers to the starting node as the origin, and the level of the right is 0 degrees, increasing counterclockwise, angle of orientation between 0 and 360. (Figure 1)

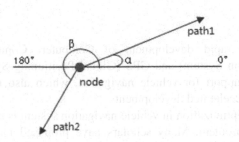

Fig. 1. Node's path angle

User-defined structure is used to store node and path information in road network.(Figure 2)

edge	nextEdge	node
⌐nodeSource : node*	⌐pEdge : edge*	⌐roadName : string
⌐nodeDirect : node*	⌐edgeAngle : signed int	⌐edges : nextEdge[]
⌐edgeLength : unsigned long		

Fig. 2. User-defined store structure

During the path search, first get the angle α from source node to target node, and then find the path which path's angle is between α-90 and α+90 in all paths of source node, and find all the next node, taking which as the source node to find the shortest path in the same way.

4   Description of Algorithm

(1). determining the starting node N_{res} and the ending node N_{des}, using set S to store passed node, and storing N_{res} to S and copying N_{res} to N_{temp},

(2). connecting the coordinates of N_{temp} and N_{des} with line, calculating the angle α, and using set G to store all the node related to the process of path optimization, and storing N_{temp} to T,

(3).starting from N_{temp}, if N_{temp}==N_{des}, entering step 5, or storing all the node between angle from α-90 to α+90 to G, and taking all the qualified adjacent node as N_{temp}, repeating step 3 until reaching N_{des};

(4).S={N_{res}}, G={other nodes}, setting up array D[Num] to store the shortest distance information from N_{res} to all the node, all of distances initialized to MAX_LENGTH

(5).Selecting a node N_{temp} angled between α-90 and α+90 from G to store in S and modifying the correspond shortest path in D[Num],

(6).Taking N_{temp} as the midpoint to modify the shortest path in D[Num]; supposing the distance between N_{res} and N_{temp} is E_t, between the node in G and N_{temp} is E_i; if E_t+E_i < D[i], then D[i] = E_t+E_i,

(7).Repeating step 5 and 6 until S including all the node;

(8).Marking all the nodes passed by the shortest path from N_{res} to N_{des}, and marking the whole path on the map and presenting the total path length and search time.

Before improving algorithm, filter the node with the opposite direction to the target one and store the real passing node selected from the layer information of map, which could save memory. During the process of path optimization, filter part of adjacent nodes among origin starting node and midpoint using angle between α-90 and α+90 to further decreasing search time.

5   Simulation

In order to compare the improved algorithm with the traditional Dijkstra algorithm on the search time, this paper achieves the two algorithms' simulation.

5.1   Simulation Environment

Develop Tools: Microsoft Visual Studio 2005, Super Map 2008
    Operation System: Windows XP.
    Take map information of part of Heping district in Tianjin. Road network's node data and path information using the default layer information of Super Map.

5.2   Simulation Result

Using the traditional Dijkstra algorithm and the improved algorithm mentioned in this paper to search the shortest path between the same source node and target node respectively and the result is as follows:

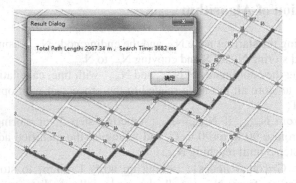

Fig. 3. Dijkstra algorithm's search result

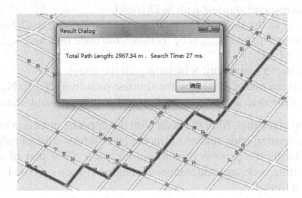

Fig. 4. Improved algorithm's search result

The shortest path length got from the two algorithms both are 2967.34m, but the time cost by Dijkstra algorithm is 3682ms, and the improved algorithm costs 27ms, achieving a great advance in search time, because the improved algorithm limits all the angles within 180°directing the target node, namely to save half of the time for every search to achieve great improvement in search time in the end.

Using user-defined data store structure to store the information of road network, takes up small space. Limiting the path angle speeds up the path optimization search time. And the programming is not difficult, which could satisfy the basic request for current shortest path optimization. The improved algorithm is a comparatively good improvement for Dijkstra.

References

1. Yu, D., Zhang, Q., Ma, S., Fang, T.: Optimized Dijkstra Algorithm. Computer Engineering 22, 145–146 (2004)
2. Xiao, J., Xie, J.: Technology of Dynamic Routing Based on Dijkstra Algorithm and Simulation. Computer Measurement & Control. 18, 1669–1672 (2010)
3. Xia, S., Han, Y.: An Improved Implementation of Shortest Path Algorithm in GIS. Bulletin of Surveying and Mapping 9, 40–42 (2004)
4. Sun, Q., Shen, J., Gu, J.: An Improved Algorithm of the Dijkstra Algorithm. Wan Fang Data 03, 99–101 (2002)
5. Li, J., Li, Q., Yang, L.: Modeling of Urban Traffic System Based on Dynamic Stochastic Fluid Petri Net. Journal of Transportation Systems Engineering and Information Technology 01, 134–139 (2010)
6. Dong, J., Gao, H.: Design and Implementation of Path Optimization Algorithm in Dynamic Networks based on PDA. Journal of Zhengzhou University of Light Industry 24, 93–95 (2009)
7. Zhang, Z., Hu, H., Zhong, G., Zheng, C.: SuperMap GIS Application and development tutorial. Wuhan University Press (2006)
8. Program Fan BBS, http://bbs.pfan.cn/

Research and Design of Communications Protocol in APTS Based on a Hybrid Model

Xiangbao Su, Weiming Wu, and Yuning Zheng

Computer Science Department, Beijing University of Posts and Telecommunications, China

Abstract. With the development of communications technology, recently, lots of attempts related to Advanced Public Transport System have come forth at home and abroad. Through the exploration of wireless communications, this paper puts forward a hybrid model of GPRS and WCDMA for the APTS. It introduces a UDP-based communications protocol named RTalk, which isolates data application from specific data transportation. The protocol can keep the efficiency and credibility of the communications process in wireless network environment. It has been used in the actual project and has achieved certain results.

Keywords: Advanced Public Transport System, Wireless communications, application layer communications protocol, RTalk.

1 System Framework

Generally speaking, the APTS (Advanced Public Transport System) is mainly composed of dispatch center system, scheduling branch system, vehicle terminal and wireless network. Based on the functions of the vehicle terminal, APTS can be constructed in two different ways, the central framework and the distributed

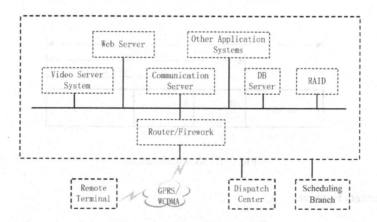

Fig. 1. System framework

D. Jin and S. Lin (Eds.): Advances in MSEC Vol. 2, AISC 129, pp. 209–213.

framework. The terminals in distributed framework system are more intelligent and can make some decisions. While terminals in the later system have limited functions of collecting data and executing the schedule. The dispatch center system commits computing and policy decision. And the later system has been widely used in practical applications for lower construction cost and higher reliability.

Based on the central framework system, we designed a protocol named RTalk. The actual system networked in a hybrid model with GPRS and WCDMA. In this way, the system utilized the economy of GPRS and the high-bandwidth of WCDMA, as well as enhanced the system's reliability with a redundant channel. We can see the system framework in figure 1.

2   Protocol Design

2.1   Protocol Architecture

In the architectural, the features related to communications details are virtualized to data transport layer, while the parts bounded up with specific applications are belonged to data application layer. Data control layer locates between the two layers, provided a division from application service and data delivery. In figure 2 and 3, we can get the relation between the system model and the classical TCP/IP architecture.

APTS   TCP/IP

Data Application Layer		Application Entity	Application Layer
Data Control Layer		RTalk	
		UDP	Transport Layer
Data Transport Layer		IP	Internet Layer
		Bearer	Network Access Layer

Fig. 2. Network architecture

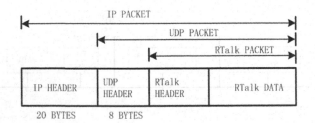

Fig. 3. RTalk encapulations

2.2   Packets Format

RTalk has two categories of packets: data packets and control packets. Data packets carry the actual payload from application, while control packets are used to maintain the communications process. Control packets are separated into response packets and hello packets. Meanwhile the response packets include communications response packets and business response packets.

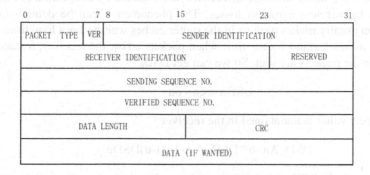

Fig. 4. RTalk packets format

2.3   Communications Process

While building the connection, server starts firstly. And the client will send hello packet when it need data service. Before hello response packet arrives and timer is stimulated, the client will send hello packets periodically. The server gets ready for data transport after sending hello response packet. Every hello packet must get feedback to avoid response packet missing. Either of the sides can send a close packet to quit the delivery. And the heartbeat packets are sent for the close connection mechanism. This process is described in figure 5. The sending and receiving entities share the same UDP port. The receiver is responsible for processing the trigger events.

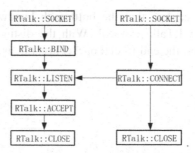

Fig. 5. Communications process

2.4   Congestion Mechanism

RTalk utilizes a closed circle algorithm to resolve congestion problems. The algorithm can track the dynamic status of the network and accommodate sending rate. The basic receiving and sending mechanism was demonstrated as follows:

The receiver gathers the information about network status, checks the losses and judges the different losses. Then feedback is delivered to the sender. The sender accommodates the sending rate according to the feedback.

Generally the delay of single direction (DSD) caused by congestion is bigger than it caused by wireless error bit losses. The phenomenon can be demonstrated that congestion usually makes packets in the router caches waiting for a long time.

RTalk makes ξ stand for the time when packets arrive the receiver, δ stand for the time when the packets are sent. So we can get DSD_n :

$$DSDn = \xi n - \delta n . \tag{1}$$

An expect value is maintained in the receiver:

$$DSD\_An = \alpha * DSD\_A \, n\text{-}1 + (1-\alpha)DSDn. \tag{2}$$

In this formula DSD is a parameter measured in the communications process. The parameter 'a' equals to 0.95. Now we can get the ratio:

$$\Theta_n = DSD_n/DSD\_A_{n-1}. \tag{3}$$

RTalk uses a filter to smooth the noise occured in the network:

$$\tau_n = \beta * \tau_{n-1} + (1-\beta)\theta_n. \tag{4}$$

β is equivalent with 0.125. When the system checked out packets losses, if $\tau_n > 1$, congestion occurred. The sending rate should be drop. Otherwise, the losses were caused by wireless link error and the sending rate needn't be regulated.

2.5   Test Results

Figure 6 and Figure 7 explain that the bottleneck bandwidth utilization remained above 90% while using RTalk protocol. With the distinction between congestion losses and wireless losses, the end-to-end packet loss rate is higher compared with the pure TCP environment.

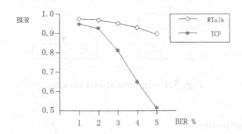

Fig. 6. Bottleneck bandwidth utilization rate(BUR) and wireless Bit Error Rate

BER PLR	0.01	0.02	0.03	0.04	0.05
TCP	0.022	0.026	0.033	0.040	0.050
RTalk	0.042	0.044	0.046	0.049	0.050

Fig. 7. Wirless Bit Error Rate (BER) and end-to-end packet loss rate (PLR)

3   Conclusions

With the flexibility in application layer, RTalk is proved to be a reliable data transfer protocol based on UDP. The protocol effectively utilizes the bandwidth in wireless network environment. With the application in Xi'an Advanced Public Transport System, RTalk fully meets the system demands in video scheduling. After one-year's trial operation, the system proved to be effective, stable, and reliable.

References

1. 3GPP TS 25.321, MAC protocol specification, V4.7.0 (December 2002)
2. Holma, H., Toskala, A.: WCDMA for UMTS, 2nd edn. Wiley (2002)
3. Song, C., Cosman, P.C., Voelker, G.: End-to-end Differentiation of Congestion. IEEE/ACM Transaction on Networking 11(5), 703–717 (2003)
4. IBM, The Globalization of Traffic Congestion: IBM 2010 Commuter Pain Survey. IBM (June 2010)
5. Hofmann, C., Weindorf, M., Wiesel, J.: Integration of GIS as a Component in Federated Information Systems. In: International Archives of Photogrammetry and Remote Sensing, vol. 33, pp. 1173–1180 (2000)
6. Gurtov, A., Passoja, M., Aaito., O., Raitola, M.: Multi-Layer Protocol Tracing in a GPRS Network. IEEE Fall VTC, 124–130 (2004)
7. Hannah, M.: Integrated Technology Services Publications And Communications strategy (2007)
8. Bettstetter, C., Hans-Jorg, Eberspacher, V.J.: GSM Phase 2+General Packet Radio Service GPRS: Architecture, protocols, and Air Interface

Fig. 7. Wirless Bit Error Rate (BER) and end-to-end packet loss rate (PLR)

5   Conclusions

With the flexibility in application layer, RTalk is proved to be a reliable data transfer protocol based on UDP. The protocol effectively utilizes the bandwidth in wireless network environment. With the application in XTan Advanced Public Transport System, RTalk fully meets the system demands in video-scheduling. After one-year's trial operation, the system proved to be effective, stable and reliable.

References

1. 3GPP TS 25.321 MAC protocol specification, V6.2.0 (December 2002)
2. Holma H., Toskala A.: WCDMA for UMTS, 2nd edn. Wiley 2007.
3. Song, C. Gossen, P.C. Voelker, C.: End-to-end Differentiation of Congestion. IEEE/ACM Transaction on Networking 13(5), 703–717 (2003).
4. IBM: The Globalization of Traffic Congestion. IBM 2010 Commuter Pain Survey. IBM (June 2010).
5. Hofmann C., Wendorff A., Wloka. J.: Integration of GIS as a Component in Federated Information Systems. The International Archives of Photogrammetry and Remote Sensing, vol. 33, pp. 1274–1150 (2000).
6. Gurtov A., Passoja, M., Aatto, O., Raitola, M., Multer-Laser Protocol Tracing in a GPRS Network. IEEE Fall VTC, 129–169 (2004).
7. Haumin, M.: Integrated Technology, Services, Publications, And Communications Strategy (2003).
8. Brismeister C., Haw-John, Ferenbacher, V.I.: GSM Phase 2: General Packet Radio Service GPRS, Architecture, protocols, and Air Interface.

The Optimum Professional Liability Insurance Contract Design for Supervising Engineer

Bao-Long Li[1,2]

[1] The Engineering Management Department at Luoyang Institute of Science and Technology,
Luoyang City, 471023, China
[2] Management School of Tianjin University, Tianjin City, 300072, China
libaolong@yahoo.com.cn

Abstract. Because there is a principal-agency relationship in professional liability insurance for supervising engineer, there is a need to introduce principal-agency theory into this field. We used the common hypothesis in principal-agency theory for reference to establish model structure, and we treated the compensation of insurance company as the incentive contract. The conclusion is that, when compensation amount takes place, the optimum effort level would lead to that insurance company need pay insurance premium as compensation under information symmetry. However, due to the universal existence of information asymmetry, the optimum effort level cannot be realized, so the compensation amount of insurance company under information asymmetry should be less than the compensation amount of insurance company under information symmetry.

Keywords: Professional liability insurance, Incentive contract, Supervising engineer, Principal-Agent.

1 Introduction

According to [1], in 1800 Ocotillo v. wlb Group, a real estate developer (Ocotillo), commenced development of a town house project in Phoenix on a parcel of property bounded on one side by the Arizona Canal. Ocotillo agreed to pay a design firm (WLB) for surveying, engineering, and landscape architecture services on the project pursuant to a written agreement between Ocotillo's design/build contractor and WLB. However, professional liability insurance (PLI) for supervising engineer appeared lately in USA until 1948. In China, the engineering supervising was begun in 1984[2], and, in 2002, Shanghai local government made experiments about the PLI for supervising engineer. Because there is a principal-agency relationship in PLI for supervising engineer, there is a need to introduce principal-agency theory into this field. From one side, we pay attention to the generality of the principal-agency theory; from another side, we pay attention to the PLI for supervising engineer. Therefore, we focus on the principal-agency problem in the field of PLI for supervising engineer.

D. Jin and S. Lin (Eds.): Advances in MSEC Vol. 2, AISC 129, pp. 215–220.
springerlink.com   © Springer-Verlag Berlin Heidelberg 2011

2   The Hypothesis of Model

In this principal-agency problem, the utility maximum problem faced by insurance company is as following:

$$\text{Max}:\overline{V} = \text{Max}:\{P\!\int V\left(W_1 - S(X(a,\theta)) + Z\right)g(\theta)d\theta + (1-P)V\left(W_1 + Z\right)\}$$

$$\text{IR: } \overline{U} \geq \overline{U}_1 \text{, i.e.}$$

$$P\!\int U\left(W_2 - Y(a,\theta) + S(X(a,\theta)) - Z\right)g(\theta)d\theta + (1-P)U\left(W_2 - Z\right)$$

$$\geq P\!\int U\left(W_2 - Y(a,\theta)\right)g(\theta)d\theta + (1-P)U\left(W_2\right) \tag{1}$$

$$\text{IC: } Max:\overline{U} \text{, i.e.}$$

$$\text{Max}:\overline{U} =$$

$$\text{Max}:\{P\!\int U\left(W_2 - Y(a,\theta) + S(X(a,\theta)) - Z\right)g(\theta)d\theta + (1-P)U\left(W_2 - Z\right)\}$$

$\overline{V}$ is the expected utility of insurance company, $\overline{U}$ is the expected utility of supervising engineer with insurance coverage, and $\overline{U}_1$ is the expected utility of supervising engineer without insurance coverage. a is the effort level of supervising engineer. The bigger a is, the higher the effort level is. θ is the natural state. The bigger θ is, the higher the natural state is. $G(\theta)$ is the distribution function of θ, and $g(\theta)$ is the density function of θ. a and θ are invisible, but provide a visible result $X(a,\theta)$. We believe that incentive contract $S(X)$ (i.e. $S(X(a,\theta))$) is a kind of design for PLI contract, which means that insurance company makes decision about compensation according to the visible result $X(a,\theta)$ and this kind of compensation could be treated as incentive. From the perspective of insurance company, when compensation does not happen, $S(X)$ is equal to 0; when compensation takes place, the amount of compensation could be treated as relative 'incentive contract'. Therefore, we can get $0 \leq S(X) \leq Z/\pi$. Z is insurance premium, $\pi(0 < \pi < 1)$ is insurance price, and Z/π is insurance amount needed by supervising engineer. V $(V' > 0, V'' < 0)$ and U $(U' > 0, U'' < 0)$ respectively represent $v - N - M$ expected utility function of insurance company and supervising engineer. W_1 and W_2 respectively represent the initial wealth level of insurance company and supervising engineer. $P(0 < P < 1)$ is the probability that the loss of supervising engineer takes place. Y is possible compensation amount that supervising engineer has to pay to the project owner.

3   The Optimum PLI Contract Design for Supervising Engineer under Information Symmetry

Under information symmetry, in the case of the optimum effort level, the problem faced by insurance company is how to select a and $S(X(a,\theta))$, so (1) could be rewritten as following:

$$\begin{cases} \text{Max} \quad : \overline{V} \\ \text{s.t.(IR)} \overline{U} \ge \overline{U}_1 \end{cases} \tag{2}$$

We create Lagrange function as following:

$$\begin{aligned} L(S(X(a,\theta)),a) &= \overline{V} - \lambda(\overline{U}_1 - \overline{U}) = \overline{V} + \lambda(\overline{U} - \overline{U}_1) \\ &= P\!\int V(W_1 - S(X(a,\theta)) + Z)g(\theta)d\theta + (1 - P)V(W_1 + Z) \\ &+ \lambda P\!\int U(W_2 - Y(a,\theta) + S(X(a,\theta)) - Z)g(\theta)d\theta + \lambda(1 - P)U(W_2 - Z) \\ &- \lambda P\!\int U(W_2 - Y(a,\theta))g(\theta)d\theta - \lambda(1 - P)U(W_2) \end{aligned} \tag{3}$$

Solving equation (3), we get:

$$S(X(a,\theta)) = Z \tag{4}$$

And

$$\begin{aligned} V'(W_1 - S(X(a,\theta)) + Z) &= \lambda U'(W_2 - Y(a,\theta)) \\ &= \lambda U'(W_2 - Y(a,\theta) + S(X(a,\theta)) - Z) \end{aligned} \tag{5}$$

Based on equation (5), conclusion could be drawn that Pareto optimum effort level could be realized under information symmetry. With or without insurance coverage, when compensation takes place, the effort level of supervising engineer will make his marginal utility equal, and also make the relation between his marginal utility and marginal utility of insurance company to be fixed. Nevertheless, because of $S(X(a,\theta)) = Z$ (Z is insurance premium and $S(X(a,\theta))$ is compensation amount), when compensation amount takes place, the optimum effort level would lead to that insurance company need pay insurance premium as compensation under information symmetry.

4   The Optimum PLI Contract Design for Supervising Engineer under Information Asymmetry

4.1   The Solution of IC

In order to solve IC, we add several constraint conditions to transfer IC as following:

$$\begin{cases} \text{Max:} \overline{U} = \\ \text{Max:} \{P\!\int U(W_2 - Y(a,\theta) + S(X(a,\theta)) - Z)g(\theta)d\theta + (1-P)U(W_2 - Z)\} \\ \text{s.t.} \quad Z/\pi < \int Y(a,\theta)g(\theta)d\theta \\ \quad P\!\int S(X(a,\theta))g(\theta)d\theta \le Z/\pi \\ \quad P\!\int S(X(a,\theta))g(\theta)d\theta \ge 0 \end{cases} \tag{6}$$

In equation (6), due to the existence of adverse selection, partial insurance must require $Z/\pi < \int Y(a,\theta)g(\theta)d\theta$. $P\!\int S(X(a,\theta))g(\theta)d\theta$ is the expected compensation of insurance company. Due to $0 \le S(X) \le Z/\pi$, it would lead to $0 \le P\!\int S(X(a,\theta))g(\theta)d\theta \le Z/\pi$. We create Lagrange function as following:

$$\begin{aligned} L_1(S(X(a,\theta)),a) &= P\!\int U(W_2 - Y(a,\theta) + S(X(a,\theta)) - Z)g(\theta)d\theta \\ &+ (1-P)U(W_2 - Z) - \lambda_1(Z/\pi - \int Y(a,\theta)g(\theta)d\theta) \\ &- \lambda_2(P\!\int S(X(a,\theta))g(\theta)d\theta - Z/\pi) + \lambda_3 P\!\int S(X(a,\theta))g(\theta)d\theta \end{aligned} \tag{7}$$

Solving the equation (7), we get:

$$U'(W_2 - Y(a,\theta) + S(X(a,\theta)) - Z) = \lambda_2 - \lambda_3 \tag{8}$$

And

$$PU'(W_2 - Y(a,\theta) + S(X(a,\theta)) - Z) = P(\lambda_2 - \lambda_3) = \lambda_1 \tag{9}$$

4.2   The Improvement and Solution of the Model

Based on equation (8) and (9), (1) could be rewritten as following:

$$\begin{cases} \text{Max} : \overline{V} \\ \text{s.t.(IR)} \overline{U} \ge \overline{U}_1 \\ \text{(IC)} U'(W_2 - Y(a,\theta) + S(X(a,\theta)) - Z) = \lambda_2 - \lambda_3 \\ \text{(IC)} PU'(W_2 - Y(a,\theta) + S(X(a,\theta)) - Z) = P(\lambda_2 - \lambda_3) = \lambda_1 \end{cases} \tag{10}$$

We create Lagrange function as following:

$$L_2(S(X(a,\theta)),a) = P\int V(W_1 - S(X(a,\theta)) + Z)g(\theta)d\theta + (1-P)V(W_1 + Z)$$
$$+ \lambda_4 P\int U(W_2 - Y(a,\theta) + S(X(a,\theta)) - Z)g(\theta)d\theta + \lambda_4(1-P)U(W_2 - Z)$$
$$- \lambda_4 P\int U(W_2 - Y(a,\theta))g(\theta)d\theta - \lambda_4(1-P)U(W_2)$$
$$+ \lambda_5(U'(W_2 - Y(a,\theta) + S(X(a,\theta)) - Z) - \lambda_2 + \lambda_3)$$
$$+ \lambda_6(PU'(W_2 - Y(a,\theta) + S(X(a,\theta)) - Z) - \lambda_1)$$

(11)

Solving the equation (11), we get:

$$\frac{V'(W_1 - S(X(a,\theta)) + Z)}{U'(W_2 - Y(a,\theta) + S(X(a,\theta)) - Z)} =$$

$$\lambda_4 + \left(\frac{\lambda_5}{P} + \lambda_6\right)\frac{U''(W_2 - Y(a,\theta) + S(X(a,\theta)))}{U'(W_2 - Y(a,\theta) + S(X(a,\theta)) - Z)G(\theta)}$$

(12)

Based on equation (12), we find that the Pareto optimum risk sharing does not realized. Also, based on equation (12), conclusion could be drawn that, under information asymmetry, the effort level selected by supervising engineer is lower than Pareto optimum effort level. Because the effort level of supervising engineer and the natural state are invisible from the standpoint of insurance company, as to compensation $S(X(a,\theta))$, the compensation amount of insurance company under information asymmetry should be less than the compensation amount of insurance company under information symmetry. Based on equation (4), the optimum effort level would lead to that insurance company need pay insurance premium as compensation under information symmetry. Therefore, under information asymmetry, when compensation takes place, the compensation amount $S(X(a,\theta))$ of insurance company should be less than insurance premium Z, which means $S(X(a,\theta)) < Z$. Therefore, insurance company need control the compensation amount. Under this precondition, insurance company would get profit from supervising engineer and also prevent possible loss from supervising engineer.

5   Conclusion

This paper introduces principal-agency theory into this field of the PLI for supervising engineer. In part 2, we use the common hypothesis in principal-agency theory for reference, and the point of creativity is that we treat the compensation of insurance company as the incentive contract. In part 3, we concentrate on the Pareto optimum effort level under information symmetry, and the conclusion is that, when compensation amount takes place, the optimum effort level would lead to that insurance company need pay insurance premium as compensation under information symmetry (i.e. $S(X(a,\theta)) = Z$). However, due to the universal existence of information asymmetry, the effort level of supervising engineer and the natural state are invisible from the standpoint of insurance company, which will lead to $S(X(a,\theta)) < Z$. This ($S(X(a,\theta)) < Z$) means that insurance company need control the compensation amount. Under this precondition, insurance company would get profit from supervising engineer and also prevent possible loss from supervising engineer.

References

1. Loulakis, M.C., McLaughlin, L.P.: Enforceability of Limitation of Liability Provisions. Civil Engineering 78(4), 92 (2008)
2. Wang, X.J., Huang, J.: The relationships between key stakeholders' project performance and project success: Perceptions of Chinese construction supervising engineers. International Journal of Project Management 24, 253–260 (2006)

The Influence of Project Owner on Professional Liability Insurance for Supervising Engineer

Bao-Long Li[1,2]

[1] The Engineering Management Department at Luoyang Institute of Science and Technology,
Luoyang City, 471023, China
[2] Management School of Tianjin University, Tianjin City, 300072, China
libaolong@yahoo.com.cn

Abstract. In the principal-agency problem of the professional liability insurance (PLI) for supervising engineer, because supervising engineer is hired by the project owner, both project owner and insurance company are the principals of supervising engineer. Therefore, there is a need to study this kind of principal-agency structure and the influence of project owner on PLI for supervising engineer. We used the common hypothesis in principal-agency theory for reference to establish model structure. The conclusion is that, due to the influence of project owner, the compensation of insurance company under the existence of project owner should be less than the compensation under the absence of project owner. Also, because project owner cannot force supervising engineer to select effort level, the effort level of supervising engineer would further decrease.

Keywords: Professional liability insurance, Incentive contract, Supervising engineer, Project Owner, Principal-Agency.

1   Introduction

In China, the engineering supervising was begun in 1984[1]. Due to short history of engineering supervising, there are many problems in practice. Making sure that as many risks as possible are recognized and managed is good practice in any project[2]. Professional liability insurance (PLI) for supervising engineer is a way to manage risk. Because principal-agency relation is involved in PLI for supervising engineer, there is a need to introduce principal-agency theory into the field of PLI for supervising engineer. Principals and agents pursue cooperative relationships, yet they have differing goals and attitudes toward risk[3]. In this paper, the principal-agency relation is between insurance company and supervising engineer. However, supervising engineer is hired by the project owner, so there are two principals in this field. We would concentrate on this kind of principal-agency relation structure, and study the influence of project owner on PLI for supervising engineer.

2   The Hypothesis of Model

In this principal-agency problem, the utility maximum model structure is as following:

$$\text{Max}: \overline{V}_1 = \text{Max}: \{\int V_1(W_3 + T - S_1(Q(a,\theta)))g(\theta)d\theta\}$$

$$\text{Max}: \overline{V} = \text{Max}: \{P\int V(W_1 - S(X(a,\theta)) + Z)g(\theta)d\theta + (1-P)V(W_1 + Z)\}$$

$$\text{IR: } \overline{U} \geq \overline{U}_1 \text{, i.e.}$$

$$P\int U(W_2 - Y(a,\theta) + S(X(a,\theta)) - Z + S_1(Q(a,\theta)))g(\theta)d\theta$$
$$+ (1-P)\int U(W_2 - Z + S_1(Q(a,\theta)))g(\theta)d\theta \geq \qquad (1)$$
$$P\int U(W_2 - Y(a,\theta) + S_1(Q(a,\theta)))g(\theta)d\theta + (1-P)\int U(W_2 + S_1(Q(a,\theta)))g(\theta)d\theta$$

$$\text{IC: Max}: \overline{U} \text{, i.e.}$$

$$\text{Max}: \overline{U} = \text{Max}: \{P\int U(W_2 - Y(a,\theta) + S(X(a,\theta)) - Z + S_1(Q(a,\theta)))g(\theta)d\theta$$
$$+ (1-P)\int U(W_2 - Z + S_1(Q(a,\theta)))g(\theta)d\theta\}$$

$\overline{V}_1$ is the expected utility of project owner, $\overline{V}$ is the expected utility of insurance company, $\overline{U}$ is the expected utility of supervising engineer with insurance coverage, and $\overline{U}_1$ is the expected utility of supervising engineer without insurance coverage. a is the effort level of supervising engineer. The bigger a is, the higher the effort level is. θ is the natural state. The bigger θ is, the higher the natural state is. $G(\theta)$ is the distribution function of θ, and $g(\theta)$ is the density function of θ. a and θ are invisible, but provide a visible result $X(a,\theta)$ for insurance company and another visible result $Q(a,\theta)$ for project owner. We believe that incentive contract $S(X)$ (i.e. $S(X(a,\theta))$) is a kind of design for PLI contract, which means that insurance company makes decision about compensation according to the visible result $X(a,\theta)$ and this kind of compensation could be treated as incentive. From the perspective of insurance company, when compensation does not happen, $S(X)$ is equal to 0; when compensation takes place, the amount of compensation could be treated as relative 'incentive contract'. Therefore, we can get $0 \leq S(X) \leq Z/\pi$. Z is insurance premium, $\pi(0 < \pi < 1)$ is insurance price, and Z/π is insurance amount needed by supervising engineer. Following the same idea, the incentive contract of project owner is $S_1(Q(a,\theta))$, which means that project owner makes decision about payment according to the visible result $Q(a,\theta)$ and this kind of payment could be treated as incentive.

$V_1(V_1'>0, V_1''<0)$, $V(V'>0, V''<0)$ and $U(U'>0, U''<0)$ respectively represent v-N-M expected utility function of project owner, insurance company and supervising engineer. W_1, W_2 and W_3 respectively represent the initial wealth level of insurance company, supervising engineer and project owner. $P(0 < P < 1)$ is the probability that the loss of supervising engineer takes place. Y is possible compensation amount that supervising engineer has to pay to the project owner. T $(T>0)$ is the production of engineering supervising to project owner. In Chinese practice, when supervising engineer's mistakes bring loss to project owner, the highest compensation amount is the income of engineering supervising, which means $\int Y(a,\theta)g(\theta)d\theta < \int S_1(Q(a,\theta))g(\theta)d\theta$. $\int Y(a,\theta)g(\theta)d\theta$ is the total loss undertaken by supervising engineer. $Max[\int Y(a,\theta)g(\theta)d\theta$ is the highest compensation amount undertaken by supervising engineer. Therefore, the loss exceeding $Max[\int Y(a,\theta)g(\theta)d\theta$ would be undertaken by project owner.

Obviously, the expected utility of insurance company is not only decided by insurance company itself and supervising engineer, but also decided by project owner. In addition, both insurance company and project owner are the principals of supervising engineer, but insurance company and project owner have different objectives and different utility maximum problem. Therefore, it is difficult to solve the utility maximum model structure showed in (1). In order to solve this problem, we adopt a two-step method to solve the utility maximum model structure showed in (1). The first step is to solve the principal-agency problem between project owner and supervising engineer, i.e. the utility maximum problem of project owner. Then, the result of first step is treated as constraint condition in second step to solve the principal-agency problem between insurance company and supervising engineer, i.e. the utility maximum problem of insurance company.

3   The Solution of IC

In order to solve the utility maximum model structure showed in (1) step by step, the IC has to be solved firstly. Therefore, we add several constraint conditions to transfer IC as following:

$$
\begin{cases}
\text{Max}: \overline{U} = \text{Max}: \{P\int U\left(W_2 - Y(a,\theta) + S(X(a,\theta)) - Z + S_1(Q(a,\theta))\right)g(\theta)d\theta \\
+ (1-P)\int U\left(W_2 - Z + S_1(Q(a,\theta))\right)g(\theta)d\theta\} \\
\text{s.t.}\quad Z/\pi < \int Y(a,\theta)g(\theta)d\theta \\
\qquad P\int S(X(a,\theta))g(\theta)d\theta \le Z/\pi \\
\qquad P\int S(X(a,\theta))g(\theta)d\theta \ge 0 \\
\qquad \int Y(a,\theta)g(\theta)d\theta < \int S_1(Q(a,\theta))g(\theta)d\theta \\
\qquad \int S_1(Q(a,\theta))g(\theta)d\theta < T \\
\qquad 0 < \int S_1(Q(a,\theta))g(\theta)d\theta
\end{cases}
\tag{2}
$$

In (2), due to the existence of adverse selection, partial insurance must require $Z/\pi < \int Y(a,\theta)g(\theta)d\theta$. $P\int S(X(a,\theta))g(\theta)$ is the expected compensation of insurance company. Due to $0 \le S(X) \le Z/\pi$, it would lead to $0 \le P\int S(X(a,\theta))g(\theta)\ d\theta \le Z/\pi$. The highest compensation amount is the income of engineering supervising, which would lead to $\int Y(a,\theta)g(\theta)d\theta < \int S_1(Q(a,\theta))g(\theta)d\theta$. Also, the income of supervising engineer must be less than the income of project owner, which would lead to $0 < \int S_1(Q(a,\theta))g(\theta)d\theta < T$. Solving (2), we get:

$$U'(W_2 - Y(a,\theta) + S(X(a,\theta)) - Z + S_1(Q(a,\theta))) = \lambda_2 - \lambda_3 \qquad (3)$$

And

$$PU'(W_2 - Y(a,\theta) + S(X(a,\theta)) - Z + S_1(Q(a,\theta))) = \lambda_1 - \lambda_4 \qquad (4)$$

And

$$(1 - P)U'(W_2 - Z + S_1(Q(a,\theta))) = -\lambda_1 + \lambda_5 - \lambda_6 \qquad (5)$$

4   The Solution of First Step Principal-Agency Problem

In first step, the utility maximum problem of project owner is showed as following:

$$\begin{cases} \text{Max}: \overline{V}_1 = \text{Max}: \{\int V_1(W_3 + T - S_1(Q(a,\theta)))g(\theta)d\theta\} \\ \text{s.t.(IC)}U'(W_2 - Y(a,\theta) + S(X(a,\theta)) - Z + S_1(Q(a,\theta))) = \lambda_2 - \lambda_3 \\ \text{(IC)}PU'(W_2 - Y(a,\theta) + S(X(a,\theta)) - Z + S_1(Q(a,\theta))) = \lambda_1 - \lambda_4 \\ \text{(IC)}(1 - P)U'(W_2 - Z + S_1(Q(a,\theta))) = -\lambda_1 + \lambda_5 - \lambda_6 \end{cases} \qquad (6)$$

Solving (6), we get:

$$\begin{aligned} &\int V_1'(W_3 + T - S_1(Q(a,\theta)))g(\theta)d\theta = \\ &\lambda_7 U''(W_2 - Y(a,\theta) + S(X(a,\theta)) - Z + S_1(Q(a,\theta))) \\ &+ \lambda_8 PU''(W_2 - Y(a,\theta) + S(X(a,\theta)) - Z + S_1(Q(a,\theta))) \\ &+ \lambda_9 (1 - P)U''(W_2 - Z + S_1(Q(a,\theta))) \end{aligned} \qquad (7)$$

Equation (7) is the result of first step principal-agency relation (project owner and supervising engineer), and the constraint condition in second step.

5   The Solution of Second Step Principal-Agency Problem

In second step, the utility maximum problem of insurance company is showed as following:

$$\text{Max}: \overline{V} = \text{Max}: \{ P\int V(W_1 - S(X(a,\theta)) + Z)g(\theta)d\theta + (1-P)V(W_1 + Z)\}$$

$$\text{s.t.(IR)} P\int U(W_2 - Y(a,\theta) + S(X(a,\theta)) - Z + S_1(Q(a,\theta)))g(\theta)d\theta$$

$$+ (1-P)\int U(W_2 - Z + S_1(Q(a,\theta)))g(\theta)d\theta \geq$$

$$P\int U(W_2 - Y(a,\theta) + S_1(Q(a,\theta)))g(\theta)d\theta$$

$$+ (1-P)\int U(W_2 + S_1(Q(a,\theta)))g(\theta)d\theta$$

$$\text{(IC)} U'(W_2 - Y(a,\theta) + S(X(a,\theta)) - Z + S_1(Q(a,\theta))) = \lambda_2 - \lambda_3$$

$$\text{(IC)} PU'(W_2 - Y(a,\theta) + S(X(a,\theta)) - Z + S_1(Q(a,\theta))) = \lambda_1 - \lambda_4 \qquad (8)$$

$$\text{(IC)}(1-P)U'(W_2 - Z + S_1(Q(a,\theta))) = -\lambda_1 + \lambda_5 - \lambda_6$$

$$\int V_1'(W_3 + T - S_1(Q(a,\theta)))g(\theta)d\theta =$$

$$\lambda_7 U''(W_2 - Y(a,\theta) + S(X(a,\theta)) - Z + S_1(Q(a,\theta)))$$

$$+ \lambda_8 PU''(W_2 - Y(a,\theta) + S(X(a,\theta)) - Z + S_1(Q(a,\theta)))$$

$$+ \lambda_9(1-P)U''(W_2 - Z + S_1(Q(a,\theta)))$$

Solving (8), we get:

$$\frac{V'(W_1 - S(X(a,\theta)) + Z)}{U'(W_2 - Y(a,\theta) + S(X(a,\theta)) - Z + S_1(Q(a,\theta)))} =$$

$$\lambda_{10} + (\frac{\lambda_{11}}{P} + \lambda_{12})\frac{U''(W_2 - Y(a,\theta) + S(X(a,\theta)) - Z + S_1(Q(a,\theta)))}{U'(W_2 - Y(a,\theta) + S(X(a,\theta)) - Z + S_1(Q(a,\theta)))G(\theta)} \qquad (9)$$

$$- \lambda_{14}(\frac{\lambda_7}{P} + \lambda_8)\frac{U'''(W_2 - Y(a,\theta) + S(X(a,\theta)) - Z + S_1(Q(a,\theta)))}{U'(W_2 - Y(a,\theta) + S(X(a,\theta)) - Z + S_1(Q(a,\theta)))G(\theta)}$$

Due to $U' > 0$ and $U'' < 0$, we get $U''' > 0$. Also, λ_7, λ_8, λ_{11}, λ_{12} and λ_{14} are bigger than 0, we get:

$$(\frac{\lambda_{11}}{P} + \lambda_{12})\frac{U''(W_2 - Y(a,\theta) + S(X(a,\theta)) - Z + S_1(Q(a,\theta)))}{U'(W_2 - Y(a,\theta) + S(X(a,\theta)) - Z + S_1(Q(a,\theta)))G(\theta)} < 0 \qquad (10)$$

And

$$-\lambda_{14}(\frac{\lambda_7}{P}+\lambda_8)\frac{U'''(W_2-Y(a,\theta)+S(X(a,\theta))-Z+S_1(Q(a,\theta)))}{U'(W_2-Y(a,\theta)+S(X(a,\theta))-Z+S_1(Q(a,\theta)))G(\theta)}<0 \quad (11)$$

Obviously, (11) could be viewed as the influence of project owner on PLI for supervising engineer. Therefore, under information asymmetry, due to the influence of project owner on PLI for supervising engineer, the compensation of insurance company under the existence of project owner should be less than the compensation under the absence of project owner. Because project owner cannot force supervising engineer to select effort level, the effort level of supervising engineer would further decrease.

6   Conclusion

In the principal-agency problem of the PLI for supervising engineer, because supervising engineer is hired by the project owner, both project owner and insurance company are the principals of supervising engineer.   Therefore, project owner plays a very important role. This paper is a study about the influence of project owner on PLI for supervising engineer. In part 2, we use the common hypothesis in principal-agency theory for reference to establish the utility maximum model structure, and the point of creativity is that we treat the compensation of insurance company as the incentive contract and also teat the payment of project owner as the incentive contract. In part 3, we add several constraint conditions to transfer IC so that IC can be solved properly. In part 4 and 5, we adopt a two-step method to solve the principal-agency problem of the PLI for supervising engineer, and find out the influence of project owner on PLI for supervising engineer. The conclusion is that, due to the influence of project owner on PLI for supervising engineer, the compensation of insurance company under the existence of project owner should be less than the compensation under the absence of project owner.   Also, because project owner cannot force supervising engineer to select effort level, the effort level of supervising engineer would further decrease.

References

1. Wang, X.J., Huang, J.: The relationships between key stakeholders' project performance and project success: Perceptions of Chinese construction supervising engineers. International Journal of Project Management 24, 253–260 (2006)
2. Zaghloul, R., Hartman, F.: Construction contracts: the cost of mistrust. International Journal of Project Management 21, 419–424 (2003)
3. Dahlstrom, R., Ingram, R.: Social networks and the adverse selection problem in agency relationships. Journal of Business Research 56, 767–775 (2003)

The Long Term Incentive Mechanism of the Professional Liability Insurance for Supervising Engineers

Bao-Long Li

The Engineering Management Department at Luoyang Institute of Science and Technology,
Luoyang City, 471023, China
Management School of Tianjin University, Tianjin City, 300072, China
libaolong@yahoo.com.cn

Abstract. In terms of seeking long term development direction and finding the influential factors of the professional liability insurance for supervising engineers, this paper would establish the long term incentive mechanism of the professional liability insurance for supervising engineers. The result is that, the effective incentive condition (both the credit level of the supervising engineers and the possibility that insurance company finds supervising engineers adopting hidden action) would be the influential factors towards whether supervising engineers adopt hidden action or not. If this effective incentive condition can be founded, incentive mechanism would function in the long term development of PLI for supervising engineers.

Keywords: Professional liability insurance, Supervising engineers, Credit, The long term incentive mechanism.

1 Introduction

In China, the engineering supervising was begun in 1984, but until 1996, the engineering supervising became popular in Chinese construction industry.[1-3] In the field of engineering supervising, the supervising engineers play a very important role. Due to the vital liability of the supervising engineers, it will lead to enormous loss of human life and wealth that they make mistakes. In order to help supervising engineers better work and transfer the huge risk undertaken by supervising engineers, in 2002, Shanghai local government made experiments about the professional liability insurance (PLI) for supervising engineers. In terms of seeking long term development direction and finding the influential factors of the PLI for supervising engineers, this paper would establish the long term incentive mechanism of the PLI for supervising engineers.

2 The Establishment of the Model

There are two kind of supervising engineers: one is supervising engineers obeying the rule; another is supervising engineers disobeying the rule. The supervising engineers obeying the rule do not adopt hidden action and maliciously acquire the compensation

D. Jin and S. Lin (Eds.): Advances in MSEC Vol. 2, AISC 129, pp. 227–232.

from insurance company. However, the supervising engineers disobeying the rule have the possibility to adopt hidden action and maliciously acquire the compensation from insurance company. The insurance company does not know the type which supervising engineers belong to, but, once insurance company finds that supervising engineers adopt hidden action, insurance company would believe that supervising engineers disobey the rule. When the supervising engineers adopt hidden action, the following equation is the expected utility function for supervising engineers.

$$w = b[(1 - D)\alpha - \alpha^e]$$ (1)

In equation (1), α ($0 \leq \alpha \leq 1$) is the utility level of supervising engineers through adopting hidden action. D ($0 \leq D \leq 1$) is the possibility that insurance company finds supervising engineers adopting hidden action. $(1 - D)\alpha$ represents the expected utility level of supervising engineers through adopting hidden action. Once insurance company finds supervising engineers adopting hidden action, the supervising engineers would not get the utility through adopting hidden action. α^e ($0 \leq \alpha^e \leq 1$) is the expected utility level for supervising engineers that insurance company forecasts, when supervising engineers adopt hidden action. b is the type of supervising engineers. $b = 0$ represents supervising engineers obeying the rule; $b = 1$ represents supervising engineers disobeying the rule. The prior probability of $b = 0$ is p_0 and the prior probability of $b = 1$ is $1 - p_0$, which mean, when $t = 0$, the probability that insurance company believes that supervising engineers obey the rule is p_0 and the probability that insurance company believes that supervising engineers disobey the rule is $1 - p_0$. Therefore, p_0 can represent the credit level of the supervising engineers to some extent.

With the development of PLI for supervising engineers, multiple stage game takes place. According to $\alpha_t^e (1 - D)$, supervising engineers select the α_t, then obtain the utility $w_t = b[(1 - D)\alpha_t - \alpha_t^e]$ at t stage. Once α_t is selected, based on α_t, D and α_t^e, insurance company would forecast α_{t+1}^e to influence the selection and utility of supervising engineers at $t + 1$ stage. Therefore, reasonable supervising engineers can decide the choice at very stage according to total utility maximization. If multiple stage game repeats in T stages, y_t is the probability that supervising engineers do not adopt hidden action, and x_t is the probability that insurance company believe that supervising engineers do not adopt hidden action. Then, under equilibrium, $x_t = y_t$. Therefore, if supervising engineers do not adopt hidden action at t stage, posterior probability that insurance company believe supervising engineers obeying the rule is as following:

$$p_{t+1} = p_{t+1}(b = 0|\alpha_t = 0) = \frac{p_t \times 1}{p_t \times 1 + (1 - p_t)x_t} \geq p_t$$ (2)

In equation (2), p_t is the probability that supervising engineers obey the rule at t stage, and 1 is the probability that supervising engineers do not adopt hidden action. Based on equation (2), if supervising engineers do not adopt hidden action, the probability that insurance company believe supervising engineers obeying the rule will be increased. If $x_t < 1$, equation (2) can be rewritten as following:

$$P_{t+1} = P_{t+1}\left(b = 0 \middle| \alpha_t = 0\right) = \frac{p_t \times 1}{p_t \times 1 + (1 - p_t)x_t} > p_t \tag{3}$$

If $x_t = 1$, equation (2) can be rewritten as following:

$$P_{t+1} = P_{t+1}\left(b = 0 \middle| \alpha_t = 0\right) = \frac{p_t \times 1}{p_t \times 1 + (1 - p_t)x_t} = p_t \tag{4}$$

If insurance company finds supervising engineers adopting hidden action, posterior probability that insurance company believe supervising engineers obeying the rule is as following:

$$P_{t+1} = E \times P_{t+1}\left(b = 0 \middle| \alpha_t = 1\right) = \frac{D \times p_t \times 0}{p_t \times 0 + (1 - p_t)x_t} = 0 \tag{5}$$

Based on equation (5), conclusion could be drawn that, if insurance company finds supervising engineers adopting hidden action, insurance company believe supervising engineers disobeying the rule.

3   The Design of Incentive Mechanism

At T stage (the final stage), there is no mean whether supervising engineers obey the rule or not. Therefore, supervising engineers must select to adopt hidden action and maliciously acquire the compensation from insurance company. The optimum choice of supervising engineers is $b = 1$ and $\alpha_T = 1$. When supervising engineers adopt hidden action, the expected utility level of supervising engineers that insurance company forecasts is $\alpha_t^e = 1 - p_t$. The expected utility level of supervising engineers is as following:

$$W_T = (1 - D)\alpha_T - \alpha_T^e = (1 - D) \times 1 - (1 - p_T) = p_T - D \tag{6}$$

Solving equation (6), we can get $\partial W_T / \partial p_T = 1 > 0$. p_T is the probability that insurance company believes supervising engineers obeying the rule at T stage, and p_T can represent the credit level of the supervising engineers to some extent. Therefore, $\partial W_T / \partial p_T = 1 > 0$ shows that the utility level of supervising engineers disobeying the rule is the increase function of the credit level, so that supervising engineers disobeying the rule have enough incentive to establish their market credit.

At $T - 1$ stage, if supervising engineers disobeying the rule did not adopt hidden action before $T - 1$ stage, $p_{T-1} > 0$. The expected utility level of supervising engineers disobeying the rule forecasted by insurance company is as following:

$$\alpha_T^e = 1 \times (1 - p_{T-1})(1 - x_{T-1}) \tag{7}$$

In equation (7), 1 is the optimum selection of α for supervising engineers disobeying the rule, $(1 - p_{T-1})$ is the probability that supervising engineers disobey the rule, and $(1 - x_{T-1})$ is the probability that insurance company believes supervising engineers adopting hidden action. In order to simplify the problem, we only consider the situation of $y_t = 0$ (i.e. supervising engineers disobeying the rule do not adopt hidden action at t stage) and $y_t = 1$ (i.e. supervising engineers disobeying the rule adopt hidden action at t stage). This is because that, supervising engineers can only select $0 < y_t < 1$, when the utility is equal under the situation of $y_t = 0$ and $y_t = 1$. If supervising engineers disobeying the rule adopt hidden action at $T - 1$ stage ($y_{T-1} = 0$, i.e. $\alpha_{T-1} = 1$), we can get following equation.

$$p_T = \frac{(1 - D)p_{T-1}}{p_{T-1} + (1 - p_{T-1})x_{T-1}} \tag{8}$$

The expected utility level of supervising engineers adopting hidden action forecasted by insurance company is α^e_{T-1}, then the total utility level of supervising engineers disobeying the rule at $T - 1$ stage is as following:

$$W_{T-1}(1) + W_T(1) = [(1 - D) \times 1 - \alpha^e_{T-1}] + p_T - D = 1 - 2D - \alpha^e_{T-1} + p_T$$
$$= 1 - 2D - \alpha^e_{T-1} + \frac{(1 - D)p_{T-1}}{p_{T-1} + (1 - p_{T-1})x_{T-1}} \tag{9}$$

Following the same analysis, if supervising engineers disobeying the rule do not adopt hidden action at $T - 1$ stage ($y_{T-1} = 1$, i.e. $\alpha_{T-1} = 0$), the total utility level of supervising engineers disobeying the rule is as following:

$$W_{T-1}(0) + W_T(1) = -\alpha^e_{T-1} + p_T - D = -D - \alpha^e_{T-1} + \frac{p_{T-1}}{p_{T-1} + (1 - p_{T-1})x_{T-1}} \tag{10}$$

Considering equation (9) and (10), if supervising engineers disobeying the rule do not adopt hidden action at $T - 1$ stage (i.e. $\alpha_{T-1} = 0$ is more optimum than $\alpha_{T-1} = 1$), the following inequation must be founded.

$$1 - 2D - \alpha^e_{T-1} + \frac{(1 - D)p_{T-1}}{p_{T-1} + (1 - p_{T-1})x_{T-1}} \leq -D - \alpha^e_{T-1} + \frac{p_{T-1}}{p_{T-1} + (1 - p_{T-1})x_{T-1}} \tag{11}$$

Solving inequation (11), we can get:

$$D \geq \frac{p_{T-1} + (1 - p_{T-1})x_{T-1}}{2p_{T-1} + (1 - p_{T-1})x_{T-1}} \tag{12}$$

Under equilibrium, the expectation of insurance company is equal to the choice of supervising engineers, i.e. $x_{T-1} = y_{T-1}$. Therefore, if supervising engineers disobeying the rule do not adopt hidden action at $T-1$ stage (i.e. $y_{T-1} = 1$), we can get $x_{T-1} = 1$ and $p_T = (1-D)p_{T-1}$. The inequation (12) could be rewritten as following:

$$D \geq \frac{1}{p_{T-1}+1} \quad 或 \quad p_{T-1} \geq \frac{1-D}{D} \tag{13}$$

Based on inequation (13), if insurance company believes that the probability of supervising engineers obeying the rule satisfies inequation (13) at $T-1$ stage, supervising engineers disobeying the rule would pretend to be obeying the rule. This means that, if the market credit level of supervising engineers is high enough, supervising engineers would have great incentive to keep market credit level. Whereas, supervising engineers with low market credit level have no incentive to maintain and improve their market credit level. Also, if $p_{T-1} = (1-D)/D$, no matter supervising engineers how to select (i.e. $y_{T-1} \in [0,1]$), the result is optimum. Due to $p_{T-1} \leq 1$, we can get $1/2 \leq D \leq 1$. Nevertheless, when D becomes smaller, p_{T-1} becomes bigger. This means that, when the possibility that insurance company finds supervising engineers adopting hidden action is smaller, the credit level of the supervising engineers must be higher.

4   Conclusion

Through above demonstration, both the credit level of the supervising engineers (i.e. p_{T-1}) and the possibility that insurance company finds supervising engineers adopting hidden action (i.e. D) would be the influential factors towards whether supervising engineers adopt hidden action or not. Therefore, $(1-D)/D \leq p_{T-1} \leq 1$ and $1/2 < D < 1$ can represent the effective incentive condition. If this effective incentive condition can be founded, incentive mechanism would function in the long term development of PLI for supervising engineers. Nevertheless, if the credit level of the supervising engineers is high enough or, the possibility that supervising engineers adopt hidden action goes down. And, if the possibility that insurance company finds supervising engineers adopting hidden action is high enough, the possibility that supervising engineers adopt hidden action goes down. According to this effective incentive condition, insurance company can decrease the damage caused by supervising engineers under information asymmetry, and prevent that supervising engineers maliciously acquire the compensation from insurance company. Obviously, at T stage (the final stage), incentive mechanism of PLI for supervising engineers has no effect, and supervising engineers would definitely adopt hidden action. Also, if $p_0 < (1-D)/D$, the supervising engineers disobeying the rule would adopt hidden action to increase their utility at every stage.

From the perspective of insurance company, from one side, increasing the credit level of the supervising engineers can help realize the self incentive of supervising engineers and realize the purpose that insurance company motivates supervising engineers. From another side, even if the credit level of the supervising engineers is high enough, insurance company still need increase the possibility that insurance company finds supervising engineers adopting hidden action. This is because that $(1-D)/D \leq p_{T-1} \leq 1$ and $1/2 \leq D \leq 1$ must be met at the same time. When $D = 1/2$, $p_{T-1} = 1$. This means that, when the possibility that insurance company finds supervising engineers adopting hidden action is the lowest, the credit level of the supervising engineers must be the highest, which does not make sense in real economic life. When $D = 1$, $p_{T-1} = 0$. This means that, when the possibility that insurance company finds supervising engineers adopting hidden action is the highest (i.e. all hidden action of supervising engineers can be found), there is no meaning whether supervising engineers have credit or not. This ($D = 1$ and $p_{T-1} = 0$) could be viewed as the situation on perfect market. Therefore, insurance company not only increases the credit level of the supervising engineers, but also increases the possibility that insurance company finds supervising engineers adopting hidden action.

Moreover, due to that incentive mechanism of PLI for supervising engineers has no effect at T stage (the final stage), insurance company has to reconsider the PLI for supervising engineers. For example, if supervising engineers are inclined to be engaged in engineering supervising for a short term or near to be retired, insurance company should not satisfy their PLI requirements. In PLI practice, insurance company should establish history recording files for every supervising engineer, then assess the credit level of the supervising engineer, and finally establish the evaluation system for the credit level of the supervising engineers. Following that, insurance company should redesign PLI contracts for the different credit level of the supervising engineers, in order that the supervising engineers with high credit level can get insurance coverage at lower cost.

References

1. Wang, C., Ren, H.: Sustainable Development Problems and Corresponding Suggestions of Construction Supervision Sector. Journal of Chongqing Jianzhu University 29(3), 145–149 (2007)
2. Wang, X.J., Huang, J.: The relationships between key stakeholders' project performance and project success: Perceptions of Chinese construction supervising engineers. International Journal of Project Management 24, 253–260 (2006)
3. Yang, X.-F.: Discussions on the garden greening engineering standards in the application of supervision quality control. Journal of Beijing Forestry University 26(3), 71–75 (2004)

Research on the Externality of the Real Estate Market in Social Security Sector

Bao-Long Li[1,2]

[1] The Engineering Management Department at Luoyang Institute of Science and Technology, Luoyang City, 471023, China
[2] Management School of Tianjin University, Tianjin City, 300072, China
libaolong@yahoo.com.cn

Abstract. Due to the existence of the externality, it is difficult to develop the real estate market in social security sector. Therefore, there is a need to study the externality of the real estate market in social security sector. We used the Cobb-Douglas production function for reference to establish the model. We get conclusion that, because the price that people want to pay is less than social optimum price, the quantity (or the quality) of real estate bought by people has to be less than social optimum quantity (or quality). Due to that the quantity (or the quality) of real estate bought by people decreases, the quantity (or the quality) of real estate provided by real estate developer must decrease.

Keywords: Externality, Real estate, Social security, Cobb-Douglas production function.

1   Introduction

In recent years, the price of the real estate goes up rapidly in China, and this kind of phenomenon becomes extremely serious throughout China. The increase of the price of the real estate not only influences the stability of Chinese society, but also leads to a large number of homeless people. In order to build better society and make people live and work in peace and contentment, the decision makers of Chinese government started to develop the real estate market in social security sector. However, as a kind of public good, due to the existence of the externality, it is difficult to develop the real estate market in social security sector. Therefore, there is a need to study the externality of the real estate market in social security sector.

2   The Hypothesis of Model

The production function of one real estate developer is Cobb-Douglas production function, which is showed as following:

$$Q = AL^\alpha K^\beta \tag{1}$$

D. Jin and S. Lin (Eds.): Advances in MSEC Vol. 2, AISC 129, pp. 233–237.
springerlink.com   © Springer-Verlag Berlin Heidelberg 2011

A ($A>0$) is the scale parameter. L is the input of labor of this real estate developer, and K is the related input of capital of this real estate developer. α ($0<\alpha<1$) is the production elasticity of labor, β ($0<\beta<1$)is the production elasticity of capital. According to [1-4], we hypothesize that E ($E=L^{a-\alpha}K^{b-\beta}$, $\alpha<a<1$, $\beta<b<1$) represents the production externality. Therefore, the production function of this real estate developer could be written as following:

$$Q_1 = AL^\alpha K^\beta E = AL^\alpha K^\beta L^{a-\alpha} K^{b-\beta} = AL^a K^b \tag{2}$$

Generally speaking, production and cost are two sides of one problem in supply theory, so the cost function of this real estate developer could be hypothesized as following:

$$C = f(Q)+FC = wL+rK+FC \tag{3}$$

$f(Q)$ ($f(Q)=wL+rK$) is the variable cost. FC is the fixed cost. w represents the price of the input of labor L, and r represents the price of the input of capital K.

Therefore, without considering the externality, the profit of this real estate developer could be showed as following:

$$\pi = Q-C = AL^\alpha K^\beta - f(Q)-FC = AL^\alpha K^\beta - wL-rK-FC \tag{4}$$

With considering the externality, the profit of this real estate developer could be rewritten as following:

$$\pi_1 = Q_1 - C = AL^a K^b - f(Q_1)-FC = AL^a K^b - w_1 L - r_1 K - FC \tag{5}$$

3   Solution of Model

Without considering the externality, in order to get maximum profit, we get:

$$\partial\pi/\partial L = \alpha AL^{\alpha-1}K^\beta - w = 0 \Rightarrow w = \alpha AL^{\alpha-1}K^\beta \tag{6}$$

And

$$\partial\pi/\partial K = \beta AL^\alpha K^{\beta-1} - r = 0 \Rightarrow r = \beta AL^\alpha K^{\beta-1} \tag{7}$$

Without considering the externality, in order to get maximum profit, we get:

$$\partial\pi_1/\partial L = aAL^{a-1}K^b - w_1 = 0 \Rightarrow w_1 = aAL^{a-1}K^b \tag{8}$$

And

$$\partial\pi_1/\partial K = bAL^a K^{b-1} - r_1 = 0 \Rightarrow r_1 = bAL^a K^{b-1} \tag{9}$$

Due to $\alpha < a < 1$ and $\beta < b < 1$, we get:

$$\frac{w_1}{w} = \frac{aAL^{a-1}K^b}{\alpha AL^{\alpha-1}K^\beta} = \frac{aL^{a-\alpha}K^{b-\beta}}{\alpha} > 1 \tag{10}$$

And

$$\frac{r_1}{r} = \frac{bAL^a K^{b-1}}{\beta AL^\alpha K^{\beta-1}} = \frac{bL^{a-\alpha}K^{b-\beta}}{\beta} > 1 \tag{11}$$

Base on (10) and (11), conclusion could be drawn that, the price of the input of labor L should be higher with considering externality than without considering externality, the price of the input of capital K should also be higher with considering externality than without considering externality.

4   The Influence of Externality on the Real Estate Developer

In order to better understand the result showed in part 3, we use the input of labor L and its related price as an example to explain in Figure 1. With considering externality, the input of labor is L_1, and its related price is w_1. Without considering externality, the input of labor is L, and its related price is w.

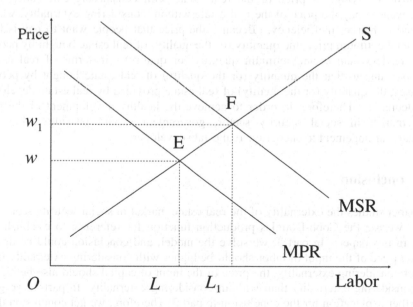

Fig. 1. The demand curve when deciding the labor input

When this real estate developer makes decision about the input of labor, the fact is that the increase of labor can increase the output of this real estate developer. If considering all the revenue (including external revenue), this real estate developer will make decision based on marginal social revenue (MSR). L_1 is the optimum labor input, and w_1 is the related optimum price. However, in reality, this real estate developer only considers private revenue, and makes decision based on marginal private revenue (MPR). Therefore, L is the input of labor, and w is the related price. From social perspective, the input of labor is too small, and the effective input of labor should be the social optimum labor input L_1 based on marginal social revenue (MSR) and should not be L based on marginal private revenue (MPR). Due to the existence of externality, the internal production and sales of this real estate developer will present an insufficient state. The input of labor is less than the social optimum labor input, and the related price of labor is lower than the reasonable price, which will produce no efficiency. The analysis of capital input is similar to the analysis of labor input.

The input of labor and its price should be higher with considering externality than without considering externality, and the input of capital and its price should also be higher with considering externality than without considering externality. This result means that, in the real operation of this real estate developer, the financial return for the labor and the capital of this real estate developer is obviously lower than the return that the whole society should 'pay' to this real estate developer. However, when people make decision about buying the real estate in social security sector, people do not consider paying the price of the real estate with considering externality, and people want to pay the price of the real estate without considering externality, which will lead to market inefficiency. Because the price that people want to pay is less than social optimum price, the quantity (or the quality) of real estate bought by people has to be less than social optimum quantity (or quality). In terms of real estate developer, due to that the quantity (or the quality) of real estate bought by people decreases, the quantity (or the quality) of real estate provided by real estate developer must decrease. Therefore, in order to achieve the healthy development of the real estate market in social security sector, government should provide a kind of systematic arrangement to encourage real estate developer.

5   Conclusion

This paper studies the externality of the real estate market in social security sector. In part 2, we use the Cobb-Douglas production function for reference to establish the model in this paper. In part 3, we solve the model, and conclusion could be drawn that, the price of the input of labor should be higher with considering externality than without considering externality, the price of the input of capital should also be higher with considering externality than without considering externality. In part 4, we give the further explanation for the conclusion in part 3. Therefore, we get conclusion that, because the price that people want to pay is less than social optimum price, the quantity (or the quality) of real estate bought by people has to be less than social optimum quantity (or quality). In terms of real estate developer, due to that the quantity (or the quality) of real estate bought by people decreases, the quantity (or the quality) of real estate provided by real estate developer must decrease.

References

1. Benhabib, J., Farmer, R.E.A.: Indeterminacy and increasing returns. Journal of Economic Theory 63, 19–41 (1994)
2. Itaya, J.: Can environmental taxation stimulate growth? The role of indeterminacy in endogenous growth models with environmental externalities. Journal of Economic Dynamics & Control 32, 1156–1180 (2008)
3. Meng, Q., Yip, C.K.: On indeterminacy in one-sector models of the business cycle with factor-generated externalities. Journal of Macroeconomics 30, 97–110 (2008)
4. Wang, X.-Q., Li, B.-L., Fan, Z.-Q.: Analysis of the Externality of Engineering Supervision. Soft Science 24(7), 6–8 (2010)

The Research on Urban Planning Data Storage System

Xiaosheng Liu[1] and Tingli Wang[2]

[1] School of Architectural and Surveying Engineering,
JiangXi University of Science and Technology
[2] School of Applied Science, JiangXi University of Science and Technology
lxs9103@163.com, 228679192@qq.com

Abstract. We study the physical truth of urban planning and choose the regulatory planning data as research object. First, we have system design. Second, we write code which adapt to the CAD data based on industry rules in favor of the change between different data formats; then base on .NET platform and C# language, through the ArcGIS Engine and Objects ARX2007 component, we change the data format by the rules. The factor is put the dwg data into personal geodatabase feature class and insert into ArcSDE Geodatabase, so we can manage it by ArcSDE. The function of data preprocessing and transfer from CAD data format to GIS format intact and GIS data management are achieved. The system not only can reduce the complex of urban planning work greatly, but also make marked progress in efficiency and accurate of urban planning.

Keywords: Urban Planning, CAD, GIS, Geodatabase.

1   Introduction

In the traditional way of urban planning and management, urban planning departments usually use planning data files or popular database to handle data. The former is the CAD data file formats, but the disadvantage is that planning data attribute can not connect with spatial entity and also difficult in the data update, maintenance, query and so; the latter is the lack of database information can not be defined in accordance with the corresponding entity. Both management methods do not have the spatial query and analysis capabilities so that the information can not be shared, which will result in some data confusion and loss. This will not only waste a lot of manpower and resources, but also affected the smooth progress of urban planning. Therefore, the establishment of a "planning results a map" is the industry trend of urban planning, urban planning and the data storage is to achieve "a map" the management of the basic method. In view of this situation, the authors conducted a data storage system of urban planning research and development.

2   System Development Technology and Overall Designs

As the data in the current urban planning mainly accounted for the majority of CAD data which is different from the GIS format. Although ArcGIS software provides the

D. Jin and S. Lin (Eds.): Advances in MSEC Vol.2, AISC 129, pp. 239–243.

functions for CAD data format conversion, but in the actual conversion process, whether it is the conversion of graphical data or attribute data, we can not be fully converted that would result in some loss of information so that aliasing. In view of this situation, we used the method in the city planning data storage system is that: First, we write code which adapt to the CAD data base on industry rules in favor of the change between different data formats; then based on the .NET platform and C# language, through the ArcGIS Engine and Objects ARX2007 component, we change the data format by the rules. The factor is put the dwg data into personal geodatabase feature class and import into ArcSDE Geodatabase, so we can manage it by ArcSDE.

Taking into account the urban planning graphic data and attribute data in the data storage need to be encoded, pre-processing, conversion, storage and management, so urban planning data storage system can be divided into the planning data processing module and GIS data storage module. The overall structure of the system is shown in figure 1.

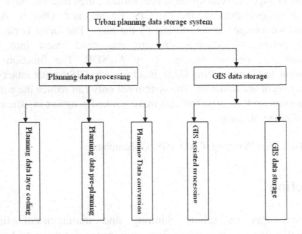

Fig. 1. System Structure

3   The Planning Data Processing

3.1   CAD Planning Data Coding

First we must encode the data in the CAD before the planning data preprocessing,. The system uses five digits to encode, the first and second digits stand for the land use classification code, the third and forth digits stand for the different layer code, the fifth digit stands for the feature type(0—point,1—line,2—polygon,3—note). According to the principles of this code, all the planning CAD data is encoded.

3.2   Urban Planning Data Pre-processing

Urban planning data including topology check and physical editing, topology check is a plan which checked topology errors and marked for the user to modify. Physical editing is edit planning data, user-friendly editing of topological errors entity.

3.3   Conversion between CAD and GIS Data

1) Conversion of basic geometric elements

The process is that according to the own characteristics coded of points, polylines, arcs, ellipses, circles and other geometric entities transform into GIS point, line and polygon feature.

(1) Conversion of point feature

Conversion of point feature includes CAD point feature changed into GIS point feature and CAD block changed into GIS point feature. The implementation process is basically the same, the different is the conversion factor to obtain the data is the point where the coordinates, and the block is to obtain the coordinates of its insertion point.

(2) Conversion of line feature

① Conversion of line feature includes CAD lines, circles, arcs, polylines, ellipses and elliptical arcs changed into a GIS-line feature. Because of a variety of graphic elements and their own characteristics, so their methods of the conversion and achieve the convert data are different.

② Conversion of line feature : just need to get a straight line graph elements start and end coordinate values will be able to achieve the conversion.

③ Conversion of Not closed polylines: Need to get the coordinates of multi-line start, end point and each node, and then create a Geodatabase line feature under the coordinates of these points.

④ Conversion of circles and arcs: Need to obtain the circle and arc starting angle, central angle, the center coordinates and radius for the conversion of these data. In fact, the conversion of circles and arcs are basically the same, but the circle is a central angle equal to 360 degrees special circumstances, they used the same conversion method.

⑤ Conversion of ellipse and ellipse arc: Need to obtain the coordinates of the center, start angle, central angle, rotation angle, long elliptical axis ratio of ellipse and ellipse arc, except rotation, other conversion data are available get ellipse and ellipse arc elements directly through ObjectARX. As the rotation angle of the graphic elements do not provide the data, and therefore to calculate the rotation of ellipse and ellipse arc. The method of conversion of ellipse and ellipse arc are the same, ellipse is just the starting angle of zero degrees in special circumstances.

(3) Conversion of polygon

Conversion of polygon includes the closed polylines, rectangles, and polygons in CAD are changed into polygon elements in GIS. As in CAD polylines, rectangles, and regular polygon are Polyline type, and all the elements to get are the coordinates of the vertices of each polygon, the methods are the same. There are two problems in CAD polygon converted : ① determine whether the polyline is closed.② determine whether the polygon is clockwise.

2) Conversion of note

The planning system will change text annotation data into the GIS point feature class, then label the feature class to label elements of the object.

3) Conversion of attribute data

Conversion of attribute data includes the conversion of planning data basic geometric attributes and expansion of data. The conversion of planning data basic geometric attributes includes layer, line, line width, color, material, etc ; The conversion of expansion of data includes according to the encoded value given to the GIS attributes and stored in the entity's XData.

4   GIS Data Storage

4.1   GIS Assisted Processing

GIS assisted processing is to display, query, edit and spatial analyze the data after CAD data format changed into GIS data format, it includes four areas: GIS data editing, symbolic, spatial query and spatial analysis.

4.2   Data Storage

The converted data are stored in Oracle Enterprise database based on SDE database engine for further planning applications. It includes: Connection and release of the database engine, database management of data sets and the non-planning data clean-up, version management, data storage, data output.

5   System Development and Implementation

The urban planning data storage system based on .NET and ArcGIS 9.3 platform, we use C# language as the development language, through the ArcGIS Engine and Objects ARX2007 component. The system implements the encoding pre-processing and conversion of planning data, as well as editing, query, analysis and storage management of GIS data conversion.

Click the planning data processing module, the system will automatically start CAD. At the same time loaded secondary development of planning storage items and planning storage toolbar menu. After data pre-processing and other operations, we implement the data conversion and then add the data which are already converted into GIS and process the data. They are shown in the figure 2 and figure 3:

Fig. 2. CAD data loading

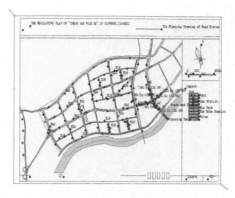

Fig. 3. The GIS data after converted

6 Conclusion

This system is the industry from the current reality of urban planning and in the understanding of user needs, we select. NET as development platform, C # as development language, with the ArcGIS Engine and OA component technology, the use of a powerful spatial database storage management functions to achieve the CAD data encoding and preprocessing, conversion of CAD data and GIS data and the management of storage. The system is already used in a city urban planning and construction bureau of geographic information systems, through the practical application we find that the system not only reduces the complexity of the work of planners and improve work efficiency, but also completes the CAD data to GIS data lossless conversion to achieve a dynamic, electronic data management to promote the development of urban planning geographic information systems.

Acknowledgement. The project is supported by National Natural Science Foundation of China (Grant No.41061041) and Natural Science Foundation of Jiangxi Province, China (Grant No.2010gzs0084).

References

1. Yun, C., Wei, Z.: The conversion technology and methods of AutoCAD data into GIS spatial data. Anhui Agricultural Sciences (2006)
2. Huang, J., Chen, W.: The cadastral data storage system based on AutoCAD platform. Geospatial Information (2006)
3. Wen, X., Zhong, W., Zhu, F., Lu, X.: The research of CAD to ARCGIS data storage system based on Geodatabase. Surveying and Mapping (2006)
4. Zhou, X., Jiao, D.: The solution of conversion of CAD data to GIS based on Geodatabase. Modern Surveying and Mapping (2004)
5. Chen, N., Shi, B.: The research and application of AutoCAD topographic map data into GIS spatial data. Mapping Bulletin (2005)
6. Zhong, S., Zheng, G.: Data conversion between AutoCAD and MAPGIS. Surveying and Mapping (2005)

Fig. 2. The GIS data after converted

6. Conclusion

This system is the industry, from the current reality of urban planning and in the understanding user needs, we select .NET as development platform, C # as development language, with the ArcGIS Engine and OA component technology, the use of a powerful spatial database storage management functions to achieve the CAD data encoding and preprocessing, conversion of CAD data and GIS data and the management of storage. The system is already used in a city urban planning and construction bureau of geographic information systems, through the practical application, we find that the system not only reduces the complexity of the work of planning and improve work efficiency, but also completes the CAD data to GIS data location conversion to achieve a dynamic, electronic data management to promote the development of urban planning geographic information systems.

Acknowledgement. The project is supported by National Natural Science Foundation of China (Grant No. 41061044) and Natural Science Foundation of Jinhgxi Province, China (Grant No. 20119a034x).

References

1. Yin, G., Wei, Z.: The conversion technology and methods bt AutoC AD data into GIS spatial data. Anhui Agricultural Sciences (2009)
2. Huang, J., Chen, W.: The archetral data storage system based on AutoCAD platform. Geomatic Information China
3. Nga, X., Zhou, W., Zhu, B.J. et.: The research of CAD to ARCGIS data storage system based on Geodatabase. Surveying and Mapping (2009)
4. Zhou, X., Han, D.: The solution to conversion of CAD data to GIS based on Geodatabase. Modern Surveying and Mapping (2007)
5. Chen, N., Shi, L.: The design and application of AutoCAD topographic map data into GIS application. Mapping Bulletin (2008)
6. Zhu, S., Yin, H.: Data conversion between AutoCAD and MAPGIS. Surveying and Mapping (2005)

Web3D Based Multimedia Software of Safety Knowledge for Children

Zhen Liu and YanJie Chai

Faculty of Information Science and Technology, Ningbo University, Ningbo 315211
liuzhen@nbu.edu.cn

Abstract. Safety education for children is still a urgent task in china, teaching children to learn safety knowledge in life is very important. With the development of Internet, there is a growing concern in web based e-learning software. Based on the psychological characteristic of children and constructivism learning theory, web3d is a good tool for developing education software for children. The design method of web3d software is introduced with traffic education and virtual digestive system, a finite state machine expresses the behavior state of a virtual character, and NPC characters are introduced to increase the interest of the scene. The Web3d based education software can run both on local PC and on the Internet, a child user can explore safety knowledge in interactive method. The demo system can make children safety education more vivid, interactive and interesting.

1 Introduction

Children's dangerous behavior can easily result in injury, there were a lot of cases about children's unexpected injury in china, In 2010, there were more that 1.92 hundred million vehicles in China. The Ministry of Public Security Traffic Management published that there were 219521 road traffic accidents with 65225 people killed and 254075 injured. Traditional safety education manner for children were not interesting, children could learn abstract safety regulation or watched safety education movies in schools. These methods seldom were suitable for the psychological characteristic of a child [1], educational effect was rarely visual and vivid, and was hard to arouse the notice of a child. In general, the ability of logic thought of a child has not growth completion, children usually like visual and interesting things by their psychological characteristic. There are many books on safety education, and the contents are too easy. Some other VCD and Flash animation for safety education have no interaction. In accordance with psychologist Jean Piaget [2], children's knowledge mainly come from cognitive construction of the objective world, and creating a learning situation is the key to increase learning interests.

There is growing on virtual reality based education software in recent years [4-5], and computer animation is playing a very important role. For examples, chemical education is set up by Web3D technology, in the virtual space, multinational people, who have different cultural backgrounds and educational levels, can communicate one another in Cyberspace [6]. Haller et al. developed a safety training system with virtual reality [7], Yoon et al. explored ecological food webs, and used the information to

D. Jin and S. Lin (Eds.): Advances in MSEC Vol. 2, AISC 129, pp. 245–250.
springerlink.com   © Springer-Verlag Berlin Heidelberg 2011

create a direct display on VRML/X3D browsers [8]. In design of modern city, visualization was presented to serve as an aid to marketing an urban commercial park. The visualization featured the ability to add new objects to the visualization and to view the visualization from different perspectives [9]. Belfore et al. simulated a garden landscape at Chiba University by using VRML. With three media representing the landscape (Photographs, Static images of VRM Land Walk-Through images by VRML), The results showed that the VRML images could be used as an effective stimulus for landscape assessment both on the Internet and in the laboratory [10]. Roussou et al. researched the interactive manner in education software, they thought that interactive manner enhanced the interests of users [11].Tomaz et al. developed some examples on biology kwnoledge by VRML and Web3D, he used Terragen tool to model virtual hills and   blue sky with clouds [12]. Cai et al. presented the research on VR-enhanced bio-molecular education using a gaming approach Learning of molecular biology may benefit from the technology of VR X games, which could stimulate the students' engagement and motivation in molecular biology learning [13]. Virtual oceanarium is set up in paper [14], in the oceanarium, a lot of fishes can move freely, users can learn more ocean knowledge. Liu developed 2D cartoon games for traffic education [15], Padgett et al. built fire education software for children [16].

On the basis of topology psychology [3], behavior is related to environment, children's dangerous behavior occurs in a given environment, a safety education software should show dangerous areas.

A believable virtual character should act well with animation scenario; the design of character modeling should suit to safety storytelling. An animation scenario has individuality that include personal living environment in a given region and culture. In a word, a virtual character should integrate with scenario as a whole system, which can express an individual safety story.

In this paper, virtual safety cases are shown by Interactive animation. we explore to construct a three-dimensional virtual scene, a child user can control his avatar in a virtual environment, this method offers enough interactions, and makes a child user to learn the correct safety knowledge easily.

2   Key Technology of the Education Software

Web3d is a PC based virtual reality, there are many Web3D software tools, such as VRML, VIRTOOLS, Shockwave3D, EON reality. In these tools, only VRML is open-source and free for users. VRML is the abbreviation of Virtual Reality Modeling Language. In 2001, a new international standard of VRML is named as X3D. VRML or X3D includes many concepts in virtual reality, they can run on any PC. X3D is a considerably more mature and refined standard than its VRML predecessor. VRML is an ASCII-based open source language; it includes a set of nodes that serves as utilization of geometry, light, viewpoint, sensor, etc. VRML integrates animation, spatial sound, collision detection and scripting together.

3DS MAX is one of the most popular three-dimensional modeling software, which can create many kinds of three-dimensional models in a virtual three-dimensional scene. If we finish a virtual world in 3DSMax, the virtual word can be exported as a VRML(VIRTOOLS) format file, we can explore the virtual world in Microsoft IE

Explorer, users can move in the virtual world. In 3DS MAX, some of VRML nodes are integrated, these nodes are: TouchSensor, ProxSensor, TimeSensor, SoundAudio,TouchSensor, Ancho, TimeSensor, Fog, LOD, InLine, NavInfo, Billboard. We can use these nodes to create a interactive animation. If we want to create advanced animation, we can use script programming to realize interactive functions. A script is as the following:

```
Script
    { URL []
            directOutput    FALSE
            mustEvaluate   FALSE
            eventIn      eventTypeName eventName
            eventOut eventTypeName eventName
            fieldTypeName fieldName initialValue
    }.
```

The URL specifies the program script to be executed. The program languages currently supported are JAVA, JavaScript, and VRML Script.

Virtools is another Web3d tool with powerful interactive functions. The most important feature of Virtools is the visual and flexible interactive design with behavior blocks. Each behavior block is a program block packaged with Specific role and function. We can drag Building Block and put it on the appropriate object or virtual characters, or using flow chart to determine the order of BB. In these ways, we can realize visual design of interactive scripts, and gradually compile a complete interactive virtual world.

We can use 3DS Max to model a virtual scene for a education software, all the geometry model of objects are drawn beforehand, and exported as VRML97 or Virtools format file. In recent years, many new functions are integrated in 3DS Max, complex 3D geometry models are easily created in 3DS Max that supports a rich modeling techniques. The steps of design a virtual landscape can be summarized as follows:

(1)Drawing of static objects with irregular shape: These objects are realized by adding noise effect on a regular object, an appropriate texture is added for representing the superficial vision effect.

(2)Drawing of a dynamic object in 3DS Max beforehand, if it is converted to VRML animation file, a dialog will appear on the screen, we should can select "PositionInterpolator" menu to create a animation VRML file.

(3) Drawing of creatures: a landscape can include some creature (for example, a virtual dog); a bone system is bound up in the geometry model of a virtual creature. The movement of a virtual creature is realized by adjusting position in different key frame, and the swinging of a creature is realized by setting up movements of skeleton.

(4) Programming of interactive animation: different viewpoints can be selected in a virtual landscape by viewpoint syntax in VRML97 language. Interaction in a VRML sample can be realized by adding VRML helper objects in 3DS Max development environment.

3　Two Examples of Safety Education Software

In the example, we use Virtools to realize a traffic education software, a virtual scene of traffic is set up, a user can control his avatar to move in the virtual scene (see Fig.1). The interactive animation integrates the sound effects in a real traffic environment, and a NPC character is a tutor. Virtual tutor can react to user's commands, such as answering user's questions. In particular situations, a virtual tutor can direct virtual environment and to tell some typical traffic accidents. For example, when a virtual tutor has perceived the avatar is violating traffic rule, he will send message and a car that is moving rapidly toward the avatar.

Fig. 1. A virtual avatar walks on a road and knocked by a car

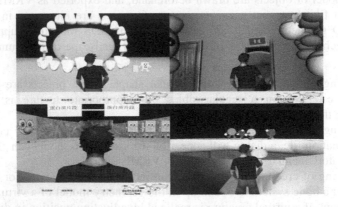

Fig. 2. The avatar walks into a digestive System

In another example, we develop a virtual digestive system in VRML, a user can act a avatar, he can walk into a digestive system, and some NPC character can tell the functions of digestive system.(see Fig.2).

4   Conclusion

The technique details of a teaching case are introduced in the paper, we illustrate the case to students on class, they like the case teaching method (see Fig.3). From the case, virtual reality is one of multimedia technology, and it will play an important role in safety education; Web3D technology is a PC based virtual reality. The relative researches on education software with Web3D tools are proposed. We use 3Ds Max to model a virtual landscape, in the case demo system, a user can interact with the virtual world, and he can control his avatar move in the virtual environment, and watch the virtual natural world. We test the software for a lot of children, they like to explore the demo system, and the virtual ecological landscape is vivid for children's learning. The preliminary result shows that the Web3D based software is useful for children's education, it can be put on a website, many users can interact with the virtual world.

Fig. 3. We test the software for pupils

Acknowledgements. The work described in this paper was co-supported in part by the National Natural Science Foundation of China (Grant no: 60973099), the Natural Science Foundation of Zhejiang Province (grant no: Y1091158), Zhejiang Public Technology Application Project(2011C23027).

References

1. Bernstein, D.A., Stewart, A.C., Roy, E.J., Wickens, C.D.: Psychology, 4th edn., pp. 360–361. Houghton Miffin Company, New York (1997)
2. Leslie, P.S., Steff, G., Jerry, E.G.: Constructivism in education. Lawrence Erlbaum Associates, Inc., Hillsdale (1995)
3. Gibson, J.J.: The ecological approach to visual perception. Lawrence Erlbaum Associates, Inc., Hillsdale (1986)
4. William, R.S., Alan, B.C.: Understanding Virtual Reality, Interface, Application, and Design. Elsevier Press, Amsterdam (2003)
5. Geroimenko, V., Chen, C. (eds.): Visualizing Information Using SVG and X3D. Springer-Verlag London Limited, London (2005)

6. Hiroshi, Y.: Virtual chemical education-agent oriented global education system for chemistry. Chemical Education 4(1) (2000)
7. Haller, M., Holm, R., Volkert, J., Wagner, R.: omVR - A Safety Training System for a Virtual Refinery. In: Proceedings of ISMCR 1999, Workshop of Virtual Reality and Advanced Human-Robot Systems, Tokyo, Japan, pp. 291–298 (1999)
8. Yoon, I., Yoon, S., Williams, R.J., Martinez, N.D., Dunne, J.A.: Interactive 3D visualization of highly connected ecological networks on the WWW. In: ACM Symposium on Applied Computing (SAC 2005), Multimedia and Visualization Section, pp. 1207–1217 (2005)
9. Belfore, L.A., Vennam II, R.: Vrml for urban visualization. In: Proceedings of the 1999 Winter Simulation Conference, pp. 1454–1459 (1999)
10. Lima, E.-M., Honjo, T., Umekia, K.: The validity of VRML Images as a Stimulus for Landscape Assessment, vol. 77(1-2), pp. 1–2 (2006)
11. Roussou, M., Oliver, M., Slater, M.: The virtual playground: an educational virtual reality environment for evaluating interactivity and conceptual learning. Virtual Reality 10(3-4), 227–240 (2006)
12. Tomaz, A.: Teaching biology in primary and secondary schools with the help of the dynamic HTML and web virtual reality (web3D) projects. Interactive Educational Multimedia (4), 89–98 (2002)
13. Cai, Y., Lu, B., Fan, Z., Indhumathi, C., Lim, K.T., Chan, C.W., Jiang, Y., Li, L.: Bioedutainment: Learning life science through X gaming. Computer & Graphics 30, 3–9 (2006)
14. Torsten, F.: The virtual oceanarium. Communications of the ACM 43(7), 94–101 (2000)
15. Liu, Z.: Design of a Cartoon Game for Traffic Safety Education of Children in China. In: Pan, Z., Aylett, R.S., Diener, H., Jin, X., Göbel, S., Li, L. (eds.) Edutainment 2006. LNCS, vol. 3942, pp. 589–592. Springer, Heidelberg (2006)
16. Padgett, L.S., Strickland, D., Coles, C.D.: Case study: using a virtual reality computer game to teach fire safety skills to children diagnosed with fetal alcohol syndrome. Journal of Pediatric Psychology 31(1), 65–70 (2006)

The Research on Viewpoints Requirements Description Method under Time-Varying

FuHua Shang, LiJuan Liu, RuiShan Du, and Ye Yuan

Computer and Information Technology Department,
Northeast Petroleum University, DaQing City,
Heilongjiang Province, China
{shangfh,liulijuankd,ruishan_du,yuanye2005king}@163.com

Abstract. In this paper, using viewpoint template we define the viewpoints requirements description method under time-varying, and add requirement identifiers to viewpoint template, to record in detail the changing requirements with development of the target software system, and to effectively reduce the degradation of requirements information. This method can comprehensively extract requirements from different project stakeholders in different period of times, form complete requirement specification and preferably guide the software development. We also propose the process of viewpoints integration, based on viewpoint templates with requirement changing with time.

Keywords: Viewpoint Template, Time-varying, Requirements Engineering, Viewpoints.

1 Introduction

Traditional software requirements analysis methods analyze and handle requirements source through how software engineers understand user demands in specific domain and through their own knowledge and experience, which can lead to results that important requirements information are lost, requirements specification are formed not completely, software systems are always re-developed, and the delivered software system cannot meet users' real needs.

The viewpoints method employs the principle of seperation of concerns [1]. To obtain requirements of software systems fully and accurately, we apply viewpoints' thought into requirements engineering to fulfill the requirements analyis of software systems. First viewpoint was proposed in the SADT method [2], but it was put forward formally in CORE method [3] after that. Henceforth, viewpoints are being applied more and more broadly in requirements engineering.

In contrast to traditional requirements engineering methods, the advantages of requirements engineering with viewpoints methods are:

(1)   They allows to collect software requirements from various concerns of the system, which can effectively lessen the possibilty of losing key requirements messages of the system, and can improve the communication with users and clients;

D. Jin and S. Lin (Eds.): Advances in MSEC Vol. 2, AISC 129, pp. 251–255.
springerlink.com   © Springer-Verlag Berlin Heidelberg 2011

(2) The adoption of viewpoints can be useful for achieving requirements information from all aspects of the system dispersedly, independently and comprehensively, then the difficulty of gainning and describing requirements can be lowered effectively, which helps improve the quality of whole software requirements engineering process.

With the development of mutiple viewpoints requirements engineering, many viewpoints models occur which are appropriate for different software developing phases, and most of researchers use viewpoint templates and frameworks to definde viewpoints models, such as:

Finkelstein defines viewpoint as five parts, which are style, domain,specification, work plan and work record, and viewpoint template only elaborates style and work plan of these five parts[4]. Nuseibeh then appoints BNF or entity-relation-attribute can be metalanguage of the style slot[5], and BNF is more suited to represent text message[6]. There are other methods which devide viewpoints and define corresponding viewpoint templates[7,8]. There are a common point of these viewpoint templates, that is, they define serveral different viewpoint templates, which are reusable and can generate many viewpoints. But these viewpoint templates do not define time in them. Pandy considers that we need to manage changing requirements over time[9].

As the software system develop, the responsibilites and concerns of stakeholders involved would also change with it. Requirements are formulated gradually with a long time, and may evolve continually in the life cycle of the system. Needs changing over time is a universal law, that is the time-varing of requirements. Boehm and others's studies suggest that the potential impact of not articulating requirements is huge[10]. And requirements from the knowledge, belief and intentions of stakeholders are closely linked to time. So time plays an important role in requirements elicitation and description, then we need to take it into consideration the relationship between the change of requirements and time. Preview method proposes the significance of traceability to viewpoints requirements analysis methods[11,12]. Time is a critical factor which embodys the traceability of requirements. In this paper, we adopt the structure of viewpoint and viewpoint template from Finkelstein to express stakeholders' requirements, we take time slot as part of template to describe requirements, and add requirement identifiers into viewpoint template.

2   Definition of Viewpoint Template

2.1   Composition of Requirement Identifiers

When specifying requirements by viewpoints method, we need to map raw requiremnts into different viewpoints, to get seperated viewpoints requirements of that raw requirements. Then these viewpoints requirements should be integrated to synthesized requirements, through conflicts or contriodictions resolution between different viewpoints requirements(as shown in Figure 1).

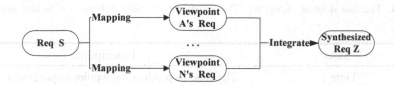

Fig. 1. Req S(requirement S) is mapped into Viewpoint A's Req(the requirement on Viewpoint A) to Viewpoint N's Req(the requirement on Viewpoint N). Then these viewpoints requirements are integrated to Synthesized Req Z(Synthesized Requirements Z), and Req Z is the final requirement specification.

The *Req S* in figure 1 is not just one identifier, but is comprised by serveral identifiers, such as module name, version number and so on. We introduce the structure decomposition method in SADT to analyze what *Req S* is constituted of. The whole system is devided into modules, which also are devided into sub-moduls seperately, which then are devided into functions, until appropriate. Hence, we assume that identifiers in the *Req S* include module name, sub-module name, function tabs and requirement tabs(as shown in Figure 2).

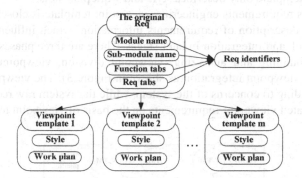

Fig. 2. The original Req(the raw requirements) has five Req identifiers(requirement identifiers): Module name, Sub-module name, Function tabs, Req tabs(requirement tabs). The original Req should be described by various Viewpoint templates(from Viewpoint template 1 to Viewpoint template m). Viewpoint template elaborates Style and Work plan slots.

2.2   Definition of Viewpoint Template and Requirement Description

We describe requirements by viewpoint template. The viewpoint is composed by five slots(as shown in Table 1), these are:

Table 1. The five slots of viewpoint. The viewpoint template elaborates style and work plan slots.

Slots	Description
Time :	The time to produce viewpoint requirements
Style :	The notations to describe specification
Work Plan :	Corresponding raw requirements, represented by requirement identifiers
Specification :	Specific requirements information
Pre-planning Content :	Some requirements to be changed

We improve the five slots of viewpoint Finkelstein defined. The aim adding time is to record the producing time of viewpoint requirements, then we can trace how requirements change. The work plan slot is different from Finkelstein's, which in this article is defined as corresponding raw requirements, and can be represented by requirement identifiers. Pre-planning content shows some requirements to be changed. And viewpoint template only describes style and work plan slots.

In viewpoints requirements engineering, viewpoint template is closely related with acquisition and description of requirements information, which influences viewpoint consistency check and integration in later stages. There are three phases in viewpoints requirements obtaining and analyzing: viewpoint division, viewpoint requirements description and viewpoint integration(as shown in Figure 3). The viewpoint templates is divided according to concerns of the system, while the system raw requirements are generated to related viewpoint requirements on the basis of viewpoint templates.

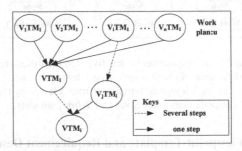

Fig. 3. V_j indicates viewpoint requirements of the jth viewpoint template; TM_i indicates time i; V_jTM_i indicates viewpoint requirements of of the jth viewpoint template in *work plan u* at time i. And VTM_i indicates that synthesized requirements from different viewpoints requirements under the *work plan u*. VTM_i can conduct subsequent software development.

3   Conclusions

This paper improves viewpoints template, adds time into template and treate it as an important part, defines requirement identifiers and uses them to compose work plan slot, promotes the concerns of time-varing requirements, and proposes the viewpoints requirements description method under time-varing. This method can trace requirements which are changed over time, and effectively controlls the affects of changing requirements to software requirements description and software developments. And we put forward the composition of viewpoint and the viewpoint integration process.

References

1. Lago, P., Avgeriou, P., Hilliard, R.: Guest editors' introduction: Software architecture: Framing stakeholders' concerns. IEEE Software 27, 20–24 (2010)
2. Ross, D.: Structured analysis (SA): A language for communicating ideas. IEEE Trans. Software Engrg. 3, 16–34 (1977)
3. Mullery, G.: CORE - a method for controlled requirements expression. In: Proceedings of 4th International Conference on Software Engineering(ICSE-4), pp. 126–135. IEEE Computer Society press (1979)
4. Finkelstein, A., Goedicke, M., Kramer, J.: ViewPoint oriented software development. In: Proc. 3rd Int. Workshop on Software Eng. and Its Applicat., Toulouse, France, Cigref EC2, pp. 337–351 (December 1990)
5. Nuseibeh, B., Finkelstein, A.: ViewPoints: A Vehicle for Method and Tool Integration. In: Proceedings of 5th International Workshop on Computer-Aided Software Engineering (CASE 1992), Montreal, Canada, July 6-10, pp. 50–60. IEEE Computer Society Press (1992)
6. Nuseibeh, B., Kramer, J., Finkelstein, A.: Expressing the Relationships Between Multiples Views in Requirements Specification. In: Proc. 15th Int. Conf. on Software Engineering, Baltimore, pp. 187–196 (1993)
7. Clark, R., Moreira, A.: Constructing Formal Specifications from Informal Requirements. Software Technology and Engineering Practice, 68–75 (1997)
8. Araujo, J., Coutinho, P.: Identifying aspectual use cases using a viewpoint-oriented requirements method. In: Early Aspects 2003: Aspect Oriented Requirements Engineering and Architecture Design, Workshop of the 2nd Intl. Conference on Aspect-Oriented Software Development, Boston, MA (2003)
9. Dhirendra Pandey, U., Suman, A.K.: An Effective Requirement Engineering Process Model for Software Development and Requirements Management. IEEE Xplore, 287–291 (2010)
10. Boehm, B.W.: Software Engineering Economics. IEEE Trans. Software Eng. 10(1), 4–21 (1984)
11. Sommerville, I., Sawyer, P.: Requirements Engineering - A Good Practice Guide. John Wiley and Sons (1997)
12. Sommerville, I., Sawyer, P., Viller, S.: Viewpoints for Requirements Elicitation: a Practical Approach, pp. 74–81 (1998)

Cost-Effective Charge Sharing Address Driver with Load Adaptive Characteristic

Hyun-Lark Do

Department of Electronic & Information Engineering,
Seoul National University of Science and Technology, Seoul, South Korea
hldo@seoultech.ac.kr

Abstract. A cost-effective charge sharing address driver for plasma display panel (PDP) is proposed. The proposed address driver utilizes an auxiliary circuit to reduce the voltage across the data driver IC when its output stages change their status. The auxiliary circuit absorbs some of the energy stored in the panel and provides the energy absorbed to the panel. Therefore, it can reduce the power consumption and relieve the thermal problems of the driver ICs. Moreover, it has load adaptive characteristics (LAC). Compared to LC resonant driving method, the component count is significantly reduced and the light load efficiency is improved due to LAC.

Keywords: Charge sharing, address driver, load adaptive, plasma display panel.

1   Introduction

A plurality of address driving ICs are required to drive the address electrodes in PDP. In the case of a 42-in XGA PDP, there are 3072 address electrodes. 16 data drive ICs of 192 outputs with single scan method and double with dual scan are required to drive these electrodes. Generally, the power consumption in the address driver circuit increases according to the screen size and the resolution of panel. The address voltage source V_a is directly applied to the data driver ICs. The stored energy in the panel capacitances are dissipated at the inevitable parasitic resistances including the distributed resistances of data lines and the equivalent resistances of switches. Also, in the sustain electrodes, there are a lot of power loss when square voltage waveforms are applied to the panel through full-bridge circuit with out any other circuits. In order to recycle the energy, an additional circuit called an energy recovery circuit (ERC) is required. The conventional LC resonant address driver is in [1]. It utilizes the resonance between the panel capacitance and an external inductor. Conceptually, it is identical to the conventional sustain driver [1], [2]. It provides high efficiency at heavy data loads which have a lot of data switchings. However, it shows relatively low efficiency at light loads such as full white image which has no data switching. It is because it has no LAC. Moreover, it has a lot of components and complex control circuits are required. Also, a resonant time interval T_{ER} should be much longer to achieve high efficiency. It shortens the effective data pulse width ($=T_{scan}-T_{ER}$) and can not guarantee an adequate addressing.

D. Jin and S. Lin (Eds.): Advances in MSEC Vol. 2, AISC 129, pp. 257–260.
springerlink.com   © Springer-Verlag Berlin Heidelberg 2011

In order to solve these problems, a cost-effective charge sharing address driver with LAC is proposed and it is shown in Fig. 1. Components count is significantly reduced. In addition, the cost of the gate drivers for power switches is reduced. In the conventional LC resonant driver, two gate drivers with independent high and low side referenced output channels are required. However, the proposed address driver requires only one gate driver with 2 ground referenced output channels by utilizing p-channel MOSFET which can be driven by a capacitor coupled driving method. Since it has intrinsic LAC, low efficiency problem at light load disappears. In addition, since the loss of the effective data pulse width is limited only by R_{eq} which represents the ohmic resistance of the printed circuit board patterns and the MOSFETs in the address driver, and inevitable parasitic resistances including the distributed resistances of data lines, high speed addressing can be achieved.

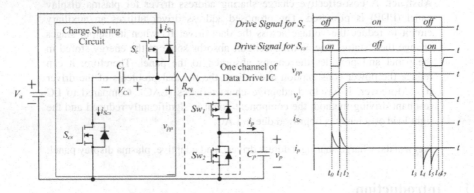

Fig. 1. Circuit diagram and key waveforms of proposed charge sharing address driver

2   Analysis of the Proposed Address Driver

Before t_0, the panel voltage v_p is maintained as zero with turn-on of S_{w2}. The switch S_{Cs} is turned on and ready for charge sharing operation.

Mode 1 $[t_0, t_1]$: At t_0, S_{w2} is turned off and S_{w1} is turned on and the panel capacitance C_p is charged through S_{Cs}, C_s, R_{eq}, and S_{w1}. During this operation, the energy stored in the charge sharing capacitor C_s is transferred to the panel. At the end of this mode, the gate signal for S_{Cs} is removed and the body diode of S_{Cs} takes over the current.

Mode 2 $[t_1, t_2]$: At t_1, v_p arrives at V_{Cs}. Then, the clamp switch S_c is turned on. In this mode, C_p is charged by the address voltage source V_a.

Mode 3 $[t_2, t_3]$: At t_2, v_p arrives at V_a.

Mode 4 $[t_3, t_4]$: At t_3, the gate signals for S_c and S_{w1} are removed. Since the address discharge is already done, there is no current flow and v_p is maintained as V_a.

Mode 5 $[t_4, t_5]$: At t_4, S_{Cs} is turned and the charge sharing operation starts. The energy stored in C_p is transferred to C_s through the body diode of S_{w1}, R_{eq}, C_s, and, S_{Cs}.

Mode 6 $[t_5, t_6]$: At t_5, S_{w2} is turned on and v_p starts to decrease to zero. At the end of this mode, v_p arrives at zero.

The power loss P_L of the address driver without charge sharing circuit is given by

$$P_L = \frac{N_{SF} N_{add}}{T_{frame}} \cdot N_{data\_chg} V_a^2 C_p \,, \tag{1}$$

where N_{SF} = number of subfields, N_{add} = number of address electrodes, N_{data_chg} = number of address data switchings, and T_{frame} = time interval of one TV frame. The power loss $P_{L,proposed}$ of the proposed address driver is given by

$$P_{L,proposed} = \frac{N_{SF} N_{add}}{T_{frame}} \cdot \left(N_{data\_chg} V_{Cs}^2 + N_{data} \left(V_a - V_{Cs} \right)^2 \right) C_p \,, \tag{2}$$

where N_{data} = the number of address data according to one scan electrode. The voltage V_{Cs} is determined by the charge-balancing condition of C_s during the charge sharing operation and it is derived as follows:

$$V_{Cs} = \frac{N_{data}}{N_{data} + N_{data\_chg}} V_a \,. \tag{3}$$

At heavy load, N_{data_chg} approaches to N_{data}. Then, V_{Cs} approaches to $V_a/2$. At light load, N_{data_chg} approaches to zero. Then, V_{Cs} is around V_a and the charge sharing operation stops. From (1), (2), and (3), the power loss relation between P_L and $P_{L,proposed}$ is given by

$$P_{L,proposed} = P_L \frac{N_{data}}{N_{data} + N_{data\_chg}} \,. \tag{4}$$

At heavy load where N_{data_chg} is large, the power loss is significantly reduced. At light load where N_{data_chg} is almost zero, V_{Cs} is around V_a and the charge sharing operation stops.

3   Experimental Results

To verify the performance of the proposed address driver, the prototype has been built and tested on a single-scan 42-inch XGA ac PDP with 16 data drivers of 192 outputs. The proposed charge sharing circuit block is designed to control two data driver ICs. Totally, 8 charge sharing circuit blocks are used. The address voltage V_a is set to 60V. T_{frame} is 16.67ms and 10 subfields are used. The scan period T_{scan} is 1.2μs. Fig. 2 shows the experimental waveforms and power consumption comparison. Since N_{data} is equal to N_{data_chg} at one-dot on/off pattern which is the heavy loads, V_{Cs} is around $V_a/2$. At full white pattern where there is no data changings, V_{Cs} goes up to V_a and the charge sharing operation stops. Power consumption of the proposed address driver is reduced by 48% at the heaviest load.

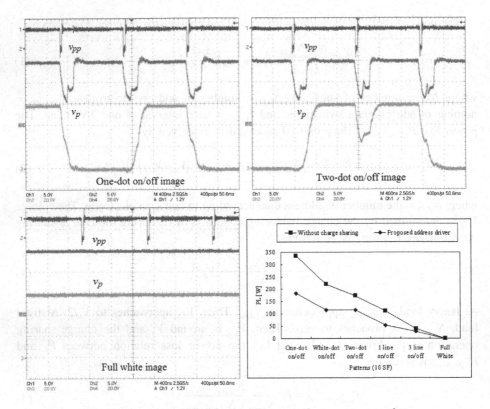

Fig. 3. Experimental waveforms and measured power consumption

4   Conclusion

A cost-effective address driver for PDP has been proposed. It utilizes a charge sharing capacitor which recycles the energy stored in the panel capacitance. Also, it has LAC. Its component count is significantly reduced compared with the conventional LC resonant driving method. The proposed address driver requires only one gate driver with 2 ground referenced output channels by utilizing p-channel MOSFET as a clamp switch.

References

1. Weber, L.F., Warren, K.W.: Power efficient sustain drivers and address drivers for plasma panel. U.S. Patents 5081400 (1992)
2. Do, H.L.: Energy recovery sustain driver with low circulating current. IEEE Trans. Ind. Elec. 57, 801–804 (2010)

A Method of Mechanical Faults Feature Extraction Based on Timbre Feature

Yungong Li[1], Jinping Zhang[2,*,**], and Li Dai[1]

[1] School of Mechanical Engineering & Automation, Northeastern University,
Shengyang 110004
[2] School of Mechanical Engineering, Shenyang University of Chemical Technology,
Shenyang 110142

Abstract. The human auditory system has remarkable ability to identify different signal depending on pitch, loudness and timbre. And timbre is a most important feature in above three features. In this paper, timbre is introduced in filed of mechanical faults diagnosis and a calculation method is proposed. This method firstly processes mechanical vibration signal using ZCPA model. The second step is to calculate six elements of timbre, which are adjusted according to the characteristics of vibration signal in this paper. To test the performance of the proposed method, rotor faults experiments are carried out, and the results show that timbre feature can distinguish different fault with high stability and small data quantity.

Keywords: auditory model, timbre, ZCPA model, mechanical faults diagnosis.

1 Introduction

When identifying the type of sound source, the auditory system is usually based on such feature information as pitch, loudness and timbre[1][2], and the timbre is a important feature that is related with frequency structure of a signal. If sound sources with different structure and materal, the sound timbres will different, even they have same pitch and loundness. Therefore, to provide effective features for equipment condition monitoring and fault diagnosis, the timbre feature might be introduced to mechanical vibration analysis and identification.

It has been clearly that timbre can be constructed by some elements such as spectrum centroid, spectrum flux, spectrum irregularity[4]. In the aspect of timbre calculation, mainly have two approaches, one[4] is to establish a complete mathematical model of the auditory system; the other way[5] is to calculate in the based of conventional signal analysis methods.

In this paper, a calculation method of timbre for vibration signal is proposed. ZCPA mode is employed as a frond-end processor to generate timbre, and the calculation method of timbre is adjusted for vibration signal. The test results indicate that the proposed method can effectively distinguish different rotor faults and has satisfactory stability of feature extraction.

---

* Corresponding author.
** This project is supported by National Science Foundation of China (50805021), Fundamental Research Funds for the Central Universities (N100403004), Science and Technology Research Project of Department of Education of Liao Ning Province(L2010442).

D. Jin and S. Lin (Eds.): Advances in MSEC Vol. 2, AISC 129, pp. 261–266.
springerlink.com   © Springer-Verlag Berlin Heidelberg 2011

2   Principle of Proposed Method

The implementation process of the proposed method is shown as fig.1. The signal x(t) is firstly processed by ZCPA model[6][7] to get the auditory spectrum. And then the timbre feature consi- sting of spectral centroid (SC), spectral centroid bandwidth (SBW), spectral flux(SF), spectral Roll-off (SRO),spectral irregularityes (SI), spectral flatness measurement(SFM)   are calculated based on the data of auditory spectrum.

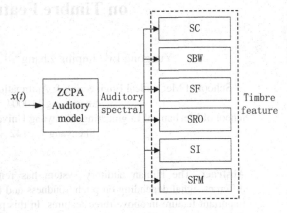

Fig. 1. The prime calculate process of timbre feature

3   ZCPA Auditory Model

The implementation process of ZCPA is shown as fig.2, it generally includes three main steps, namely basilar membrane bandpass filter, inner hair cells and auditory nerve feature extract,and feature information synthesis.

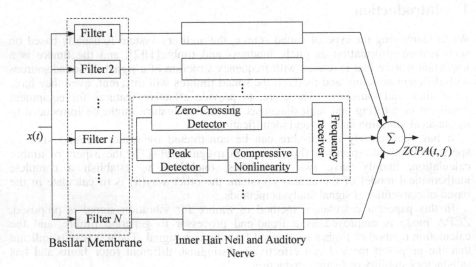

Fig. 2. Principle diagram of ZCPA model

In ZCPA, Gammatone filterbank [8][9] is used to simulate the bandpass filter characteristics of basilar membrane, every filter center frequency in the frequency axis is logarithmically evenly distributed. Given N-filter, No. i filter center frequency is f_i, Gammatone filter time-domain expression is

$$h(t,i) = B^n t^{n-1} e^{-2\pi Bt} \cos(2\pi f_i t + \phi_i) u(t) \tag{1}$$

Where $u(t)$ is step function and n is filter rank, when n=4 can well simulate the characteristics of basilar membrane. Phase ϕ_i is usually taken as zero, B can be calculated as

$$B = 1.019 \cdot \mathrm{ERB}(f_i) = 1.019 \cdot (24.7 + 0.108 f_i) \tag{2}$$

Where $\mathrm{ERB}(f_i)$ is the equivalent rectangular bandwidth of filter.

Let $y_1(t,i)$ be the output of NO.i channel fuliter. It is conveyed to zero-crossing detector and peak detector. Zero-crossing detector is in charge of checking all zero-crossing points in $y_1(t,i)$ during this period of time from given time t, and calculating the time interval between neighboring zero-crossing, let time interval between l th and $(l+1)$ th zero-crossings of $y_1(t,i)$ be ΔT_{il}. Peak detector fulfills the job of detecting the maximum peak between two zero-crossings, let peak between l th and $(l+1)$ th zero-crossings of $y_1(t,i)$ be p_{il}. After normalization processing, p_{il} will be compressed nonlinearly through S function described by

$$g(p_{il}) = \frac{2}{1 + \exp(-\gamma \cdot p_{il})} - 1 \tag{3}$$

The contact ratio in frequency domain is usually great between basilar membrane filters, then M frequency intervals are divided, named that as frequency box, the frequency range of each box is increasing according to geometric progression form. The receiver's work corresponding to i th filter is shown as follows

$$y_2(t,m,i) = \sum_{l=1}^{Z_i-1} \delta_{mil} g(p_{il}) \quad (m = 1, 2, \cdots, M) \tag{4}$$

Where Z_i is the total zero-crossing number, m is the order number of frequency box, each m is corresponding to a frequency range fb_m, δ_{mil} is Kronecker operator, if $f_{il} = 1/\Delta T_{il}$ falls into mth frequency box, then $\delta_{mil} = 1$, else $\delta_{mil} = 0$.

Last, the summing of each frequency receiver output is the final result of ZCPA,

$$zcpa(t, f_m) = \sum_{i=1}^{N} y_2(t,m,i) \quad (m = 1, 2, \cdots, M) \tag{5}$$

Where f_m is the center frequency of m th frequency box, $zcpa(t, f_m)$ is called as auditory spectral. In this paper the stationary signal is the prime extract feature, so auditory spectral can be expressed as $zcpa(t, f_m)$.

4   The Elements of Timbre Feature

In order to make the timbre features suitable for the equipment faults working under different frequency, it is nessecery to find the maximum amplitude Z_{max} and its frequency f_{max} in $zcpa(f_m)$, then do normalization processing $zcpa(f_m)$ by Z_{max}, let the result still be $zcpa(f_m)$. Obtained based on auditory spectrum $zcpa(f_m)$, in this paper, timbre is compose of six elements, each element calculate method and are stated as follows.

$$SC = [\sum_{m=1}^{M} f_m \cdot zcpa(f_m)] \Big/ [f_{max} \sum_{m=1}^{M} zcpa(f_m)] \tag{6}$$

$$SBW = \frac{2}{f_s}[SC_h - SC_l] \tag{7}$$

$$SF = \sum_{m=1}^{M} b_m \cdot zcpa(f_m) \tag{8}$$

$$\sum_{m=1}^{K} zcpa(f_m) = c \sum_{m=1}^{M} zcpa(f_m) \tag{9}$$

$$SI = [\sum_{m=2}^{M} [zcpa(f_m) - zcpa(f_{m+1})]^2] \Big/ [\sum_{m=1}^{M} zcpa(f_m)] \tag{10}$$

$$SFM = 10\log_{10}\{[\prod_{m=1}^{M} zcpa(f_m)]^{1/M} \Big/ \frac{1}{M} \sum_{m=1}^{M} zcpa(f_m)\} \tag{11}$$

Where SC_h is the spectral centroid within $[SC, f_M]$, SC_j is the spectral centroid within $[0, SC]$, f_s is the sampling frequency. b_m the bandwidth of mth frequency box. Given constant c, let $SRO = f_K / SC$, K should meet usually, given c as 0.85.

5   Experimental Test

In order to test the actual effect of the proposed method in the feature extraction of mechanical faults, there are four faults (misalignment, rotor-to stator rubbing, oil film whirl and pedestal looseness) to be detected by using rotor test-bed , sampling frequency is 2000Hz, the using signal is shaft radial displacement signal.The experimental system is shown as fig.3.

Fig. 3. Rotor experiment system

In ZCPA model, given $\gamma = 6$, the number of Gammatone filter is 60, the frequency box number is 16, the frequency range of each frequency box is increasing with 1.2 times. Then, using the results of ZCPA model, six elements of timbre feature are calculated. In order to test the stability of timbre, each fault is tested for five times, and get five groups data, shows as tab.1.

Table 1 shows that different fault signal has different timbre feature, and different perceptual element of timbre has different sensitive degree, for example, SRO is almost invariably for one fault value, but others change slightly. Then, timbre is stable for one fault but different for different fault and is propitious to intelligent recognition.

Table 1. Timbre feature data of four faults

Timbre feature	SC	SBW	SF	SRO	SI	SFM
Misalignment	1.832	0.054	0.051	0.776	1.842	-0.528
	1.829	0.053	0.050	0.776	1.839	-0.522
	1.838	0.050	0.059	0.776	1.846	-0.533
	1.825	0.049	0.048	0.775	1.793	-0.525
	1.840	0.061	0.056	0.776	1.910	-0.532
Rotor-to-stator rubbing	3.135	0.087	0.066	1.606	1.832	-0.321
	2.959	0.080	0.069	1.606	1.829	-0.330
	3.189	0.085	0.058	1.606	1.837	-0.319
	3.170	0.079	0.067	1.606	1.728	-0.317
	3.267.	0.093	0.070	1.607	1.880	-0.328
Oil film whirl	1.757	0.042	0.107	0.809	1.325	-0.586
	1.763	0.044	0.101	0.809	1.330	-0.581
	1.742	0.036	0.110	0.810	1.322	-0.593
	1.755	0.041	0.103	0.809	1.329	-0.577
	1.753	0.039	0.100	0.809	1.326	-0.579
Pedestal looseness	2.202	0.075	0.078	1.482	1.613	-0.459
	2.195	0.071	0.070	1.482	1.619	-0.460
	2.207	0.079	0.069	1.482	1.598	-0.451
	2.197	0.070	0.083	1.482	1.620	-0.444
	2.200	0.074	0.077	1.482	1.617	-0.457

6   Conclusion

The method of extracting mechanical faults features based on ZCPA auditory model and timbre feature is proposed, and the calculation method of timbre is improved and adjusted combining the characteristic of mechanical vibration signal. Experimental verification shows that the auditory spectrum of ZCPA model of proposed method has good distinction, and timbre feature can further condense the information, and has smaller data quantity, more higher stability. Therefore the proposed is conducive to further intelligent recognition work and has practical applicability.

References

1. Aucouturier, J.-J., Pachet, F., Sandler, M.: "The way it sounds": timbre models for analysis and retrieval of music signals. IEEE Transactions on Multimedia 7(6), 1028–1035 (2005)
2. Cao, X.-Z., Meng, H.-L., Xu, J.-C.: Timbre model of software musical instrument based on sine interpolation. In: International Conference on Digital Object Identifier, pp. 358–361 (2009)
3. Miyasaka, E.: Timbre of complex tone bursts with time varying spectral envelope. In: IEEE International Conference on ICASSP, pp. 1432–1465 (1982)
4. Wang, N., Chen, K.-A.: Applications of auditory perceptual features into targets recognition. Journal of System Simulation 21(10), 3128–3132 (2009)
5. Ma, Y., Chen, K.-A., Wang, N., Zheng, W.: Application of auditory spectrum-based features into acoustic target recognition. Acta Acustica 34(2), 142–150 (2009)
6. Kim, D.-S., Lee, S.-Y., Kil, R.M.: Auditory processing of speech signal for robust speech recognition in real word noisy environments. IEEE Transactions On Speech And Audio Processing 1(7), 55–68 (1999)
7. Serajul, H., Roberto, T., Anthony, Z.: Perceptual features for automatic speech recognition in noisy environments. Speech Communication 51(1), 58–75 (2009)
8. Schluter, R., Bezrukov, L., Wagner, H., Ney, H.: Gammatone features and feature combination for large vocabulary speech recognition. In: ICASSP 2007, pp. 679–652 (2007)
9. Li, Y., Zhang, J., Dai, L., Zhang, Z., Liu, J.: Auditory-Model-Based Feature Extraction Method for Mechanical Faults Diagnosis. Chinese Journal Of Mechanical Engineering 21(3), 391–397 (2010)

Appendix: Springer-Author Discount

LNCS authors are entitled to a 33.3% discount off all Springer publications. Before placing an order, they should send an email to SDC.bookorder@springer.com, giving full details of their Springer publication, to obtain a so-called token. This token is a number, which must be entered when placing an order via the Internet, in order to obtain the discount.

Yungong LI, born in 1976,is currently a associate professor in School of Mechanical Engineering and Automation,Northeastern University, China. His research interests include mechanical faults diagnosis and engineering signal analysis.
Tel:+86-24-83680540
E-mail:ygli@mail.neu.edu.cn
Corresponding Author: Jinping Zhang, born in 1977, is currently a lecture in School of Mechanical Engineering, Shenyang University of Chemical Technology, China. She received her PhD degree from Northeastern University, China, in 2008.
Tel: +86-24-89383281 E-mail: jinping7707_cn@sina.com

Based ADL on the 3d Virtual Characters Based Action of the Research and Application

FuHua Shang, Bo Li, and Hong Zhou

School of Computer and Information Technology, Northeast Petroleum University, Daqing City HeilongJiang Province China 163318
Shangfh@163.com, libo6679969@yahoo.com.cn, zh_hlj@126.com

Abstract. This paper, based on the rehabilitation medicine field ADL, Create accord with human body model of virtual minimum basic skeleton model, Extraction and split the human daily life basic action, Obtained based ADL on the 3d virtual characters latched basic action, 17 molecular action and 37 atomic movement, According to the software engineering layered development and atom model thought, In view of the small and medium-sized virtual reality, the three dimensional animation, Put forward the application based ADL on the 3d characters latched based movement efficient development method of complicated movements, Effectively improve the 3d characters movement development efficiency.

Keywords: ADL, 3d virtual characters, Skeletal animation, Sports generate.

1 Introduction

Activities of daily living (ADL) from the concept of rehabilitation medicine field, the first produced by American Deaver physicians and Brown physicians [1,2]. ADL is to show people to live independently and must be repeated every day, the most basic have common body action group, namely for food, clothing, shelter and transportation, personal health and other activity basic action and skills [3].

Virtual people are people in the computer generated space geometry and behavior characteristics of the realistic said [4]. At present, the research of the technology of virtual people can be divided into the geometry of the virtual human expression, virtual a people's movement to create and control, virtual behavior expression and virtual the people's cognitive express four main aspects [5].

This paper makes a study of the virtual simulation system of 3d virtual a people's movement is generated. Analysis 3d body bone structure characteristics, structuring accord with human body model of virtual minimum basic skeleton model, extraction separation based on the 3d latched the virtual characters basic action. 17 molecular action and 37 atomic movement, the paper puts forward how to through movement design table 17 molecules called action joining together, and according to need to modify the molecular movement parameters of the atoms in action, concise and efficient method of developing complex movement.

D. Jin and S. Lin (Eds.): Advances in MSEC Vol. 2, AISC 129, pp. 267–272.
springerlink.com

2   Basic Action Extraction and Separation

Literature [3] to human body all daily life activities summary. Preliminary analysis got 40 items daily life basic activities, and carry on the experiment, through the action to decompose, the extraction and filtering, draw a group of typical basic action. As long as the human body can do put forward the basic action can realize the daily life for, these movements alternative daily life activities. After 40 items human daily life activities, get a action decomposition 17 human daily life basic action. Eventually got as long as can do these basic action, can satisfy the human body daily life activities, body joints can meet all freedom if these basic action requirements, we can be successful in daily life activities[6,7].

Table 1. A human body 17 daily life basic action

Serial number	Daily life basic action	Daily life example
1	Hand to head	Comb my hair, wear a hat
2	Hand to the occipital	Comb my hair
3	Hand to nape	Bathing, and comb my hair
4	Hand to the side shoulder	A bath, backpacks
5	Hand to the side shoulder	Backpack, a coat on
6	Hand to the waist	Bath
7	Hand to the waist with side	Bath
8	Hand to the sacral	The toilet
9	Hand to the perineal	Physiological health
10	Forearm rotation	Open the door
11	Palms stretch	Brush my teeth
12	Hand back to the central	Wear underwear
13	Hand to foot	From the ground up objects
14	The swivel	Fluctuation bed
15	Ducks	From the ground up objects
16	SLR to go to bed	Fluctuation bed
17	Feet into his shoes in	Wear shoes

3   Create 3d Body Bone Structure Model

We know from the literature [8] with stratified virtual human representation of basic skeleton, virtual people by the joint determine its state layer, decided the basic sports virtual attitude, such, can directly control of framework to achieve the virtual characters generate movement[8].Therefore, the study of virtual simulation system of 3d virtual a people's movement before the generation, we need to describe and build a virtual character basic skeleton.

VRML(Virtual Reality Modeling Language) has a special child standard used to describe virtual human model, that is, H-Anim. Virtual people said to focus for the human body for the root node, with joint for child nodes, and for the attachment to bone tree. Through the limbs of the geometric model of attachment to the corresponding bone, to form a complete virtual person model[9].

Through the literature[10] we know that, according to the human body bone structure characteristics and the anatomy the research focus, from sports joints and exercise two on the character of freedom Angle bone movement system is simplified, can get a 16 joints, 37 degrees of freedom, namely the simplified model of virtual people minimum model of the human body. This article is based on the above analysis, the building used for 3d virtual character bone action generated skeleton model, that is, the 3d virtual character of basic skeleton layer. As shown in figure 1 shows, namely the minimum for virtualization human body model skeleton structure, the graph of the Numbers noted each joint, of which the human body gravity Numbers for 0, their sports freedom for six.

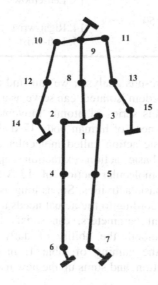

Fig. 1. Virtualization of human body model skeleton minimum

4   Based on ADL 3d Virtual Character Base Motion Resolution

ROM(Range of Motion) is when the knuckles movement through sports arc[11], use the virtual characters skeleton model building, completed in table 1 item 17 of human body daily life basic action, determination of the completion of each movement involves the joint activities. After finishing analysis, get 16 joints in a degree of freedom that the movement of the Angle range, the experimental results such as shown in table 2.

Table 2. 16 Motions of joint angle range determination result

Joint name	Axis	Sports angle range	Joint name	Axis	Sports angle range
0:Root	X	-180 — 180	8: Spine	X	-50 — 100
	Y	-180 — 180		Y	-50 — 55
	Z	-180 — 180		Z	0 — 0
1:Waist	X	-45 — 90	9: Neck	X	-50 — 90
	Y	-50 — 50		Y	-80 — 80
	Z	-55 — 55		Z	-60 — 60
2:Right-groin 3:Left-groin	X	-165 — 45	10:Right-shoulder 11:Left-shoulder	X	-80 — 180
	Y	-20 — 20		Y	-80 — 160
	Z	-120 — 20		Z	-90 — 90
4:Right-knee 5:Left-knee	X	0 — 165	12: Right-elbow 13: Left-elbow	X	-90 — 40
	Y	0 — 0		Y	0 — 120
	Z	0 — 0		Z	0 — 0
6:Right-ankle 7:Left-ankle	X	-45 — 50	14: Right-wrist 15: Left-wrist	X	-45 — 45
	Y	0 — 0		Y	0 — 0
	Z	0 — 35		Z	-90 — 90

Above the determination results analysis, we can find that as long as we will each joint action in a degree of freedom isolated, can serve as a series of atomic movement. By the analysis we know, this series of atomic movement through the on-demand combination, can realize the need of human body 17 daily life basic action, we will this 17 human daily life basic action called molecular action. Through the human body to 17 of daily life basic action extraction separation principle, we can demonstrate, complete the molecular action of 17 virtual people can complete skeleton 40 items daily life basic activities. So, as long as through a molecular action the joining together of 17, according to the actual needs a molecular action within the adjustment atomic movement parameters, can satisfy the human body daily life activities, then can have smooth the ability of daily life activities. This paper molecular action continue the naming of table 1, in order to see named name meaning, 37 atoms known action, and sums up the new name as shown in table 3.

Table 3. 37 atomic movement

Action name	Finish joint	Angle range	Action name	Finish joint	Angle range
Root-X	Root	-180 — 180	Spine-Y	Spine	-50 — 55
Root-Y	Root	-180 — 180	Neck-X	Neck	-50 — 90
Root-Z	Root	-180 — 180	Neck-Y	Neck	-80 — 80
Waist-X	Waist	-45 — 90	Neck-Z	Neck	-60 — 60
Waist-Y	Waist	-50 — 50	Right-shoulder -X	Left-shoulder	-80 — 180
Waist-Z	Waist	-55 — 55	Right-shoulder -Y	Left-shoulder	-80 — 160
Right-groin-X	Right-groin	-165 — 45	Right-shoulder -Z	Left-shoulder	-90 — 90
Right-groin-Y	Right-groin	-20 — 20	Left-shoulder-X	Left-shoulder	-80 — 180
Right-groin-Z	Right-groin	-120 — 20	Left-shoulder-Y	Left-shoulder	-80 — 160
Left-groin-X	Left-groin	-165 — 45	Left-shoulder-Z	Left-shoulder	-90 — 90
Left-groin-Y	Left-groin	-20 — 20	Right-elbow-X	Right-elbow	-90 — 40
Left-groin-Z	Left-groin	-120 — 20	Right-elbow-Y	Right-elbow	0 — 120
Right-knee-X	Right-knee	0 — 165	Left-elbow-X	Left-elbow	-90 — 40

Table 3. (*continued*)

Left-knee-X	Left-knee	0 — 165	Left-elbow-Y	Left-elbow	0 — 120
Right-ankle-X	Right-ankle	-45 — 50	Right-wrist-X	Right-wrist	-45 — 45
Right-ankle-Z	Right-ankle	0 — 35	Right-wrist-Z	Right-wrist	-90 — 90
Left-ankle-X	Left-ankle	-45 — 50	Left-wrist-X	Left-wrist	-45 — 45
Left-ankle-Z	Left-ankle	0 — 35	Left-wrist-Z	Left-wrist	-90 — 90
Spine-X	Spine	-50 — 100			

5   Based on ADL 3d Virtual Character Base Motion Applications

In the 3d virtual characters of the practical developing action according to actual needs, we designed a series of actions by the table design table, realizing complex movements of the management, and the formation and use. Used to split the current action and complicated movements for simple actions design table, such as table 4 shows, is a drilling workers turned round, pick a brush, clean up the drill bit, then mother buckle brush, rose up and went to the resolution process.

Table 4. Down for the current action

Time	1-2s	3-6s	7-15s	16-18s	19-20s
Action	Workers turned	Brush up	Brush bit	Put brush	Workers to stand up

Molecular action design table 1 table, the resolution will be with the molecular action movement design, then according to call out the design has been made complete molecular action.

Table 5. Molecular action design table for example

Time	Action	Molecular action	Explain
1-2s	Workers turned	Root	
3-3s	Brush up	Ducks	
4-6s		Hand to foot	Reciprocating
7-8		Forearm rotation	Parameters adjustment
9-10		Forearm rotation	Parameters adjustment
11-12	Brush bit	Forearm rotation	Parameters adjustment
13-15		Forearm rotation	Parameters adjustment
16-18s	Put brush	Hand to foot	Reciprocating
19-20s	Workers to stand up	Ducks	Inverse process

The atom movement design table 4 and 5, table to table, need to adjust the parameters of the molecular action in view of the actual need to design the atom movement parameter modification. This paper here, pick apart, for example, only such as shown in table 6.

Table 6. Atomic design table for example: Brush bit

Joint	X	Y	Z
Right-elbow	-30—30	——	——
Right-winst	——	45	——

6 Summary

This paper based on the extraction separation out latched 3 d virtual character basic action, i.e. 17 molecular action and 37 atomic movement, and but according to need to modify the molecular movement parameters of the atoms in action, concise and efficient method of developing complex movement.

Through the drilling of a complete set of three dimensional simulation training system research and development, including 22 independent operation regulations, involving 325 3d virtual drilling workers complicated movements of the development of the generation, and tests the application based on the 3d characters latched based movement development complex action can be efficient overall finish the actual demand.

Acknowledgement. National Natural Science Foundation of China (Grant No. 61170132).

References

1. Qin, L., Tong, L.: Low latched the care of the sick self care deficiency. Foreign Medical Nursing 17(4), 156–159 (L998)
2. Lawton, M.P.: Aging and performance of home tasks. Human Factors 32(5), 527–536 (1990)
3. Song, H.Y., Zhang, J.G.: Human daily life activities of basic action research extraction. China Rehabilitation Medicine Journal 24(5), 451–453 (2009)
4. Badler, N.I., Phillips, C.B., Webber, B.L.: Simulating Humans: Computer Graphics, Animation, and Control. Oxford University Press, London (1999)
5. Luo, G., Hao, C.Y., Zhang, W.: Virtual human technology research and reviewed in this paper. Computer Engineering 31(18), 7–9 (2005)
6. Jin, Y., Meng, D.H., Jiang, Z.L.: Stroke patients daily life activities of the dynamic analysis ability recovery level. Practical Geriatric Medicine (06) (2010)
7. Zhang, A.L.: Based on the MATLAB human upper limbs motion analysis and simulation, pp. 19–28. Tianjin university of science and technology (2003)
8. Thalmann, D., Shen, J., Chauvineau, E.: Fast Realistic Human Body Deformations for Animation and VR Applications. In: Proc. Computer Graphics International 1996, pp. 166–174. IEEE Computer Society Press (1996)
9. Wang, Z.Q.: Virtual people in synthesis. Graduate School of Chinese Academy of Sciences Journal 17(2) (2000)
10. Luo, G., Hao, C.Y., Zhang, W.: Virtual human bone structure of multi-rigid-body system modeling method research. Computer Aided Design and Graphics of Journals 17(6), 1354–1359 (2005)
11. Magermans, D.J., Chadwick, E.K., Veeger, H.E., et al.: Requirements for upper extremity motions during activities of daily living. Clinical Biomechanics 20(6), 591–599 (2005)

Vehicle Scheduling for Transports in Large-Scale Sports Meeting

Lei Wang and Chunlu Wang

Computer Science Department, Beijing University of Posts and Telecommunications, China

Abstract. This paper proposes a vehicle scheduling model for transports in large-scale sport meeting with the aim to reduce the operating costs. The problem is constrained by various practical operational constraints, travel time, and route time restrictions. A constrained local search method is developed to find better vehicle schedules. The model was implemented and tested on real data sets from Universidad Vehicle Scheduling System. The results indicate that a significant reduction of the operational costs can be achieved by optimization of constrained local search method for vehicle schedule with multi-depot with line change.

Keywords: Large-scale Sport Competitions, Line change, Vehicle scheduling.

1 Introduction

Vehicle scheduling system has become an essential part of many large-scale sports meeting to meet the goals of improving mobility, protecting the environment and saving energy. Developing a convenient and efficient vehicle scheduling system depends heavily on an efficient and fast scheduling process. The vehicle scheduling system usually includes 4 steps: designing routes, setting frequencies and building timetables, vehicle scheduling, and crew scheduling (Figure 1). In this paper, we focus

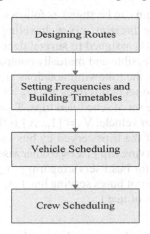

Fig. 1. Process of vehicle scheduling system

D. Jin and S. Lin (Eds.): Advances in MSEC Vol. 2, AISC 129, pp. 273–277.

on the vehicle scheduling problem of services with multi-depot and line change operations. We also focus on developing a flexible algorithm that can accommodate additional side constraints commonly found in real cases. Since the multi-depot vehicle scheduling problem is NP-hard, we propose a local search method that uses a simple local move to obtain a good quality solution within a viable time. The proposed model and algorithm are evaluated with the real data from the vehicle scheduling system in Universidad SHENZHEN 2011.

2   Literature Review

Moreover a lot of research has been on multi-depot vehicle scheduling. A column generation technique was used by Ribeiro et al. (Ribeiro and Soumis, 1994) to solve the problem to optimality for sizes up to 300 trips and 6 depots. (L"obel, 1999) solves an ILP formulation derived from a multi-commodity flow formulation by column generation and is able to handle large real-life instances. A recent approach from (Kliewer et al., 2006) extended in (Gintner et al., 2005) shows good results. The problem is represented as a time-space network with covering constraints. In (Kliewer and Bunte, 2007) vehicle scheduling is combined with the possibility to make small changes to the timetable, i.e. by shifting the departure times of trips that have been identified as critical within a limited time window. This is achieved by extending the time-space network. Their computational results are encouraging.

In this paper, we propose a different algorithm for MDVS. The method is based on a special type of local search algorithm which operates simple local moves to improve the solution. The algorithm proposed is arguably more flexible, and can easily accommodate additional side constraints. The users can easily introduce and/or modify the constraints without any changes to the algorithm structure.

3   Problem Definition

The vehicle scheduling problem can be stated as follows. For a given set of timetabled trips within a fixed planning horizon, given traveling times between all pairs of locations, and a fleet of vehicles assigned to several depots, find minimum cost sets of vehicle such that vehicle are feasible and mutually compatible.

To facilitate the discussion, the following notations are used: $D = \{1,..., d\}$ denotes a set of depots, $L = \{1,...,l\}$ is the set of bus lines and $L'_d = \{1,..., L_d\}$ is the set of bus lines for depot d, $T = \{1,...,t\}$ is the set of scheduled trips, $T_1 = \{1,...,T_1\}$ is the set of trips of line 1, $V = \{1,...,v_l\}$ is the set of vehicle, $V'_1 = \{1,...,v\}$ is the set of vehicle of line l, $Si = \{1,..., s\}$ is the ordered set of the trips served by bus i s, c_{ij} is the fixed cost of bus i serving trip j, $x_{ij} = 1$ if bus i serving trip j ; $x_{ij} = 0$ otherwise, e_{ij} is the starting time of trip j for bus i, t_{ij} is the trip time for bus i servicing trip j, t_{pq} is the transfer time between depots p and q , l_s is the number of buses serving line l , l_w is the total trips of line l , l_k is the number of dedicated buses of line l.

Constraints are following:

Constraint (1) ensures that the bus can only start the trip after finishing the previous assigned trip.

$$\sum_{i \in V} \sum_{j \in D_i} \left[\left(e_{i,j-1} + t_{i,j-1}\right) - e_{ij}\right] \geq 0 \tag{1}$$

Constraint (2) is associated with vehicle range (due to fuel limit), maintenance period, and maximum driver's working hours which can be written as:

$$\sum_{i \in V} \left(e_{i,m^*+1} - \left(e_{im^*} + t_{im^*} + bk\right)\right) \leq 0 \tag{2}$$

Constraint (3) ensures that the vehicle cannot run longer than the allowable maximum working hours.

$$\sum_{i \in V} \left| \sum_{j \in T} x_{ij} - A_i \right| \geq 0 \tag{3}$$

Line change and vehicle transfer operations are proposed to reduce the operating costs by efficiently utilizing the current resources. Typically, the vehicles are assigned to specific lines (routes) and depots. With this fixed plan, some vehicles may be underutilized and some lines may be lack of vehicles at some periods due to the fluctuation of travel times and demands during a day. The line change operation is the assignment of a vehicle belonging to one line to serve a trip of another line. The multi-depot operation (vehicle transfer) allows the vehicles to arrive or depart from different depots from their original depots. Thus, the vehicle scheduling with multi-depot and line change operations can potentially increase the efficiency of the bus utilization. Nevertheless, some buses may not be available for this operation due to some contractual/practical issues. The violation for the constraints can be formulated as:

$$\sum_{l \in L} \left[\sum_{i \in V} \left(\sum_{j \in T_l} x_{ij} \right) m_{il} \right] \leq 0 \tag{4}$$

$$\sum_{d \in D} \sum_{i \in V} \left[\sum_{l \in L_d} \left(\sum_{j \in T_l} x_{ij} \right) n_{id} \right] \leq 0 \tag{5}$$

4   Solution Approach

The CLS is used to solve the vehicle scheduling problem. The constraints are represented in a constraint matrix. CLS employs the random variable selection strategy and simple variable flip as a local move. The move quality is assessed by its total constraint violation. The violation scheme and redundant constraints are also used to guide the search into more promising regions of the search space. In addition, the constraint propagation is maintained, i.e. once vehicle is assigned to a particular scheduled trip the flip operations associated with infeasible trips (e.g. those overlapping with the current assigned trip) will be banned. This reduces the size of the search space and speed up the computation time.

```
// Constrained local search (CLS)
// INPUT: constraints
// OUTPUT: a best feasible solution found
A = initial assignment
A_b = best assignment
V_b = best total violation
while (iteration < stopping criterion) do
        nc = Random (numcol)
        V_m = MAX_INTEGER
        for i =1 to nc
                C_fst = select-columns(A)
                P = select -variable(C)
                V_m = MAX_INTEGER
                for j=1 to numcol
                        A' = flip(C_fst, C_i, P)
                        V' = total violation (A')
                        if V' < V_b then
                                A_b = A'
                                V_b = V'
                        end if
                        if V' <V_l then
                                A_l = A'
                        end if
                end for
                IF l g V <V THEN
                        A_g = A_l
                        V_g = V_l
                end if
        end for
        A = A_g
end while
```

The procedure of CLS can be summarized into three steps:

- Step 1: Constraint selection. After the initial assignment A, the nc columns are randomly selected in accordance with the violated columns are given the priority.
- Step 2: Variable selection. CLS selects the row (variable assigned to 1) to perform a trial flip with other variables in the same row.
- Step 3: Move acceptance. CLS chooses the best V' amongst the flipped variables in every single iteration and assigns the best V' to the current solution, $A = A'$.

5 Conclusion

The computational experiments were performed to evaluate the performance of the proposed method. The data sets from Universidad Vehicle Scheduling System were used for the case study. By scheduling nine bus lines with 829 scheduled trips, the results indicated the potential operating cost savings by around 8.84 percent compared to the current schedule (manually prepare).

References

1. Bertossi, A.A., Carraresi, P., Gallo, G.: On some matching problems arising in vehicle scheduling models. Networks 17, 271–281 (1987)
2. Clement, R., Wren, A.: Greedy Genetic Algorithms, Optimizing Mutations and Bus Driver Scheduling. In: Daduna, J.R., Branco, I., Paixao, J.M.P. (eds.) Computer-Aided Transit Scheduling. Lecture notes in Economics and Mathematical Systems, vol. 430, pp. 213–235. Springer, Heidelberg (1995)
3. Ribeiro, C.C., Soumis, F.: A column generation approach to the multiple-depot vehicle scheduling problem. Operations Research 42, 41–52 (1994)
4. Haghani, A., Banihashemi, M.: Heuristic approaches for solving largescale bus transit vehicle scheduling problem with route-time constraints. Transportation Research Part A 36, 309–333 (2002)
5. Lobel, A.: Vehicle scheduling in public transit and Lagrangian pricing. Management Science 4, 1637–1649 (1998)
6. Hadjar, A., Marcotte, O., Soumis, F.: A branch and cut algorithm for the multiple depot vehicle scheduling problem. Operations Research 54, 130–149 (2006)

Collaborative Filtering Recommendation Algorithm Based on User Interest Evolution

Dejia Zhang

Wenzhou Vocational and Technical College, Wenzhou 325035, China
zhangdejiaei@sina.com

Abstract. Personalized recommendation systems provide personalized item recommendations during a live user interaction, and they have achieved widespread success in electronic commerce nowadays. In many personalized recommender systems, collaborative filtering algorithm is the most famous technique and especially in collaborative filtering methods, neighborhood formation is an essential algorithm component. In order to make a recommendation in collaborative filtering algorithm, it is required to form a set of users sharing similar interests to the target user. But traditional collaborative filtering recommendation algorithm does not consider the evolution of user interest when finding the nearest neighbors in different time periods. And the recommendation results can not reflect the user's true interests. For this reason, a personalized collaborative filtering recommendation algorithm based on user interest evolution is given. This recommendation approach takes into account the important factor that user interests changes over time.

Keywords: personalized recommendation, collaborative filtering, algorithm, user interest evolution, time weight.

1 Introduction

With the rapid development of Internet, personalized recommendation has been an important application. At the same time, more and more researchers pay attention on personalized recommendations. Personalized recommendation refers to the characteristics of the user's interest. Personalized recommendation currently used technologies are: collaborative filtering, content-based recommendation, the recommendation based on association rules, recommendations based on Bayesian network technology, recommendation based on the Horting map technology. Collaborative filtering is the most extensive and most successful recommendation techniques. But the existence of traditional collaborative filtering high-dimensional sparse data and interest in the issue of failure often leads to inaccurate recommendations.

Collaborative filtering technique has been successfully used in commerce, science and other fields, such as Amazon, CDNow, eBay and so on. However, collaborative filtering-based recommendation system has encountered two basic challenges: how to effectively improve the collaborative filtering algorithm scalability and how to provide users with high-quality recommendations. To address these two fundamental

D. Jin and S. Lin (Eds.): Advances in MSEC Vol. 2, AISC 129, pp. 279–283.

challenges, in recent years, there has been a variety of improvement strategies. In this paper, a personalized collaborative filtering recommendation algorithm based on user interest evolution is given. This recommendation approach takes into account the important factor that user interests changes over time.

2   Collaborative Filtering

In real life, for the problem or do not understand things, people tend to consult their friends or some people with similar interests to make their own judgments based on their choices. Collaborative filtering is to simulate the process, according to the history of behavior between users.

In general, collaborative filtering algorithms can be divided into 3-steps: building user-item rating matrix, finding target user's neighbors based on the similarity of users, and generating the recommendation employing the neighbors.

2.1   Building User-Item Rating Matrix

On the assumption that a recommended system has m users and n items, then the system can be expressed as an m × n matrix. In the matrix, each user i rates on item j. Definition of user i on item j's score is as Rij, and if the user i did not rate item j, then Ri,j = 0. Collaborative filtering algorithm can be seen to predict Rij that is not rated.

2.2   Finding Neighbors

Finding similar neighbors of the target user, mainly it is based on the similarity of users. In general, the traditional method of similarity measure has the following three kinds.

Cosine similarity: User rating is seen as the n-dimensional vector space program, if the user did not rate the item, the value will be set to 0. The formula is shown as following.

$$sim(i, j) = \cos(\overset{v}{i}, \overset{v}{j}) = \frac{\overset{v}{i} \cdot \overset{v}{j}}{\left\| \overset{v}{i} \right\| \cdot \left\| \overset{v}{j} \right\|}$$

Modified cosine similarity: In the cosine similarity measure, it does not consider the problem of different users rating scale, and modified cosine similarity measure considers this problem. The formula is shown as following.

$$sim(i, j) = \frac{\sum_{c \in I_{ij}} (R_{ic} - \overline{R}_i)(R_{jc} - \overline{R}_j)}{\sqrt{\sum_{c \in I_i} (R_{ic} - \overline{R}_i)^2} \sqrt{\sum_{c \in I_j} (R_{jc} - \overline{R}_j)^2}}$$

Correlation similarity: Let users rate over i and j together a collection of items with Iij that the user i and user j is the similarity between the sim (i, j) measured by Peason correlation coefficient as following.

$$sim(i, j) = \frac{\sum_{c \in I_{ij}} (R_{ic} - \overline{R_i})(R_{jc} - \overline{R_j})}{\sqrt{\sum_{c \in I_{ij}} (R_{ic} - \overline{R_i})^2} \sqrt{\sum_{c \in I_{ij}} (R_{jc} - \overline{R_j})^2}}$$

2.3   Predicting

The traditional recommendation prediction algorithm is as follows:

$$P_{ui} = \overline{R_u} + \frac{\sum_{a=1}^{n} (R_{ai} - \overline{R_a}) \times sim(u, a)}{\sum_{a=1}^{n} |sim(u, a)|}$$

3   The Personalized Recommendation Algorithm Based on User Interest Evolution

3.1   Analysis

Traditional collaborative filtering methods can not change the interest. For example, the user's interest is constantly changing. In order to improve the sensitivity of prediction for new items, and achieved the desired effect in real time, we propose a time-weighted collaborative filtering algorithm to obtain new information on interest changes.

3.2   The Algorithm

Traditional collaborative filtering algorithm uses similar neighbors' interested users interested in a project the size of the project predicted the current user preference. But usually the algorithm does not take into account user interests' change, thus affecting the accuracy of the algorithm. Collaborative filtering algorithm needs for the current user selection and his recent interest in similar neighbor users, while out of his past only to users interested in a similar neighborhood.

Therefore, the traditional collaborative filtering algorithm is improved. First, by the time given to each score weight decay, weights of the recent scores, score past weight is small, and then press the weighted score to determine the similarity between users. Algorithm is as follows:

Step1 gradually forgotten function calculation.

Recommended system is a Web Intelligent Systems, and in order to improve its intelligence, recommender systems need to machine learning, data mining and learn from different disciplines such as psychology, valuable ideas. Recommended system is consistent with the role of forgotten psychology theory. In order to quickly capture changes in user interests, the choice should be fully learn the function of psychology forgotten people forget about the results of the theory. Forgotten strategy calculated as follows.

$$f(t) = m \times \frac{t - MIN}{MAX - MIN} + (1-m) \qquad \begin{array}{l} MIN \le t \le MAX \\ 0 \le m \le 1 \\ 1 - m \le f(t) \le 1 \end{array}$$

Step2 calculated similarity between users using Pearson correlation coefficient.

$$w(u,v) = \frac{\sum_{i \in Com} (R_{ui} \times f(t) - \overline{R_u})(R_{vi} \times f(t) - \overline{R_v})}{\sqrt{\sum_{i \in Com} (R_{ui} \times f(t) - \overline{R_u})^2 \sum_{i \in Com} (R_{vi} \times f(t) - \overline{R_v})^2}}$$

Step3 According to the similarity between users calculated by Pearson correlation similarity, we choose the highest similarity of the user as the target user's nearest neighbor set.

Step4 Recommend using following formula.

$$P_{ai} = \overline{R_a} + \frac{\sum_{j \in Na} sim(a, j)(R_{ji} - \overline{R_j})}{\sum_{j \in Na} sim(a, j)}$$

4   Summary

In this paper, we give the personalized collaborative filtering recommendation algorithm based on user interest evolution. This recommendation approach takes into account the important factor that user interests changes over time.

Acknowledgment. This work was supported by the Wenzhou Vocational and Technical College Foundation, China (Grant No. WZY2010010).

References

1. Gong, S.: A Personalized Recommendation Algorithm on Integration of Item Semantic Similarity and Item Rating Similarity. Journal of Computers 6(5), 1047–1054 (2011)
2. Min, S.-H., Han, I.: Detection of customer time-variant pattern for improving recommender systems. Expert Systems with Applications 28, 189–199 (2005)
3. Gong, S.: Privacy-preserving Collaborative Filtering based on Randomized Perturbation Techniques and Secure Multiparty Computation. International Journal of Advancements in Computing Technology 3(4), 89–99 (2011)
4. Lee, T.Q., Park, Y., Park, Y.-T.: A time-based approach to effective recommender systems using implicit feedback. Expert Systems with Applications (2007)
5. Gong, S.: Employing User Attribute and Item Attribute to Enhance the Collaborative Filtering Recommendation. Journal of Software 4(8), 883–890 (2009)
6. Gong, S.: A Collaborative Filtering Recommendation Algorithm Based on User Clustering and Item Clustering. Journal of Software 5(7), 745–752 (2010)

7. Cho, Y.B., Cho, Y.H., Kim, S.H.: Mining changes in customer buying behavior for collaborative recommendations. Expert Systems with Applications 28, 359–369 (2005)
8. Yu, L., Liu, L., Li, X.: A hybrid collaborative filtering method for multiple-interests and multiple-content recommendation in E-Commerce. Expert Systems with Applications 28, 67–77 (2005)
9. Gong, S.: An Efficient Collaborative Recommendation Algorithm Based on Item Clustering. In: Luo, Q. (ed.) Advances in Wireless Networks and Information Systems. Lecture Notes in Electrical Engineering, vol. 72, pp. 381–387. Springer, Heidelberg (2010)

Wavelet TRANSFORMATION for Content-Based Image Retrieval COMBINE G-Regions Of Interest

Xugang, Fuliang Yin, and Chaorong Wei

A333, Building of CHUANG XIN YUAN, Computer Science Institute,
Dalian University of Technology, 116024, China
xugang@dlut.edu.cn

Abstract. This paper in the research of Region-Based characteristic image retrieval method foundation, proposed a new kind of the color image retrieval method based on wavelet transformation G-Regions Of Interest (GROI). We first use HVS(Human Visual System) characteristic to choose the color space which fit for the visual characteristics, then use K-means clustering to extract the areas of interest in the wavelet transform domain, and using the local energy of the wavelet coefficients in the areas of interest as the texture feature, color's mean and variance as the color feature, the barycentric coordinates as the position feature. We calculated the similarity between the image content and retrieval. The simulation results show that this GROI method which combined color characteristic, texture characteristic and position characteristic can more accurately find the necessary content of the images to users, significantly improve the accuracy in color image retrieval.

Keywords: Image Retrieval, GROI, Wavelet Transform.

1 Introduction

We proposed one kind image retrieval method based on wavelet transformation G-Regions Of Interest, this method have characteristics as below:

(1) It unions human eye vision sensation characteristic, accepted the extraction strategy based on the wavelet transformation regions of interest.
(2) It not only quite accurately described the regions of interest texture characteristic (to get interest area by the wavelet coefficient partial energy achievement of texture characteristic), simultaneously considered the interest area position characteristic.
(3) It has consummated similarity computation model.

In this method ,we use K-means clustering to extract the area of interest in the wavelet transform domain, use the local energy of the wavelet coefficients in the areas of interest as the texture feature, the color's mean and variance as the color feature, the barycentric coordinates as the position feature,then calculate the similarity of images for content retrieval. [1-2]

D. Jin and S. Lin (Eds.): Advances in MSEC Vol. 2, AISC 129, pp. 285–291.
springerlink.com   © Springer-Verlag Berlin Heidelberg 2011

2   Region of Interest Characteristic Extraction

2.1   Color Characteristic

One kind extremely simple and the effective color characteristic can be obtained by the color moment method which is proposed by Stricker and Orengo and Dimai et al.. This method mathematics foundation lies that any color distribution in the image can be expressed by its moment .In addition, because that the color distribution information mainly concentrates in the low-order moment, in order to reduce the computation load, this article only uses the color the first moment (average value), the second moment (variance) to express the image the color distribution.

We assume that the interest region is R (the size is $m \times n$), then it's color average value and the color variance respectively are:

$$u_i = \frac{\sum\limits_{x=0}^{m-1}\sum\limits_{y=0}^{n-1} I_i(x,y)}{m \times n} \tag{1}$$

$$\sigma_i = \sqrt{\frac{\sum\limits_{x=0}^{m-1}\sum\limits_{y=0}^{n-1} (I_i(x,y)-u_i)^2}{m \times n}} \tag{2}$$

In which $I_i(x,y)$ is the i th color component the picture element to select (x,y) in interest region R .We select HSV color space which can more conform to the human eye vision sensation characteristic, then the color characteristic of the interest region R obtained is:

$$C = (\mu_H, \sigma_H, \mu_S, \sigma_S, \mu_V, \sigma_V) \tag{3}$$

2.2   Texture Characteristic

We extract the texture characteristic of the region of interest in the wavelet transformation territory, in order to make full use of the wavelet coefficient characteristics that are not related to the image translation, revolving, reproduce by pantograph and so on. the region of interest texture characteristic extraction method based on the wavelet transformation is as follows:

First from the interest region R we withdraw its brightness component, and carry on 2 level of wavelet transformations to brightness R^V . Then in the wavelet transformation territory, we calculate the wavelet coefficient partial energy of the interest region brightness component R^V .

We assume that wavelet coefficient $\{d_{r,s}^{\theta,l}\}$ partial energy which is in the coordinates position (x,y) is $E_l^\theta(x,y)$ (among this, $l \in \{1,2\}$ expressed wavelet transformation layer, $\theta \in \{LL, LH, HL, HH\}$, indicated in wavelet territory innertube image direction, r, s expressed motion factor), then

$$E_l^\theta(x, y) = \sum_{r,s} (d_{r,s}^{\theta,l})^2 K(2^{-j}x - r, 2^{-j}y - s) \tag{4}$$

Here, the $K(x, y)$ selection is the *Gaussian* essence function:

$$K(\vec{x}, \vec{y}) = \exp(\frac{-\|\vec{x} - \vec{y}\|^2}{2\sigma^2}) \tag{5}$$

In order to reduce the computation order of complexity, we select brightness component R^V various innertubes image following energy characteristic (partial energy to value) as the texture characteristic T of interest region R, namely:[3-4]

$$T = \{\log E_1^{LH}, \log E_1^{HH}, \log E_1^{HL}, \log E_1^{LL},$$
$$\log E_2^{LH}, \log E_2^{HH}, \log E_2^{HL}, \log E_2^{LL}\} \tag{6}$$

2.3   Position Characteristic Withdraws

Here we use the existing center of gravity formula directly. The two-dimensional center-of-gravity position remaining to distinguish objects is calculated by the following formula: [3-4]

$$\bar{x} = \frac{\sum\limits_{y=1}^{N}\sum\limits_{x=1}^{M} xg(x, y)}{\sum\limits_{y=1}^{N}\sum\limits_{x=1}^{M} g(x, y)} \tag{7}$$

$$\bar{y} = \frac{\sum\limits_{y=1}^{N}\sum\limits_{x=1}^{M} yg(x, y)}{\sum\limits_{y=1}^{N}\sum\limits_{x=1}^{M} g(x, y)} \tag{8}$$

$$g(x, y) = \begin{cases} 1 & (x, y) \in Interested \\ 0 & (x, y) \in Background \end{cases} \tag{9}$$

Considered the convenient of computation, position characteristic O of interest region R is indicated with its center of gravity normalized coordinates, namely:

$$O = [\frac{\bar{x}}{M}, \frac{\bar{y}}{N}] \tag{10}$$

Here $\bar{x}$, $\bar{y}$ expresses areal coordinates, M and N are the image affable high.

Finally, the i th interest region R_i characteristic with the vector representation is $F_i = (C_i, T_i, O_i)$, but the interest region the image characteristic vector including m is expressed as[10-12]:

$$\{F_0, F_1, F_2, ..., F_{m-2}, F_{m-1}\} \tag{11}$$

3   Similarity Measure

It is convenient to normalize the different dimension characteristic variable through adjusting various characteristics variance σ_i, and the similarity is always between [0,1]. When $s \to 1$, the two interest regions are most similar, when $s \to 1$ the two interest region are least similar. The similarity formula shows as follows: [5-7]

(1) color similarity

$$s_C = \exp[-\frac{\sum\limits_{g \in \{H,S,V\}}[(\mu_g(R_i) - \mu_g(R_j))^2 + (\sigma_g(R_i) - \sigma_g(R_j))^2]}{6\sigma_C^2}] \tag{12}$$

(2) texture similarity

$$s_T = \exp[-\frac{\sum\limits_{l \in \{1,2\},\theta \in \{LL,LH,HL,HH\}}[\log E_l^\theta(R_i) - \log E_l^\theta(R_j)]^2}{7\sigma_T^2}] \tag{13}$$

(3) position similarity

$$s_O = \exp[-\frac{(\bar{x}(R_i) - \bar{x}(R_j))^2 + (\bar{y}(R_i) - \bar{y}(R_j))^2}{2\sigma_O^2}] \tag{14}$$

Therefore, the similarity between any two interest regions the R_i and the R_j in the image is each characteristic similarity weighted averages, namely

$$s_R(R_i, R_j) = w_C s_C + w_T s_T + w_O s_O \text{ , and}$$
$$w_C + w_T + w_O = 1 \tag{15}$$

There is m interest regions in demonstration image Q, n interest regions in the database any image I, then similarity of image Q in relation to I is:

$$s(Q,I) = \sum_{i=0}^{m-1} W_i s_R(R_i, P_I(R_i)) \quad \sum_{i=0}^{m-1} W_i = 1 \tag{16}$$
$$\text{,} \quad \text{且}$$

Among them, W_i is the weight interest region R_i. Because of the difference area of interest region in the image, the important degree is various. This article initialize W_i into the percentage of the interest region R_i to occupy the entire picture area, thus causes weight W_i to be proportional with the area of the interest region R_i. $P_I(R_i)$ $(i = 0,1,..., m-1)$ is expressed the mapping relations between interest region R_i of image Q and the interest region in the image I, namely it will return in the interest region which in image I is most similar with the ith interest region R_i of the image Q.[8-10]

4   Experiments Result

Here we select 5 kinds of image in the Corel image database (indigenous life, flower, dinosaur, building, automobile), extract a sample image from each kind of images at random as the demonstration image, and select the before 30 most similar images to be the retrieval result in each time inquire, total 25 times inquire. Select the before 12 most similar images as the image retrieval result shown in Fig.1(a-b). The numeral above each picture represents the similarity factor.Number 1 represents the selected sample images.

The standard accuracy ratio and the standard recall mean value of its 5 times inquiry results shown in Table 1.Compareing with the other image retrieval method such as NID and Gabor filter,the GROI method show more perfect in Table 2. Comparing with the retrieve data in the data sheet, we find the Average accuracy ratio and the Average recall ratio of the construction and the automobile image is highter than the other images,because of their texture characteristics are more regular. Above experiments show that the retrieve result are ideal.[11-12]

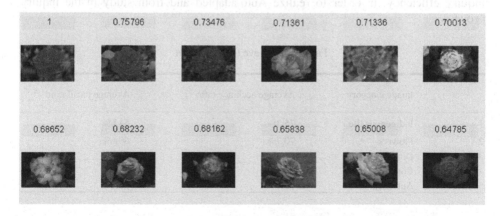

Fig. 1(a). Flower retrieval effect

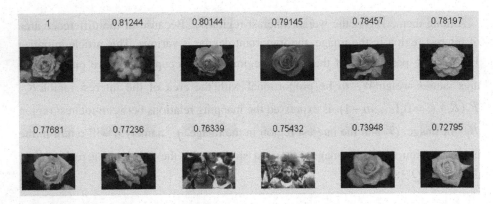

Fig. 1(b). Flower retrieval effect

5   Conclusion

This article proposes a new kind of the color image retrieval method based on wavelet transformation G-Regions Of Interest. The experimental result indicated that not only this algorithm has the better retrieval performance, but also its retrieval result can well close to person's visual sensation effect.

The deficiency of this article lies in that because we divide the image into certain sub-blocks when image processing, and the operand can be very great when we compare the similarity, thus it would cause the running time of procedure to be long.

The further research work includes: (1) Improving algorithm, reducing the procedure running time; (2) Selecting small image database to train, withdrawing the best characteristic value coefficient; (3) Exploring more effective region of interest extraction method (specially regarding background complex image); (4) Combining the correlation feedback and the machine learning technology, further enhancing the inquiry efficiency, in order to realize Auto-adapted and from study in the inquiry process.

Table 1. Retrieve the data sheet

	Image category	Average accuracy ratio	Average recall ratio
1	Indigenous life	75.3%	64.6%
2	Flowers	70.7%	60.6%
3	Dinosaur	73.3%	62.9%
4	Construction	76.7%	65.7%
5	Automobile	88.7%	76%

Table 2. Compareing with NID and Gabor method

method	Image category	Average accuracy ratio	Average recall ratio
NID	Automobile	82.3%	73.6%
Gabor	Automobile	81.7%	71.3%
GROI	**Automobile**	**88.7%**	**76%**

References

1. Chan, Y.-K., Ho, Y.-A., Liu, Y., Chen, R.: A ROI image retrieval method based on CVAAO. Image and Vision Computing 26(11), 1540–1549 (2008)
2. Gondra, I., Heisterkamp, D.R.: Content-based image retrieval with the normalized information distance. Computer Vision and Image Understanding 111(2), 219–228 (2008)
3. Doyle, S., Hwang, M., Naik, S., Feldman, M., Tomaszeweski, J., Madabhushi, A.: Using manifold learning for content-based image retrieval of prostate histopathology. In: MICCAI 2007 Workshop on Content-based Image Retrieval for Biomedical Image Archives: Achievements, Problems, and Prospects, pp. 53–62 (2007)
4. Quellec, G., Lamard, M., Cazuguel, G., Cochener, B., Roux, C.: Wavelet optimization for content-based image retrieval in medical databases. Medical Image Analysis 14(2), 227–241 (2010)
5. Jhanwar, C.N., Chaudhuri, S., Seetharaman, G., Zavidovique, B.: Content based image retrieval using motif cooccurrence matrix. Image and Vision Computing 22(14), 1211–1220 (2004)
6. Qiu, G.: Embedded colour image coding for content-based retrieval. Journal of Visual Communication and Image Representation 15(4), 507–521 (2004)
7. Mejdoub, M., Fonteles, L., BenAmar, C., Antonini, M.: Embedded lattices tree: An efficient indexing scheme for content based retrieval on image databases. Journal of Visual Communication and Image Representation 20(2), 145–156 (2009)
8. Müller, H., Michoux, N., Bandon, D., Geissbuhler, A.: A review of content-based image retrieval systems in medical applications—clinical benefits and future directions. International Journal of Medical Informatics 73(1), 1–23 (2004)
9. Liu, Y., Zhang, D., Lu, G., Ma, W.-Y.: A survey of content-based image retrieval with high-level semantics. Pattern Recognition 40(1), 262–282 (2007)
10. Cheng, S.-C.: Content-based image retrieval using moment-preserving edge detection. Image and Vision Computing 21(9), 809–826 (2003)
11. Chang, H., Yeung, D.-Y.: Kernel-based distance metric learning for content-based image retrieval. Image and Vision Computing 25(5), 695–703 (2007)
12. Arevalillo-Herráez, M., Ferri, F.J., Domingo, J.: A naive relevance feedback model for content-based image retrieval using multiple similarity measures. Pattern Recognition 43(3), 619–629 (2010)

An Optimal Algorithm for Resource Scheduling in Cloud Computing

Qiang Li

College of Computer Science, Sichuan University,
Chengdu, Sichuan, China
liq@scu.edu.cn

Abstract. Resource scheduling based on SLA (Service Level Agreement) in cloud computing is NP-hard problem. There is no efficient method to solve it. This paper proposes an optimal algorithm for solving the problem by applying Gröbner bases theory and stochastic integer programming technique. The experimental results of the implementation for the algorithm are also presented.

Keywords: Gröbner Bases, SLA, Stochastic Integer Programming, Resource Scheduling, Cloud Computing.

1   Introduction

Cloud computing is a resource delivery and usage model to get resource (hardware, software, applications) via network "on-demand" and "at scale" as services in a multi-tenant environment. The network of providing resource is called Cloud. All resources in the cloud are scalable infinitely and used whenever as utility. In practice of cloud computing, providing an optimal/appropriate resource to user becomes more and more important.

In cloud computing, each application is often designed a business process which includes a set of abstract services. Each abstract service encapsulates the function of an application component using its interface, and a concrete service(s) or resource(s) is selected (bound) at runtime to fulfill the function. A Service Level Agreement (SLA) in cloud computing is defined upon a business process as its end-to-end Quality of Service (QoS) constraints since a business process defines how abstract services interact to accomplish a certain business goal. Since different concrete services may operate at different QoS measures, an appropriate/optimal set of concrete services/resources may be selected so that it guarantees the fulfillment of SLA and cost is minimal. Such problem, the QoS-aware service composition problem, is a combinatorial optimization problem which ensures the optimal mapping between each abstract service and available resources [1], [2]. Since the problem is known as NP-hard[3], it takes a significant amount of time and costs to find optimal solutions (optimal combinations of resources) from a huge number of possible solutions.

This paper defines a resource composition model based on stochastic integer programming which address the SLA-aware resource composition problem. We propose an algorithm to solve stochastic integer programming problem and the results

D. Jin and S. Lin (Eds.): Advances in MSEC Vol. 2, AISC 129, pp. 293–299.

of the implementation for the algorithm are also presented. This paper is organized as follows. Section 2 shows the problem of resource composition and section 3 proposes a model of a resource composition based on stochastic integer programming technique. Section 4 describes the solution for stochastic integer programming based on Gröbner Bases theory [4], [5]. Section 5 introduces an optimal algorithm for solving the resource scheduling problem in cloud computing and presents the simulation results to evaluate the algorithm. Sections 6 conclude with some discussion on related work.

2   SLA-Based Resource Composition Problem

We define the SLA-based resource composition problem with the following assumptions:

(1) A SLA between user and cloud provider: a service agreement on throughput, latency and cost.
(2) A business process instance: it realizes users request.
(3) A series of abstract services: it executes the business process.
(4) Concrete service/Resource: it implements an abstract service.
(5) QoS for each concrete service/resource: it has three attributes: throughput, latency and cost, while throughput and latency can vary at runtime.

The problem is how to select concrete service/resource to realize abstract service in business process instance that satisfies SLA, while the overall cost is minimal.

3   Model For SLA-Based Resource Scheduling

Applying stochastic integer programming technique, we give a model for SLA-based resource schedule. With this model, we can find the optimal resource schedule to realize a business process and satisfies SLA.

Model 3.1 Optimal Resource Scheduling Model

$$Min \sum_{i=1}^{|\alpha|} \sum_{j=1}^{|\beta_i|} c_{ij} x_{ij} \text{ subject to} \tag{1}$$

$$\sum_{i=1}^{|\alpha|} \sum_{j=1}^{|\beta_i|} c_{ij} x_{ij} \leq C_{SLA} \tag{2}$$

$$\sum_{j=1}^{|\beta_i|} x_{ij} = 1, i \in \{1,...,|\alpha|\} \tag{3}$$

$$p \left\{ \begin{matrix} \min\{\xi_{ij}^t x_{ij} : i \in \{1,...,|\alpha|\}, j \in \{1,...,|\beta_i|\}, x_{ij} \neq 0\} \geq T_{SLA}, \\ \sum_{i=1}^{|\alpha|} \sum_{j=1}^{|\beta_i|} \xi_{ij}^l x_{ij} \leq L_{SLA} \end{matrix} \right\} \geq \gamma \tag{4}$$

$$x_{ij} \in \{0,1\} \tag{5}$$

Where

α : a set of abstract services

$|\alpha|$: number of elements in set α

β_i : a set of resource available to implement i-th abstract service

$|\beta_i|$: number of elements in set β_i

$x_{ij} = \begin{cases} 1{:}select \quad jth \quad resource \quad for \quad ith \quad abstract \quad service \\ 0{:}otherwise \end{cases}$

c_{ij} : cost of j-th resource for implement i-th abstract service

C_{SLA} : cost of SLA

T_{SLA} : throughput of SLA

L_{SLA} : latency of SLA

ξ_{ij}^t : random variable for resource's throughput

ξ_{ij}^l : random variable for resource's latency

γ : probability of fulfillment of a given SLA

Formula (1) means that the overall cost is minimal under the solution of variables ψ x_{ij} . Formula (2) means that overall cost is less than or equal to SLA's cost.

Formula (3) means that only one resource is selected to implement an abstract service and formula (4) means that probability of fulfillment of SLA's two attributes is great than or equal to γ . The solution of above model is discussed in next two sections.

4   Gröbner Bases for Stochastic Integer Programming

In this section we first introduce Gröbner Bases for integer programming (IP) and then extend it to solve stochastic integer programming (SIP).

Consider the following model of IP problems:

$$IP_{A,C}(b) = \min\{Cx : Ax = b, x \in N^n\}$$

where C is an n-vector of real numbers, A is an $m \times n$ matrix of integers and b is an m-vector of integers. This model means that we solve variables x under the constraints $Ax = b$ so that the value of Cx is minimal. We use $IP_{A,C}$ ψ to denote a generic IP problem.

The method for solving IP via Gröbner Bases was firstly introduced by Conti et al. in commutative algebra [6] and by Thomas in geometry [7] independently. The key idea is to encode an IP problem into a special ideal associated with the constraint matrix A and the cost (objective) function Cx. An important property of such an encoding is that its Gröbner bases correspond directly to the test sets of the IP problem. Thus, by employing an algebraic package such as MACAULAY or MAPLE, the test sets of the IP problem can be directly computed. Using a proper test set (such as the minimal test set which corresponds directly to the reduced Gröbner basis of the encoded ideal), the optimal value of the cost function can be computed by constructing a monotonic path from the initial non-optimal solution of the problem to the optimal solution.

In practical IP problem, the size of Gröbner basis increases quickly and the computation for it becomes expensive as the number of variables in IP becomes large. To overcome the disadvantage, Qiang et.al proposed a new algorithm for solving IP called *Minimised Geometric Buchberger Algorithm* (MGBA)[9], which improves the computation of Gröbner Basis for IP problem by truncating the basis with the fixed right hand b. The experimental results of the implementing BGBA state that MGBA shows significant performance improvement comparing to Conti[6] and Thomas[7]'s algorithms.

Now we consider the class of SIP as the following form.

Min h(x) subject to

$$P\{Tx \leq \xi\} \geq \gamma$$

$$Ax = b$$

$x \in N^n$, ξ is a vector of random variables.

The above model means that we solve variables x under the probabilistic constraint $P\{Tx \leq \xi\} \geq \gamma$ and the constraints $Ax = b$ so that the value of $h(x)$ is minimal.

Based on S.R.Tayur et al.'s idea[10], we apply MGBA to solve SIP problem with the following process.

(1) Divide the model into two parts

One is composed of the probabilistic constraints and some complicated constraints, called membership oracle; and the other is a simple IP after removing the membership oracle from the SIP problem, called Reduced IP (RIP).

(2) Compute the test set of RIP

We compute RIP's test set simply by using MGBA.

(3) Compute the test set of SIP

The test set of RIP provides a set of directions that can be used to trace paths from every nonoptimal solution to the optimal solution of the RIP. So, for any feasible solution of the RIP, we get an optimal solution by searching the set of directions. Simultaneously, we can also walk back from the optimal solution to every other feasible solution of the RIP by simply reversing these paths. By reversing all directions of RIP's test set, we get the test set of SIP, which was proved by S.R.Tayur et al.[10].

(4) Compute the optimal solution of SIP

We compute the optimal solution of the SIP by walking back from the optimal solution of RIP to other feasible solutions and querying the membership oracle to check whether the reached point is feasible or not. If the reached point is feasible for the membership oracle, it is the optimal solution of SIP. Same as the test set of IP, we can prove that the search (walking back) terminates with either the optimal solution of SIP or all paths are searched, i.e., the SIP is infeasible.

5   Optimal Algorithm and Implementation

Algorithm 5.1. Algorithm for   Model 3.1

1. Decompose Model3.1 into RIP (including formula(1),(2),(3) and(5)) and membership oracle (including formula (4)).
2. Compute a test set $\wp$ of RIP by using MGBA
3. Compute an optimal solution x^R from any feasible solution of   RIP
4. Derive the test set ϖ of   Model3.1   by reversing all directions of vectors in $\wp$
5. Compute the optimal solution of Model3.1 with test set ϖ and optimum x^R of RIP by using feasible checking method

We have implemented above algorithm and have down several simulation by finding optimal resource schedule for business process with different abstract services and the different SLA. The experiment result is shown in Table 1. From the result, we can see that the computation of optimal resource schedule finished in a reasonably short time. But the time is growing exponentially with the increasing of the numbers of the services and resources. The reason is that the computation of Gröbner basis is still suffer from the fact that the size of Gröbner basis grows exponentially with the increasing of the variables.

Table 1. Experimental Results

Problems	γ	Optimal resource schedule	Time(sec.)
$\lvert \alpha \rvert = 2, \lvert \beta_i \rvert = 3$ $i = 1,2$	0.84	$x_{11} = 0, x_{12} = 1, x_{13} = 0, x_{21} = 0, x_{22} = 1, x_{23} = 0$ Overall QoS: C=44, T=100, L=10	2.36
	0.88	$x_{11} = 1, x_{12} = 0, x_{13} = 0, x_{21} = 0, x_{22} = 0, x_{23} = 1$ Overall QoS: C=97, T=130, L=8	182.98
$\lvert \alpha \rvert = 4, \lvert \beta_i \rvert = 3$ $i = 1,...,4$	0.5	$x_{11} = 0, x_{12} = 1, x_{13} = 0, x_{21} = 0, x_{22} = 0, x_{23} = 1$ $x_{31} = 1, x_{32} = 0, x_{33} = 1, x_{41} = 1, x_{42} = 0, x_{43} = 1$ Overall QoS: C=93, T=250, L=30	235.59
$\lvert \alpha \rvert = 6, \lvert \beta_i \rvert = 3$ $i = 1,...,6$	0.62	$x_{11} = 0, x_{12} = 0, x_{13} = 1, x_{21} = 0, x_{22} = 1, x_{23} = 0$ $x_{31} = 1, x_{32} = 0, x_{33} = 0, x_{41} = 1, x_{42} = 0, x_{43} = 0$ $x_{51} = 0, x_{52} = 1, x_{53} = 0, x_{61} = 0, x_{62} = 1, x_{63} = 0$ Overall QoS: C=122, T=450, L=40	433.64
$\lvert \alpha \rvert = 7, \lvert \beta_i \rvert = 3$ $i = 1,...,6$ $\lvert \beta_7 \rvert = 5$	0.47	$x_{11} = 1, x_{12} = 0, x_{13} = 0, x_{21} = 0, x_{22} = 0, x_{23} = 1$ $x_{31} = 0, x_{32} = 1, x_{33} = 0, x_{41} = 0, x_{42} = 0, x_{43} = 1$ $x_{51} = 0, x_{52} = 1, x_{53} = 0, x_{61} = 1, x_{62} = 0, x_{63} = 0$ $x_{71} = 1, x_{72} = 1, x_{73} = 0, x_{74} = 0, x_{75} = 0$ Overall QoS: C=212, T=650, L=80	1219.18

6   Related Works and Conclusion

In [11], H.Wada et al. develop a multi-objective optimization model to tackle SLA-aware service composition problem. They consider multiple SLAs simultaneously and provided a set of solutions of equivalent quality. But their model is based on the heuristic genetic algorithm in which the performance cannot be expected. In [12], S.Chaisiri et al. apply stochastic integer programming for resource provision optimization problem. The algorithm minimizes the total cost of resource provision in a cloud computing environment. The optimal solution is obtained by formulating and solving stochastic integer programming with two-stage recourse. However, they do not consider the notion of SLA, which is one of the most important business notion in cloud computing.

In this paper, we define a model for optimization of SLA-based resource schedule in cloud computing based on stochastic integer programming technique. By applying Gröbner bases theory, we give an algorithm by extending MGBA to solve the optimal model of SLA-based resource schedule problem. The simulation results show that the optimal solution is obtained in a reasonable short time by implementing the optimal algorithm. In future, we plan to speed up the performance of the algorithm by improving the computation of Gröbner bases (test set) for integer programming.

References

1. Anselmi, J., Ardagna, D., Cremonesi, P.: A QoS-based Selection Approach of Autonomic Grid Services. In: ACM High Performance Distributed Computing, Workshop on Service-Oriented Computing Performance (June 2007)
2. Yu, T., Lin, K.J.: Service Selection Algorithms for Composing Complex Services with Multiple QoS Constraints. In: Int'l Conf. on Service-Oriented Computing. Addison-Wesley Professional (December 2005)
3. Canfora, G., Penta, M.D., Esposito, R., Villani, M.L.: An Approach for QoS-aware Service Composition based on Genetic Algorithms. In: Genetic and Evolutionary Computation Conference (June 2005)
4. Adams, W.W., Loustaunau, P.: An Introduction to Gröbner bases. American Mathematical Society, vol. 3 (1994)
5. Buchberger, B., Winkler, F. (eds.): Gröbner bases and applications. Cambridge University Press (1998)
6. Conti, P., Traverso, C.: Buchberger Algorithm and Integer Programming. In: Mattson, H.F., Rao, T.R.N., Mora, T. (eds.) AAECC 1991. LNCS, vol. 539, pp. 130–139. Springer, Heidelberg (1991)
7. Thomas, R.R.: A geometric Buchberger algorithm for integer programming. Mathematics of Operations Research 20, 864–884 (1995)
8. Kall, P., Wallace, S.W.: Stochastic Programming. John Wiley & Sons Ltd., Chichester (1994)
9. Li, Q., Guo, Y.K., Ida, T.: Minimised Geometric Buchberger Algorithm for Integer Programming. Annals of Operations Research 108, 87–109 (2001)
10. Tayur, S.R., Thomas, R.R., Natraj, N.R.: An algebraic geometry algorithm for scheduling in the presence of setups and correlated demands. Mathematical Programming 69(3), 369–401 (1995)

11. Wada, H., Oba, K.: Multi-objective Optimization of SLA-aware Service Composition. In: Proc. of IEEE Workshop on Methodologies for Nonfunctional Properties in Services Computing, Honolulu (2008)
12. Chaisiri, S., Lee, B.S., Niyato, D.: Optimal Virtual Machine Placement across Multiple Cloud Providers. In: IEEE APSCC 2009, Singapore (December 2009)

An Empirical Study on Governance Characteristics and Performance of High-Tech Corporations

Ye Liu, Lili Zeng, Duo Wang, Fanyun Sun, and Liang Feng

School of Business Administration, Northeastern University
Shenyang, P.R. China, 110004
Yliu@mail.neu.edu.cn

Abstract. This paper carries out an empirical study on corporate governance characteristics and performance of high tech corporations. Our conclusion shows that among the severely competitive high-tech corporations, ROE is significantly positively correlated with Tobin's Q; the average salary of executives, the stock ownership of executives and institutions are significantly positively correlated with ROE. And the degree of equity restriction is significantly positively correlated with Tobin's Q. In order to improve corporate performance, we need to consummate the corporate governance.

Keywords: High-tech, Corporate Governance, Empirical study, Human capital.

1   Introduction

In the past 40 years, scholars have analyzed many different models of corporate governance and found that the increasing path and life cycle of common corporations is different from high-tech corporations. The development features of high-tech corporations are quite different from common ones. All those factors show that high-tech corporations need an appropriate corporate governance model. The key point for corporations to grow from small to large and become from closed to open is to take advantages of the period of the next life cycle. Lerner (2000) expresses some views on corporate governance of Silicon Valley. Firstly, the purpose of establishing a new corporation is to develop new and high risk products. Actually, it's difficult for the former corporation to carry out a thorough revolution due to the lack of an appropriate incentive system to take the risk. So, it's easier for the new corporation to develop new products. Secondly, because of the high cost of developing new technologies and the uncertainty of future, in order to avoid the risk, it requests to supplement external capital. The early venture investment may meet this request. In order to make uncertain profits, the venture capitalists prefer to take the high risk.

2   An Overview and Hypothesis

2.1   The Influence of Corporate Ownership Structure on Corporate Performance

How does the degree of centralized equity influence corporate performance? It still remains an argument now. In some scholars' opinion, the degree of centralized equity

D. Jin and S. Lin (Eds.): Advances in MSEC Vol. 2, AISC 129, pp. 301–307.
springerlink.com   © Springer-Verlag Berlin Heidelberg 2011

has nothing to do with corporate performance. For example, Demsets and Leth (1985) researched 511 large corporations in the USA, and found that there is no relationship between the degrees of centralized equity with the accounting index of business performance (ROE). Holderness (2003) argued that there are few evidences supporting the conclusion that corporate ownership structure has a significant influence on corporate value. But, there are many scholars holding the view that it is correlated with the corporate performance too. The degree of equity restriction mainly reflects the top major shareholders' power balance. All those researches have accepted the effect of the degree of equity restriction in corporate governance. But some other scholars hold the opposite views.

H1a: The degree of centralized equity is significantly positively correlated with corporate performance;
H1b: The degree of equity restriction is significantly positively correlated with corporate performance.

2.2   The Influence of Incentive Mechanism of Executives on Corporate Performance

Jensent and Meckling (1976) argued that if executives are allowed to hold shares, their interests will coincide with the outside shareholders' interests, and that will reduce their incentives of in-service consumption, corroding shareholders' wealth and other non-value-maximizing behavior. The increasing of shares held by executives may reduce the agency cost caused by the separation of ownership and control, and then improve the corporate performance. Mehran's epirical research (1995) showed that corporate performance is positively correlated with the proportion of the shares held by CEO and of their rewards based on equity. But Gang Wei (2000) proved that there is no significant relationship between executives holding shares and corporate performance which is represented by ROA.

H2a: The corporate performance is significantly positively correlated with the summation of the proportion of shares held by executives in the top ten shareholders;
H2b: The corporate performance is significantly positively correlated with the average salary of executives.

2.3   The Influence of Board of Director on Corporate Performance

Shukun Wu, Jie Bo and Youmin Xi's research (1998) found that in Chinese listed corporations, whether the executives both work in the board and corporation or not, corporate performance won't change. Obviously, CEO can't obey the decisions which may corrode his/her interests effectively, and he/ she has the power to pursue his own interests rather than shareholders'. It is even harder for small shareholders' voice to be heard on some important issues, so the opportunistic behavior of controlling shareholders is likely to happen, and then it will influence corporate performance.

H3a: If CEO and chairman of the board is the same person, it will have a significantly negative correlation to corporate performance;

H3b: The proportion of independent directors is positively correlated with corporate performance;

H3c: The size of board is positively correlated with corporate performance.

2.4 The Influence of Human Capital on Corporate Performance

Many high-tech corporations are founded by the people who master the core technology and most of the employees have relatively higher diplomas. The corporate key capability is the "key resources", and the core technologists are the most important key resource of the corporations. For the corporations, equity capital is scarce, but the R&D capability is much scarcer than that. The core technologists' capability directly affects the R&D capability and the creation of corporate surplus value. So we hypothesize:

H4: The proportion of the employees who has a diploma above college is significantly positively correlated with corporate performance.

2.5 The Influence of Institutional Investors Holding Shares

Some scholars argue that the scale of shares held by institutional investors is significantly positively correlated with corporate performance. But some other scholars have different views. They argue that it is impossible for all the institutional investors to take part in corporate governance actively. In their opinion, the institutional investors just hold the shares. They are not able to have the practical experience and ability as the inside executives do, and they can't get involved in making specific decisions. So their behavior has nothing to do with corporate performance.

H5: The proportion of shares held by institutional investors is significantly positively correlated with corporate performance.

3 Data and Methodology

3.1 Data

The research object of this paper is the cross-sectional sample of all high-tech corporations in Shanghai Stock Exchange and Shenzhen Stock Exchange, 2006. (When we chose the sample, we excluded all the H-shares, B-shares corporations and ST corporations. We chose 104 high-tech corporations as research sample). We measure corporate performance by ROE. Furthermore we'll use Tobin's Q to measure corporate vertical value-increase capability. The definition and descriptive statistics is as shown in Table 1.

Table 1. Basic instance statistics of the samples

Variable	Definition	Max	Min	Mean	Standard deviation
ROE	Return on equity (%)	35.12	-10.66	8.15	6.84
Q	Tobin's Q	4. 17	0.81	1.45	0.60
CR	Proportion of the first largest shareholders' shares (%)	73.97	0.77	33.41	14.29
DR	Degree of equity restriction	2.50	0.03	0.74	0.60
MR	Proportion of the shares held by executives in the top ten largest shareholders (%)	57.22	0	5.20	13.00
MS	Average salary of executives	151.10	3.44	17.25	18. 90
CEO	Having positions both in board and corporation	1	0	0.19	0.40
ID	Proportion of independent director (%)	0.50	0.2	0.33	0.06
B	Members of board	17	6	10.82	2.31
HC	Proportion of employees with a diploma above college (%)	98.04	12.13	53.25	24.00
F	Proportion of institutional investors' shares in the top ten largest shareholders (%)	50.97	0	10.41	10.56

3.2 Methodology

By SPSS12.0, we analyze the corporate value and the corporate increase-value capability by regression in the least square method. We use the standard parametric test (T test and F test) to check the significance of correlation.

$$ROE= a_0 + a_1CR+ a_2DR+ a_3MR+ a_4MS+ a_5CEO+ a_6ID+ a_7B+ a_8HC+ a_9F+\varepsilon \quad (1)$$

$$Q= a_0 + a_1CR+ a_2DR+ a_3MR+ a_4MS+ a_5CEO+ a_6ID+ a_7B+ a_8HC+ a_9F+\varepsilon \quad (2)$$

4 Empirical Results

4.1 Correlation Analysis

Analyzing the performance index — corporate value (ROE) and corporate increase-value capability (Tobin's Q) and all the influencing factors of corporate governance by multiple linear regressions. (See Table 2)

Table 2. All variable correlativity matrixes

	ROE	Q	CR	DR	MR	MS	CEO	ID	B	HC	F
ROE	1.000	0.431(***)	0.049	0.115	0.182(*)	0.463(***)	-0.034	0.034	0.049	-0.175	0.239(**)
		0.000	0.641	0.270	0.078	0.000	0.746	0.742	0.636	0.191	0.020
Q		1.000	-0.101	0.218(**)	0.097	0.146	0.076	-0.014	0.126	0.076	0.162
			0.333	0.035	0.355	0.159	0.469	0.896	0.228	0.466	0.119
CR			1.000	-0.814(***)	-0.259(**)	0.163	-0.135	-0.097	0.101	-0.031	-0.222(**)
				0.000	0.012	0.117	0.196	0.350	0.335	0.764	0.032
DR				1.000	0.405(***)	-0.068	0.158	0.169	-0.080	0.018	0.283(***)
					0.000	0.515	0.129	0.104	0.443	0.860	0.006
MR					1.000	0.039	0.077	0.012	-0.145	0.041	-0.107
						0.711	0.460	0.905	0.163	0.694	0.305
MS						1.000	-0.068	0.016	0.054	0.066	-0.032
							0.513	0.878	0.605	0.529	0.757
CEO							1.000	0.062	-0.126	-0.133	0.054
								0.552	0.224	0.200	0.603
ID								1.000	-0.092	-0.057	0.071
									0.377	0.584	0.498
B									1.000	0.035	0.219(**)
										0.739	0.034
HC										1.000	0.150
											0.129
F											1.000

** * Correlation is significant at 0.01 level(2-tailed). ** Correlation is significant at 0.05 level(2-tailed). * Correlation is significant at 0.10 level(2-tailed).

4.2 Multiple Regression Analysis

All the explained variables are regressed to ROE and Tobin's Q as shown in Table 3, Table 4, Table 5 and Table 6.

Table 3. The regression analysis of explained variable to ROE

Model		Sum of Squares	Df	Mean Square	F	Sig.
1	Regression	1586.800	9	176.311	5.368	.000(a)
	Residual	2758.759	94	32.842		
	Total	4345.559	103			

Table 4. The regression coefficient of explained variable to ROE

Model		Unstandardized Coefficients		Standardized Coefficients	t	Sig.
		B	Beta			
1	(Constant)	1.569	5.755		.273	.786
	CR	.109	.074	.228	1.473	.144
	DR	2.307	1.920	.203	1.202	.233
	MR	.094	.053	.178	1.783	.078
	MS	.162	.032	.449	5.033	.000
	CEO	-.965	1.548	-.056	-.623	.535
	ID	-1.415	10.570	-.012	-.134	.894
	B	-.033	.272	-.011	-.120	.904
	HC	-.051	.026	-.178	-1.989	.050
	F	.158	.063	.244	2.491	.015

Table 5. The regression analysis of explained variable to Tobin's Q

Model		Sum of Squares	df	Mean Square	F	Sig.
2	Regression	4.027	9	.447	1.284	.258(a)
	Residual	29.264	94	.348		
	Total	33.292	103			

Table 6. The regression coefficient of explained variable to Tobin's Q

Model		Unstandardized Coefficients		Standardized Coefficients	t	Sig.
		B	Std. Error			
2	(Constant)	.548	.593		.925	.358
	CR	.007	.008	.176	.968	.336
	DR	.338	.198	.339	1.710	.091
	MR	.001	.005	.017	.147	.884
	MS	.004	.003	.138	1.313	.193
	CEO	.116	.159	.077	.730	.468
	ID	-.549	1.089	-.053	-.505	.615
	B	.029	.028	.110	1.018	.311
	HC	.002	.003	.085	.803	.424
	F	.006	.007	.100	.869	.388

5   Counter Measurements on Corporate Governance of High-Tech

Venture capital is a joint process of investment and financing. Venture Capitalists raise funds from the investors then invest them into several venture business, so a double principal-agent risk of investors versus venture capitalists and venture capitalists versus venture business forms. The realization of principal's profits will rely on the agent's behavior. Usually, the investors won't get involved in the management of venture capital. Though the venture capitalists participate in the management of venture capital, they are not able to participate as much as the venture business managers do. So, the agent has the opportunity to conceal the information from the principal. With the advantage of asymmetric information, the agent may be likely to corrode the principal's interests when their interests conflict. Then, the moral hazard emerges. The equity restriction relationship of the shareholders should become the goal of equity structural optimization .We find the degree of equity restriction has a positive correlation to corporate performance. And it tells us, in the process of equity structure optimization of high-tech corporations in our country, if we want to control the principal-agent cost, we should try to reduce the concentration of ownership structure and enhance the degree of equity restriction. The goal of equity structure optimization is to form the equity restriction relationship of the shareholders.

Acknowledgments. This paper is supported by the humanities and social sciences fund of the ministry of education in China (10YJA630103).

References

1. Van, L.A.A., Abigail, L.: The Role of the Venture Capitalist as Monitor of the Company: A Corporate Governance Perspective. Corporate Governance 10(3), 124–135 (2002)
2. Sigurt, V.: Frankfurt's Neuer Market and the IPO Explosion: is Germany on the Road to Silicon Valley? Economy and Society 30(4), 553–564 (2001)
3. Sheu, H.J., Yang, C.Y.: Insider Ownership Structure and Firm Performance: A Productivity Perspective Study in Taiwan's Electronics Industry. Corporate Governance (3), 326–337 (2005)
4. Yoshikawa, Y., Phillip, H.P.: The Performance Implications of Ownership-driven Governance Reform. European Management Journal 21(6), 698–706 (2003)
5. Chiang, H.: An Empirical Study of Corporate Governance and Corporate Performance. The Journal of American Academy of Business (3), 95–101 (2005)
6. Xiang, C., Xie, M.: The Empirical Analysis on the Relationship between the Corporate Governance Structure and Corporate Performance of the Listed Corporations in Our Country. Management World 5, 117–124 (2003)
7. Song, L., Han, L.: Characteristic Analyses of Ownership Structure of Private Listed Firms. Nankai Business Review (4), 69–71 (2004)
8. Shi, D., Situ, D.: The Empirical Research on the Influence of the Corporate Governance Level on Chinese Listed Corporation Performance. World Economy (5), 69–79 (2004)

Research of Node Location Algorithm in Wireless Sensor Network

Jianqiang Zhao[1], Ge Yao[2], Jie Zhang[2], and Sufen Yao[1]

[1] Institute of Information Engineering, Tianjin University of Commerce
Tianjin, China
[2] Nanjing University
Nanjing, China
zhjq@tjcu.edu.cn

Abstract. To solve the location problem of WSN nodes, an improved MCL algorithm is used. Phase velocity and motion direction of the node are estimated by Newton's difference method, and the energy consumption in the stage of the node's position forecasting and filtering is reduced. Recursion type importance sampling method is used to forecast the posterior probability distribution of node location in this algorithm, which improves locating accuracy and optimizes the positioning performance of MCL algorithm.

Keywords: WSN, position, MCL, optimization.

1 Introduction

In recent years, Wireless Sensor Network has been more and more popular as a new platform of information collection. And the monitoring and tracking of moving target is one of the most typical ones, which has been a hot issue on research for the parts of port and storage.

Position technology is one of important technologies of appliance of WSN. The general location algorithm has been divided into range-based location algorithm and range-free positional algorithm. Taking cost and energy into consideration, using rang-free note location technology decreases the measurement between notes or location. In order to solve the location problem of WSN, this paper uses MCL algorithm to follow location, and improve the limited appliance of MCL algorithm.

2 MCL Location Algorithm

The core idea of MCL algorithm is that the location of the target may appear in the form of a weighted sample set by expressing as posterior probability distribution function:

$$p(x_t \mid z_1, \cdots, z_t) \approx \sum_{i=1}^{N} \omega_i \delta(x_t - x_i) \tag{1}$$

Where, x_i is the nodes of current forecast, n ω_i is the node weight. Algorithm includes two stages of prediction and location update at every step.

D. Jin and S. Lin (Eds.): Advances in MSEC Vol. 2, AISC 129, pp. 309–313.
springerlink.com   © Springer-Verlag Berlin Heidelberg 2011

2.1 Position Estimation

Nodes initially do not have any own prior knowledge of N positions. They need to be initialized. (N is the number of samples by the process of implementation in the algorithm). M_0 [node deployment in the region may be randomly selected N-position].

According to M_{t-1}, on the possible location of a period of time sequence of nodes and the new observations n_t possible to the location of the new node m_t is calculated. While (size(M_t)<N)do $R = [m_t^i | m_{t-1}^i]$ is selected from $p(m_t^i | m_{t-1}^i)$, $m_{t-1}^i \in m_{i-1}$, for all [$m \leq i \leq N$].

$$Rfilter = \left[m_t^i \in R, p(n_t | m_t^i)\right]$$

$$M_t = choose(M_t \cup Rfilter \cdot N)$$

Phase velocity of the node and the direction of motion are estimated and predicted by Newton's difference method. At the same time, the structural anchor of the box ways is utilized to improve the accuracy of prior estimates. Node is usually a smooth trajectory. Assuming the position of the first 3 times, respectively (x_{t-3}, y_{t-3}), (x_{t-2}, y_{t-2}), (x_{t-1}, y_{t-1}). For x, y direction, the data is available to 2 times Newton interpolation.

$$x_t = x_{t-3} + 3 \cdot (x_{t-2}, x_{t-3}) + 3 \cdot (x_{t-1} - 2 \cdot x_{t-2} + x_{t-3})$$

$$y_t = y_{t-3} + 3 \cdot (y_{t-2}, y_{t-3}) + 3 \cdot (y_{t-1} - 2 \cdot y_{t-2} + y_{t-3})$$

Then, the current speed and direction of the moment can be estimate

$$v_t = \min(\sqrt{(x_t - x_{t-1})^2 + (y_t - y_{t-1})^2}, v_{max}) \tag{2}$$

$$\alpha_t = \arctan(\frac{y_t - y_{t-1}}{x_t - x_{t-1}}) \tag{3}$$

The rectangular anchor structure location information box was formed within using the received hop anchor K nodes. The box area is decided by Eq. 4.

$$BoxAnchor = \left\{(x_{min}, x_{max}); (y_{min}, y_{max})\right\} \tag{4}$$

After determining the initial position in each period of time, position sequence will be updated by the movement of nodes and new observational information.

2.2 Position Prediction

The node is calculated from the previous phase position of a group of M_{t-1}. The sample value of each application node mobility model obtains a new set of samples M_t. If $d(m_1, m_2)$ expresses two Euclidean geometry interval between m_1 and m_2. Node speed is in the interval [0, v_{max}] on the uniform distribution. Then, the node position based on the current position of the previously estimated and the probability distribution can be given in the form of uniform distribution.

$$p(m_t \mid m_{t-1}) = \begin{cases} \dfrac{1}{\pi \, v_{\max}} & if \quad d(m_t, m_{t-1}) < v_{\max} \\ 0 & if \quad d(m_t, m_{t-1}) \geq v_{\max} \end{cases} \tag{5}$$

Therefore, in the forecast period, the probable location of the node is calculated in a point set of sequences $R\,M_{t-1}$ at any point in the center of the circle and the radius of a circular area v_{max} .

2.3   Location Filter

In the filtering stage, the nodes need to obtain new observations which can not filter out the location information.

Fig. 1. Node location filter conditions

Fig. 1 depicts the location of the node filtering conditions. In the figure, S describes node N can listen to all of the beacon node group. T describes neighbor node N, node N can listen to the whole itself can not listen to the beacon node. Therefore, the node location l of the filter conditions can be Eq. 6 said.

$$filter(l) = \forall s \in S, d(l,s) \leq r \wedge \forall s \in T, r < d(l,s) \leq 2r . \tag{6}$$

After filtering, the collections of observations that are inconsistent with the position are deleted from the node to all possible locations. Guessing process and the filtering process will be not constantly repeated until the node may be at least N position.

2.4   Importance Sampling

The ultimate goal is to estimate the algorithm may be the location of the node posterior probability distribution $p(l_t \mid o_0, o_1, \cdots, o_t)$. Typically, the location directly from the node samples the posterior probability distribution is unlikely, therefore, the importance of a recursive algorithm function is utilized in this paper.

$$\pi(l_t \mid o_0, o_1, \cdots, o_t) = p(l_0) \prod_{k=1}^{t} p(l_k \mid l_{k-1}) \tag{7}$$

$$\omega_t^i = \tilde{\omega}_{t-1}^i \, p(o_t \mid l_t^i) \tag{8}$$

The node continuously updated and adjusted based on observations obtained samples of the new weight values. By Eq. 8 weight $\tilde{\omega}_t^i$ is normalized values, and ω_t^i is obtained. Weight sequence (l_t^i, ω_t^i) of the node location is used to estimate the posterior probability distribution.

3   MCL Algorithm Optimization

Sensor nodes in the computing resources and storage are generally more scarce resources and so on. The first polygon approximation point test method is utilized to determine the direction of the node. The method is based entirely on the connectivity between nodes. The nodes need only hop beacon location broadcast messages, and so no additional power consumption of nodes and hardware are required. Fig. 2 shows the polygon point test method for a case.

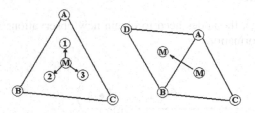

(a) the location of nodes   (b) the direction of nodes

Fig. 2. Points within the triangle test method

In the Fig. 2 (a), the number of hops between A and M is 4. The number of hops between node A and 1 is 3. The number of hops between node A and 2 is 5, The number of hops between node M and B is 3, The number of hops between node 2 and B is 2, The number of hops between node B and 1 is 4. When M moves to the node 1, you can come close to the node A node M after an absence of a node B, node M can not both come close to or after an absence of three vertices must be inside the triangle. When the nodes near the same time or after an absence of three M vertices. It should be outside the triangle.

In Fig. 2 (b), when a node M left the triangle ABC into the triangle ABD, they can generally get the direction of the node. Then Eq. 6 filtering conditions can be simplified as

$$filter(l) = \forall s \in S, d(l,s) \leq r \wedge \forall s \in T', r < d(l,s) \leq 2r . \qquad (9)$$

Where, T' is ABD in the region within the node N nodes can listen to the neighbors and the node N itself can not listen to all the beacon nodes.

Nodes in the direction of the MCL algorithm can greatly reduce the estimated location of speculation and location of the computational filter stage, saving the energy consumption of the node. And more easily filter out positions inconsistent with the observations.

4   Conclusion

In the simulation MCL, Location algorithm, the improved MCL and centroid location algorithm are compared. In simulation experiment, wireless sensor networks, nodes and the algorithm parameters are constantly changing. The simulation experiment can be seen, Improved MCL localization algorithm accuracy in the initial stage will increase rapidly over time and thus into the stabilization phase.

The position estimation error will be the ultimate stable at minimum fluctuations. MCL algorithm can be seen that the positioning accuracy compared with the centroid location algorithm has a large upper hand.

Positioning system of communications technology, sensor technology, computer technology, highly integrated signal processing applications, is the core of ITS. In this paper, the MCL algorithm for wireless sensor networks, logistics and transport vehicle location, given some of the algorithms and models. Random motion in the case of the node, with no additional hardware facilities, we can obtain a higher accuracy to solve, with mobile wireless sensor network node localization problems. By estimating the direction of the node, simplify algorithm is executed to optimize the performance of MCL localization algorithm. For the establishment of logistics tracking and scheduling planning service system to provide technical support.

References

1. Smith, T.F., Waterman, M.S.: Identification of Common Molecular Subsequences. J. Mol. Biol. 147, 195–197 (1981)
2. May, P., Ehrlich, H.C., Steinke, T.: ZIB Structure Prediction Pipeline: Composing a Complex Biological Workflow through Web Services. In: Nagel, W.E., Walter, W.V., Lehner, W. (eds.) Euro-Par 2006. LNCS, vol. 4128, pp. 1148–1158. Springer, Heidelberg (2006)
3. Bahl, P., Padmanabhan, V.N.: RADAR: An in-building RF-based user location and tracking system. In: Proceeding of IEEE INFOCOM 2000, Tel Aviv, Israel, March 26-30, pp. 775–784 (2000)
4. Bulusu, N., Heidemann, J., Estrin, D.: GPS-less low-cost outdoor localization for very small devices. IEEE Personal Communications 7(5), 28–34 (2000)
5. He, T., Huang, C., Brian, M.B., et al.: Range-free localization schemes for large scale sensor networks. In: Proceedings of MobiCom 2003, San Diego, CA, USA, September 14-19, pp. 81–95 (2003)
6. Doherty, L., Pister, K.S.J., Ghaoui, L.E.: Convex position estimation in wireless sensor networks. In: Proceedings of 20th Annual Joint Conference of the IEEE Computer and Communications Societies, vol. 3, pp. 1655–1663 (2001)
7. Sheu, J.P., Hsu, C.S., Li, J.M.: A Distributed Location Estimating Algorithm for Wireless Sensor Networks. Wireless Networks 1, 218–225 (2006)

The Study of Bus Superstructure Strength Based on Rollover Test Using Body Sections

Guosheng Zhang, Xuewen Zhang, and Bin Liu

Key Laboratory of Operation Safety Technology on Transport Vehicles Ministry of Communication, PRC, Research Institute of Highway Ministry of Communications, Beijing 100088, China
gs.zhang@rioh.cn

Abstract. Taking a full load bus as the research object, the finite element analysis theory is applied to build the finite element model of the bus and the numerical simulation environment of the structural strength of the superstructure. According to the ECE R66 equivalent authentication method, gravity center position of the bus is calculated, the rollover test of bus body section is carried out. The deformation of the superstructure and its invasion to residual space of passengers are evaluated. Comparing the rollover test result with the numerical simulation result, it can be found that there is a good agreement between the two results. On this basis, the energy-absorbing situation of the side wall pillars during the rollover process is studied and evaluated. The results show that the body section of bus is complied with the regulatory requirements; its structural safety characteristic is good. This design method of the rollover crash safety has important significance to research and development of manufacturer.

Keywords: Bus, Superstructure strength, Body section, Residual space.

1 Introduction

With the prosperity of the domestic market and the rapid development of highways, the number of bus continues to grow rapidly, following by the frequent occurrence of traffic accident, especially occurrence of road accidents such as "cluster killed and cluster wounded" and so on, resulting in extremely bad social impacts. Relevant government departments and people concern much about this situation.

Although the topic is the research focus currently, and it has important practical significance, because of fierce commercial competition among manufacturers, various reasons such as technology related to secrecy .etc, the relevant research output is still rare. In this paper, according to ECE R66 regulations equivalent authentication method, a finite element model of nine-meter full load bus is built. Using the method of combining body section rollover test with numerical simulation, energy absorbing of the side wall pillars during the rollover test is studied, and the deformation of the superstructure and its invasion to residual space of passengers are evaluated. The results have important guiding significance and application value on superstructure strength design of bus.

D. Jin and S. Lin (Eds.): Advances in MSEC Vol. 2, AISC 129, pp. 315–322.
springerlink.com   © Springer-Verlag Berlin Heidelberg 2011

2   The Superstructure Strength of Bus

Many countries take the bus rollover test as a mandatory certification program of imported bus, requiring the superstructure of bus have sufficient strength. The ECE R66 regulation offers five ways to detect the superstructure strength when the bus rollovers: (1) rollover test; (2) rollover test using body sections; (3) quasi- static loading test of body sections; (4) quasi-static calculation based on testing of components; (5) computer simulation of rollover test of complete vehicles.

The fundamental idea of the regulation is that bus manufacturers can choose among the above-mentioned five test methods, because the last four methods are equivalent to the first one, the standard rollover test. GB/T 17578-1998 provisions on strength of the superstructure of bus, Australia ADR59/00 and South Africa SANS 1563, are cited from the ECE R66 the bus rollover test part. The ECE R66 regulation requires that the superstructure of the vehicle shall have the sufficient strength to ensure that the residual space during and after the rollover test on complete vehicle is unharmed. That means no part of the vehicle which is outside the residual space at the start of the test shall intrude into the residual space during the test. No part of the residual space shall project outside the contour of the deformed structure. Specification requirements are detailed in document [1].

3   Determination of the Centre of Gravity of the Vehicle

In this paper, the nine-meter full load bus in the stage of research and development of company serves as research object. The reference and the total energy to be absorbed in the rollover test directly depend on the position of the vehicle's centre of gravity position. Therefore, its determination should be as accurate as practical [1].

The longitudinal position (l1)of the centre of gravity relative to the centre of the contact point of the front wheels is given by:

$$l_1 = \frac{(P_3 + P_4) \cdot L_1 + (P_5 + P_6) \cdot L_2}{P_{total}} \tag{1}$$

where: P_i is reaction load on the load cell; P_{Total} is unladen kerb mass or total effective vehicle mass, as appropriate L_i is the distance from centre of wheel on 1st axle to centre of wheel on (i+1)th axle, if fitted.

The transverse position (t) of the vehicle's centre of gravity relative to its longitudinal vertical centre plane is given by:

$$t = \left((P_1 - P_2)\frac{T_1}{2} + (P_3 - P_4)\frac{T_2}{2} + (P_5 - P_6)\frac{T_3}{2} \right) \cdot \frac{1}{P_{total}} \tag{2}$$

where: T_i is distance between the centers of the footprint of the wheel(s) at each end of the ith axle.

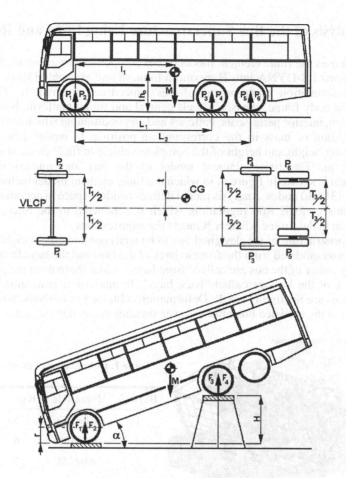

Fig. 1. The centre gravity

The height of the centre of gravity (h_0) shall be determined by tilting the vehicle longitudinally and using individual load-cells at the wheels of two axles. The support structures shall be sufficiently tall to generate a significant angle of inclination α (> 20°) for the vehicle. The height of the vehicle centre of gravity is given by [1][2]:

$$h_0 = r + \left(1 \middle/ \tan\left(\arcsin\left(\frac{H}{L_1}\right)\right)\right)\left(l_1 - L_1\frac{F_3 + F_4}{F_{total}}\right) \qquad (3)$$

where, r is height of wheel centre (on first axle) above the load cell top surface; F_{total} is unladen kerb mass or total effective vehicle mass.

The transverse position (t) is 23.4 mm, the longitudinal distance from the centre line of front axle (a) is 2903.4 mm, the height of the vehicle center of gravity (H_s) is 1036 mm.

4   Analysis of the Bus Superstructure Using FEM and Results

This paper uses the finite element model of bus components based on the CAD model, which imports LS-DYNA into Hypermesh to mesh and assemble [3][4], building the numerical simulation environment of bus superstructure strength. The structure includes the body frame, chassis, wheels general and tilting platform, body parts such as glass, skin, interior parts, seats, batteries and air conditioning which are equivalent to the distribution of mass in the corresponding position, to ensure that deviation of gravity center height and height of the complete vehicle in finite element model is less than 0.02 m. The finite element model of the bus and numerical simulation environment is shown in figure 4, in which the finite element model including 133,431 elements, 131,080 nodes. Analysis model defines residual space in accordance with test requirements of ECE R66 regulation, which is conducted to the reference of body deformation to determine whether it meets the requirements.

The superstructure of the designed bus to be analyzed consists of eight closed bays that are accommodated from the front to back of the bus, and the bays from the front to the gravity center of the bus are called "front bays", while those from the gravity centre to the back of the bus are called "back bays". In analysis it is assumed that elastic deformations are negligibly small. Deformations obtained in analysis, which occurred in all bays of the modeled bus after rollover simulation, are determined as follows:

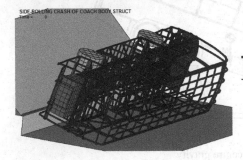

SIDE ROLLING CRASH OF COACH BODY STRUCT

Fig. 4. Numerical simulation environment

Table 1. Deformation of closed bays unit: m

Bays	Front side	Bays	Rear side
1	d1=0.70	5	d5=0.93
2	d2=0.68	6	d6=1.00
3	d3=0.74	7	d7=1.04
4	d4=0.83	--	----------

5   Rollover Test of Bus Body Section

In this paper, the superstructure section of the bus to be tested is abutted on the third and the fourth frame bearer. The simulation analysis shows that the total energy absorbed by body section is 5.46kJ. The rollover test of a body section is carried out on the same way as the standard rollover test with a complete vehicle. The body section have the same position, same the centre of gravity's height on the tilting platform as that of the complete vehicle. The body section is equipped with residual space indicator to prove that during the rollover test residual space is unharmed[5][6].

The rollover test of real section structure is executed by a bus manufacturer according to the conditions which are defined by ECE R66 regulation. The body section is located on a rotational tilt platform, and the platform is lifted up by a crane. The body section structure has the weight of 746 kg about 8.14 % of bus mass, and the free dropping begins after the center of gravity moves out of the hinge point.

As the result of rollover test, the deformation of the superstructure doesn't invade the residual space of passengers. The fall of the center of gravity of body section is about 0.83 m at the end of the roll-over. By means of this value, the energy absorbed by the section rolled-over in the test is determined by equation (4) as E_{F3}=6.08KJ.

$$E_{F3} = mgh_f \qquad (4)$$

Comparing the rollover test result with the numerical simulation result, it can be found that there is a good agreement between the two results. The absorbed energy of other bus section can be calculated by equation (5):

$$E_j = E_3 \times (dj/d3) \quad j=1,2,4,5,6,7 \qquad (5)$$

Fig. 5. Body section rollover test

The energy that would be absorbed by every body section was determined as in Table 2.

Table 2. The absorbed energy of every body section unit: KJ

Front side	Rear side
E_{F1}=5.75	E_{R5}=7.62
E_{F2}=5.62	E_{R6}=8.20
E_{F3}=5.46	E_{R7}=8.57
E_{F4}=6.84	——
$E_{F,Total}$=24.27	$E_{R,Total}$=24.27

therefore, $E_{Total} = \sum E_{Fi} + \sum E_{Ri}$ =48.653 KJ.

6   Energy Absorption of Bus Structure

According to ECE R66 regulation, assuming suspension system is rigid fixed without considering other energy loss, the total energy to be absorbed in a roll-over test can be determined by equation (6):

$$E^T = 0.75M \cdot g \left[\sqrt{\left(\frac{W}{2}\right)^2 + Hs^2} - \frac{W}{2H}\sqrt{H^2 - 0.8^2} + 0.8\frac{Hs}{H} \right] \quad (6)$$

Where, M is the unladen kerb mass of vehicle (kg); W is the overall width of the vehicle (m); H is the height of the vehicle (m), H_s is the height of the centre of gravity of the unladen vehicle (m). Therefore, E^T=45.4KJ [7][8][9].

The primary body deformation shows the deformation of the pillar. The energy is absorbed by the pillar as the deformation energy. In accordance with the technical requirements of bus, the energy absorbed by the pillar of superstructure during the rollover should be evaluated by equation (7).

$$E_{Total} = \sum_{i=1}^{n} E_i > E^T \quad (a)$$

$$E_{F,Total} = \sum_{i=1}^{n} E_{iF} \geq 0.4E^T \quad (b)$$

$$E_{R,Total} = \sum_{i=1}^{p} E_{iR} \geq 0.4E^T \quad (c)$$

$$\hspace{10cm} (7)$$

$$L_F = \frac{\sum_{i=1}^{n}(E_{iF} \times L_{if})}{\sum_{i=1}^{n} E_{iF}} \geq 0.4l_F \quad (d)$$

$$L_R = \frac{\sum_{i=1}^{p}(E_{iR} \times L_{ir})}{\sum_{i=1}^{p} E_{iR}} \geq 0.4l_R \quad (e)$$

Where, E_i is the absorbed energy by the ith pillar; E_{iF} is the declare amount of energy that can be absorbed by the ith bay forward of the centre of gravity of the vehicle; E_{iR} is the declare amount of energy that can be absorbed by the ith pillar to the rear of the centre of gravity of the vehicle ; L_{if} is the distance from the centre of gravity of the vehicle of the ith pillar forward of the centre of gravity ; L_{ir} is the distance from the centre of gravity of the vehicle of the ith pillar rearward of the centre of gravity.

Table 4. Checking of technical requirement

Conditions	Checking	Regulatory requirements
a	48.653 kJ > 45.40 kJ	Yes
b	24.27 kJ > 18.16 kJ	Yes
c	24.383 kJ > 18.16 kJ	Yes
d	2.61 m > 1.96 m	Yes
e	2.20 m > 1.57 m	Yes

From the results we can be seen, the energy absorbed in all superstructure section is 7 to 36% higher than the required values and thus the superstructure of the designed bus fulfils the Regulation conditions. In addition, the calculated values L_F and L_R meet the requirements.

7   Conclusion

In this paper, the rollover test of the body section of a bus is performed according to ECE R66 regulation and the approximate values of the displacements at the main pillars are obtained. Then, with these values the energy absorbed by main pillars is calculated. Additionally, the modeled superstructure of the bus is analyzed so that the energy every bus section absorbs is determined. By means of the results obtained, it is concluded that the results of the rollover test of the bus body section can be approached by finite element analyses.

Acknowledgement. This research is supported by National Key Technology R&D Program (Grant No.2009BAG13A04) and Ministry of Transportation Western Construction Projects (Grant No.2009318000043) and Research Institute of Highway of Transport Basic Science Research Project (Grant No.2010-1004, 2010-1026).

References

1. United Nations Economic Commission for Europe ECE/R66/01, Uniform Technical prescriptions concerning the Approval of Large Passenger Vehicles with Regard to the Strength of Their superstructure. United Nations Economic Commission for Europe, Geneva (2006)
2. Koppel, S., Charlton, J., Fildes, B.: How important is vehicle safety in the new vehicle purchase process? Accident Analysis and Prevention 40, 944–1004 (2008)

322   G. Zhang, X. Zhang, and B. Liu

3. Zhao, H.: LS-DYNA Guide to Dynamic Analysis. The publish of weapon industry, Beijing (2003)
4. General Administration of Quality Supervision, Inspection and Quarantine of the People's Republic of China, Standardiaztion Administration Of The people's republic of china. GB 13094-2007 the safety requirements for bus construction. China Standards Publishing House, Beijing (2007)
5. Yin, H.: Study on Strength of Bus Superstructure and side-rolling Crash Test. HeFei University of Technology, HeFei (2006)
6. Liu, D.: Analysis of Strength of Bus Body Structure and Its Side-rolling Crash Simulation. Chang'an University, Xian (2007)
7. General Administration of Quality Supervision, Inspection and Quarantine of the People's Republic of China, standardiaztion administration of the people's republic of china. GB/T17578-1998 Provisions of Strength for the Superstructure of Bus. China Standards Publishing House, Beijing (1998)
8. Gürsel, K.T.: Analysis of the superstructure of a designed bus in accordance with regulations ECE R66. G.U Journal of Science 23(1), 71–79 (2010)
9. Belingardi, G., Gastaldin, D., Martella, P.: Multibody Analysis of M3 Bus Rollover: Structure Behavior and Passenger Injury Risk. In: The 18th International Technical Conference on the Enhanced Safety of Vehicle(ESV), Paper Number: 288 (2002)

Strengthen the Experiment and Training Management, Construction, and to Improve Teaching and Research

Yandong Song

School of Mechanical Engineering, Nanjing Institute Industry Technology,
Nanjing, 210046, China
damy-s@163.com

Abstract. From the experimental and training management, experiment and practical teaching management, the building and management of the team of training and experiment, the equipment of experimental and training management and open, strengthen the security management et cetera discusses how to strengthen the management and construction of experimental and training. Improve experimental and training for teaching quality and research standards. Pointed out that the experimental training room management must be adapted to the development of modern education. Only by fully understanding the experiment and practical teaching in the process of training personnel in the position and role, to the ideological, fundamental construction of the experimental training room should be placed in the position. Training team also must improve building and management; improve equipment utilization and efficiency of experimental training of students. Training room to play better in the experimental work of College Teaching an important role.

Keywords: Experimental and training management, Teaching and research, Team, Equipment management.

1 Introduction

With the development of profession technical education, Vocational and Technology Institutes experimental and training room building have been expanding, Capital and equipment investment increased year by year. Experimental and training Management in Higher Vocational Colleges management position in the increasingly prominent.

By teaching how to build a new experimental training room management system, in order to improve the experimental quality of teaching and research level, to improve the overall efficiency of experimental training room has become the focus of higher education. Experimental and training room of colleges is both a for practice teaching students to develop intellectual capacity and development sites, but also teachers and laboratory technicians engaged in scientific research base. Construction and management of experiment training room is good or bad, will directly affect the teaching and research in progress. Experimental training room level is a measure of College teaching quality, the level of scientific research and the management of important flag.

D. Jin and S. Lin (Eds.): Advances in MSEC Vol. 2, AISC 129, pp. 323–328.
springerlink.com   © Springer-Verlag Berlin Heidelberg 2011

2   Experimental Training Room Management Must Adapt to the Development of Modern Education

Modern construction is from traditional teaching pattern in higher vocational colleges are teaching and scientific research development, increasing emphasis on cultivation of postgraduates in higher vocational colleges and scientific research of teachers.

In order to meet the needs of higher vocational education in the development of modern scientific research and, based on the course or subject to Division of experimental training room management mode, establish a new management mechanism of experimental training room, becoming the development trend of the research work in schools.

Rationalizing the management system, centralized management devices consolidated provisioning, the experimental and practical room bear the ability of teaching and research tasks, increase equipment utilization, on the experimental teaching reform of experimental teaching contents and methods, strengthen practice, promoting the construction of experimental training room more higher level.

Fully opening professional teaching of experimental teaching of basic and course-wide experiment at the same time, concentrate construction of new professional experimental training room, purchase advanced equipment, ensure that the requirements for teaching. Strengthening the construction of the number of regular instruments and equipment, improving experimental ability, ensure the intact rate of experimental equipment. Experimental training room management modes adapt to the development of modern education, creating conditions for personnel training and scientific research, to shape the economy requires innovative talents to provide a strong guarantee.

3   Standardizing the Management of Experimental and Practical Teaching, Strengthen the Experimental and Practical Teaching Position in the Entire Teaching Process

Experimental teaching in cultivating the practice ability of students, experimental skill, analysis of observations and ask questions, and problem-solving skills, inspire creative thinking, and so has a theory course irreplaceable important role. But for a long time, think of experimental and practical teaching theory course teaching of subsidiary, master books of experimental and practical teaching as an auxiliary teaching of knowledge, making experimental and practical arrangements of the contents of the main principle to verify that the structure, as described in the book, rather than experimental and practical teaching pattern of teaching itself and its effects on cultivation of students ability.

To enhance their ideological understanding, clearly recognizing the experimental and practical teaching is an important means of science and technology research, training; is the transformation of achievements in scientific research to the production prototype; is the agriculture and industry, defense industry of the birth of new product base. To enhance their ideological understanding, clearly recognizing the experimental and practical teaching is an important means of science and technology research, training; is the transformation of achievements in scientific research to the production prototype; is the agriculture and industry, defense industry of the birth of new product base. In personnel training plan and curriculum standards, gradually increasing comprehensive training course training course.

To embody the entire system of teaching theoretical teaching and experimental and practical teaching in the teaching of two different but equal ways, changes of experimental and practical teaching of subsidiary status.

4   Construction and Management of Experimental and Practical Teaching Staff

4.1   Problems in the Experimental and Practical Teaching Staff Building

Due to experimental training room management system, experimental training room construction investment irrationality of the system, resulting in experimental and practical teaching staff in higher vocational college internal status is not respected, experimental and practical teaching units with responsibilities and rights are not well defined, a series of problems such as inadequate attention to professional training of laboratory staff.

Allows many people with high academic standards in the experimental training room kept, coupled with the effects of environmental factors, treatment, equipment, experimental training room to introduce the people, unable to form a rational team structure. Low overall quality of the experimental and practical personnel could not be better to complete experimental and practical teaching; scientific research tasks, there is no way for technology development. Full-time teachers of experimental and practical directly engaged in the management of experimental training room, but because of capacity constraints, if you are experiencing other staff when needed in the management, "have responsibilities but not the powers" enabling difficulties in carrying out their work.

Experimental and practical teaching staff's lack of professional training, device usage and utilization is not high, most experimental training room only able to do the experimental and practical teaching of General, General, structural, the original reason of experimental and practical occupy the subject status of experimental and practical teaching, when some of the more deep-seated, high difficulty of the experiment teaching and scientific research projects often gain.

4.2   Experimental and Practical Teaching Staff Construction of Main Ideas

4.2.1   School Leaders at All Levels Attaches Great Importance to Experimental and Practical Teaching Staff Construction

For the construction of experiment teaching staff in training room problem, school leaders at all levels should attach great importance to teachers doing experimental training room for strengthening discipline construction of one of the team. Combination of short-term goals and long-term planning approach, cultivating a good experimental practice of teachers and staff. In the formulation of appropriate policies, you should consider the proper tilt of experimental training room, to actively introducing high-level capable personnel to experimental and practical work, including undergraduate students, and even master's, strengthen scientific research capacity and management level in experimental and practical room.

4.2.2   Good Policy and Incentive Mechanism

To break down the artificial limits of the experimental training and full-time teachers, encouraging two-way communication, swaps positions, both in terms of promotion, professional title evaluation, training teachers and other school personnel should be treated in the same way, enjoy the same powers. Schools can require new young teacher, at least in experimental training room after 1-2 years of experience, and then enter the teacher's role, this can not only increase the experimental training room fresh blood, and on the teaching of young teachers in the future work is of great help.

In addition, require full-time teachers to participate in the experimental and practical development work, and ensure that support distribution policy and other related policies, make sure to have a certain number of teachers are actively engaged in experimental and practical teaching, scientific research and management work.

5   Experimental and Practical Device Management and Open, Increase Equipment Utilization

Specific measures can be taken are the following:

5.1   Lower Occupancy Rate, Improve Utilization

Occupation equipment responsibility, right and benefit the three are combined from the units on the management mechanism of changing the past to buy light phenomena. Implementation of system of instrument and equipment paid occupation, establish an effective mechanism for device management, with market economy rules and economic instruments to strengthen management, improving investment efficiency and utilization of instrument and equipment, promoting the development and utilization of large-scale instrument and equipment, makes resource allocation more reasonable and more science.

To fully justify the necessity of large apparatus and equipment investment, resource optimization, remnant assets fundamentally address the phenomenon of nonstandard equipment management, waste of resources effectively.

5.2   Use of Social Forces Experimental Training Centers

At present, the scientific research management departments at all levels strongly supported the establishment of experimental training center, many companies based on market changes necessary to set up experimental study on the training center, to speed up the development of new products.

To take advantage of existing human and material resources in higher vocational colleges, mature, clear directions for conditions, to promote the development of scientific research in higher vocational colleges and enterprises project, going out, introduce to ways to set up joint research center for experimental and practical, to enhance experimental practice training equipment for rapid development of higher vocational colleges.

6   Strengthen the Management of Safety

Safety management of experimental training room is the important part of the construction and development of experimental training room, related to the school teaching, scientific research, experimental and practical work, such as the successful completion of prerequisite conditions.

Experimental training room should be pay sufficient attention to aspects of the management. Science and engineering experimental real training room has many precious instrument device and important technology information, and most using type range of chemical drug, and flammable easy explosion articles and highly toxic articles, some experimental real training to in high temperature, and high pressure or vacuum, and strong magnetic, and microwave, and radiation, and high voltage and high speed, special environment Xia or conditions Xia for, some experimental will emissions toxic material, coupled with high vocational college experimental real training room most has using frequently, and personnel concentrated and liquidity large of features, security management on is particularly important.

As the intensity of running school in higher vocational colleges in China opening up to the strengthening and deepening of the reform of internal management of schools, experiment in higher vocational college training room use, mobility and internal management of many new situations and new problems have emerged, in the experimental and practical room accident occurred more and more.

Therefore, careful analysis of higher vocational college experimental training room security in the new situation, deep human analysis of the cause of the accident, study on experimental training in higher vocational colleges under the new situation room security management solutions.

For higher vocational college experimental status of security characteristics and management of the training room, experiment in higher vocational college training room safety management countermeasures to highlight "people foremost, prevention first" security thinking, focus on building the safety culture on campus atmosphere, enhanced full sense of security, and improve the security management system, clear security responsibilities, focus on the basic of safety management, strengthening the standardization construction of security and other aspects of the work and take effective measures.

Experimental training room in safety work, should seriously implement the "safety and prevention first," policy and "whoever is in charge, who is responsible for", "who, who is responsible for", "who used, who is in charge of" principle, get down to base security education. Sound experimental training room security management system, gradual signing safety responsibility, and clear implementation of the system, both by taking punishment approach to security management. Attention to the experimental and practical security implications of governance, strengthening daily security checks, increase the intensity of supervision and inspection.

References

1. Zheng, H., Xinnian, L.: Talking about laboratory management. Study on College Laboratory Work 1, 48–49 (2006)
2. Wei, C., Changyun, C.: Strengthen the experimental training room management in higher vocational colleges. Experimental Study on the Training Room 1, 46–47 (2004)
3. Xiaojun, X., Chao, Y.: Present situation and countermeasures of University's experiment technical staff. Higher Education Research in China 7, 50–51 (2005)
4. Yi, Y.: Some thoughts on teaching instrument and equipment work. Experimental Study on the Training Room 1, 53–54 (2006)

The Optimization on the Multiperiod Mean-VaR Portfolio Selection in Friction Market

Peng Zhang and Lang Yu

School of Management Wuhan University of Science and Technology
Wuhan P.R. China 430081

Abstract. The mean-VaR approach is extended in this paper to multiperiod portfolio selection. Considering the transaction costs and the constraints on trade volumes, the paper proposes the multiperiod mean-VaR portfolio selection model. An efficient algorithm-the discrete approximate iteration method is also proposed for finding an optimal portfolio policy to maximize a utility function of the expected value and average absolute deviation of the portfolio selection. At last, the paper proves the linear convergence of the algorithm.

Keywords: Multiperiod portfolio selection, Mean-VaR, Utility function, Discrete approximate iteration.

1 Introduction

Portfolio theory deals with the question of how to find an optimal distribution of the wealth among various assets. While the mean-variance approach [1,2] provides a fundamental basis for modern portfolio selection in single-period, the mean-VaR approach proposed by Alexander&Baptists [3] represents another school of thinking in single-period. The problem of multiperiod portfolio selection has been studied by Smith [4], Merton [5], Samuelson [6]. Enormous difficulty was reported by Chen et al [7] in finding optimal solutions for multiperiod mean-variance formulation. The literature in multiperiod portfolio selection has dominated by the results of maximizing expected utility functions of the terminal wealth and/or multiperiod consumption. Specifically, investment situations where the utility functions are of power form, logarithm function, exponential function, or quadratic form have been extensively investigated in the literature.

Many researchers use different approaches to study the multiperiod portfolio models. Markus Leippold et al use the geometric methods to solve the assets and liabilities the multiperiod portfolio selection [8]. Michael W brandt et al use the augment states space and simulating regression methods to solve the multiperiod portfolio selection models, and get a result that the approximate approach reduces calculation [9]. Li and Ng consider the mean-variance formulation in multiperiod portfolio selection and determine the optimal portfolio policy and an analytic expression of mean-variance efficient frontier by dynamic programming [10]. U. Celikyurt and S. O zekici use dynamic programming to solve the multiperiod portfolio models and get the ananlytical optimal solution. [11]. Martin Haugh solves the models

D. Jin and S. Lin (Eds.): Advances in MSEC Vol. 2, AISC 129, pp. 329–335.
springerlink.com

by the martingale [12]. Dimitris Bertsimas and Dessislava Pachamanova use the equilibrium controlling to solve the models with linear transaction costs [13]. Wei and Zhong formulate the multiperiod portfolio selection considering the bankruptcy controlling and using Markovian to describe the assets return [14]. Wang et al formulate estate insurance dynamic portfolio selection and use simulating method to solve it [15]. Li and Wang use non-arbitrage methods to optimize the dynamic consumption and investment [16]. Guo and Hu propose the multiperiod portfolio selection with the uncertain stopping time, and use the dynamic programming to solve [17].

The organization of this paper is as follows. In Section 2, the mean- VaR formulation for multiperiod portfolio selection is discussed, and the transaction costs and constraints of trade volumes are considered. The numerical solution to the multiperiod mean-VaR formulation is got by the new algorithm- the discrete approximate iteration method. The algorithm is proved linearly convergent in Section 3. One case is studied in Section 4 and the paper concludes in Section 5 with a suggestion for further study.

2   The Multiperiod Mean Average Absolute Deviation Formulation

The returns of assets are random. The exact distributions of the returns are not known, but their means and average absolute deviation are known. These factors change randomly on a periodic basis and form our stochastic market. As the state of the market changes over time, the returns also change accordingly. In short, We have a model where asset returns are modulated by the stochastic market.

We consider a capital market with risky n securities. An investor joins the market at time 0 with an initial wealth s_0. s_t is the initial wealth of the period t. The wealth can be reallocated among the n assets at the beginning of each of the following T consecutive time periods. The rates of returns of the risky securities of period t within the planning horizon are denoted by $R_t = [R_{1t}, \cdots, R_{nt}]$, where R_{it} is the random return for security i of period t. The return R_t has a known mean $r_t = E(R_t)$ for $t = 1, 2, \cdots, T$. Let $x_{it}, i = 1, 2, \cdots, n$ be the rate of the possession of the ith risky asset at the beginning of period t. An investor is seeking the best investment strategy, $x_t = (x_{1t}, x_{2t}, \cdots, x_{nt})$ for $t = 1, 2, \cdots, T$. Let a_{it} and v_{it} be respectively the pursing and selling volumes, so $x_{it} + a_{it} - v_{it}$ is the rate invested in the ith risky asset at the beginning of period t. Let $c(a_{it})$ and $c(v_{it})$ be respectively pursing and selling transaction costs. $R_{pt} = R_{it}x_t$ is the rate of return of the portfolio of period t. Its mean and VaR are respectively

$$r_{pt} = r_{1t}(x_{1t} + a_{1t} - v_{1t}) + \cdots + r_{nt}(x_{nt} + a_{nt} - v_{nt}) \tag{1}$$

$$f(r_{pt}) = \Phi^{-1}(c)\sqrt{x_t'Gx_t} - r_t'x_t \tag{2}$$

It can be seen that the expected return r_{pt} and average absolute deviation $f(r_{nt})$ are linear functions of $x_{it} + a_{it} - v_{it}$.

Let I_{pt} be the net expected return of portfolio at time of period t, so

$$I_{pt} = r_{pt} - \sum_{i=1}^{n}[c_{it}(a_{it})a_{it} + c_{it}(v_{it})v_{it}] \tag{3}$$

Let$U(I_{pt}, f(r_{pt}))$be the expected utility function of the investor at the period t, so

$$U(I_{pt}, f(r_{pt})) = \beta^{t}(I_{pt} - \omega f(r_{pt})) \tag{4}$$

where β is discount coefficient, and $0 < \beta \leq 1$; ω is the risk preference coefficient, and $\omega \geq 0$. Clearly, $U(I_{pt}, f(r_{pt}))$is concave function of $x_{it} + a_{it} - v_{it}$.Since investors always would like to maximize their wealth with a low risk level, Utility function $U(I_{pt}, f(r_{pt}))$ is assumed to satisfy the following:

$$\frac{\partial(U(I_{pt}, f(r_{pt})))}{\partial I_{pt}} > 0 \tag{5}$$

$$\frac{\partial^{2}(U(I_{pt}, f(r_{pt})))}{\partial(r_{pt})^{2}} < 0 \tag{6}$$

Considering the transaction costs and trade volumes, the following multiperiod mean- VaR portfolio problem is formulated:

$$max \sum_{t=1}^{T} \beta_{t}(I_{pt} - \omega f(r_{pt}))$$
$$s.t. \begin{cases} S_{T} = (1 + I_{pt})S_{t-1} \\ I_{pt} \geq r_{0} \\ 0 \leq a_{it} \leq \sum_{j=1, j \neq i}^{n} x_{jt}, i = 1, 2, \cdots, n \\ 0 \leq v_{it} \leq x_{it} \\ a_{it}v_{it} = 0 \end{cases} \tag{7}$$

where the first constraint is the state transfer equation. The second denotes that the net expected return of portfolio I_{pt} is not less than the smallest return r_0. The third denotes that the pursing volume is less than the amount of the possession of other risky asset. It means that the short buying is not allowed. The forth denotes that the selling volume is less than the possession of other risky asset. It means that the short selling is not allowed.

Investment situations where there exists a riskless asset can be regarded as a special case in the general multiperiod mean- VaR formulation discussed above.

The state transfer equation is transformed into

$$I_{pt} = \frac{S_{t}}{S_{t-1}} - 1 \tag{8}$$

Theorem 1. Assume that there are transaction costs of model (7), the fifth constraint $(a_{it}v_{it} = 0)$ is useless.

Proof. If the average absolute deviation $f(r_{pt})$of the objective function$\beta^{t}(I_{pt} - \omega f(r_{pt}))$ doesn't change, the rate invested in the risky asset also doesn't change. It means that the expected rate of return of the portfolio $r_{pt} = \sum_{i=1}^{n} r_{it}(x_{it} + a_{it} - v_{it})$ doesn't change. In order to optimize the objective function, the amount of transaction costs $\sum_{i=1}^{n}[c_{it}(a_{it})a_{it} + c_{it}(v_{it})v_{it}]$ must minimize. The transaction costs that $a_{it} \neq 0$and$v_{it} \neq 0$ are not less than the transaction costs that $a_{it} = 0$ and $v_{it} \neq 0$, or $a_{it} \neq 0$ and $v_{it} = 0$. So $a_{it}v_{it} = 0$, it means that the fifth constraint is useless.

3   The Discrete Approximate Iteration Method

A.   Transform the model into the multiperiod weighted digraph
The state variable of the period t is discretized into four same internals among the smallest and the biggest values. It means that I only study five discrete values of the state variable every time. So the model (7) is transformed into the multiperiod weighted digraph 1. Where the period of the weighted digraph is investment period. The point is the discrete values of the state variable. The edge is the utility of the investor.

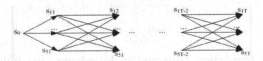

Fig. 1. The multiperiod weighted digraph

Let m be very small number, the smallest value of every state variable in t period of the model (1) is got by the following method:

$$max\beta_t((1 - \omega)I_{pt} - \omega f(r_{pt}))$$
$$s.t \begin{cases} I_{pt} \geq m \\ 0 \leq a_{it} \leq \sum_{j=1, j\neq i}^{n} x_{jt}, i = 1, 2, \cdots, n \\ 0 \leq v_{it} \leq x_{it} \end{cases} \tag{9}$$

Using the sequence quadratic programming and the pivoting algorithm to get the optimal solution of model (9) [9,10,11]. At the same time, $a'_t x_t$ and b_t (the smallest value of every state variable) can be got.

B.   Calculate the longest path of the multiperiod weighted digraph
The longest path of the multiperiod weighted digraph from the starting point to the ending point can be got by the max-plus algebra[18,19].
According to the max-plus algebra, the longest path U of the figure 1 is

$$U = \beta U_1 \otimes \beta^2 U_2 \otimes \cdots \otimes \beta^T U_T \tag{10}$$

Let the longest path U_0 be $S_0 \to S_{1i_1} \to S_{2i_2} \to \cdots \to S_T$. The optimal solutions of the longest path of figure 1 are also admissible solutions of the multiperiod mean- average absolute deviation portfolio selection model. Based on $(S_0, S_{1i_1}, \cdots, S_T)$, I respectively discretize the state variable from period 1 to period$T$ – 1into four same internals. For example, I respectively discretize $S_{11} and S_{1i_1}, S_{1i_1} and S_{15}$into two same internals. So $(S_{11}, S'_{12}, S_{1i_1}, S'_{13}, S_{15})$ are five discrete values of S_1. Based on $(S_{2i_2}, \cdots, S_{T-1, i_{T-1}})$, I respectively discretize the state variable from period 2 to period $T - 1$ into four same internals by the same way. The utility of period t can be got by the pivoting algorithm. At the same time, the multiperid weighted digraph can be got. The longest path and another admissible solution can be got by the (12). If the two admissible solutions are near, we can get the optimal solution of model (7). Otherwise, continue iterating.

C.   The convergence of the discrete approximate iteration method

Theorem 2. Let the $U(I_{pt}, f(r_{pt}))$ be concave function, the discrete approximate iteration is linearly convergent.

Proof. Let the longest side in the period 1 be $U_1(0, j_1)$,the longest side in the period t be $U_t(i_{t-1}, j_1), t = 2, \cdots, T-1$,and the longest side in the period T be $U_T(i_{T-1}, T)$, so the upper bound of the solution in model (7) is $U_1(0, j_1) + U_2(i_1, j_2) + \cdots + U_T(i_{T-1}, T)$.

Let the optimal value of model (7) be U_0^*, the optimal value of period t of model (7) be U_t^*, the optimal value of iteration $k+1$ and k epectively be $U_t^{(k+1)}$ and $U_t^{(k)}$. Because the algorithm is convergent and the objective function of model (7) is concave, $0 \le \frac{|U_t^{(k+1)} - U_t^*|}{|U_t^{(k)} - U_t^*|} \le 1$ It means $0 \le \frac{\sum_{t=1}^T |U_t^{(k+1)} - U_t^*|}{\sum_{t=1}^T |U_t^{(k)} - U_t^*|} \le 1$. So the algorithm is linearly convergent.

4   Numerical Example

In this section, we suppose that an investor chooses six componential stocks of Shanghai 50 index for his investment. We collect historical data of the six kinds of stocks which exchange Codes are 600005, 600016, 600050, 600104, 601318, 600601 from April, 2006 to March, 2008. The data are downloaded from the web site www.stockstar.com. Then we use three months as a period to obtain the historical rates of return during fourteen periods in Table 1. Let $\beta_t = 1, \varepsilon = 10^{-6}, c_{it}(x_{it}) = 0.008|x_{it}|$, $t = 1, \cdots, 3$. When $T = 3$, what are the optimal investment policy?

Table 1. The quarter rates of six stocks returns (%)

period \ stock	S_1	S_2	S_3	S_4	S_5	S_6
1	6.4500	10.4400	3.4500	4.2900	10.0000	4.5100
2	12.4030	14.5600	5.4000	6.3900	10.6000	7.1100
3	17.5170	19.8400	7.3500	4.3700	15.5000	9.2100
4	6.0290	5.4100	2.2000	2.9500	4.4000	1.8200
5	14.5950	6.5600	4.9700	6.5000	8.3205	3.4100
6	19.9560	9.4700	6.7000	9.5800	11.0000	5.3800
7	24.5830	12.5500	7.8900	11.9200	13.8600	7.5200
8	7.4190	5.0600	2.2000	3.2700	5.2000	1.7300
9	17.9120	11.0600	4.7500	5.1000	8.8000	3.3300

The paper uses the moving average method to get the future return of two periods and uses the sequence quadratic programming and the pivoting algorithm to get the smallest and the biggest value of every state variable, ie $S_{1max} = 0.1410, S_{1min} = 0.1337$.

The state variable of the period 1 is discretized into four same internals among the S_{1min} and the S_{1max}. The optimal policies and utilies are caculated, the result is the following:

Table 2. The optimal policies of the mulitiperiod mean-VaR portfolio in period 1

r_p	The optimal policies						σ_p	U
	x_1	x_2	x_3	x_4	x_5	x_6		
0.1337	0.7953	0.2047	0.0000	0.0000	0.0000	0.0000	0.0573	0.0734
0.1345	0.8178	0.1822	0.0000	0.0000	0.0000	0.0000	0.0581	0.0733
0.1363	0.8686	0.1314	0.0000	0.0000	0.0000	0.0000	0.0600	0.0732
0.1381	0.9193	0.0807	0.0000	0.0000	0.0000	0.0000	0.0620	0.0730
0.1410	1.0000	0.0000	0.0000	0.0000	0.0000	0.0000	0.0654	0.0726

Where x_1, x_2, x_3, x_4, x_5 and x_6 denote the investment rate of stock S_1, S_2, S_3, S_4, S_5 and S_6. $U_1^{(1)} = (0.0734, 0.0733, 0.0732, 0.0730, 0.0726)$.

Using the sequence quadratic programming and the pivoting algorithm, we can get $I_{p2max} = 0.1410, I_{p2min} = 0.1358, I_{p3max} = 0.1417, I_{p3min} = 0.1396$. The optimal policy of first period is the following: $a_{11} = 0.7953, a_{21} = 0.2047, a_{i1} = 0, i \neq 1, 2, v_{i1} = 0, i = 1, \cdots, 6;$

The optimal policy of second period is the following: $a_{12} = 0.0584$, $v_{22} = 0.0584, a_{i2} = 0, i \neq 1, v_{i2} = 0, i \neq 2$

The optimal policy of last period is the following: $a_{13} = 0.0874, a_{i3} = 0, i \neq 1$, $v_{23} = 0.0874, v_{i3} = 0, i \neq 2$.

5   Conclusions

The Feinstein mean- VaR approach has been extended in this paper to multiperiod portfolio selection problems. The numerical solution has been derived for the multiperiod mean-average absolute deviation formulation by the new algorithm: the discrete approximate iteration method. It is proved linearly convergent and can avoid the curse of dimensions. The derived numerical solutions for the multiperiod portfolio selection will definitely enhance investors' understanding of the trade-off between the expected return and the average absolute deviation. At the same time, the derived optimal multiperiod portfolio policy provides investors with the best strategy to follow in a dynamic investment environment. A future research is investigation of the optimal policy for the multiperiod mean-different risk measure formulation using the discrete approximate iteration algorithm.

References

1. Markowitz, H.: Portfolio selection. The Journal of Finance 7(1), 77–91 (1952)
2. Markowitz, H.: The optimization of a quadratic function subject to linear constraints. Naval Research Logistics Quarterly 3, 111–133 (1956)

3. Alexandre, G., Baptisa, A.: Economic implications of using mean-VaR model for portfolio selection comparison with mean-variance analysis. Journal of Economic Dynamic and Control 26, 115–126 (2002)
4. Merton, R.C.: An Analytic Derivation of the Efficient Frontier. The Journal of Finance. Quant. Anal. 7, 1851–1872 (1972)
5. Smith, K.V.: A transition model for portfolio revision. The Journal of Finance 22, 425–439 (1967)
6. Merton, R.C.: Lifetime portfolio selection under uncertainty: the continuous-time case. The Review of Economics and Statistics 51, 247–257 (1969)
7. Samuelson, P.A.: Lifetime portfolio selection by dynamic stochastic programming. The Review of Economics and Statistics 50, 239–246 (1969)
8. Chen, A.H.Y., Jen, F.C., Zionts, S.: The optimal portfolio revision policy. Journal of Business 44, 51–61 (1971)
9. Leippold, M., Trojani, F., Vanini, P.: A geometric approach to multiperiod mean variance optimization of assets and liabilities. Journal of Economic Dynamics & Control 28, 1079–1113 (2004)
10. Brandt, M.W., Clara, P.S.: Dynamic Portfolio Selection by Augmenting the Asset Space. The Journal of Finance 10, 2187–2217 (2006)
11. Li, D., Ng, W.L.: Optimal dynamic portfolio selection: Multiperiod mean-variance formulation. Mathematical Finance 10, 387–406 (2000)
12. Celikyurt, U., Zekici, S.O.: Multiperiod portfolio optimization models in stochastic markets using the mean–variance approach. European Journal of Operational Research 179, 186–202 (2007)
13. Haugh, M.: Martingale Pricing Applied to Dynamic Portfolio Optimization and Real Options. Financial Engineering: Discrete-Time Asset Pricing, 1–12 (Fall 2005)
14. Bertsimas, D., Pachamanova, D.: Robust multiperiod portfolio management in the presence of transaction costs. Computers & Operations Research 3, 1–15 (2006)
15. Wei, S.-z., Ye, Z.-x.: Multi-period optimization portfolio with bankruptcy control in stochastic market. Applied Mathematics and Computation 7, 1–12 (2006)
16. Wang, C., Yang, J., Jiang, x.l.: Multistage Stochastic Programming Model for the Portfolio Problem of a Property Liability Insurance Company. Transactions of Tianjin University 8(3), 203–206 (2002)
17. Li, Z., Wang, S.: Optimal consumption portfolio selection in frictional markets. Journal of Systems Science and Mathematical Sciences 7, 406–416 (2004) (in Chinese)
18. Zhang, P., Zhang, Z.-z., Zeng, Y.-q.: The Optimization of The Portfolio Selection with The Restricted Short Sales. Application of Statistics and Management 27, 124–129 (2008) (in Chinese)
19. Heidergott, B., Olsder, G.J., Van der Woude, J.: Max Plus at Work—Modeling and Analysis of Synchronized Systems: A Course on Max-Plus Algebra and its Applications. Princeton University Press (2006)

Analysis and Program Design on Time Effect of Foundation Settlement

Wei Lei

Chang chun Institute of Technology
Changchun130012, China
lwlwlw2086@sina

Abstract. Nowadays, infiltrating solidification theory of saturated soil is adopted to analytically calculate the relation between building subsidence and time. It is inevitable to employ hypotheses in quantified calculation. Calculation error is personally added and the calculation is excessively detailed because different relation curve between solidification and time factor are adopted. The method is out of date. In order to promote the development of rock mechanics, a program is drawn up with VB language. The problem of subsidence and time can be easily solved with the program, which is of significance in theory and engineering circles.

Keywords: subsidence of foundations, solidification degree, time factor, solidification coefficient.

1 Introduction

The foundation settlement of engineering design should be predicted during construction and use period of the buildings, course of foundation settlement refers to relation between settlement and time, which aims to design clearance between relevant parts of obligate buildings and consider connection method and construction sequence; especially, buildings with accidents as crack and slope etc needs knowing about its settlement and development of settlement in future, namely relation between settlement and time, an important ground for accident handling proposal.

Analysis on foundation settlement of buildings is built upon unidirectional penetration consolidation theory of saturated soil; this theory has three basic assumptions:

The load is inflicted once instantaneously and it is in uniform distribution along the soil depth z.

The soil is homogeneous and saturated. The osmotic coefficient K and compressibility coefficient in compression process are both constants.

The soil layer only causes compression and drainage in vertical direction.

In practical engineer, load situation and drainage condition of the foundation are quite complex, which can be basically divided into 5 situations, seen in Figure 1.

Applicable conditions under various situations are approximately as follows:

Circumstance 0 is suitable for the situation that the foundation soil is consolidated by self weight, the base area is large and the compressed soil layer is quite thin.

D. Jin and S. Lin (Eds.): Advances in MSEC Vol. 2, AISC 129, pp. 337–343.
springerlink.com © Springer-Verlag Berlin Heidelberg 2011

Circumstance 1 is suitable for the situation that self weight stress of soil equals to compression stress in large area of newly-deposited soil layer when the downside side of the soil layer is incompletely consolidated.

Circumstance 2 is suitable for the situation that the compression stress caused by basic load at the bottom of the compressed soil layer is approximate to zero when the foundation soil is consolidated by self weight, the basement area is small and the compression soil layer is rather thick.

Circumstance 3 is amount to the situation that the soil layer is not consolidated under self weight, which suffers from continuous even load or partial load function of the foundation.

Circumstance 4 is amount to the situation that the soil layer has been consolidated under self weight, but the compression soil layer is not thick enough, the compression stress of the bottom of the compression layer is still quite large and cannot be regarded as zero.

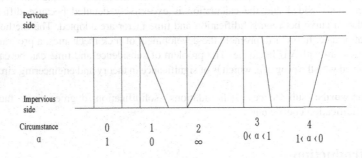

Fig. 1. Load Situation of the Foundatio

It still needs pointing out that the compression stress allocation plan along the basic center line is often used as the calculation graph in calculation process of the foundation consolidation if the bottom area of the foundation is not large.

All the above-stated are single-sided drainage situation, if it is two-sided drainage, the consolidation degree of the homogeneous soil layer can be calculated according to circumstance 0 as long as it is in linear distribution no matter which circumstance the compression stress distribution in the soil layer is in accordance with, which is concluded from superposition principle.

2   Establish Mathematical Model of Foundation Consolidation Degree under Various Circumstances

Ratio between settlement St and the final settlement S of the foundation in process of consolidation at any time point is called the consolidation degree of the foundation at t, indicated with U, namely: $U=St/S$.

Under the situation that the compression stress, property of the soil layer and drainage condition have been fixed, the consolidation degree is a function of time, the positive and negative functional relation between the consolidation degree and time of the foundation under various loads (meeting requirements of accuracy) can be seen from complex mathematical deduction (omitted here).

Functional relation under circumstance 0

$$Uto=1-8/ (\pi^2 *E^{(\pi2 * tv/4)})$$
$$Tv=-4/\pi^2 *(Ln (1-Ut0) +Ln\pi^2 -Ln8)$$

Functional relation under circumstance 1

$$Ut1=1-32/ (\pi^3 *E^{(\pi2 * tv/4)})$$
$$Tv=-4/\pi^2 *(Ln (1-Ut1) +Ln\pi^3 -Ln32)$$

Functional relation under circumstance 2

$$Ut2=1- (16/\pi^2 -32/ \pi^{3)}/ (E^{(\pi2 * tv/4)})$$
$$Tv=-4/\pi^2 *(Ln (1-Ut2) - Ln(16/\pi^2 -32/ \pi^3))$$

Functional relation under circumstance 3 and 4

$$Ut =1- 1 / (1+\alpha)*(16*\alpha/\pi^2 + (32-32*\alpha)/ \pi^{3)} / (E^{(\pi2 * tv/4)})$$
$$Tv=-4/\pi^2 *(Ln (1-Ut) +Ln (1+\alpha) - Ln (16*\alpha/\pi^2 + (32-32*\alpha)/ \pi^3))$$

The above-established functional relations pay foundation for computer programming.

3   Compile Program Flow Chart

The program flow chart can be complied according to the above calculation formula (seen figure 2).

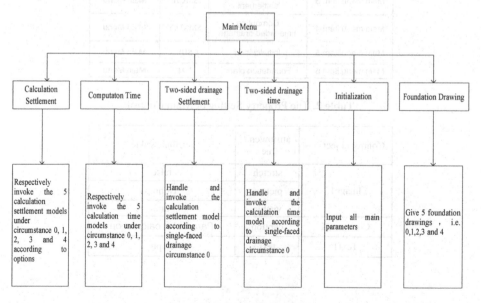

Fig. 2. Procedure Flow Chart

4   Rrlevant Programs Can Be Complied By the Flow Chart

Program structure as shown in Table 1 and Table 2。

Table 1. Menu Structure of Application

Menu item	Title	Name	Level
Main menu item 1	Foundation Settlement	C j L	Main Menu
Sub-menu item 1	The settlement of the case 0	CJ0	A sub-menu
Sub-menu item 2	The settlement of a case	CJ1	A sub-menu
Sub-menu item 3	2, the settlement of the case	CJ2	A sub-menu
Sub-menu item 4	3, the settlement of the case	CJ3	A sub-menu
Sub-menu item 5	4, the settlement of the case	CJ4	A sub-menu
Main menu item 2	Computing time	SJ	Main Menu
Sub-menu item 1	0, calculate the time	SJ0	A sub-menu
Sub-menu item 2	1, calculate the time	SJ1	A sub-menu
Sub-menu item 3	2, calculate the time	SJ2	A sub-menu
Sub-menu item 4	3, calculate the time	SJ3	A sub-menu
Sub-menu item 5	4 calculate the time	SJ4	A sub-menu
Main menu item 3	Double-sided water settlement	SMCJL	Main Menu
Main menu item 4	Computing time-sided drainage	SMSJXY	Main Menu
Main menu item 5	Initialization	CSHCS	Main Menu
Main menu item 6	Foundation plans	DJT	Main Menu

Table 2. The Property Setting of Main Control

Control object	attributename	attribute value
Image1	stretch	false
	picture	T8.bmp
	visible	ture
Command1	caption	Parameter calculation
Text1	text	empty

The initial subprogram can be seen as follows:

```
Private Sub Command1_Click()
s = Val(Text1.Text)
h = Val(Text2.Text)
e1 = Val(Text3.Text)
rw = Val(Text4.Text)
av = Val(Text5.Text)
k = Val(Text6.Text)
cv = k * (1 + e1) / (av * rw)
Text7.Text = cv
End Sub
```

among :

```
s  : Final Settlement
h  : Drainage layer
e1 : Void ratio
rw : Bulk density of water
av : Compressibility
k  : Permeability coefficient
cv : Coefficient of consolidation
```

Calculate the settlement of a given time, the subroutine is as follows:

```
Private Sub Command2_Click()
t = Val(Text8.Text)
tv = cv * t / (h * h)
ex = -(pi * pi) / 4 * tv
ut = 1 - 8 * Exp(ex) / (pi * pi)
st = ut * s
Text9.Text = st
End Sub
Private Sub Command1_Click()
t = Val(Text8.Text)
tv = cv * t / (h * h)
ex = -(pi * pi) / 4 * tv
ut = 1 - 32 * Exp(ex) / (pi * pi * pi)
st = ut * s
Text9.Text = st
End Sub
Private Sub Command1_Click()
t = Val(Text8.Text)
tv = cv * t / (h * h)
ex = -(pi * pi) / 4 * tv
ut = 1 - (16 / (pi * pi) - 32 / (pi * pi * pi)) * Exp(ex)
st = ut * s
```

```
Text9.Text = st
End Sub
Private Sub Command1_Click()
t = Val(Text8.Text)
tv = cv * t / (h * h)
ex = (pi * pi) / 4 * tv
ut = 1 - 1 / (1 + a) * (16 * a / (pi * pi) + (32 - 32 * a)
/ (pi * pi * pi)) / Exp(ex)
st = ut * s
Text9.Text = st
End Sub
among :
t   : time
st  : The settlement of a certain time
```

Achieve a given amount of computing time required for settlement, the following subroutine :

```
Private Sub Command1_Click()
st = Val(Text8.Text)
ut = st / s
tv = -4 / (pi * pi) * (Log(1 - ut) + Log(pi * pi) - Log(8))
t = tv * h * h / cv
End Sub
Private Sub Command1_Click()
st = Val(Text8.Text)
ut = st / s
tv = -4 / (pi * pi) * (Log(1 - ut) + Log(pi * pi * pi) - Log(32))
t = tv * h * h / cv
Text9.Text = t
End Sub
Private Sub Command1_Click()
st = Val(Text8.Text)
ut = st / s
tv = -4 / (pi * pi) * (Log(1 - ut) - Log(16 / (pi * pi) -
32 / (pi * pi * pi)))
t = tv * h * h / cv
Text9.Text = t
End Sub
Private Sub Command1_Click()
q2 = Input Box("Please input compressive stress on a
permeable surface", "q2 Input parameters ", 1)
q3 = Input  Box("Please input compressive stress on a
impermeable surface ", "q3" Input parameters", 1)
a = q2 / q3
st = Val(Text8.Text)
ut = st / s
```

```
tv = -4 / (pi * pi) * (Log(1 - ut) + Log(1 + a) - Log(16 *
a / (pi * pi) + (32 - 32 * a) / (pi * pi * pi)))
 t = tv * h * h / cv
Text9.Text = t
End Sub
```

5   Examples

Thickness of the saturated claypan in the first foundation is 10m with sand layer at the upper and bottom. The subsidiary stress caused by the base pressure (distributed uniformly) in the claypan: the top surface is 235.2kpa, while the under surface is 156.8kpa. The physical property index of the soil layer is knows as: the initial void ratio e1=0.8, the compressibility coefficient a_v =0.225*10^{-3}1/kpa, and the osmotic coefficient k=2.0cm/a. Then try to calculate

  a. Settlement of the foundation one year after adding load.
  b. Time needing for the foundation settlement reaching 25cm.
  Answer:

a. Average value of the subsidiary stress upon the claypan is σ z= (235.2+156.8)/2=196(kpa)
   The final settlement of the foundation is S=a_v*σz*H/ (1+ e1) = (0.225*10^{-3}*196*1000)/ (1+0.8) =27.8 cm
   It can be seen by putting the know parameters into consolidation coefficient and time factor formula that Cv=1.44*10^5(cm^2/a)   ,Tv=0.576
   The subsidiary stress of the soil layer takes on ladder-shaped, however, it is calculated according to circumstance 0 (namely α=1) due to drainage on the upward and downward side. It can be seen by checking up table of consolidation degree and time factor that: degree of consolidation U=0.802.
   Settlement one year after adding load St:

  St=U*S=0.802*27.8=22.3cm
  b. Consolidation degree of St reaching 25cm: U=St/ S=25/27.8=0.9
   It can be seen from α=1, U=0.9 and table of consolidation degree and time factor that Tv=0.85
   t=Tv*H^2/Cv=0.85*500^2/1.44*10^5=1.48 (a)

   It can be calculated by inputting all parameters into the program that: 1. St=22.34cm 2. t=1.47(a)
   The result is accurate from calculation angle, which reduces errors of artificial table check-up and saves a lot of time.

References

1. Qing, Y.: Experts of Development Tool—Visual –Basic6.0. Electronic Industry Press
2. Feng, G.: Soil Mechanics. Water Resources and Electricity Press. Wuhan Water Electricity
3. Chen, X. f.: Settlement Calculation Ethods and Research Progress. Civil Engineering (6) (2004)

Close Eye Detected Based on Synthesized Gray Projection

Ling Lu, Xing Ning, Ming Qian, and YongKe Zhao

Jiangxi Province Key Lab for Digital Land
East China Institute of Technology
56 Xuefu Road, Fuzhou, JiangXi, P.R. China
luling@ecit.cn

Abstract. A Method, to detect close eye in image by using synthesized gray projection, is presented. We utilize the gray differential projection and gray integral projection for the facial image to locate eye horizontal position, and utilize the gray integral projection to locate eye vertical position. According to the change of horizontal differential projection, we can detect whether the eye is close. The method is available and its calculation is less. The experiment results show that close eyes can be detected accurately by using gray projection.

Keywords: eye location, gray projection, close eye detected.

1   Introduction

Eye is an important character in face. It acts part of important on face recognition and face detection. Eye location is a first work in face recognition algorithm. Many scholars have presented many methods to locate eye in image. Kawaguchi et al. introduced a method base on edge detection and Hough transform[1].They utilize edge detection operator to draw character edge and use the result of Hough transform to determinate area with optimum matching. Yuille[2], Shi[3] and Fei[4] et al. represented a method base on template matching and its improve. The template matching is mode identify technology. It utilizes the similar extent between having already eye template and an area in image to locate eye. The key of the method is how make selection the eye template. Ai[5] and Zhu[6] represented the eye location method base on blocks. According to the gray character of eye, they defined different parameters to find eye block position in image. Some method that was used on computer vision has been used on getting face character. Reisfeld utilized generalized symmetry transform to get face characteristic and to realize face recognition [7]. Zhou defined a directional symmetry transform to used for eye accurately location [8]. Liu combine above two transform to represent generalized symmetry transform [9].

Projection is an effective method for getting image character. A two-dimension image may be analysis by two orthogonal a dimension projection function. The project function may decrease dimension, decrease calculation, and convenient for analysis image characteristic. So project function has been used for location face character by a lot of scholar successfully. Kanade[10] used integral projection function for face

D. Jin and S. Lin (Eds.): Advances in MSEC Vol. 2, AISC 129, pp. 345–351.

recognition at earliest. Brunelli and Poggio[11] improved Kanade's method. They applied integral project function to edge figure analysis and determinate each characteristic position on face. Feng and Yuen [12] represented variance projection function at the earliest and used this function for location eye. The method is simple. Geng et al. combine integral and variance projection function and represented a hybrid projection function [13]. Zhang et al. gave differential projection [14] and combine integral projection to locate eye.

At present, most methods about close eye detected are using the length ratio of to wide on minimum circumscribed rectangle. The key of the method is how calculate minimum circumscribed rectangle [16].

We synthesize gray differential and integral projection to locate the horizontal l position of eye for upright face, and utilize the gray integral projection curve to locate the vertical position of eye. Finish, we may determine eye close status.

2   Projection Method

2.1   Integral Projection

The integral projection is a popular projection method. Supposed $f(x, y)$ mean a gray value at the point (x, y), x is taken in the range $x_1 \leqslant x \leqslant x_2$., y is taken in the range $y_1 \leqslant y \leqslant y_2$. The function for horizontal integral projection and for vertically integral projection is defined as following.

$$S_h(y) = \sum_{x=x_1}^{x_2} f(x, y) \qquad S_v(x) = \sum_{y=y_1}^{y_2} f(x, y)$$

The integral projection may reflect total gray case on horizontal direction or vertical direction. The average integral projection M_h and M_v are following.

$$M_h(y) = \frac{1}{x_2 - x_1} \sum_{x=x_1}^{x_2} f(x, y) \qquad M_v(x) = \frac{1}{y_2 - y_1} \sum_{y=y_1}^{y_2} f(x, y)$$

2.2   Differential Projection

When x is taken in the range $x_1 \leqslant x \leqslant x_2$, y is taken in the range $y_1 \leqslant y \leqslant y_2$. The function for horizontal differential projection and for vertically differential projection is defined as following.

$$D_h(y) = \sum_{x=x_1+1}^{x_2} | f(x, y) - f(x-1, y) | \qquad D_v(x) = \sum_{y=y_1+1}^{y_2} | f(x, y) - f(x, y-1) |$$

The differential projection may reflect gray change case on horizontal direction or vertical direction. The average differential projection A_h and A_v are following

$$A_h(y) = \frac{1}{x_2 - x_1} \sum_{x=x_1+1}^{x_2} | f(x, y) - f(x-1, y) | \qquad A_v(x) = \frac{1}{y_2 - y_1} \sum_{y=y_1+1}^{y_2} | f(x, y) - f(x, y-1) |$$

3   Eye Location

The accurate position of eye is applied in intellective system widely. The premise about our method is that the face area has been detected.

Eye area has two characteristics. One is that its area is darker than surrounding area and its gray value is lower. Another is that its gray change is bigger. So we may use integral projection and differential projection to location eye.

3.1   Delimiting Eye Area

We may delimit eye area and eliminate interference information.

(1) The range of eye horizontal area

General, the shape of a front face is long circle. The gray of the face is bigger than ambient. In vertical gray projection, a high gray area is face's horizontal range(see Fig. 1). On horizontal direction, the average integral projection value is $p_i(i=0,1,2,....W-1)$, there W is image width. The maximum of p is $pmax$ and minimum of p is $pmin$. The left and right edge at $(pmax-pmin)/2$ are the horizontal range of eye.

(2)The range of eye vertical area

A eye's position on face is represented with h/H (see Fig.1). Upper bound of the eye is h'/H and lower bound is h''/H. Using BJUT-3D face database [15], we can get eye's position proportion on face that eye' average proportion on face is 0.26, the maximum is 0.45, the minimum is 0.1. So the range of eye's position proportion is from 0.05 to 0.5. Fig.1 shows the results of delimiting eye area.

3.2   Horizontal Position of Eye

(1) The feature about horizontal differential projection of eye

The eye's gray change is high frequency in the horizontal direction. Fig.2 shows gray horizontal projection curve that closed image are average differential projection and differential projection has extrema at eye and eyebrows. If eye area includes nose, the extrema maybe occur at nose. (see Fig.2b) and it is smaller than at eye. We may locate candidate for election horizontal position of eye according to average differential projection.

(2) The feature about horizontal integral projection of eye

The eye's gray value is low in the horizontal direction. In Fig.2, the curve that stay out image are average integral projection and there is extremum at eye. The extremum is minimum or second small. In Fig.2a, the average gray on eyebrows position is minimum. In Fig.4b, the average gray on eyebrows position is closed by eye position and there is extramum at nose but it is larger than at eye and eyebrows.

Therefore, we synthesize differential and integral projection to decide the horizontal position of eye.

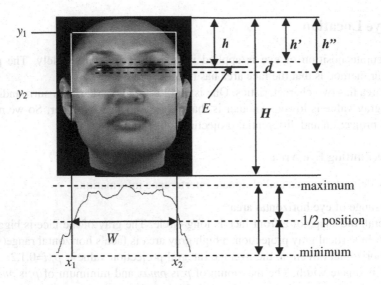

Fig. 1. Eye's range

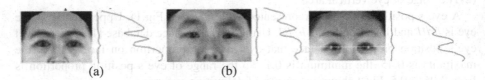

(a)   (b)

Fig. 2. Differential and integral projetion   **Fig. 3.** Eye horizon tal position

(3) The method step

① Calculated extrema series d_i ($i=0,1,\ldots m-1$)in horizontal average differential projection, there m is number of extrema. ②Amalgamated extrema with closed and getting d_i' ($i=0,1,\ldots n-1$),there $m \geq n$. ③Sort by extrema and getting maximum $dmax1$ or second big $dmax2$. ④Calculated extrema series s_i ($i=0,1,\ldots j-1$) in horizontal average integral projection, there j is number of extrema. ⑤Amalgamated extrema with closed and getting s_i' ($i=0,1,\ldots k-1$),there $j \geq k$. ⑥Sort by extrema value and getting minimum $smin1$ or second small $smin2$. ⑦Compared the positions error about $dmax1$, $dmax2$, $smin1$ and $smin2$.If having an error is smaller than getting error, it is eye position. Else if having two errors are smaller than getting error, the eye position is .below (see Fig.3). Else location is fail.

3.3   Horizontal Area of Eye

The key of decision horizontal area of eye is eye's height. Fig.1 shows that the eye's height is represented by d. d/H represent the ratio between eye's height and face's height. Using BJUT-3D face database [15], we can get the ratio between eye's height and face's height. The average of d/H is 0.08, a maximum is 0.1, a minimum is 0.06. When the horizontal position of eye is p, the eye's area is from $p+0.04H$ to $p-0.04H$ (see Fig.4).

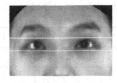 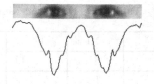

Fig. 4. Eye's area   **Fig. 5.** Vertical integral projetion

3.4   Vertical Position of Eye

We utilize above horizontal area of eye and make vertical integral projection (see Fig.5). In projection curves, there are two clear extrema that are eye's vertical position. In regard to eye's area, two eye lie two side area and every side area has a minimum that is eye's vertical position.

4   Detecting Close Eye

Making discovery on experiment, we may judge close eye from horizontal project curve. Fig.6 shows that the horizontal gray average differential at eye's extremum lessen when eye from open to close.

Fig. 6. The difference curve of different eye state

The curve about horizontal gray average integral at eye extrema change is little. So we may use it to locate eye in closed eye.

4.1   Setting Parameter

Supposed: D_{max} is maximum value about horizontal gray average differential at eye, D_{ave} is average value about horizontal gray average differential in face. D_m is defined as following.

$$D_m = D_{\max} - D_{ave}$$

where(see Fig.1)

$$D_{\max} = \frac{1}{W} \sum_{x=x_1+1}^{x_2} | f(x,E) - f(x-1,E) | \qquad D_{ave}(y) = \frac{1}{WH} \sum_{y=y_1}^{y_2} \sum_{x=x_1+1}^{x_2} | f(x,y) - f(x-1,y) |$$

According to the image about same person with different attitude, we may get D_m as Table 1. Table 2 shows the result of different person. Tab.1–2 show that D_m value descends above half when eye is from open to close.

Table 1. D_m value about same person with different attitude

	I	II	III	IV	V	VI	VII
Open eyes	8	11	8	10	7	7	8
Closed eyes	1	2	2	2	2	1	3

Table 2. D_m value about different person

	I	II	III	IV	V
Open eyes	4	5	11	7	5
Closed eyes	1	1	2	3	2

4.2   The Method Step about Close Eye Detecting

(1) Using open eyes image to locate eye's position E and calculated $D_{m.}$ Recording the extrema position about horizontal average integral projection at eye. (2) Getting next frame image. According to the extrema position about horizontal average integral projection at eye to get eye's position E' and calculated D'_m. (3) If $D_m / D'_m > 2$, the eye may be close. (4) Repetition (2) and (3).

5   Conclusion

We have described our proposal for eye location and close eye detecting. Our method is based on gray projection. Using differential projection and integral projection may locate eye and detect close eye. The method's special characteristic is (1) Before location eye, delimiting eye area may delete some interference information. (2)Synthesizing differential projection and integral projection and making result's accuracy is high. (3)The quantitative method is given and can detect close eye. It is simple and calculation fast.

When a person wears dark edge glasses, it can affect the result of location eye. If a person's eye is small, it can affect the result of close eye detecting.. Our next work is improving the method and making it broad-spectrum.

Acknowledgment. This work was funded by Jiangxi Province Key Lab for Digital Land (DLLJ201003).

References

1. Kawaguchi, T., Hidaka, D., Rizon, M.: Detection of eyes from human faces by hough transform and separability filter. In: Proceedings of IEEE International Conference on Image Processing, vol. (1), pp. 49–52. IEEE Computer Society, Vancouver (2000)
2. Yuille, A.L., Cohen, D.S., Hallinan, P.W.: Feature extraction from faces using deformable templates. International Journal of Computer Vision 8(2), 99–111 (1992)

3. Shi, H.R., Zhang, X.S., Liang, Y., et al.: An Improved Template Matchig Eye Location Method. Computer Engineering and Applications 40(33), 44–45 (2004)
4. Fei, J.L., Yu, W.X., Wang, Z.Z.: An Improved Template Matchig Eye Character points Location Method. Computer Engineering and Applications 430(32), 207–209 (2007)
5. Ai, J., Yao, D., Guo, Y.F.: Eye Location Based on Blocks. Journal of Image and Graphics 12(10), 1841–1844 (2007)
6. Zhu, X.J., Wand, X., Li, B.W.: Algorithm of eye location in face recognition. Journal of Circuits and Systems 12(2), 98–100 (2007)
7. Reisfeld, D., Yeshurun, Y.: Preprocessing of face images: detection of features and pose normalization. IEEE Trans. Computer Vision and Image Understanding 3(71), 413–420 (1998)
8. Zhou, J., Lu, C.Y., Zhang, C.S., et al.: Human face location based on directional symmetry transform. Acta Electronica Sinica 27(8), 12–15 (1999)
9. Liu, W.Y., Pan, F.: Application of Discrete Symmetry Transform in Eye Mage Features location. Infrared Millim. Waves 20(5), 375–379 (2001)
10. Kanade, T.: Picture processing by complex and recogition of human faces, Ph.D.Thesis. Kyoto University (1973)
11. Brunelli, R., Poggio, T.: Face recognition: Feature versus templates. IEEE Transactions on Pattern Analysis and Machine Intelligence 15(10), 1042–1052 (1993)
12. Feng, G.C., Yuen, P.C.: Variance projection function an its application to eye detection for human face recognition. Pattern Recognition Letters 19(9), 899–906 (1998)
13. Geng, X., Zhou, Z.H., Chen, S.F.: Eye Lcation Based on Hybrid Projection Function. Journal of Software 14(8), 1394–1399 (2003)
14. Zhang, L., Jiang, J.G., Qi, M.B.: Eye lcation algorithm Based on differential and integral projection. Jounral of Hefei University Technology 29(2), 182–185 (2006)
15. The BJUT-3D large-scale Chinese face database [EB/OL] (August 2005), http://www.bjut.edu.cn/sci/multimedia/mul-lab/3dface/face-database.htm
16. Yan, G.L., Yan, W.X., Cheng, L.: Fatigue Driving Detection System Based on Skin-color Model and Gray Integral Projection. Journal of Chongqing Institute of Technology 22(12), 10–13 (2008)

Healthy Record and Hospital Medical Record System Intercommunication Based on ESB

Guang Dong, Honghua Xu, and Weili Shi

Changchun University of Science and Technology,
7089 Weixing Road, Changchun, 130022, China
light@cust.edu.cn, honghuax@126.com, shiwl@cust.edu.cn

Abstract. According to the requirement of data intercommunication of the resident healthy records and existing mainstream hospital medical record systems, this paper presents a solution of community health and the hospital information system intercommunication. Based on the ESB, SOA and XML technology in this paper, we construct a exchanges platform of community health service center and medical and health institutions at various levels. In addition, we design the massage middleware and data exchange adapter based on HL7 and DICOM3.0 standard in order to solve technical problems of the resident healthy records and medical and health information system data intercommunication. Consequently, this paper realizes the information exchange of the resident healthy records and hospital medical record system such as HIS, LIS, CIS, PACS, RIS, and so on.

Keywords: healthy records, hospital medical records, exchanges platform, massage middleware.

1   Introduction

Western countries attach great importance to developing the information system related technical research and application, especially strengthening community-based health information system construction. England, the United States, Canada, Australia, France and other countries have spent a huge sum of money on carrying out the regional health information construction of national and local level in the more than 10 years, for the core to the electronic healthy records and electronic medical records data sharing.

It has been fully verified that the effect of improving the medical service efficiency, quality, and accessibility and reducing health-care costs and medical risk through the health information sharing, and acknowledged as the future development direction of health information construction.

At present, HIS system application in the domestic hospital has already become more and more mature, but community health information system is a new system implemented in recent years. However, because the hospital information system was usually developed by different developers, the standards and specifications of each hospital information system are not unified and the information formats among each medical institution are also not same. Consequently, there are several problems when

D. Jin and S. Lin (Eds.): Advances in MSEC Vol.2, AISC 129, pp. 353–358.
springerlink.com   © Springer-Verlag Berlin Heidelberg 2011

community health information system exchanges and shares data with the hospital, namely we can not get a very good solution to the problem of system interconnection and resources sharing. So this makes each system independent and causes a lot of information isolated points, not really to realize information exchanging and sharing[1,2]. To solve this problem, it needs to shield the concrete technology used by the hospital information system and its realization ways, and utilizes the intercommunication system to realize the resources sharing between the resident health records[3] and hospital medical records[4,5].

2   Method

Enterprise Service Bus (ESB) is a group of basic architecture implemented by the middleware technology, and supports Service-Oriented Architecture (SOA). In addition, ESB supports service and message in the heterogeneous environment and the interaction based on events, and ESB is provided with the appropriate levels of service and manageability.

In order to realize the linkage of the medical health business data and community health information platform, the first thing to do is to realize the standardization of heterogeneous data between system, which needs to arrange the lead service component of data exchange in the medical and health institutions. Each application system realizes the information interaction with the news interactive center by installing the corresponding software adapter in the business system part through the service bus ESB. The theory of the resident healthy records and hospital medical records intercommunication is shown in Fig.1 below.

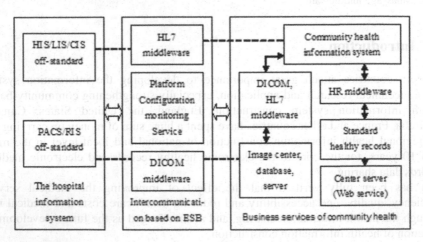

Fig. 1. The model of the resident healthy records and hospital medical record system intercommunication

Based on message middleware technology and the routing function of news content, this system integrates the workflow service. And we can realize the integration access of each regional medical and health information system such as HIS, LIS, CIS, PACS,

RIS, and so on, in the way of data exchange adapter. The integrated system need to encapsulate the functional components and data components interacting with the data exchanges platform into "service", and to shield the concrete technology used by integrated system and its realization ways, and finally to realize the link with exchanges platform in the way of standard interface.

In the integral designing structure of the news bus system, various specific business systems can transmit and receive business data through the adapter connected to information exchange platform. The adapter plays a role of coupling news exchange platform and specific business system.

2.1 Middleware Technology Based on HL7 and DICOM3.0 Standard, to Realize the Information Transmission of Medical Records between Systems

In this part, combined with HL7, DICOM3.0 standards and XML language, we design message middleware, and realize the information intercommunication of the resident healthy records and diagnostic information domain, the drug prescription information domain, clinical examination information domain and medical imaging information domain in the hospital medical record systems such as HIS, LIS, CIS, PACS, RIS, and so on.

(1) Utilizing the middleware based on HL7 to realize HL7 standardization of heterogeneous data

HL7 standardization of heterogeneous data firstly transforms data of each system into a standard HL7 message format, and then in accordance with the consultative communication rules sent to the receiving system. The receiving system resolves the received HL7 message and transforms into the application data, so as to realize the data exchange between the systems.

HL7 middleware consists of three modules. The first module is used to achieve HL7 message construction, namely to transform the information in the database which will be sent into the text data of HL7 format through function. Another module is to achieve the HL7 message analytical function, namely, to extract structure based on the HL7 message structure and field structure according to paragraph structure, which will be processed into the corresponding information, and store or update the corresponding data in the database. The third module is responsible for sending and receiving messages. The system structure of HL7 middleware is shown in Fig 2 below.

The system structure of HL7 middleware has adopted the C/S mode and is divided into the server and the client. The client extracts the corresponding fields from the local database, which are processed to comply with the data structure of HL7 through the HL7 middleware, and then transforms the corresponding data into HL7 messages and sends them to the server. It is considered that producing a standard format or XML format, according to the situation of applicant. After the server receives news, the server will resolve the HL7 messages through the HL7 middleware, and which will be reverted to data structure which is consistent with this system.

(1) Realize the DICOM standardization of heterogeneous data through DICOM middleware

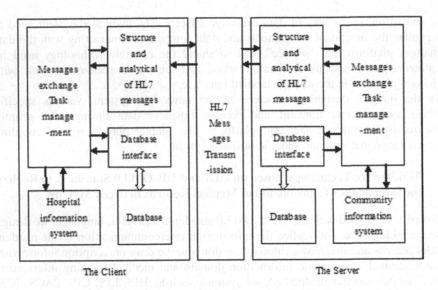

Fig. 2. The system structure of HL7 middleware

The DICOM standardization of heterogeneous data should have two functions which are conversion and transmission of data. For the medical equipments and medical images which do not accord with DICOM3.0 standards, we can convert the non-DICOM data into DICOM data through the DICOM middleware, and transmit in accordance with the standards of DICOM. After the conversion from the non-standard data to the DICOM data is completed, we should send the DICOM data to the server PACS and complete archiving and storing.

The communication mechanism of DICOM middleware is shown in Fig.3 below[6]. Compared with the traditional ESB accessing mode, The DICOM sending equipment is the outside visitor, and the service cell of receiving DICOM on ESB is the service provider. When accessing ESB in the outside, we can use the custom protocol interface by way of a communication tool.

This communication tool has two aspects of the requirements.

① It must comply with the DICOM communication format, namely, must include the necessary elements information such as the IP address, port, entity title, and so on.

② It must be able to establish a connection with ESB bus, because this is the basis and premise of accessing the service unit.

In the sending end on DICOM, after the service of ESB bus is opened, the DICOM sender only need to send the request of establishing a connection to the receiving entity with specify IP address, port and entity title, and doesn't have to consider the realization of the receiving end. This way reduces the coupling coefficient between components in practical application, and in the condition that the sending end on DICOM does not make any changes, the traditional pattern which is from the sending equipment to the receiving equipment will be promoted to the new model which is from the sending equipment to ESB bus.

According to the normal protocol of DICOM communication, the work which the internal functional components of service unit will accomplish is to establish a connection with the DICOM sender based on the analytic entity title, IP, port and other information, and to complete the task of file transmission. After the service unit completes internal functions, it will send the ESB message to the bus for being used by other service unit.

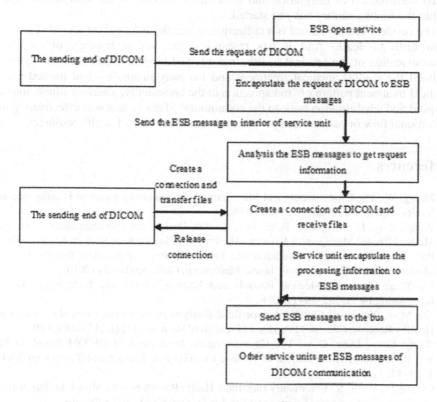

Fig. 3. The communication mechanism of DICOM middleware

2.2 Realize the Reliability and Integrity of the Data Transmission Based on the Data Cache Service Technology

In this part, we should optimize and deploy the lead database according to the requirement, exchange the lead cache of data, and provide the caching mechanism of medical institutions in the process of uploading the original medical data. In the process of uploading mass original data, we will adopt the uploading forms of the quasi real-time and batch, which can ensure the service response time and stability and improve the transmission efficiency of system.

3 Conclusion

The information-based construction in health has experienced from scratch, from local to global and from hospitals to other businesses which is a process of unceasing infiltration. Health Information has gradually become an indispensable part of the medical and health service system. At present, the information system of medical institutions at or above the county level have basically gained popularity, but 85% of HIS systems still make financial accounting as the center. However, the information-based construction of integration and information sharing of the information system within the health industry has just started.

The resident healthy record is a collection of healthy information records for a man from birth to death. And it has researched the key technology of the inter-communication of the resident healthy records and hospital medical record systems, realized the grading medical treatment and two-way diagnoses, and formed a new medical treatment pattern of "indisposition in the community, a serious illness into the hospital and rehabilitation back to the community". This system will effectively guide the rational flow of patients and promote the reasonable use of health resources.

References

1. Zhang, W.: The Establishment and Management of Community Resident Healthy Records. Health of Mount Taishan 33, 29–30 (2009)
2. Wang, B.-q., Lv, J.-l., Hui, H.-f., Tian, J.: The Design and Implementation of Hospital Medical Record Management Systems. The Project of Information System, 52–55 (2009)
3. Pan, B.-k., Cai, G.-q.: Exploitation and Use of Community Resident Healthy Records. Modern Distance Education of Chinese Medicine in China 8(02), 124 (2010)
4. Li, Y.-q.: Electronic Medical Records and Related Information Technology. Medical Information 14(7), 382–384 (2001)
5. Xu, M.-y., Yu, H.-x., Chi, Y.-j.: Superficial Analysis of the Construction about Electronic Healthy Records. Chinese Magazine of Preventive Medicine 11(2), 215–216 (2010)
6. Wang, C.-w., Ding, G.-y.: The Communication Mechanism of DICOM Based on ESB and Project Integration. Journal of Beijing University of Science and Technology 30(11), 1315–1319 (2010)
7. Guang, D., Weili, S.: Community Electronic Health Record System Model. In: International Conference on Computer Science and Service System, CSSS 2011 (2011)

Improving Oral Fluency through Readers Theater in the EFL Classroom—A GET Innovation in China

Man Cao* and Lan Huang

School of Foreign Languages, Wuhan University of Technology
430070 Wuhan, China
appletreeleaf@126.com, whuthuang@163.com

Abstract. This study initiated the effectiveness of Readers Theater (RT) in improving oral fluency of graduate English (GE) of Chinese students after one semester's experiment of using RT in graduate English teaching (GET) as a foreign language (EFL). The pretest and the posttest of GE oral fluency were analyzed via SPSS 13. Students' learning attitudes towards RT were surveyed and analyzed. The other three oral tasks, reciting, reading aloud and pronunciation correcting, were also surveyed to suggest their correlations with after-class learning. Future research is suggested in applying RT in reading and writing abilities for GET.

Keywords: Readers Theater, oral fluency, GET.

1   Introduction

In 1992 China's State Education Commission launched "English Teaching Syllabus for Non-English Major Graduate Students (GET Syllabus)" [1], providing teaching regulations for universities. The syllabus hasn't been updated ever since, but graduate education of the 21st century is popularized. Such rapid growth has witnessed wider admission and bigger class size; therefore, graduate English teaching (GET) will face greater challenges when nearly forty students share one classroom.

With more multimedia classrooms, one direction of oral GET reform is "movie/video teaching". Zhao [2] pointed that movie teaching should be important as extensive reading and listening. Long [3] stressed the authenticity of movie teaching in setting and language. Ou [4] indicated two benefits: to motivate students' interest and to provide cultural learning. Li [5] introduced five auxiliary functions of playing original English movies in class. Teaching methods targeted at oral English are: brainstorming, group discussion and dubbing [3, p.50]; questioning, unsynchronized input and discussion [6]; imitation, conversation and recitation [7].

Another form is to bring "drama teaching" into GET. Back in 1994, Zhang [8] introduces Carol Cox's Drama method, which aims to engage students into language learning, to enlarge their vocabulary and to cultivate their critical reading skills. Xu [9] constructed theoretical foundations for drama: educational psychology, art education, memory rules and commutative mode. It should be noticed that drama here

---

* Man Cao: lecturer at Wuhan University of Technology; researcher in applied linguistics.

D. Jin and S. Lin (Eds.): Advances in MSEC Vol. 2, AISC 129, pp. 359–364.
springerlink.com   © Springer-Verlag Berlin Heidelberg 2011

does not involve the aspect of literature; it emphasized the aspect of acting out plays on stage.

In classroom teaching, the above methods are less popular for inadequate multi-media classrooms. Most drama or mini-plays should be played on stage. More often, a drama night might stage ten shows, while there are more shows in the list. The audition failure may discourage weak teams. Meanwhile, a successful drama would pose higher demands on performance, line-memorization, acting, and costumes. Even if a team wins, it needs time in preparing the scene with props rather than improving their lines.

2   Theoretical Framework

Readers Theater (RT), or Reader's Theater, or Readers' Theater, is minimal theatre in support of literature and reading. It is an oral presentation of scripts by two or more readers, which does not require stage, costumes, setting or memorizing. Therefore, RT is a simple but efficient way of practicing in EFL classrooms. Each student reads aloud one role of the scripts with gestures, facial expressions and simple setting.

The scripts may focus on literature like story, drama, poem but not be confined to it [10]. Researchers have found that RT can also be used in nonfictions, such as history [11], science, and social studies. An important feature of the scripts is repeated reading in different roles, which may lower the anxiety of low performance learners. RT has been popular from universities to middle schools for it can help improve reading fluency [12], writing performance [13], oral discourse and listening skills.

Fluency in language output can target at three of the four skills: reading, writing and speaking, and this paper deals with oral fluency specifically. For GET, oral fluency can be defined as accuracy in linguistic level, quick response in communicative level and appropriate intonation in supra-segmental level.

However, literature on RT as applied to EFL classrooms at Chinese college level or above is still scarce. Most of the research was done by Taiwan researchers with focus on English teaching in middle school and primary school settings [14]. College language classrooms are left behind. Using CNKI (an online database of published academic journals in China) as a search tool, the phrases "Reader's Theater/ Readers Theater/ Readers' Theater/ RT" are keyed for related literature. Only four papers were found, two of which deal with RT in English teaching and learning: Mei [15] and Chen [16]. None of them was designed at college level or above.

To address the issue of oral fluency, this paper proposes an innovation of applying RT to GET in mainland China. It is the first research that examines the effectiveness of RT experiment in GET with pretest and posttest in oral fluency.

3   Research Methodology

This study was carried out by a quasi-experimental design for a semester of 12 weeks. The participants were measured by pretest-posttest in oral fluency in Week One and Week Twelve. Each student was evaluated by two teachers in three features of oral fluency: accuracy, response and intonation. The experiment of RT teaching lasted from Week Two to Week Eleven. In Week Twelve, a questionnaire was given to participants on their opinions or feelings in RT and their attitudes about oral tasks.

The participants consisted of three classes of Grade One graduate students in a Chinese university. The total number is 102, with 28, 36, and 38 respectively. They were taught by the same instructor in an extensive English course for the whole academic year. Among them 25 were female and 77 male. The distribution of gender was influenced by their science majors.

4 Results

The software Statistical Package for Social Science (SPSS 13.0) was used to analyze the collected data in this study. Performances in pretest and posttest of oral fluency were analyzed by paired-sample t tests to see if there was a significant improvement. The questionnaire was also analyzed by frequency to evaluate students' attitude toward RT in GET. Meanwhile, the other three oral tasks were analyzed by correlation with students' after class learning.

4.1 Paired-Sample T Test Results on Oral Fluency

Table 1 showed that the students' oral fluency reached the significant difference with value of sig. (2 -tailed) at .000, less than 0.01. It showed that students had made significant progress in oral fluency after the one-semester experiment with RT in GET.

Table 1. Paired Sample Test

		Paired Differences							
		Mean	Std. Deviation	Std. Error Mean	95% Confidence Interval of the Difference		t	df	Sig. (2-tailed)
					Lower	Upper			
Pair 1	Pre-test Post-te	-1.53922	.99176	.09820	-1.73402	-1.34441	-15.674	101	.000

4.2 Questionnaire Results on RT

The main focus of the questionnaire (Table 2) was to ascertain students' experience in RT. Their response to the first question 54.8% agreed that RT made English class more interesting. 48% of them strongly agreed that RT is a good way of learning oral English, and 44.1% agreed that they were more confident speaking English due to RT. To ascertain if the students liked working in a group, 54.9% agreed and 31.4% strongly agreed.

Only 6.9% indicated that they preferred to learn English alone than in a group. It is interesting to note that most students were motivated to improve their English after watching their friends perform (45.1% agreed and 38.2% strongly agreed). This could mean extrinsic motivation plays a major role in their learning.

Table 2. Questionnaire on Readers Theater and other oral tasks
1(strongly disagree); 2 (disagree); 3 (neutral); 4(agree); 5 (strongly agree)

		1	2	3	4	5
1	RT made English class more interesting.	0%	0%	3.9%	38.2%	57.8%
2	RT is a good way of learning oral English.	0%	1%	3.9%	47.1%	48%
3	I am more confident speaking English due to RT.	0%	2.9%	13.7%	44.1%	39.2%
4	RT is difficult in learning English.	5.9%	40.2%	46.1%	5.9%	2.0%
5	Making our own scripts was not difficult.	2.0%	22.5%	30.4%	39.2	5.9%
6	I was willing to perform RT with my classmates.	0%	3.9%	9.8%	54.9	31.4%
7.	After watching others' performance, I want to improve my English.	0%	2.9%	13.7%	45.1%	38.2%
8.	RT made no difference in the way I learn English.	32.4	49%	16.7%	1.0%	1.0%
9.	RT helped me improve my writing.	0%	7.8%	34.3%	52%	5.9%
10	I prefer to learning English alone than in a group.	6.9%	52.9%	24.5%	10.8%	4.9%
11	I found writing the script difficult.	2.9%	41.2%	38.2%	13.7%	3.9%
12	I prefer reciting than RT. (f 12)	11.8%	39.2%	44.1%	2.9%	2.0%
13	I often learn English after class. (f13)	3.9%	21.6%	41.2%	28.4%	4.9%
14	I do not like reading aloud in class. (f 14)	19.6%	48%	25.5%	4.9%	2.0%
15	I like pronunciation corrections in class. (f 15)	1.0%	2%	35.3%	29.4%	32.4%

Table 3. Correlations between after-class learning and reading aloud

			f13	f14
Kendall's tau_b	f13	Correlation Coefficient	1.000	-.225**
		Sig. (2-tailed)	.	.009
		N	102	102
	f14	Correlation Coefficient	-.225**	1.000
		Sig. (2-tailed)	.009	.
		N	102	102
Spearman's rho	f13	Correlation Coefficient	1.000	-.245*
		Sig. (2-tailed)	.	.013
		N	102	102
	f14	Correlation Coefficient	-.245*	1.000
		Sig. (2-tailed)	.013	.
		N	102	102

**. Correlation is significant at the 0.01 level (2-tailed). *. Correlation is significant at the 0.05 level (2-tailed).

Table 4. Correlations between after-class learning and pronunciation correction

			f13	f15
Kendall's tau_b	f13	Correlation Coefficient	1.000	.236**
		Sig. (2-tailed)	.	.006
		N	102	102
	f15	Correlation Coefficient	.236**	1.000
		Sig. (2-tailed)	.006	.
		N	102	102
Spearman's rho	f13	Correlation Coefficient	1.000	.268**
		Sig. (2-tailed)	.	.007
		N	102	102
	f15	Correlation Coefficient	.268**	1.000
		Sig. (2-tailed)	.007	.
		N	102	102

**. Correlation is significant at the 0.01 level (2-tailed).

4.3 Correlations between Other Oral Tasks and After-Class Learning

The correlations between after-class learning (f 13) and the other three oral tasks, namely, reciting (f 12), reading aloud (f 14), and pronunciation correction (f 15) showed: 1) no significant correlation between reciting and after-class learning (due to the space limits, the respective table was not shown); 2) significantly negative correlation between not liking reading aloud, which may be explained as reading aloud being associated with after-class learning; 3) significantly positive correlation between pronunciation correction and after-class learning.

5   Conclusion

This paper conducted the first innovative study of RT being introduced into GET in China. The quantitative results are: 1) significant differences in the oral fluency; 2) positive experiences of students using RT; 3) willingness to take reading aloud and pronunciation correction tasks for students learning English after class.

In conclusion, there is no doubt that students have a positive experience after using RT to improve their oral fluency in GET. Also, based on the collected data of the two oral tests, it shows that most students made significant progress in oral fluency. According to the questionnaire, students liked RT since it was interesting and it made learning English more fun than before. As far as group learning is concerned, RT is an economic and helpful way of engaging EFL students into GET activities. Inspired by successful experience of RT in class, students are more likely to write their own scripts, using English in authentic settings, expressing their own feelings and sharing their own stories.

However, there may be difficulties to bring RT into class, especially when students have no suprasegmental knowledge of English, that is, linking, deleting, stress, intonation and rhythm. Teachers need more knowledge about theater as a genre and performance as an art. When students write their own scripts, teachers have to proofread and edit the scripts before students are ready to practice or perform.

The limitations of this study are as follows. First, this study investigated only the oral fluency of GE, while it left out the writing ability and reading fluency. Future studies may associate RT with these two linguistic abilities. Second, the sample size of this study was small and the participants were not chosen randomly. Therefore, the results of the research might not be generalized to other contexts. For this reason, more participants and sample randomization should be considered in future research. Third, this study was a quasi-experiment. Future studies may be a longitudinal study among control groups and experimental groups to understand GET by multiple methods like classroom observation, tape-recording, video-recording and interviews with participants.

Acknowledgement. This paper was sponsored by Provincial Teaching Research Projects of Higher Institutions in Hubei, China (Project No. 2009102).

The authors are grateful to Professor Zhu Hanxiong for her insightful comments on this paper. Thanks also go to the graduate students participating in RT experiment.

References

1. Syllabus Compiling Group of Non-English Major Graduate English Teaching, English Teaching Syllabus for Non-English Major Graduate/doctoral Students. Chongqing University Press, Chongqing (1993)
2. Zhao, G.: Position and Function of Movie-viewing Class. Educational Technology for Foreign Language Teaching. 4, 23–25 (1994)
3. Long, Q.: Films and English Listening and Speaking Teaching. Educational Technology for Foreign Language Teaching 91, 48–51 (2003)
4. Ou, L.: Effect of Watching Films on English Study. Journal of Chongqing University (Social Sciences Edition) 9(6), 72–73 (2003)
5. Li, S., Liu, Y.: On Auxiliary Functions of Original English Film in College English Instruction. Shandong Foreign Language Teaching Journal 3, 63–65 (2007)
6. Some Considerations, W.Y.: about Integrating Film into College English Class. Journal of Longdong University (Social Science Edition) 17(2), 94–96 (2006)
7. Y.Y., Y, Y.D.Y.: Analysis and Design of English Movies in College English Content Teaching. Movie Literature 5, 31–32 (2007)
8. Zhang, W.: Language Teaching via Drama by Cox. Modern Foreign Languages. 4, 33–35 (1994)
9. Xu, K.: The Scientific Basis of Process-oriented Role Play in Foreign Language Teaching. Foreign Languages and Their Teaching 3, 61–63 (2001)
10. Patrick, N.C.L.: The Impact of Readers Theater (RT) in the EFL Classroom. Polyglossia 14, 93–100 (2008)
11. All the World's a Stage: Using Readers Theater to Teach History and Develop Reading Fluency, http://www.ait.net/lessons/LanguageArts_5.pdf
12. Readers Theater: Science and Social Studies. Steck-Vaughn Company (2006)
13. Liu, J.: The Power of Readers Theater: from Reading to Writing. ELT Journal 54(4), 354–361 (2000)
14. Peng, S., Peng, T.: Effects of Readers Theater on Taiwanese Fourth Graders' English Reading ability. The Journal of Educational Science 9(1), 1–28 (2010)
15. Mei, M.: A study on the Application of Drama in English Teaching—based on Readers Theater. Journal of Tangshan Teachers College 29(6), 136–138 (2007)
16. Chen, S.: Bringing into English classroom—Effects and Methods on Readers Theater in English teaching. Education Science and Culture Magazine 4, 69–70 (2010)

The Management Mechanism of Education Exchange and Cooperation Platform for Southeast and South Asia in Yunnan China

Jing Tian[1] and Yun Zeng[2]

[1] College of Educational Science and Management, Yunnan Normal University,
Kunming 650092, China
[2] College of Architectural Engineering, Kunming University of Science and Technology,
Kunming 650500, China
{jingtian2003003,zengyun001}@163.com

Abstract. The purpose of establish the education exchange and cooperation platform for the Southeast and South Asia is to promote education institutions in Yunnan develop international cooperation and education services, increase exchanges, and realize mutually beneficial and win-win cooperation. The platform of international education exchange and cooperation needs to mobilize various resources, involving multiple benefit main body, so the management mechanism becomes a key factor. This paper focus on three targets that establishes the long-effect mechanism of international education cooperation with regional characteristics, and builds the platform of education exchange; establishes the international education cooperation pattern by central government support, local dominant, school participant; expands the role of local government in international education cooperation. Based on the scientific management theory, system theory and collaborative theory, this paper designs the management and operation mechanism of education exchange and cooperation platform, and ensures the platform has the adaptive ability.

Keywords: Management mechanism, Education exchange and cooperation, Platform, Southeast and South Asia, Bridgehead strategy.

1 Introduction

The 17th national congress of the Communist Party of China put forward the national strategy that deepen coastal open, accelerate the mainland open, improve border open, and realize opening-up policy to promote each other. The border open is an important part of China open, its features different from the coastal open and mainland of China open. The President Hu Jintao had put forward building two open bridgeheads, Xinjiang Uygur Autonomous Region (Sept. 2006) and Heilongjiang province (June. 2009), for northwest, northeast border open. And in July 2009, he had put forward to build Yunnan Province to the important bridgehead of China to southwest open. The three bridgeheads respectively located in China's three important road direction, which are the whole idea of the Communist Party of China and State Council to seek

D. Jin and S. Lin (Eds.): Advances in MSEC Vol. 2, AISC 129, pp. 365–370.
springerlink.com   © Springer-Verlag Berlin Heidelberg 2011

new land exits and the implementation of China's coastal open thirty years later, based on global economic development, the national safety and steady development.

Yunnan province is located in southeast of China, and connects with the Southeast and South Asia subcontinent. Since 1980s, base on facing the Southeast of Asia's the Greater Mekong Subregion (GMS), facing the South of Asia' Bangladesh China India and Myanmar (BCIM) subregion, and connecting south Beibu Bay and Bengal, Yunnan province has basically formed unimpeded economic channel of foreign exchange cooperation and openness. Facing the southwest open bridgehead strategy is the important deployment of China to perfect opening-up, is an opportunity to Yunnan province further enhance development power and vigor, and promoting the comprehensive, coordinated and sustainable development. The education exchange and cooperation platform (EECP) to expand to the Southeast South Asia open is one of soft strength support security conditions as bridgehead transformation of Yunnan province.

2   Background of the Management Mechanism of the EECP to Southeast and South Asia

Recent years, the education exchange has been increasing between the Yunnan' all levels schools, universities and research institutions with the southeast and South Asia. It has presented various forms, such as the government support, the school official, official combination folk cooperation, and promoted the cooperation pattern by means of technical assistance, technology trade, foreign investment, cooperative development, personnel exchanges, investigate and visit. The 'going global' strategy has promoted education internationalization. The 'going global' strategy gives actively fund supports to international cooperation and exchanges of universities. It cost 1 billion Yuan RMB to construct The Base of Chinese Language Council International each year. In 2004, the Yunnan provincial government set up a 'recruit neighboring countries scholarships', which is cost 1.8 million Yuan RMB each year. In 2009, the two Promotion of Chinese Language Council International School bases, five Chinese Language International Promotion Center, two Confucius Institute, two Confucius Schools and others Chinese education centers had been built [1]. At the same time, Yunnan has expanded foreign students scale. Its oversea students figures increased from 760 in 2001 to 10,000 in 2009[1]. Another measure is Strengthen international course construction to meets the need of minor language talents. In 2010s, minor language major of universities in Yunnan keep about 20, and the Number of foreign students are about to 3,000. The mechanism of talent exchange and the platform of humanistic exchange have been established. The Number of foreign students and workers in Yunnan from the Southeast and South Asia are more than 8,000. Over the years, some educational actions have good results, such as the technological training for the Southeast and South Asia relying on the greater Mekong institute, participate in international and Asia-Pacific regional tourism teacher training project, etc [2].

The international EECP is an operation system of international cooperation in teaching, scientific research and social service aspect, in which involves various interest relationship. From the view of participants, it involves government, schools, research institutions, non-profit organizations and the social public; from the view of the organization, it needs multi-level administrative organization, such as coordinate departments, the macro management department, executive departments, intermediaries and public services department. It is obvious, the management and operation mechanism of the EECP is a complex system, its operation and management is a gambling process with multi-stage asymmetric information, and needs to overall planning, optimum combination of various management tools, so that it can achieve good performance and realize the strategic objectives of the EECP.

3 The Design of the Management Mechanism of the EECP

3.1 Main Difficulties

Review the construction and operation practice of the EECP, we can say that the management and operation mechanism is key factor whether platform can normal operation and reach goals. The EECP of Yunnan is in start level, the management and operation mechanism of platform is not studied systematic and its management system is also not perfect. Firstly, the opening policy of Yunnan is not perfect in aspect of foresight, integrity and planning. Cultural exchanges and cooperation with the Southeast and South Asia at the national level lack of overall long-term planning and overall arrangement, the government's guidance function is not fully embodied out, particularly in constructing the market environment of education exchange, guiding regulation of education exchange, thus the necessary legal safeguard system and policy system need to be strengthened. EECP is a new exploration to change government functions and realize the system innovation. But due to a lack of mature operation mode, each department basically follows the way which is 'touch stone across the river', that impact the performance of platform, even failed. More over, the EECP has three basic elements, the education resources, service system and safeguard measures, in which the integrated education resources is the foundation, the co-construction and sharing of service system is the basic purpose, the security measures is the necessary condition. Three factors involved in schools, research institutions, government and non-profit organizations and social public interest bodies etc. Because the education resources lacks of sharing regulations and incentive and restraint mechanisms, the state-owned education resources or the unit or individual lack of enthusiasm to share responsibility and obligation. In addition, there are many problems in the system, management, planning, policy, the allocation of funds. These problems make platform can't gather enough education resources to provide services.

3.2 The Framework

Jeffrey Pfeffer (1997, 192) points out, 'After all, most of us don't really borrow physics theory too often to understand organizations (chaos theory notwithstanding), but temptation to rely on rational economic models of choice and organizational behavior, regardless of their ability to development novel insights or to produce empirically valid predictions, remains compelling,' [3]. It appears to us that many management theories as well as practicing managers in their work tend to present information about organizations in quantitative form, by using tables and graphs, and to express economic information as often as possible by mathematical functions [4]. Many managers see organization as rational and structural systems. It appears as if the management deliberately wishes to rise above the complex situations of education exchange organizational by adopting a bird's-eye perspective.

By adopting this bird's-eye view, the overall operation procedure of the EECP is shown as Fig. 1.

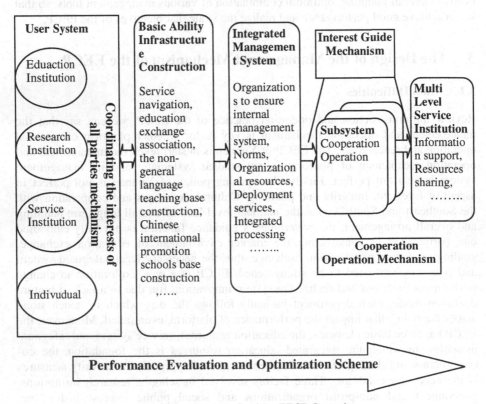

Fig. 1. The framework of the EECP Operation

It involves four parts, basic ability and infrastructure construction, integrated management system, corporately operation and service network. Schools, Education institutions, research institutions, service organizations and individuals are the user system of the platform. The aims of the basic ability and infrastructure construction is to develop the basic conditions of the colleges, schools and science and technology zone, and exert its radiation effect to the neighboring countries education and culture. The integrated management system is the platform hub, which is responsible for all education resource service in platform to deploy and manage, and offers comprehensive services, such as regulations, service guide, talent support, integrated processing, service and distribution. The corporately operation is composed by each subsystem in platform and related services institution, which can achieve multi level service cooperation and distributed service support. All above parts are not mutually separated, but fuse in together reasonably. The coordination is very difficult in the process of organizing technologic resource. Therefore, the operation mechanism characterized by sharing is the core system. In order to ensure a smooth operation platform, the core problem solved in each link need be analyzed. On the basis of overall planning, the management mechanism must be design by using the optimized combination algorithm to form complete system, which includes the cross-department interest coordination, the technical management system, organization safeguard, the internal management system, interest guide, collaborative operation, service evaluation system, etc. In addition, the whole process of platform operation need to be evaluated and optimized ongoing, and make the platform to operate perfect.

3.3   The Features

In spite of a promising start and some significant initial successes, 'systems thinking' has been marginalized in the social sciences since the late 1960s. The widespread rejection of the systems approach did not, however, stem the incorporation of a number of systems concepts into other social science theoretical traditions. Consequently, some of the language and conceptualization of modern systems theories has become part of everyday contemporary social science [5]. The EECP of Yunnan is a complex system with multi benefit body coordinated, that they should be integrated into the platform framework to balance various interactive elements, overall planning and implement step by step, so the operation system will not be suppressed. The combination algorithm of measure optimized is applied to system design process, which can solve the interest coordination difficult during the platform construction and operation process.

The management of the EECP can not be solved by management sector itself. The platform needs to integrate cross-industry, interdisciplinary and interdepartmental all kinds of science and technology resources, faces the universities, research institutions, government departments and the social public to provide, system, multi-function, multidisciplinary, convenient, efficient public services and promote communication, and must be based on various interest coordination mechanism and related system. Therefore, it is necessary to analysis the operation mechanism of platform, straighten out the interests of different participation body, master the relations of its conduction chain, explores the linkage mechanism between government and other related body to play the multiplier effect of public finance, which is the base of system effective implementation and application.

Synergetic leads to a new constructive dialogue between specialists in different disciplinary fields. The theory makes steps toward the synthesis of natural sciences and of the humanities, of the Eastern and of the Western worldviews, of the new science of complexity and the old traditions of culture, art, and philosophy [6]. The key of EECP cooperative management is to build appropriate operation mechanism, and form collaborative organizations in which participators coordination is the centre and management is the core. This operation mechanism requires that education exchange management must be established under education resources, contract and information management level, and it is an interactive system in which includes the coordinated management, personnel coordination management, information coordination management, resources system management subsystem. Although, its subsystem will increase or decrease according different subject, but the coordination management section is indispensable. For specific goal, coordinated management and operation mechanism will be based on 'the competition−cooperation−coordinated' idea of the synergetic theory, the communication collaborative department will play a role by using its privilege, and aggrandizes foresee control and guides orderly competition.

4   Conclusions

The aims of EECP is utilize regional advantages of Yunnan Province, faces the international and domestic educational markets, to develop the education communication and cooperation with the Southeast and South Asia in multilevel, and to wide field to advance education internationalization. The management and operation mechanism is the key to the outgrowth of EECP. This study explored the management and operation mechanism of the EECP. There is and needs to be continuing debate about the supporting measures and How to better improve the management and operation mechanism of the EECP with development of the times.

Acknowledgments. The research reported here is financially supported by the Humanity and Social Science Funds of Education Ministry of China (Grant: 10YJC880112). And part works is financially supported by the Science Research Foundation of Yunnan Education Department (Grant: 2010Z081).

References

1. Explore Reform Implementation of International Educational Cooperation and Exchange Platform with Regional Characteristics in Yunnan Province, China (June 6, 2011), http://china.jyb.cn/gnxw (in Chinese)
2. Li, H.Q., Wang, Y., Sun, L.: Education Research (2), 69–70 (2010) (in Chinese)
3. Pfeffer, J.: New Directions for Drganization Theory: Problems and Prospects, p. 192. Oxford University Press, New York (1997)
4. Parviainen, J., Koivunen, N.: Consumption Markets & Culture 9(2), 147–155 (2006)
5. Burns, T.R.: World Futures: Journal of General Evolution 62(6), 411–440 (2006)
6. Knyazeva, H.: World Futures: Journal of General Evolution 60(5), 389–405 (2004)

An Approach to Sound Feature Extraction Method Based on Gammatone Filter

YaHui Zhao, HongLi Wang, and RongYi Cui*

Intelligent Information Processing Lab., Department of Computer Sci & Tech
Yanbian University
Yanji, Jilin, China
{yhzhao,cuirongyi}@ybu.edu.cn

Abstract. A feature extraction method for specified sound based on Gammatone filter is researched in this paper, which can be applied to the audio security system. Firstly, audio data was carried out fourier transform, and then its spectral energy was calculated. Secondly, the inverse discrete cosine transform was computed, after the spectral energy was filtered by use of Gammatone filter, so the feature vector set was constructed. Finally, the audio data are trained and tested by making use of SVM. The experimental results show that the extracted audio features are effective and reasonable, and lead to satisfactory classification and detection performance.

Keywords: sound detection, MFCC, gammatone filter, SVM.

1 Introduction

Audio retrieval, together with image retrieval, video retrieval is present research hotspots based on content retrieval. At present, research achievements of multimedia information retrieval based on content mainly concentrates upon image and video. With the increasing information content of audio that needs processing, the information content of audios accounts for 20% of the total information content. It is of big significance to retrieve needed audio information quickly and effectively from a large amount of audio information [1]. The research of audio information retrieval technology begins the last century [2]. Currently, audio retrieval is a relatively new field, and it is still the exploratory stage in the country [3].

The generic process of audio retrieval is as follows. First, the audio feature is extracted. Then analyze the different features of different audio types, identify and classify audios according to the differences between features. Many researchers have done a lot of work in this area and proposed different features and classification methods. For instance, American Muscle Fish audio retrieval engine extracts properties of each frame of data such as tone, loudness, luminance and bandwidth. Then, calculate the typical value, variance, self-correlation value and energy of the property sequences. Next, construct 13-dimension feature as the feature vector of audio data. The domestic ARS system is an audio information retrieving and classifying system

---

* Corresponding author.

D. Jin and S. Lin (Eds.): Advances in MSEC Vol. 2, AISC 129, pp. 371–376.
springerlink.com   © Springer-Verlag Berlin Heidelberg 2011

based on content. In the actual retrieving course, it adopts 5 features such as tone, loudness, luminance, bandwidth and zero-crossing rate [1]. However, the actual audio data is diverse and complex. It is very difficult to use limited categories to cover all audio data. The study of how to classify audios only can't fully solve the problem of retrieval. As a result, according to the actual application background, aimed at different retrieval applications, study of suitable retrieval algorithm is of big significance [4].

An audio feature extraction method is researched in this paper, which can be applied to the audio security system, and this method is effectively applied to identify the sound of knocking. The sound of knocking is a common voice. The robot is carried out the command to open the door through knocking, so information technology can be better serve humanity. In addition, when there is a long time knocking, but no response, an alarm can be emitted.

MFCC feature existed some shortcomings, so the triangular filter is replaced with Gammatone filter in order to fully extract the energy of each band. In addition, the classification method based on SVM is used in this paper. SVM can learn the feature space and find the optimal separating hyperplane to overcome the shortcomings of rule-based classification algorithm, so SVM can use less data to achieve high classification accuracy. Experiments show that the audio features extracted from the text, the correct detection rate of 95.8% better performance than the MFCC features.

2   Related Work

MFCC is a kind of parameter that can make the best use of human ear, which is a special perceptual characteristic. It mainly describes the energy distribution of voice signal in frequency domain and has proved to be an important characteristic parameter. MFCC takes the nonlinear perceptual characteristic of human ear towards frequency into consideration. But, MFCC itself is a processing of homographic deconvolution, it can't well conform to the hearing physiological model and perceptual characteristic. In the method of MFCC feature extraction, a string of triangular filters that crosses and overlaps in low-frequency area are used to perform frequency domain filtering and capture the spectral information of audios. Because of the nonlinear corresponding relationship between Hz and Mel frequency, there are a large quantity of filters that be used and distribute densely in low-frequency area and there are a small quantity of filters that be used and distribute sparsely in medium-high-frequency area, which causes that the calculation accuracy of MFCC decreases and spectrum energy between adjacent frequency bands leaks significantly [5]. It is unfavorable for reflecting resonance. Furthermore, the frequency band is divided uniformly according to Mel scale based on center frequency, which doesn't conform to the concept of critical band in audio characteristic.

3   Feature Extraction Algorithm

3.1   Gammatone Filter

Gammatone filter is a standard cochlear auditory filters, and it can simulate the physiological data of hearing experiment by the use of few parameters, which reflects the sharp filtering characteristics of the basement membrane, and gammatone filter has a simple impulse response function to derived the transfer function of gammatone filter, which can analyze the performance [6][7]. Its impulse response pattern is as follow.

$$g_i(t) = At^{N-1} \exp(-2\pi ERB(f_i)t) \cdot \cos(2\pi f_i t + \phi_i)U(t) \tag{1}$$

In which, A denotes the filter gain, N denotes filter order, f_i denotes center frequency, ϕ_i denotes phase. In the simplified model, $ERB(f_i)$ is the equivalent rectangular bandwidth, and it determines the decay rate of impulse response related to the filter bandwidth, whereas the bandwidth of each filter is related to the human auditory critical band (Critical Band, CB). According to the auditory psychology, $ERB(f_i)$ is expressed as follow.

$$ERB(f_i) = 24.7(4.37\frac{f_i}{1000} + 1) \tag{2}$$

The bandwidth of each filter is determined by the above formula (2), in which the mathematical expression of center frequency f_i is given in the following formula.

$$f_i = (f_H + 228.7)\exp(-\frac{v_i}{9.26}) - 228.7 \quad 1 \le i \le N \tag{3}$$

The value N is 64, it indicates that the cochlear filter model is achieved by stacking 64 filters. The center frequency of each gammatone filter distributes between 30Hz and 4000Hz.

In which, f_H is the filter cutoff frequency, v_i is the filter overlap factor which indicates the percentage of overlap between adjacent filters. The corresponding bandwidth is calculated by equation (2), after each filter's center frequency is determined.

3.2   Description of Extracting Feature

Support vector Machine stems from the processing of data classification [8]. SVM principle is that two kinds of points were separated correctly by hyperplane, and the maximum edge was achieved.

Assuming that the training set contain N elements. Feature vector of ith training sample is represented as $\{(x_1,y_1),(x_2,y_2),...,(x_N,y_N)\}$, $x_i \in R^k$, the x_i category tags is y_i $\in \{-1,1\}$. If x_i belongs to first class, then $y_i=1$, otherwise $y_i=-1$.

The purpose of SVM training is to gain a minimum error decision function through these training data, that is to say, finding the largest interval hyperplane in the feature space. According to the different classification problem, the different kernel function classifier can be used. The linear kernel function is selected in this paper. The expression of liner kernel is as follow.

$$K(X,Y) = X \cdot Y \tag{4}$$

There are two types of audio signals, including specific audio (such as the sound of knocking) and other audio. The classification algorithm is as follow.

Step1: The short time fourier transform is carried out, and the spectrum is gained, after pre-emphasis, framing and adding hamming widow;

Step2: The spectral energy is calculated;

Step3: The spectral energy $x(k)$ is filtered by making use of gammatone filter;

Step4: The output of each filter is logarithmic, and then carried out the inverse cosine transform, so the gammatone filter coefficients is gained. The following formula was:

$$C_n = \sum_{k=1}^{N} \log x(k) \cos[\pi(k-0.5)n/N] \qquad n = 1,2,\cdots,L \tag{5}$$

Step5: The M-D feature is constructed after calculating the mean of gammatone filter coefficients;

Step6: SVM is entered as characteristic matrix Traindata of *trainS* feature vectors;

Step7: SVM is tested by use of the remaining *testS* feature vectors.

4   Experimental Results and Analysis

4.1   Experimental Data from Knocking Recognition by Using SVM

Our data set comprise 150 audio data, including the sound of knocking, voice examples, music ringtones, animal sounds, environmental sounds and transport sounds. Audio data are digitized at 8kHz sampling frequency and 16bit quantization accuracy, their lengths are 2s. The number of training set is 100, test set is 50. The ratio of knocking and non-knocking samples is 1:1 in the training and test set.

The frame length of audio data is 32ms, the frame shift is 10ms, the pre-emphasis coefficient is equal to 0.97, the number of filters is set to 24.

4.2   Experimental Results and Analysis

The 12-D feature of each sample is extracted through method of this paper, and then the SVM training is carried out. The training result is shown Fig1. After extracting the features of test samples, these audio data are tested by taking advantage of SVM, the test accuracy is 98%, and the test result is shown Fig2.

MFCC is a robust feature parameters and the most widely used. At the same time, 150 audio data are extracted the corresponding 12-D MFCC feature, then the MFCC

Fig. 1. Training Result of Proposed Feature   **Fig. 2.** Test Result of Proposed Feature

mean is calculated and entered as SVM to train and test. The MFCC training result is shown Fig3, the test result is 82%, and the test result is shown Fig4.

The blue diamonds represent the sound of knocking and the red triangles represent non-knocking in Fig1, Fig2, Fig3 and Fig4.

In order to fully demonstrate that Gammatone filter is superior to the triangle filter for extracting feature of knocking. The comparative experiments are done about the different number of filter based on the same audio data in this paper. The number of filter are set to 1, 2,....., 24 in contrastive experiments, the experimental results are shown in Fig5, in which, the recognition result of proposed feature is the curve with an asterisk, the recognition of MFCC is the curve with a box.

Fig. 3. MFCC Training Result   **Fig.4.** MFCC Test Result

From the experimental results, the average test rate of this feature is 95.8%, MFCC feature is 89.4%. The feature in this paper is better than MFCC, it can make SVM a good learn ability, so SVM can accurately identify the type of unknown samples in the testing phase, which MFCC can't done. Furthermore, it can be found by the comparing experiments that the correct recognition rate may not higher as the number of filter is more in the process of extracting proposed feature and MFCC. When the number of filter equals eight, SVM achieves optimal state after learning the proposed feature.

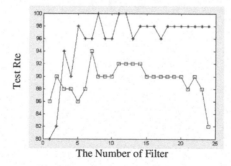

Fig. 5. Comparison of Two Features

5   Conclusion and Next Step

Audio classification is the key in content-based audio retrieval, while feature extraction is its basis. In this paper, the method of extracting the sound of knocking is proposed, and experiments show that the recognition rate of SVM based on proposed feature is better than MFCC. In application setting, the only study of feature extraction can't really solve the problem, how to design an effective detection algorithm about machine learning is the next step based on proposed feature.

Acknowledgements. Our work was supported by the Scientific Research Foundation of Yanbian University of China under Grant No. 2010-010.

References

1. Li, C., Zhou, M.Q.: Research on technology of Audio Retrieval. Computer Technology and Development 18(8), 215–218 (2008)
2. Foote, J.: An overview of audio information retrieval. Multimedia Systems 7(1), 2–10 (1999)
3. J., L, J., Mao, X.L., Wen, G.H.: Fractal feature based audio retrieval. Computer Engineering 34(11), 211–213 (2008)
4. Zheng, G.B., Han, J.Q., Lil, H.F.: Fuzzy Histogram Audio Retrieval Method Based on Principal Loudness Component. Signal Processing 22(4), 471–475 (2006)
5. Yuan, Z.W., Xiao, W.H.: Improved speech recognition algorithm based on MFCC feature. Computer Engineering and Applications 45(33), 108–110 (2009)
6. Chen, S.X., Gong, Q., Jin, H.J.: Gammatone filter bank to simulate the characteristics of the human basilar membrane. J Tsinghua Univ (Sci&Tech) 48(6), 1044–1048 (2008)
7. Ghitza, O.: Auditory Models and Human Performance in Tasks Related to Speech Coding an Speech Recognition. IEEE Trans. 2(1), 115–132 (1994)
8. Zhang, Z., Zhou, H.P.: A method of segmenting music based on support vector machine. Computer and Modernization 3, 127–129 (2009)

Explore New Countryside from the Perspective of the Rural Environmental Protection Measures

YueFen Wang

Northeast Forestry University, Harbin, 150040, Heilongjiang, China
mlbwyf@126.com

Abstract. This paper discusses the rural environment protection measures in order to promoting the construction of new countryside, and realizes the sustainable development of rural economy. It puts forward the strict control the pollution of environment, reasonable of ecological construction, strengthen environmental protection monitoring to rural environment protection measures and policies for the new rural construction provide a good social environment, the living environment and ecological environment.

Keywords: New countryside, Environment protection, Sustainable development, Ecological construction.

Good ecological environment is the base for sustainable development of rural economy and essential conditions, in turn; the rural economy sustainable development ability and level can also affect the ecological environment protection and construction level. Therefore, to make the rural environment protection work will have to develop rural productivity, not to protect the environment as an excuse, and limit the economic growth, that is unrealistic.Insist on taking economic construction as the center, the realization environmental and economic "win-win", this is the premise of the rural environmental protection do.

1   Strictly Control the Pollution of the Environment

Strengthen the prevention and control of pollution. Adhering to the control of point sources to non-point management, from end to the whole process of management control,; from the concentration control to the combination of concentration and total amount control and from pure management to adjust the industrial structure of products and reasonable layout change.

Meanwhile, according to the different degrees of pollution around and characteristic, take different targeted measures. On the ecological damage in force "Deadline to restore the management measures", rural blow down unit to focus on the "emission permits system", poor quality of regional environment of the rural area, the focus of "Centralized pollution control measures". In practice we should tightly pay special attention to the two main link: agriculture and township enterprises.

First, in agriculture, to strictly prohibited and punish chemical fertilizers, pesticides, veterinary medicine, herbicide, plastic film on rural environment caused by the pollution problem of the endogenous. Reduce the amount of chemical pesticides, doing well the non-toxic demonstration area construction, promote biological control technology, adjust the use of chemical fertilizer structure, reduce the proportion of inorganic fertilizer use, control of agricultural film cause "white pollution". Strengthening agricultural chemicals environment safety management. Strengthening the safety management, reduce the pesticide environmental harm caused by the improper use. Perfect pesticide production and use of environmental safety management regulations and standards, explore and establish high poison, high residual pesticide use notification system, strengthen in the population concentration areas, natural preservation areas use pesticides management; encourage the development and promotion of the high efficiency, low toxicity and low residual chemical pesticide, develop biological pesticide, and gradually make methamidophos, methyl parathion, monocrotophos, ammonium phosphate and high toxic organic phosphorus pesticide breed exits the main production and use of pesticide variety series. Strengthening the safety management in the application of chemical fertilizer environment, reduce agricultural non-point source pollution, and actively explore different river basin, different planting structure of chemical fertilizer seems reasonable, wastage rate and the contribution of water pollution to environment of the application of chemical fertilizer, make the environmental security of control standards; actively developing high density and slow-release fertilizer,and rationally adjust the nitrogen, phosphorus, potassium fertilizer proportion, encourage the use of the organic fertilizer. Actively promote deep application; envelope; slow release; compound formula and the measuring of the soil and the applying of fertilizer, improve chemical fertilizer utilization ratio and reduce loss rate. Strengthening agricultural films using the environment safety management, and carry out the agricultural pollution situation survey,   establish the agricultural pollution control regulations and standards, control the serious pollution area of agricultural films dosage; actively developing agricultural films recycling technology and biodegradable production technology, strictly control the ultra-thin agricultural production and use.At the same time, the governments at all levels should strengthen control of livestock waste and life rubbish strict management, the harmless treatment and comprehensive utilization. Positive development and promotion of the size of the village, biogas straw gasification to clean energy, building sewage treatment facilities, banned crop straw in airports, highways, important railway lines, high voltage transmission line and scenic spots, and nature reserve areas, population concentration and the surrounding areas open burning, promote straw comprehensive utilization. From the feed, fertilizer, fuel and industrial raw material and other fields exploit straw comprehensive utilization channels, and vigorously promote the straw returned, straw gasification technology and other comprehensive utilization measures to develop industrial use of straw, and strive to make new ways of burning the comprehensive utilization ratio of the straw be greatly improved. Ensure that the rural air environment quality is good. Strengthen the pollution control, promoting the waste of planting and raising industry to be comprehensive utilization. Establish supervision, assessment system and major accident responsibility system; control of the pollution for livestock and poultry breeding industry and fishing livestock breeding,set prevention and control policy of economic and technical standards and discharge of water pollutants, develop of livestock farming and livestock waste recycling of the waste water processing technology to greatly promote the breeding and planting of integrated ecological engineering construction, and actively guide the related industries, establish and perfect the organic fertilizer production, marketing and use of market mechanism, promote recycling of the waste; strengthen the fishery resources and fishing areas ecological protection, reasonably determine the breeding capacity and fishing intensity; preferred promoting scaled farms pollution treatment, in key river and the region, sewage discharge standards within the time limit.

Second, in the town enterprises, they should strictly control the reproduce, during, postproduce and other links, from the source minimize pollution, in the end of the production process, the most efficient handling pollution. Rural governments at all levels should through the various administrative measures to stop the city, industrial and mining enterprises to rural emissions of "three wastes", eradicate all external pollution,insist on construction projects at the same time, strict control the produce of the new pollution, implementation of "waste minimization" technology and clean process to reduce pollution emissions; implementation of pollution charge system, the one which didn't conform to the state provisions, fails to meet the relevant standards of the enterprise will be collected of sewage charges to promote pollution source management; improvement of main pollution industry, the old sources of pollution to year after year deadline governance, rural comprehensive management of the inclusion of quantitative assessment system, reduction of pollutant emissions to the environment in rural areas. Adjust the industrial structure of township enterprises, pay attention to the development of less emissions industry to contribute to the upgrading of industrial structure; environmental protection departments at various levels shall strengthen the environmental guide for township enterprises, from product design to control the generation of the waste and recycling of management, and gradually establish a perfect paid for the use of natural resources, and promote the economic compensation mechanism of development of resources to increase the order ,turn the supervision and inspection to the shut, stop, merge and steering enterprises, prevent the resurgent, prevent backward equipment and process by the east to the west transfer, while strengthening the coordination with the related departments and cooperation to protect the good rural natural ecological system.

2   Reasonable Ecological Constructions

First, adjust measures to local conditions

Our country rural region is vast, geography and climate conditions complex, these features decide in China's rural ecological environment construction is a complicated system engineering, and local resources endowment different, in the face of the environment and sustainable development is also different.

Development departments to rural environmental problems make further investigation; widely listen to rural grassroots cadres and the opinions of the farmers, the different ecological conditions of a representative and the typical significance of rural survey analysis and evaluation, according to the ecological system all over the structure characteristics and function character, establishing various different types of ecological system. At present, in the national scope are actively carry out the small town construction activities, in the process, we should according to the actual local, thorough research, reasonable planning, according to the different function of town, established enterprise industrial zone, the residents living quarters, culture and education area, the business district and other regions, and the environmental protection into the overall planning of urban construction, through comprehensive controlling the environment, promote the modernization construction of rural towns and drive around the countryside environment protection work. [1] For example, planning and construction of small towns, can establish a garbage processing center, such can solve the urban waste handling problems, and can change the current widespread situation as well. The current widespread situation is that administrative villages collect garbage be free throwed because of no suitable processing methods. In addition, with the industrial wastewater and sewage ingredients have big difference, in a small town planning and construction process to establish industrial wastewater and sewage two different emission system respectively, the sewage treatment center processing cycle after use.

Second, to plan as a whole, over all consideration, the comprehensive decision-making, reasonable development. Economic development must follow the natural rules, full consideration of the carrying capacity of ecological environment, insist on protection of development and development of protection, make the recent and long-term protection of unity, local and global unity, to sacrifice will never be allowed in the ecological environment as the price for the immediate and local interests. So to strengthen rural environmental protection is to develop ecological agriculture, further strengthen the ecological agriculture, especially of the pilot guidance should be increased in the technical content, and promote the level of improving and expanding the scale; through the typical demonstration, gradually extending to comply with local actual situation of ecological agriculture mode and the development of ecological agriculture with the strategic adjustment of agricultural structure closely combined. Pay attention to the development of forestry, fisheries, grass and tourism in rural areas is considerably prospected. Pay attention to the preservation of the agricultural production capacity, control in the use of chemicals; develop organic agriculture and ecological agriculture. Such as creating a large ecological farm, a modern vegetable plantation, special animal farms, flowers and Grass Company, etc.

To speed up the rural ecological construction, construct of a batch of ecology model district. Planting trees ,growing grass, water and soil conservation, water conservation , increase the forest vegetation coverage, strengthen the construction of the nature reserve, strengthening the protection of biodiversity ,soil and water conservation and land degradation and the comprehensive improvement of the demonstration project construction, aims to in combination with   pollution and ecological construction.

3   To Strengthen Environmental Protection Monitoring

To strengthen the environmental protection work of the rural management supervision strength. To monitor the application of the sewage to declare system, emission permits system, charge system for discharging and deadline governance system, organize implementation of pollutant concentration control; to waste water, waste gas and solid waste, noise, vibration, pollution prevention and ecological environment protection for the unified supervision and management.

Supervise the ecological environment of the influence of natural resources development and utilization of activities, important ecological environment construction and ecological destruction recovery; supervision of the various types of nature reserves, ecological function regions and scenic spots, and forest park of environmental protection; the supervision and inspection biodiversity protection, wildlife protection.Guide ecology model district and construction of ecological agriculture and biological technology environment safety management. Timely report to superior the major environmental problems; the investigation and handling environment pollution accidents and ecological destruction events in the region; carry out the environmental protection law enforcement inspection activities. Be responsible for the construction and management of the regional environment monitoring network, environmental statistics and environmental information system; carry out the environmental investigation and environmental monitoring; prepare the region environment quality organization report, in order to provide county environmental conditions to superior departments and environmental statistics. Suggest enforcing one ballot veto of environmental work, for that environmental health and environmental protection work of poor administrative villages, to cancel the advanced qualifications. [2]And at the same time, it is necessary to establish a positive, actively guide and promote the administrative villages of protecting the environment as a daily work to grasp.

Agricultural environmental protection monitoring departments should formulate and implement various environmental management system; implement environmental protection certified system; guide and promote of environmental protection industry development; guidance of cleaner production. According to the regulations, examine and develop construction activity environmental impact report, makes the environmental supervision and inspection of environmental protection administration, and to speed up the environmental information circulation, and regularly release environmental quality status; to the rural environment, regular and irregular make a survey and monitoring, and according to the agricultural environment quality evaluation standards, reported to the relevant authority in time so that establish policy as soon as possible, implement measures to address the problem. Strengthen rural environmental protection, and promote rural ecological civilization construction of the new socialist rural construction is a great strategic task. The rural environment protection relates directly to the people's quality of life and health of body and mind, related to their survival and development, in relation to the sustainable development of the whole society. Must be done to do a good job of rural environmental protection, to build a beautiful new socialist countryside providing a good social environment, living environment and ecological environment, and promote sustainable and healthy development of rural economy and society.

References

[1] Zeng, B., Su, X.: Strengthen the Protection of the Environment. Environmental Protection Industry 02 (2009)
[2] Shen, Y., Wang, W.: Xinjiang New Rural Construction and the Rural Environment Protection. The Agricultural Environment and Development 10 (2008)

Analysis on Forest Industry Enterprises Culture and the Necessity of Reconstruction

YueFen Wang

Northeast Forestry University, Harbin, 150040, Heilongjiang, China
mlbwyf@126.com

Abstract. According to the function of enterprise culture and the current new problems logging enterprises facing, this paper puts forward the logging enterprises must take place cultural recreation, implement cultural management. Logging enterprises culture reconstruction is the need of forest enterprises new development; the need of promote the sustainable development of forest enterprises; the need of improve the management level of logging enterprises; the need of constructing the harmonious society; the need of constructing boom of ecological culture system.

Keywords: Forest industry, Cultural recreation, Sustainable development, Harmonious society, Ecological culture system.

In the era of knowledge economy, the competition between enterprises, in fact, is the culture competition. Culture of enterprises, is the reflect of their comprehensive strength, is the reflect of civilization, is the source of the productivity of knowledge form into physical form. At present, ecological priority has become the world's forestry development trend; forestry development goes into a new period of strategic opportunities. In the face of new situation and task, new opportunities and challenges, we want to win in the fierce market competition and make the forest enterprise bigger and stronger, realize the sustainable development of forest enterprise, it is necessary to build up the concept of "Manage company with culture" and "Develop company with culture", integrate and innovation the original culture, reengineering cultivate advanced enterprise culture, actively promote the strategy of "enhance enterprises with culture", spare no effort promoting corporate culture with advanced forestry enterprise to reform and develop.

1 The Require for Completing New Development of Forestry Enterprises

Enterprises' culture is in the service of the strategic goal of enterprise development, when to confirm the development of the enterprise strategic, it should rely on effective enterprises' culture construction to implement. The United Nations' environment and development conference called for in the current problems for human to solve, "no any problem is more important than forestry," in sustainable development, "forestry should be given priority". With the improvement of forestry position and the change of national forestry strategic, the task of logging enterprises is changing in our country, that is, from the main wood production primarily to the protection of natural forest resources, plantation with directional cultivating and improving the eco-environment primarily. During the transition, the logging enterprises should not only change their management style, management mechanism, the more to change the management idea, values and management system, strengthen cultural construction, recreating cultural, cultivating common values, enhance cohesion, innovation and competitiveness, to complete the new historic task and the realization of logging enterprises nice and fast development provides spiritual power and cultural support.

2   The Need for Promoting the Sustainable Development of Forest Enterprises

Enterprises' culture reengineering is a profound transformation, the success of cultural recycling is the success of enterprise reform, the determinants of innovation. It is an innovative leader in forestry enterprises, is the transformation of forestry enterprise development and the driving force as well.

The current logging enterprises is also facing the decrease in the number of forest resources, the decline in the quality, forest industry in small scale, the structure unreasonable, low benefit and other problems, the development of logging enterprises is facing new challenges and test. In this case, it is need for logging enterprises culture reengineering and innovation. It can not only drive the forest enterprises ideal innovation, system innovation, technology innovation and management innovation, and through the reengineering to form advanced enterprise culture, it can make the enterprises' culture of coherence function to give full play, enterprise's organization and staff, put personal ideal faith into the whole enterprise ideal faith and form values consensus. The cohesive affinity of the enterprise culture can through establish common values, the business goal to make the staff gather around the enterprise, make the staff has the sense of mission and sense of responsibility and determine to leave their own wisdom and strength for the integral goal of enterprise, united in enterprise behavior with the employee behavior for common direction, think the same and do together, causes the enterprise to become of the consensus of the same people, so as to promote the combination of logging enterprises transition smoothly, and then promote enterprises sustainable development.

3   The Need for Improving the Management Level of Logging Enterprises

Culture management is a new stage of the development of enterprises management theory. It stresses that spiritual cultural strength, hoping to use a kind of intangible cultural force to form a kind of behavior standard, values and ethics, condenses the staff sense of belonging, enthusiasm and creativity, and guide enterprise employees try hard for the enterprise and the development of the society. Enterprises management only based on enterprises culture, using the power of the culture and the means of non-economic manage the employee, only the respect and inspire people, cultivate people as the starting point and the foothold of management can make the enterprises form the internal atmosphere of respect the rules and disciplines, courtesy and honesty, solidarity and friendship, professional dedication, healthy, positivity and vibrant. Therefore, logging enterprises should improve the level of management, should absorb advanced management concept and management ways, increasing the cultural component, and then do the culture management. And the implement of cultural management is based on the good enterprise culture. However, some problems existing in the construction of the current logging enterprises and restricted the role of culture, and it is hard to implement cultural management. Therefore, we should reshape the logging enterprises culture in order to adapt to the need of the management culture, playing the role of cultural soft power, and improving the management level.

4   The Need for Constructing the Harmonious Social

Harmonious society emphasizes man and man, man and nature, man and society are in harmony. Excellent enterprise culture is the carrier to improve the ideological and moral qualities, to promote enterprise civilization image, to promote the whole society spiritual civilization construction. Enterprise wants to be "comprehensive, coordinated and sustainable development", we must cultivate positive enterprise culture, strengthen the construction of ideology and morality with staff, carrying forward the national spirit, shape the staff of common values. The interpretation of enterprise culture and natural, especially the relationship between man and forest, contains the value orientation of people and the natural, and ethics. It is the relationship between human and the forest, man and nature, it is the interdependence, interaction and mutual fusion, and the combination of relationship which created the material civilization and the spiritual civilization. Promote the cultural recycling of the forest enterprises is to promote harmonious culture construction, the construction of harmonious society. Therefore, the reengineering of enterprise culture to make full play of their radiation function, culture radiation is continue to grow of the society, diffusion of advanced entrepreneurship, values, business ethics, and gain the recognition from the society, plays out the culture education functions of enterprises, form the situation of protecting nature, loving the forest, the harmonious development between man and nature. In short, give great impetus to enterprise culture reconstruction, construct of beautiful rivers and mountains, constructing the harmonious society are our historic mission entrusted by the times.

5   The Need for Constructing Boom of Ecological Culture System

Ecological culture is the reflection in harmony with nature ecological values. The Chinese government attached great importance to the development of construction of ecological culture and the prosperity of the ecological culture system, to make the whole society firmly establish the ecological values that human and nature are in harmony. In January 2008, the meeting of the National Forestry Office of the Secretary, Hui Liangyu recommends to push forward ecological culture system, strengthen the building of ecological civilization concept of the whole society. Forestry workers as the main body of the ecological construction and the subject of ecological civilization

386   Y. Wang

construction, undertake on construction and protection forest ecological system, protect and restore wetland ecological system, management and improve the desert ecosystem and biodiversity maintenance, carry forward the important functions of the ecological civilization. Logging enterprises should be in the construction of foresty characteristics and ecological characteristics culture system, make special contribution for constructing and enriching ecological culture system. Logging enterprises culture outstands green culture, ecological culture and harmonious culture. Therefore, constructing of ecological culture system, advocating the value concept of ecological priority and the spirit of dedicate green, proposing the harmony between human being and nature, carrying forward ecological civilization, form the good society atmosphere of respecting nature and cherishing the ecological. Therefore, the role of forest enterprise culture should be played, reform and recreate the existing enterprise culture, make the forest enterprise culture outstanding, environmental protection and green, and promote the enterprise culture development and prosperity.

References

[1] Yang, C.: Explore and Analysis of China's State-owned Enterprise Culture Construction. Economist 02 (2011)
[2] Bing, Z., et al.: Influence of the Enterprise Culture, Organization Learning to the Innovation Achievements. Soft Science 01 (2011)

Discusses the Construction of Ecological Good Yichun Forest Region

YueFen Wang and Yao Xiao

Northeast Forestry University, Harbin, 150040, Heilongjiang, China
mlbwyf@126.com

Abstract. This paper discusses the good ecological environment construction in Yichun forest region in order to promote a harmonious environment, and realize the forest sustainable economic and social development. The proposed increase ecological environment construction and the management of the sense of urgency and sense of responsibility, speed up ecological function area and ecological civilization, and promote the ecological economic development, improve the aid declining industries mechanism, form save resources's idea, green consumption consensus and other measures, make Yichun forest region good ecological environment and realize the harmony between human and nature.

Keywords: Good ecological environment, Harmonious forest region, Yichun.

To realize the harmony between man and nature, economy can continue to development. Prominent ecological protection, to strengthen and improve the ecological system. Yichun forest region will continue to carry out "Strict management of forestry" policy, consolidate and expand the ecological protection results. Continue to implement of "natural forest resource conservation project phase ii", strengthen the scientific afforestation and improve the quality of forestry, do well forest management and raising the trial work, explore the recovery of forest quantity and quality in the effective way.Continue to strengthen in Korean pine, blueberries, forest frog and other precious species protection and restoration, advancing the nature reserve construction, keep and speed up recovery biological diversity. For Yichun, forest region ecological harmony is a real harmonious society. Basic level must have a stable and equilibrium of the ecological environment, the harmonious society must be kept the development in a appropriate ecological environment, no balance of ecological environment, social politics, economy and culture can't existence and development, and harmonious interpersonal relationship will also become the castles in the air and there is no existing basis.Therefore, we must leave no stone unturned construction of ecological good forest region.

1 Strengthenen the Sense of the Ecological Environment Construction and the Management of Urgency and Sense of Responsibility

The construction of ecological good Yichun forest region, the achievement of the goal of harmony with nature.According to the target for concrete quantification spreading, the current Yichun forest region "natural forest resource conservation project" continue to promote, each forest farm has stopped cutting down trees, including Korean pine forest of childbearing age.Greater Khingan Mountains and Lesser Khingan Range are two of the eight functional areas of Heilongjiang province. Because Lesser Khingan Range as the first and biggest state-owned forest region,it has very important water conservation, water and soil conservation,flood storage, maintain functioned biodiversity and other important ecological functions;it is China's important commodity grain and livestock production base of the ecological barriers and water conservation district, to adjust the climate of northeast plain and the north China plain,alleviate global warming has indispensable ecological function; Lesser Khingan Range has irreplaceable role in protecting the ecological security of northeast Asia,maintaining regional ecological balance and security in northeast Asia to bird migration channels, raising global ecological environmental quality. So, the province, the Yichun forest region from the concept should enhance the understanding, standing in certain height recognition of the importance of the ecological view.

We should unite ideas; enhance the sense of urgency and sense of responsibility. First of all,leading cadres should carry out learning and training about knowledge of forest and forestry within the city to make them "know forest ,love forest and protect forest" , continue to deepen the understanding and awareness of the" green GDP ", the majority of the cadres should consciously join in the development of forest. Second, through the amateur remedial class, night school and other ways,goes down to the basic unit to the forestry workers on the site to communicate new spirit and a new knowledge. Enhance worker masses on the current forest ecological environment construction and the management of the sense of urgency and "forest master" consciousness; Third, through television, newspapers, network, billboards ,blackboard newspaper and other publicity approaches to all the people of the importance of ecological information,promote the sense of the importance of people in Yichun forest region ecological construction of responsibility. Resolutely change the wrong ideas of ecological environment construction and the economic construction are contradictory , and resolutely change the thought and behavior of excessive focus on economic development but not enough attention to environmental protection. Fully realize the importance of ecological environment construction and management, do the development of protection, the protection of the development.

2   Speed Up Ecological Function Areas and the Construction of Ecological Civilization

Yichun actively support for national policy, one is "natural forest resource conservation project phase ii", another is full cessation of the main cutting. The two tasks must have realized on the mechanism of system supporting and national policies support.So to increase the forest resources to protect and nurture.We can promote "lang township experience", in the scope of the city ,encourage worker masses through the "non-cutting" way of production and living, forestry bureau can open factories of the relevant collective substituted industry ,it also can do joint production by worker, individual manage should be more vigorously support, such as lang township of forestry administration "Cuihua sauerkraut" series of products has now become the famous product, such not only ensure ecological function regions, but also does not damage the construction worker masses economic benefits to the measures and ways should be priority applied, the government should establish priority long-term mechanism to support and establish the guiding thought for a long time without changing. Through the introduction and popularize advanced technology, adjust and update methods, optimize the tree structure of forest types, speeding up forest tending and forest management.By studying the Nordic countries advanced and successful forestry experience and study large equipment, such as Finland and Sweden,eliminate backward productivity, the ecological environment is the basic condition for human beings survival and development, for yichun forest region, to protect the ecological environment is to protect the productive forces, and improve the ecological environment is the development of the productive forces. General secretary Hu Jintao in the population, resources and environment symposium pointed out that "Good ecological environment is the important basis for sustainable development of social productive forces and increasing people's quality of life, is the basic conditions of human survival and development, we should firmly establish scientific concept of development that the human and the nature are in harmony. Only rational utilize and protect the ecological environment, can the economy continue to develop."The administration bureau in Yichun forest stop the felling area, to make the net growth of trees from now increased to 5 million cubic meters above of 10 million stere, strive to 20 years' time, make the ecological function regions of Lesser Khingan Range ecological environment significantly improved, forest, wetland, the sources of rivers,key waters and mines operations fully comprehensive protection and restoration, ecosystem rationally, significantly improve water conservation, keep water and soil conservation of biological diversity and maintain ecological function, the ecological environment of Lesser Khingan Range recovery to the state before.

Solve ecological and harmonious Yichun forest region, we need to continue launching ecological consciousness education propaganda. Adhere to the correct orientation of public opinion, the media to strengthen social responsibility and disclosure of the relevant case about harmonious ecological destruction, make forest people consciously to recognize the importance of guarantee ecological safety, and then make the people of forest region of the conscious action to achieve the harmony between man and nature. Through various media to continue to propaganda "Green ideas" of "natural-man-society "idea" again, it is that the whole ecological system live in harmony, strengthen the cadres and masses ecological education and ecological and ethical responsibility education in Yichun forest, cultivate people's environmental consciousness in Yichun region, arouse people's love of nature, let the forest social consensus and the core values to be idea of the environment-friendly society ,contradiction of limited resources and the unlimited human desire should be coped with through the practice of the human, production idea and consumption idea should be adapted to ecological civilization.

3   Promote of Eco-economic Development, Improve the Mechanism to Aid Declining Industries

Must find a suitable for Yichun forest region eco- economic development model, to develop environmentally friendly of circular economy, and make economic and social development in Yichun can minimize damage to the environment. Make feasible policy according to the new situation and to promote the development of ecological industry in Yichun policy system, to increase the subsidies of ecological industry development of Yichun forest region , give full play to the policy to encourage, support and guide effect, make the development of ecological industry enterprises in Yichun forest region obtain real ideal economic benefits. Yichun forest region government at all levels should be increasing investment, encourage technology innovation and promotion, vigorously developing and cultivating of new technology, new products and new energy, speed up the construction of ecological civilization in Yichun forest region with environmental protection, and gradually replace circular economy and environmental protection industry with high consumption, high pollution and high ecological damage of unsustainable development mode, go a new way in Yichun forest region of industrial ecology and eco-industrial.

Yichun forest region government initiative through exit aid mechanisms to solve the problem that only depend upon the enterprise industrial transformation and exit that can't solve the problem. For the industrial sector of high energy consumption and heavy pollution , and the poor process , considering the opportunity cost,determined to take industry limited or exit policy;for example,"Xilin Iron and Steel Company" and "Hao Lianghe cement plant", the major pollution enterprises shoule be received concerned and monitoring of relevant departments, and in protecting the ecological environment on the premise of production development, must not to damage the ecological environment for the material wealth, and also should increase the independent research and development of high efficiency energy saving and emission reduction of environmental protection product, should strengthen the introduction of new technology, multiple measures, various means, consider associating the enterprise production development benefit with environment friendly interest ; at the same time,in compression long-term industry,eliminate low-level redundant construction,we use green technology with a comprehensive reform of traditional industries,eliminate backward technology and equipment within the law,control environmental pollution from the source,thereby reducing the tragedy and transition costs, orderly development of economic and social Yichun forest region, avoid "pollution first in the governance" of the error concept.

4 Formation the Consensus of Save Resources and Green Consumption

Yichun is a big city with large numbers of foresty,the masses' strength of the consciousness of resource sense of participation directly influence on the use of resources results.Therefore, change the traditional, backward and not amnesia way of thinking and values, and based on this, formed the common recognition, consciously abide by the ecological environment of the ethics is particularly important.Based on the correct practice of Yichun,vigorously promoted the concept of moderate consumption of forest people,encourage thrift behavior, advocate life and entertainment in harmony with the environment, make forest people set up energy conservation and environmental protection concept,form green consumption way of life and consumption model,for example, According to the relevant provisions of the state,now

Yichun forest region has gradually been pushed the one-time paid for the use of plastic shopping bags, it will greatly reduce the one-time use of plastic shopping bags, so as to reduce the chance of environmental pollution, the next step in Yichun is to promote green lunch box, electric cars and other energy conservation and environmental protection products, through the idea and technical development to advance the construction of Yichun forest region environment friendly society .

According to the ecological harmony idea, Yichun forest region economic development mode is on the direction, by increasing scientific and technological innovation and establish system security to promote and form a save energy and protect the ecological environment project, it is of the industrial structure, the production method, consumption model, this is a grand project related to the future development of Yichun forest region, it needs to the joint efforts of the whole city. The building of a harmonious society should correctly handle the relationship between human and nature of Yichun forest region, make the person and the nature harmoniously, and keep good ecological environment and production development, and rich people's lives in forest region.

Research on the General Method of Round Robin Scheduling

Jian Chen[1,*] and DongFeng Dong[2]

[1] Wuhan Textile University, China
jianchenmail@sohu.com
[2] Changsha Telecommunications and Technology Vocational College, China
dongdongfeng@tom.com

Abstract. Different sports competitions use the same way of Round-Robin Scheduling, however different arrangement systems are adopted. Through the analysis, comparison and research to different arrangement systems, with a view to optimize, replenish and complete the arrangement theory and to provide the calculating basis of the computer scheduling, this article attempt to establish a general arrangement system that is suitable for all kinds of sports competitions.

Keywords: Round-robin, Odd to take the middle bye, Home/Away alternating balance, Game rounds table, Tournament scheduling, Computer scheduling.

1 Introduction

Different athletic competitions adopt different scheduling methods while the same method of round-robin. For example, the basketball adopts a fixed 1 anticlockwise round robin; the chess competitions adopt the Berger scheduling. Other competitions adopt fixed N clockwise round robin, odd to take the middle bye, snaking scheduling and so on. Currently, theoretically speaking, we haven't found a universal scheduling method for the scheduling methods of round robin.

Contemporarily, judging from the situations of software scheduling and practical application of round robin in the foreign countries, we haven't found universal scheduling software which is applicable for all of the competitive events. From the perspective of computer programming, we've found that, the scheduling methods of all competitions' round robin are able to identify a desirable result among all the possible scheduling results, that is, the universal scheduling method for all the competition events. However, there are two problems which need to be considered and solved in this respect, first, if there are too many constraint conditions, conflicts are possible to occur among different conditions. Second, we need to put the maneuverability of manual scheduling into consideration. That is to say, the scheduling shall be simple and convenient, being favorable to popularization and application.

---

* Jian Chen (1962 -) Male, Hubei Wuhan, Wuhan Textile University Associate Professor, Research Interests: sports psychology, basketball teaching and training.

D. Jin and S. Lin (Eds.): Advances in MSEC Vol. 2, AISC 129, pp. 393–399.
springerlink.com   © Springer-Verlag Berlin Heidelberg 2011

2 Review and Analysis on the Major Scheduling Methods of Round Robin

The round robbing competition is also called "round robin system", one of sport competitions. Based on some combinations, participants (individuals or groups) compete in turns and their records in all the competitions are integrated to decide the final ranking [13]. Presently, there are two types of round robin scheduling adopted in the current contact sports; serving rotate one position and serving rotate multiple positions. Serving rotate one position contains the fixed 1 (or 0) anticlockwise round robin scheduling, the fixed N (N is an even number the largest one.) clockwise round robin scheduling and the odd to take the middle bye. The typical scheduling of serving rotate multiple positions contains: Berger scheduling and snaking scheduling.

2.1 Serving Rotate One Position

2.1.1 Fixed 1 Anticlockwise Round Robin
We take the typical fixed 1 anticlockwise scheduling for example. The scheduling method makes schedules based on the even number. In the first round, U pattern is adopted in scheduling in accordance with the order of position number. Match laterally. In the following rounds, the number 1 position (up left) is fixed. As for other positions, in each round, one position is rotated anticlockwise. In case of odd position, the largest position number shall be altered into 0. The round is empty when coming upon 0. The advantage of the methods is: easy to remember, 1:2 is occurred in the last round. The event intervals are perfect. The disadvantages are: successive left and right. The even number can't balance the court. Unreasonable odd to take the middle bye.

2.1.2 Fixed N Clockwise Round Robin
The method is scheduled according to the position of even number. In the first round, U pattern scheduling is adopted based on the position number, with lateral matches. In the following rounds, the number N position is fixed (up right). In other positions, one position is clockwise rotated in each round. In case of odd position, the largest position number is altered into 0. Have the bye when come upon 0. The advantage of the method is: it is most reasonable to have the bye when there's odd number group. The disadvantage is: the seeded match is not the final round. The even number is unable to balance the court. Unable to switch.

2.1.3 Odd to Take the Middle Bye
In the scheduling method is as follows: in the first round, add 0 to the follow (N+1)/2. Adopt the U pattern in scheduling according to the order after 0 is added. Enable 0 to be in the bottom right corner. Lateral matching. Have the bye where there's 0. In the following rounds, fix the No. 0 position. As for other positions, one position is served in each anticlockwise rotate. Actually, the method is also applicable for the even number. In case of the even number, alter the No. 0 into the largest number. When there's even number, switch positions of the largest number and the opponent in turns. (Left and right). Advantages of the method: reasonable to have the bye in case of odd

number. 1: 2 occurs in the final round. The odd number balances courts. The disadvantage is: successive left and right, regardless of the application when there is participation of even numbered group.

2.2 The Scheduling Method for the Multiple Positioned Round Robin

2.2.1 Berger Scheduling

The scheduling method is that, in the first round, the U pattern arrangement is adopted according to the order of position number with lateral match. In the following rounds, fix the largest position number N (N is even number). In the other positions, in each round, anticlockwise rotate N/2-1 positions. When all the rounds are paired, switch the largest position number in the even rounds with the position of opponents, with an aim to enable the left and the right appears successively in each round. In the position of odd number, alter the largest number into 0 and have the bye when there's 0. The method has the following advantages: alternate left and right. Same method for even and odd number. Calculated, easy to remember and reasonable bye. The disadvantage is: premature contact of 1:2, unreasonable round intervals. Where there's even group, the court can't be balanced.

2.2.2 Snaking Scheduling

As shown in the picture, the scheduling method is that, first, determine the round times according to the teams. Draw vertical lines according to the detailed rounds. The bottom of vertical line stands for the left while the top stands for the right. Afterwards, whether total groups are odd number or not, we arrange according to the odd number. The scheduling method is, making snaking arrangement according to

Fig. 1. Snaking scheduling

the position number (small to big). When the largest number is arranged, we can connect the head and the tail to continue the snaking arrangement. The scheduling route is started from the left side of the first vertical line, from the top to the bottom. When reaching the bottom of the first vertical line, we start from the right side of the vertical line, from the bottom to the top, till we reach the up of the second vertical line. Then, we start from the right side of the second vertical line from the top to the bottom till we reach the bottom of the third vertical line. We make arrangement in the right side of the third vertical line from the bottom to the top. The rest may be deduced by analogy. We do this job until we finished the right side of the last vertical line. Teams in the both sides of vertical line shall be matched and equal in number. If the team is odd number, the team on the vertical line has the bye. If team is even number, the largest number collides with the team on the vertical line. Moreover, the collision occurs in the final one in each round. The advantages of method: home and away alternating balance, reasonable bye. The disadvantages are: inconvenient manipulation, unable to balance the court, unreasonable game intervals.

3 Basic Requirements and Analysis for the Universal Scheduling Method of Round Robin

By making analysis on the above five typical scheduling methods in round robin and combining the practical experiences in our realistic manipulation, we've generalized 10 basic requirements in terms of universal scheduling method for round robin and made comparative research.

Table 1. Comparison on Five Scheduling methods of Round Robin

Basic requirement of universal scheduling method for round robin	Serve rotate one position						Serve rotate multiple positions			
	Fixed 1 rotation		Fixed N rotation		Odd to take the middle bye		Berger		Snaking	
	Odd	Even	Odd	Even	Odd	Even	Odd	Even	Odd	Even
1.Balance of intervals after each two matches for teams	√	√	√	√	√	√	×	×	×	×
2.Balancing the host/guest(left/right) times of each team	√	√	√	√	√	√	√	√	√	√
3.Balance the host/guest(left/right) alternating of each team	⊙	⊙	⊙	⊙	⊙	⊙	√	√	√	√
4.Balancing the match court of each team	⊙	⊙	√	⊙	√	⊙	√	⊙	⊙	⊙
5.Each odd team's bye shall be reasonable	×		√	√	√		√		√	
6.the unified scheduling method for both odd and even teams	√	√	√	√	√	√	√	√	√	√
7.The match climax of 1:2 shall be in the last round	√	√	×	×	√	√	×	×	×	×
8.Convenient for manual scheduling	√	√	√	√	√	√	√	√	×	×
9.Able to realize the computer scheduling (arithmetic)	√	√	√	√	√	√	√	√	√	√
10. Universal for all games	×	×	×	×	☆	☆	×	×	×	×

Note: √ That they can not meet demand; × That they can not meet demand; That there is no demand for space; ⊙ Adjustable to meet the demand that; ☆ Adjusted to meet the needs of that.

From Table1, it is not difficult for us to find out that, by referring to the "10 basic requirements", there's no method which can satisfy all 10 requirement among the five methods. There are 4 methods which are far from meeting the requirement. The fixed 1 anticlockwise round robin can't meet the fifth requirement and its problem can't be solved by any additional strategy. Fixed N clockwise round robin, Berger Scheduling and snaking scheduling can't satisfy the 7th requirement and their problems can't be solved by any additional strategy. Now we only have the odd to take middle bye. In the actual practices, we often make manual adjustment to meet the 3rd and the fourth requirement. If we can find any calculable method to meet the 3rd and the fourth requirement, we can assume that, we've found the universal scheduling method for the round robin.

4 Discussion and Analysis on the Universal Scheduling Method of Round Robin

Odd to take the middle bye is a most possible prototype in becoming the universal scheduling method among the above five scheduling methods. Based on relevant research literature and material to which we've referred, we've found some solutions to improve the scheduling method of odd to take the middle bye which can't meet the third (host/guest alternating problem) and the fourth requirement(balancing the competition courts). Please refer to the black frame in Table 1.

4.1 The Chessboard Method and Its Application in Scheduling

In order to improve the odd to have the middle bye which fails to meet the third requirement (host/guest alternating balance), we have applied the chessboard based on the original scheduling method of odd to have the middle bye. By using the black and white grid, we can distinguish the host/guest alternating situations in order to solve the problem of unbalanced host and guest alternating.

Table 2. The Chessboard method and its application in scheduling

Field	Round 1	Round 2	Round 3	Round 4	Round 5	Round 6	Round 7	Round 8	Round 9	Round 10	Round 11
Field 1	1 11	11 10	10 9	9 8	8 7	7 6	6 5	5 4	4 3	3 2	2 1
Field 2	2 10	1 9	11 8	10 7	9 6	8 5	7 4	6 3	5 2	4 1	3 11
Field 3	3 9	2 8	1 7	11 6	10 5	9 4	8 3	7 2	6 1	5 11	4 10
Field 4	4 8	3 7	2 6	1 5	11 4	10 3	9 2	8 1	7 11	6 10	5 9
Field 5	5 7	4 6	3 5	2 4	1 3	11 2	10 1	9 11	8 10	7 9	6 8
Bye	6 0	5 0	4 0	3 0	2 0	1 0	11 0	10 0	9 0	8 0	7 0

Without applying the chessboard, we make that, in each battle, the team listed in the front (left) is the host while the team in the back (right) is the guest. Now we can find that, for each team, successive front (left) or successive back (right) will occur. For example, in table 2, for No.1, it is in the left in the first 6 rounds and in the right in the second 5 rounds. However, under the situation when we remain the round table of odd to take the middle bye and apply the chessboard, we find that, if we make the white stand for "left" (host), black for right (guest). Now, we turn to No.1 team and find that, No.1 has naturally formed a black and white alternating, that is, left and right alternating. The same is also true for other teams in left and right alternating. That is to say, when applying the chessboard, we can get a round table which balances the host and guest (left/right) alternating. The method is convenient in manual manipulation and can be applied universally in the method of serving rotate one position. Moreover, it provides arithmetic support to the computer scheduling.

4.2 Fixed Exchange and Its Application in Scheduling

Odd to take the middle bye is able to reach the natural balance in scheduling the odd team, but it has problem of unbalanced court when scheduling the even team. In the table 1, the 4th requirements which can't be satisfied are mentioned. From the table 2, we can find that, when altering each 0 in bye round into 12, we'll find that, team arranged with No. 12 in draw lots always appears in the last match in each round.

By introducing the table balance in the round robin of table tennis, that is, fixed exchange method, we try to solve the problem of the balanced court of even team in the odd to take the middle bye. The Table 3 shows the round order after adjustment. We take the first round of adjustment as example. The method is: before the adjustment, fix the last match (no need to exchange). Then exchange the match order of the largest N in each round one by one. (As for the match order of No.12, we need to find out the opponent A (No.6) of N (12) in this round and A's opponent B (No.8) in the last round during the exchange. Then find out the round where B (No.8) is arranged in this round. Finally, exchange the matches which contain A (6) and B (8). As for other rounds, the adjustment is carried out in the same way)

Table 3. Fixed exchange and its application in scheduling

Field	Round 1	Round 2	Round 3	Round 4	Round 5	Round 6	Round 7	Round 8	Round 9	Round 10	Round 11
Field 1	1 11	11 10	*12* 4	9 8	8 7	7 6	6 5	10 *12*	4 3	3 2	2 1
Field 2	2 10	*12* 5	11 8	10 7	9 6	8 5	7 4	6 3	9 *12*	4 1	3 11
Field 3	3 9	2 8	1 7	*3* *12*	10 5	9 4	*12* 11	7 2	6 1	5 11	4 10
Field 4	6 *12*	3 7	2 6	1 5	11 4	10 3	9 2	8 1	7 11	*12* 8	5 9
Field 5	5 7	4 6	3 5	2 4	*12* 2	1 *12*	10 1	9 11	8 10	7 9	6 8
Field 6	4 8	1 9	9 11	6 11	3 1	2 11	3 8	4 5	5 2	6 10	7 *12*

5 Discussion and Conclusion

In the field of universal scheduling theoretical research of round robin, there's little literature which touches upon the problem. Scholar Wang Pu put forward his opinion: "in the competitive games which holds the round robin as the competition methods, the fact is that, not every game adopts the same competition order. In different games, different competition methods are adopted. As for the same competition method, different competition order is adopted. In this respect, the above mentioned facts become a theoretic and practical problem which require through research and proper solving methods in terms of competition methods."

It is true that the universal scheduling method for round robin doesn't exist right now. However, from the research, we find that, based on the original odd to take the middle bye, we can alter the order of competitions' round order by additional solving method, try to meet all of those scheduling conditions as much as possible, with an aim to reach the universalism. From the research, we contend that, 1. The method of odd to take the middle bye is able to apply the scheduling of even teams. Based on the round table of odd teams, we only need to alter 0 into large number. When there's even round, we exchange the left and right position of the largest number. Now we have an adjusted round table for even teams. Due to the unchanged characteristic of serving rotate one position, the order of round table is same as the fixed 1 round robin. 2. Based on the round table of odd to take the middle bye, we apply the chessboard to distinguish the left and right (host/guest) of each team, by which we formed a competition round table which has left and right alternations to meet the scheduling requirement of host and guest. The method has replaced the previous manual adjustment and makes the scheduling has the law to abide by. Moreover, the method provides arithmetic support for future computer programming. 3. Based on the round tables of even teams in the odd to take the middle bye, we introduced the fixed exchange methods. Thereafter, we have a round table which balances the game court of each team to satisfy the scheduling requirement on the balanced court.

It is noteworthy that, although fixed exchange method is able to balance the game times of each even team in each court, it brings two problems: first, the each team can't reach balance in terms of host and guest in each court. If we reach the balance, we lose the alternating balance. Second, the original round table is able to meet the 1st requirement. However, because we have adjusted the game order of the largest number in each round, we have to break the balance of teams in having interval after two games. Therefore, in the practical application, we adopt it if when want to highlight the court of each team. Or else, we can omit this step.

All in all, conditioned that, we preserve the intrinsic advantages of "odd to take the middle bye" while we integrate the scheduling of even team. By applying the "chessboard method", we make alternating scheduling of host and guest and adopt the "fixed exchange method" to carry out the scheduling of court scheduling for the even team. We are able to get a round table which satisfies all those requirements. We contend that, the round table is applicable for all the games. Moreover, the method acts as an improvement, supplement and perfection of the round table of odd to take the middle bye.

References

1. People's Sports Publishing House, Approved by the Committee of National Sport Institutions Textbook. Basketball Senior Course (2010)
2. Lu, H.: Athleticism Theories. Tsinghua University Press (2005)
3. Xu, J.: Judge's Manual in the Chess Game. People's Sports Publishing House (2007)
4. Xuhu: Analysis and Innovation Research on the Berger Scheduling of Round Robin in the International Sport Competition Scheduling System. China Sport Science and Technology, Period 2, 38 (2002)
5. Song, G., et al.: Odd to take the middle bye in the Single Cycle Games' Scheduling. Journal of PLA Institute of Physical Education, Period 2, 20 (2001)
6. Wang, P.: Perfecting the Reform Plan of Competition Order of Round Order and its Demonstration. Journal of Tianjin Institute of Physical Education, Period 4, 22 (2007)
7. Pan, S.: A Research on the Table Tennis Table Balance in the Single Cycle match. Journal of Nanchang Hydraulic & Water Power Engineering College 13(1) (1994)

The Development of Ontology Information System Based on Bayesian Network and Learning

ZhiPing Ding[*]

Department of Computer Application, Qingyuan Polytechnics,
Qingyuan, 511510, Guangdong, China

Abstract. One main advantage of ontology is the ability to support the sharing and reuse of formally represented knowledge by explicitly stating concepts, relations, and axioms in a domain. But traditional ontology construction is time-consuming and costly procedure. In this paper we present using Bayesian network and learning for building ontology-based information systems. In order to address this research issue, this paper proposes a novel method consisting of Bayesian Network and learning probability theorem to automatically construct ontology. RDF is recommended by W3C and can deal with the lack of standard to reuse or integrate existing ontology. The experimental results indicate that this method has great promise.

Keywords: Bayesian network, ontology, Bayesian learning, Information System.

1   Introduction

As one example of the importance of ontology quality consider the area of business process modelling. The description of business processes is an important first step in the development of information systems that are intended to support and implement these processes. Ontologies are formal, explicit specifications of shared conceptualizations of a given domain of discourse. Generally, an ontology for a domain contains a description of important concepts, properties of each concept as well as restrictions and axioms upon properties [1]. A significant contribution of the knowledge engineering discipline to the software engineering one comes from techniques and formalisms for knowledge representation and ontology development. Unfortunately, constructing and maintaining ontology is a difficult task. Traditional ontology construction leans on domain experts but it is costly, lengthy, and arguable.

The Bayesian Network is constructed from the data, it is imperative to determine the various probabilities of interest from the model. The availability of large-scale high quality domain ontologies depends on effective and usable methodologies aimed at supporting the crucial process of ontology building. In general, current search engines face two fundamental problems. First, the index structures are usually very

---

[*] Author Introduce: ZhiPing Ding (1980.01-),Male,Han,Master of South China University of Technology, Research area: Data Mining, Application System Development, Intelligent algorithm.

D. Jin and S. Lin (Eds.): Advances in MSEC Vol. 2, AISC 129, pp. 401–406.

different from what the user conjectures about his problems. The success of contemporary research studies often depends on the ability to merge data of different nature.

Finally, the paper uses Bayesian Networks to reason out the complete hierarchy of terms and to construct the domain ontology intelligent systems. For this purpose, we probe into the information integration and question answering of semantic web based on pervasive ontology intelligent. This paper presents a formal and implemented generic framework for building ontology-based intelligent systems. This paper proposed ontology-based similarity measurement to retrieve the similar sub-problems that overcomes the synonym problems on case retrieval. This paper proposes a novel method consisting of Bayesian Network probability theorem to automatically construct ontology intelligent systems.

2   Ontology Intelligent Systems

As mental health disorders and psychotic illnesses represent a considerable demand on health resources, maximizing the value of mental health research data by developing a domain-specific ontology is potentially of great benefit [2]. Ontology plays a pivotal role in the development of the Semantic Web. This does not mean that a person do not have an age, it is just not known. Data instances can be interpreted correctly only if they are semantically valid with respect to the ontology they comply with.

The fitness operation is to eliminate the individuals which performed not well enough by comparing the sample value, the mutation operation is to transform the individuals into binary string and randomly chose one bit to inverse, and the crossover is randomly choose two genes to change. The relationship between the *concept layer* and the *instance layer* is "*Instance of.*" Let T_{ij} be the weight of concept c_j in the web page d_i shown in equation (1).

$$T_{ij} = \sum_{t=1}^{m} n_{ij}^t \times W_t \qquad (1)$$

RDF can be used to describe the resources of a given web page, using a meaning graph of the RDF to represent a problem. The *category layer* defines several categories, namely "category *1*, category *2*, category *3*, ..., category *k*." Each concept in the *concept layer*, contains a concept name C_i , an attribute set $\{A_{Ci1},...,A_{Ciqi}\}$, and an operation set $\{O_{Ci1},...,O_{Ciqi}\}$ for an application domain. The Web Ontology Language for Services (OWL-S) is imported to express the process ontology and process control ontology.

Search in the World Wide Web (WWW) is currently based on search engines that simply feature keyword-based search functionality. Framework requirements and architecture design need to be identified. Each node in the graph corresponds to a variable x_i and each directed edge is constructed from a variable in D_{xi} to the variable x_i. The advent of easy to use and cost-effective technologies such as the web has paved the way towards user friendly e-government systems [3].

There may be several geo-services that produce the runoff as output by using an interpolation operation or according to a formula like the following.

$$R = \frac{S \times C \times P}{160} \tag{2}$$

where S is the surface slope categorized into values of 1 (0–3°), 2 (3–6°), 3 (6–9°), or 4 (greater than 9°). X is the subset of P that is similar to p, and Y is the subset of P that is determined similar to p by our case retriever R. The W3C is currently examining various approaches with the purpose of reaching a standard for the SWS technology: OWL-S, WSMO, SWSF, WSDL-S, and SAWSDL.

Identifying the semantic view needs the knowledge of implication analysis, i.e. the dependencies and independencies between the components in the ontology. Below is how the semantic view can be generated for a data instance. Thus, the recall and precision of the ontology-based similarity measurement performed better than that of the traditional similarity measurement.

3 Bayesian Network and Learning

Bayesian methods utilize heuristic methods to search the space of DAGswhich are evaluated by scoring functions. Its feature is to search what users expect, irrelative to what they express. The Bayesian Network is constructed from the data, it is imperative to determine the various probabilities of interest from the model [4].

The variables X and Y are conditionally independent given the random variable Z if $P(x/y, z) = P(x/z)$. From the Bayesian rule, the global joint distribution function $P(x_1, x_2, ..., x_n)$ of variables $X_1, X_2, ..., X_n$ can be represented as a product of local conditional distribution functions. Different groups of methodologies for developing ontologies can be found. Requester plans to search some interested resource, however, he didn't know which kind of resources he wants.

We also propose an implicit webpage expansion mechanism oriented to the user interest to better capture the user intention. This user-oriented webpage expansion mechanism adds webpages related to the user interest for further retrieval into the website models.

3.1 Bayesian Network System

Hence a Bayesian network can be described as a directed acyclic graph consisting of a set of n nodes and a set of directed edges between nodes. Each node in the graph corresponds to a variable x_i and each directed edge is constructed from a variable in D_{xi} to the variable x_i[5]. Consequently, ontology mapping this study conducts similarity matching for concept names and considers the similarities of essential information and relationship to precisely identify the similarity between concepts.

Eq.(3) is applied for name similarity matching for two concepts. If each variable has a finite set of values, to each variable xi with parents in D_{xi} , there is an attached table of conditional probabilities $P(x_i/ D_{xi\,i})$.

$$T_{iB_1} = T'_{iB_1} + \frac{(n_b - 1) \times T_{iB}}{n_b} \qquad (3)$$

T'_{iB_1} is the weight of concept B_1 and T_{iB} is the weight of its parent concept B. n_b is the number of B's child concepts, $n_b \geqq 1$. If the sibling concepts appears, the weight of concept B_1 will be enhanced.

In general, given a network, the calculation of a probability of interest is well known as probabilistic inference, and is usually based on Bayes' theorem. The travel route recommendation agent receives the results from the context decision agent and infers the context information. They represent a specific domain and they are used as computational entities for reasoning purposes. When we compare the two concepts from web page d_1 to web page d_2, the relation of concepts in the two web pages is that of parent to child in the domain ontology. Hence, the similarity check plays a crucial role in the projected clustering of PART. In addition to the vigilance test, the PART adds a distance test to increase the accuracy of clustering. The PART algorithm is presented below.

3.2 Ontology and Bayesian Learning

Ontology extraction and similarity calculation are the two main components of the prototype system. This process has four sub-processes, namely main process, ontology mapping process, ontology merging process and sub-concept merging process.

Protégé is an open-source development environment for ontologies and knowledge-based systems. The fitness operation is to eliminate the individuals which performed not well enough by comparing the sample value, the mutation operation is to transform the individuals into binary string and randomly chose one bit to inverse, and the crossover is randomly choose two genes to change.

The most significant drawback to Bayesian methods is that they are relatively slow and they may not find the best structure due to heuristic methods. Thus, the questionnaire design contains a set of Kansei words and properties of a specific dog breed. Others introduced to search scoring functions which are used to evaluate each network for maximizing Bayesian scores, such as BDe method and Minimum Description Length Principle.

4 Using Bayesian Network and Learning to Constructing Ontology Information

In the study, an automatic ontology construction based on Bayesian Network and learning is presented. The experimental procedures presented here can serve as templates for future research to evaluate other ontologies. We are currently working on applying the method to other upper-level and domain ontologies, and augmenting the method with other techniques from cognitive research.

In virtue of the Web services semantic markup, develop the agents to support the automatic reasoning and fulfill the composition of Web Services and inter-operation is possible. The Framework employs a simple yet specialised ontology to explore and 'make explicit' the relations between.

The system consists of (a) a metadata repository, storing learning object descriptions, learner profiles and ontological knowledge for the educational domain under consideration. The flowchart as Fig.1.

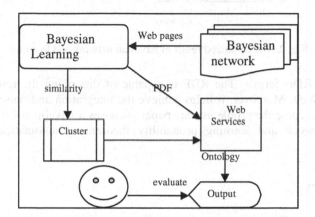

Fig. 1. The system architecture of ontology construction

To measure the performance of ontology and P2P based data and service unified discovery system and to verify the means of service matching with QoS presented in this paper, we have developed a prototype of the system. Therefore, the part of ontology extraction was implemented and there are 16 randomly chosen documents used to be evaluated the result. In this section the prototype system was implemented and 100 documents are chosen and classified into clusters to evaluate the implementation results. In the first experiment, we explore whether the quantity of data will affect the result. Bayesian reasoning is usually affected by the quantity of data. We divided the experiment into five stages and used Precision (C_P) and Precision (C_L_P) to evaluate the five ontology results.

This system has been employed to grade the students' work results. For it, an experimental system prototype that controls student's knowledge by means of ontological analyses in the URAN network was developed. After clustering, the system will select the highest term weight (TF-IDF value) of each cluster to represent the cluster, and obtain the basic hierarchical concept (PART tree). The system calculates the conditional probability table and sets the threshold θBN=0.35 in order to insert the remainder terms into the PART tree. The flowchart as Fig.2. We applied the feature selection program as described in ontology-reorganization to all collected webpages to select ontology features for each class.

Ontology has been widely applied in a variety of domains to represent information or knowledge models owing to the fact that its formal semantic can be unambiguously interpreted by humans and machines. We adopted RDF, a standard ontology web language recommended by W3C.The system utilizes the Jena package to output the

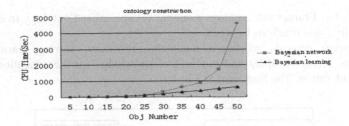

Fig. 2. The compared result of bayesian network and learning

results of the RDF format. The RDF is capable of describing the resources of the World Wide Web. Moreover, it helps achieve the integration and reuse of ontology. In order to overcome the problems, this paper proposes a novel method consisting of Bayesian Network and learning probability theorem to automatically construct ontology.

5 Summary

OWL is a new ontology language for the Semantic Web, developed by the World Wide Web Consortium (W3C) Web Ontology Working Group. The system then uses a Bayesian network to insert the terms and finish the complete hierarchy of the ontology. This is especially crucial, considering incremental changes and updates on products due to new technology advances. In the future work, we will attempt to improve the precision of term location. Our findings suggest that none of three ontologies that we examined is a good specification of the conceptualization of the domain.

References

1. Berners-Lee, T., Hendler, J., Lassila, O.: The Semantic Web, vol. 5. Scientific American (2001)
2. Denoyer, L., Gallinari, P.: Bayesian Network Model for Semi-structured Document Classification. Information Processing and Management 40, 807–827 (2004)
3. Guarino, N., Welty, C.: Evaluating ontological decisions with OntoClean. Commun. ACM 45(2), 61–65 (2002)
4. Tsai, W.T., Huang, Q., Sun, X.: A collaborative service-oriented simulation framework with Microsoft Robotic Studio. In: Proceedings of ANSS, Ottawa, April 14–16, pp. 263–270 (2008)
5. Missikoff, M., Navigli, R., Velardi, P.: Integrated Approach for Web Ontology Learning and Engineering. IEEE Computer 35(11), 60–63 (2002)

Study on the Relationship between Goal Orientation and Constitution of Collage Students in Sport

Jian Chen[1,*] and Gang Qin[2]

[1] Wuhan Textile University, Wuhan, Hubei, China
jianchenmail@sohu.com
[2] Zhejiang Lishui University, Lishui. Zhejiang, China
QinGang918@hotmail.com

Abstract. Goal orientation theory: A social-cognitive approach to the study of achievement motivation and behavioral. The theory is more concerned by sport psychologists. This investigation used "Task and Ego Orientation in sport Questionnaire" and constitution of collage students to explores the differences between goal orientation and collage students' constitution in sport.

Keywords: Goal orientation, Task orientation, Ego orientation, Constitution test.

1 Introduction

Goal orientation theory: A recent social-cognitive approach to the study of achievement motivation and behavioral. Previous studies have shown that in the academic and sports achievement situations, individuals do exist two different goal orientation states: one is task orientation (the composition of its main goals is to master the learning content and aimed at improving the skills level), the other is ego orientation (the composition of its main goals is to compare with others or aimed at victory over another).

Motivation is the starting point of behaviour, and the reason and motivation which aroused students to carry on sustainable physical exercise. In order to understand whether individual goal orientation will bring different influences to their physical health when students participating in sports activities, we tried to explore whether there are differences among the situation of physical health of students with different goal orientation, the difference source and the relationship between the difference source and its goal orientation by the test results of "Physical Fitness and Health Criteria for National Students" which can show a more comprehensive response to the situation of physical health of students.

2 Research Subjects and Methods

2.1 Research Subjects

The research subjects are the college students of Zhejiang Lishui University, and then carry on constitution and questionnaire testing by randomly selected 110 students

---

* Jian Chen (1962 -) Male, This article is the human social science problem of Hubei ProvinceDepartment. Project number: 2009y077.

whose specialty is CJ mathematics, CJ Chinese, financial management, construction and construction management. The valid data included is 96: 48 boys and 48 girls. Mean age = 20.67, SD = 0.76.

2.2 Research Methods

2.2.1 Goal Orientation Test Questionnaire

"Questionnaire of Task Orientation and Ego Orientation in Motion" which originated from (Duda, 1993 & Chi, 1995). After the questionnaire had gone through the return translation to ensure the equivalency of the language in home, the test of reliability and validity was conducted, the results are considered to have high construct validity and internal consistency reliability(Chen Jian, Si Gangyan, 1998). After the scale was tested by Liu Yujiang and other scholars (2008) by confirmatory test, it was proved that the scale had higher construct validity and internal consistency reliability once again.

2.2.2 Constitution Test Data Extraction

According to the test content of "Physical Fitness and Health Criteria for National Students", extract the constitution test data this year which is corresponding to the "Questionnaire" (including: height and weight levels; lung capacity weight level; endurance project level; flexibility, strength project level; speed, dexterity project level).

3 Research Results

Constitution test is in accordance with the scores and evaluation of students standard, we will regard the evaluation level(excellent, good, pass and fail) of each student as the research target. "Questionnaire" is based on two dimensions(task orientation and ego orientation) of each student to conduct statistical treatment, in which the average score and standard deviation of task orientation is 26.68 ± 3.88, the average score and standard deviation of ego orientation is 14.63 ± 4.54. According to goal orientation theory, task orientation and ego orientation are not the two ends of a continuum. That is, task orientation and ego orientation of a person may be very high or very low, or one orientation is very high while the other one is low (Duda, 1988). To this end, we confirmed task orientation and ego orientation whose basis of average is greater than one standard deviation as the high task and ego orientation. In that way, confirm whose basis of average is lower than one standard deviation as the low task and ego orientation. We identified the orientation state of students into 6 levels when making statistics, namely: high task orientation, higher ego orientation, and high task + high ego orientation (double high), low task + low ego orientation (double low), task orientation and ego orientation.

The research data were statistically settled by SPSS17 for Windows. In order to understand whether the student gender is related with physical level and goal orientation state, we carried out partial correlation analysis to these three subjects, the results shown in Table 1.

Table 1. Partial Correlation Analysis of Gender, Constitution Level and Goal Orientation

	Constitution level	goal orientation	Gender
Constitution level Correlation	1	-0.285	0.015
Sig (2-Tailed)		0.005	0.882
df	0	94	94
Goal Orientation Correlation	-0.285	1	-0.035
Sig (2-Tailed)	0.005		0.732
df	94	0	94
Gender Correlation	0.015	-0.035	1
Sig (2-Tailed)	0.882	0.732	
df	94	94	0

Partial correlation analysis result in Table 1 showed: gender, constitution test level and goal orientation state of the students in the sample are not significantly related (P> 0.05), while the constitution level and goal orientation have a significant negative correlation (P <0.05).

Because this correlation analysis results showed that gender and goal orientation state has no significant correlation, and therefore, under the premise of ignoring gender, we carried out single factor ANOVA in constitution level and goal orientation state, in an attempt to understand whether students with different orientation exist diverse in the constitution level. The results are as follows:

Table 2. ANOVA Homogeneity Test Results of Single Factor

Levene Statistic	df1	df2	Sig
0.348	5	90	0.883

Table 3. ANOVA of Single Factor

	Sum of Squars	df	Mean Square	F	Sig
Between Group	5.487	5	1.097	2.565	0.032
Within Group	38.603	90	0.428		
Total	43.99	95			

Table 4. Mean Multiple Comparison Test Results

(I) Orientation State		(J) Orientation State Mean	Difference(I-J)	Std.Erro	Sig
LSD	Double Low	High Task	0.639	0.256	0.014
		Task	1.25	0.401	0.002
		Ego	0.917	0.353	0.011
	Task	Double High	0.75	0.358	0.039
		High Ego	0.727	0.356	0.044

Table 5. Constitution Level * Goal Orientation State Crosstabulation

		constitution level				
		fail	pass	good	excellent	Total
Orientation State	Double Low	2	2	0	0	4
	Double High	1	13	6	0	20
	High Task	5	24	5	2	36
	High Ego	3	11	8	0	22
	Ego	2	3	1	0	6
	Task	0	2	6	0	8
Total		13	55	26	2	96

Table 2 is ANOVA homogeneity test results of single factor: t value is 0.348, two degrees of freedom is 5 and 90, two-tailed significance probability is P = 0.883> 0.05, thus accept the homogeneity of variance assumptions. Table 3 (ANOVA of single factor), the result of ANOVA test showed: interblock P=0.032<0.05, negate null hypothesis, indicates that the student of goal orientation state of six levels has significant differences among constitution level. Table 4 is mean Multiple comparison test results. LSD method showed: low task + low ego orientation and high task orientation, task orientation and ego orientation has significant differences among constitution test level(P values were 0.014,0.002,0.011, respectively. P values were all less than 0.05).Task orientation and high task+high ego orientation and high ego orientation also has significant differences among constitution test level. (P values were 0.039,0.044, respectively. P values were all less than 0.05). Table 5 is constitution level and goal orientation state crosstabulation.

4 Conclusion and Discussion

From the above test result, we can see that students with different goal orientation has significant differences among constitution test level. In the mean Multiple comparison test results of ANOVA, we can see that students with high task orientation, task orientation and ego orientation was significantly higher than students with low task+low ego (double low) state in constitution test level; while students with high task+high ego(double high) and high ego orientation was significantly higher than students with task orientation state. In the constitution level and goal orientation state crosstabulation, the population of students with high task, high ego and double high was significantly higher than orientation state of other levels in the level of pass and good. In the sex difference which is similar with the study of Liu Yujiang and other scholars, this test showed that gender, goal orientation and constitution test level does not appear the significant correlation.

"Physical Fitness and Health Criteria for National Student" (hereinafter referred to as "Criteria") has been an important tool which can change the healthy awareness of students, encourage students to actively exercise, ensure the school physical teaching can be on the rails and realize the aim of P.E. course since the full implementation of national various school in 2007.The goal of "Criteria" test is to encourage students to actively exercise, promote the integrated development of normal physiological growth and body shape, enginery (Ministry of Education, 2002 NO12). We think: although "criteria" test can reflect the dominant index of students when participating in physical activities more completely, it can't reflect the recessive index of mental stratification of students when participating in physical activities for a long time. This means the recessive index of mental stratification plays an essential role to complete or achieve the requirement of "Criteria". Cultivate lifelong physical exercise is the ultimate goal of physical education, that is also the purpose of implementing "Criteria", while the agent base of urging students to persist in participating in the long-term physical exercise is the requirement and satisfaction of mental stratification. The goal orientation theory in the motivation field provides us that how people think and how these thinking affects people's behaviour in the process of trying to understand the physical activities(ChenJian,1999). Some scholars believe: before taking intervention of changing people's behaviour, we must understand the motivation why people participate in the physical activities. Because the motivation of participate in the physical activities is an important factor of engaging in regular exercise (Shen MengYing,2010). Research also shows that a certain level of sports motivation can make physical activity participants try harder, more focused, more practice time in the motor activities(HanXu,2007).

Nicholls, Duda, and Chi and other scholars believed that, for sports participants and non participants, the meaning of sports is multidimensional, and is composed of many important beliefs and cognition. The difference among individual goal orientation will lead them to different participating motivations, behaviors, attitudes and beliefs, meanwhile, it also have a close relationship with its participating knowledge, interest, will, emotion, attention, anxiety and other psychological factors in the process of physical activities(ChenJian,1999).Just because of these different knowledge of individual ability perception and struggling methods which originated from childhood and nurture, combined with different environmental factors and motivation atmosphere, individuals will adopt different goal orientation states according to different situations in the achievement situation. Currently, a large number of theoretical and validation studies have shown that task orientation is a positive, proactive and ideal goal orientation state, participants can get more fun from goal orientation, so as to enhance internal motivation, promote the sustainability of their participating activities; while ego orientation is considered as a goal orientation states which can passively, forwardly, easily lead individual to poor fitness, participates will weaken the internal motivation till exiting activities in the case of low ability level perception. The results of this study shows that students whose orientation state is high task, high ego orientation and double high reflects higher scores on constitution health, this indicates the higher goal orientation level of students, the higher their constitution health level will be, both relevant degree attains high the level of significance. From Table 5 we can easily see that the passing population of students with high task orientation level is obviously larger than the orientation level of other levels, followed by the double high, then high ego. Ego orientation acts as a lower test situation of students with general orientation and double low. The results seems that the structure of goal orientation theory is in consistent with prediction degree. As a preliminary situation of using constitution test and the association study of goal orientation theory in sport psychology to investigate the causes of the difference of physical health of university students, this is the first time in the country. In the future research, we believe: sport psychology researchers need to make full use of the provided resource of "Criteria" test, intensively explore and study the performance of students constitution health caused by psychological effects. Goal orientation may serve as predictors of students physical exercise situation, in future studies, we need to expand the sample size, conduct the cross-regional research, so as to provide a theoretical basis which guide the students for lifelong physical activity.

References

1. Chi, L.K., Duda, J.L.: Multi-Sample Confirmatory Factor Analysis of the Task and Ego Orientation in Sport Questionnaire. Quarterly for Exercise and Sport 66(2), 91–98 (1995)
2. Chen, J., Si, G.: Goal Orientation Theory and Initial Survey of Its Questionnaire. Journal of Wuhan Institute of Physical Education 128(1), 53–56 (1999)
3. Liu, Y., et al.: An confirmatory Factor Analysis on Task Orientation and Ego Orientation of Exercising University Students. Journal of Beijing University of Physical Education Section 31, 974–975 (2008)
4. Ministry of Education, No.12 State Physical Culture Administration, A notification in Respect of Publishing "Physical Fitness and Health Criteria for Students (on trial)" and "Executive Method of Physical Fitness and Health Criteria for Students (on trial)"(Ministry of Education, State Physical Culture Administration (2002)
5. Duda, J.L.: The relationship between goal perspectives and persistence and intensity among recreational sport participants. Leisure Sciences 10, 95–106 (1998)
6. Chen, J.: Achievement Goal and Achievement Situation—Research on the Goal Orientation of Students of Wuhan Youth Athletic School and Normal School, Master Thesis of Wuhan Institute of Physical Education (1999)
7. Shen, M., et al.: Overview on Achievement Goal Orientation of National Adults Participating in Leisure Exercise and Activities. The Ninth National Article for University Exchange of Sports Psychology (September 2010)
8. Xu, H.: Research on Motivation in Sports of Henan Female Volleyball Team, Master Thesis of Henan University (2007)

Sub-block Size on Impact of Fundus Image Noise Estimate

YiTao Liang, WeiYang Lu, YuanKun Zhu, and Rui Pang

College of Information Science and Technology,
Henan university of Technology, Zhengzhou, China
liang-yt@yahoo.com.cn

Abstract. Image noise variance estimation is an important indicator to determine the image quality. In this paper, LMLSD algorithm based on Gaussian wave extraction is used for fundus image noise estimation, and it analyses the sub-block size effects for the performance of this algorithms. In the expriment, 6 different sub-block size are taken for the simulation and analysis of noise variance estimation respectively in 10 fundus images. Exprimental results show the accuracy of noise variance estimation which takes 5×5 sub-block size is better than others in 256*256 fundus images.

Keyword: Image quality assessment, LMLSD, Size of block, Noise, Fundus image.

1 Introduction

In Image acquisition, storage and transportion process, image is affected inevitably by noise pollution as the image equipment and conditions constraints[1], eventually leading to cause degradation of image quality , so there appears low contrast, image blurred, etc. Image quality is an important reference standard for further processing, but it is still no unified comprehensive evaluation method, only given quantitative indicators or parameters based on model to mesure. Image noise is an important factor affecting image quality, how to accurately estimate the amount of noise contained is an importan research work in image quality assessment.

At present, image noise estimation algorithm includes three categories. (a)Based on filtering: By subtracting from original image I the filtered image, a measure of the noise is computed. This method requires a high-performance filter, and the filtered image not only retains details in the image information, and effectively eliminates the noise. When the image contains a wealth of edge and texture information, the method is prone to obtain a larger noise estimation. (b)Based on wavelet: After wavelet tranformation, the energy of image is mostly concentrated in large-scale sub-band, the coefficients of small-scale high-frequency sub-band are smaller and energy is lower[2]. When the noise is large, the high frequency sub-band coefficients are a direct reflection of noise, thus the variance of these coefficients can be considered as noise variance estimation [3]. The easiest method is to calculate the variance of high-frequency sub-band(HH) coefficients as the noise estimation, Donoho and Johnstone proposed a noise estimate

D. Jin and S. Lin (Eds.): Advances in MSEC Vol. 2, AISC 129, pp. 413–417.
springerlink.com   © Springer-Verlag Berlin Heidelberg 2011

formula[4]: σ_n =MAD/0.6745, MAD is the median of HH sub-band coefficients. when the noise is small, this method is prone to get a biased result because the HH sub-band coefficient is not sensitive for small noise. (c)Based on blocks: the noisy image is divided into several small pieces, compute the variance of each block, then get the entire image noise estimation by statistical methods. Gao.B.C proposed a block method[5]—local mean and local standard deviation(LMLSD),this method is to divide the image into several the same size of blocks, and estimate the noise of image by the variance of these blocks. This method is more suitable for the image which contains a lot of flat areas, simple steps and facilitate the implementation of hardware and software.

2   LMLSD Based on Gaussian Wave Extraction

This paper has taken the LMLSD based on Gaussian wave extraction because fundus image contains more non-uniform blocks. When LMLSD is used for calculating image noise estiomation, the statistical curve which responses to the number of blocks in each interval should be Gaussian distribution theoretically. As fundus image contains more edge and detail information, the statistical curve will appear several peaks, the first peak represents the number uniform blocks, other peaks is reflection of non-uniform blocks. The mean of standard deviation of blocks which the first peak is corresponding to is treated as the noise estiomation of the whole image.Algorithm is shown in figure 1. As statistics curve influenced by the image content in exprement, waveform usually is not clear, so curve becomses smooth through Fourier transformation and low pass filter in order to improve the accuracy of the algorithm.

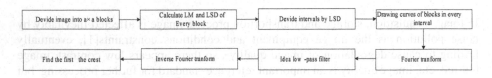

Fig. 1. Flow chart of LMLSD algorithm based on Gaussian wave extraction

This method calculates the image noise estimation by statistical characteristics of the image pixel gray level difference, image need not be completely uniform, just owns sevral uniform areas, so this algorithm has a better adaptability for fundus image.

3   Experimental Simulation and Analysis

This paper took 10 fundus images from CF-60U camera imaging instrument, and these images became digital by high-definition HP photo scanner. In the experiment, this paper took the target region split from the original image for reducing the amount of data operations. Author selected partition where the size of the target 256*256 image information according to priori knowledge, and selected Matlab 7.1 as experimental platform.

Figure 2 shows the fundus1 has obtained the scatter of local mean and local standard deviation by different block size(a,...,f represent repectively 3×3,...,8×8 block). As figure 2 shown, local standard deviation are gathered in a more concentrated area, and the values don't change with the local mean , so it illustrates there is no abvious correlation between the local standard deivation and local mean. It can be seen that fundus image noise is mainly constituted by the additive noise.

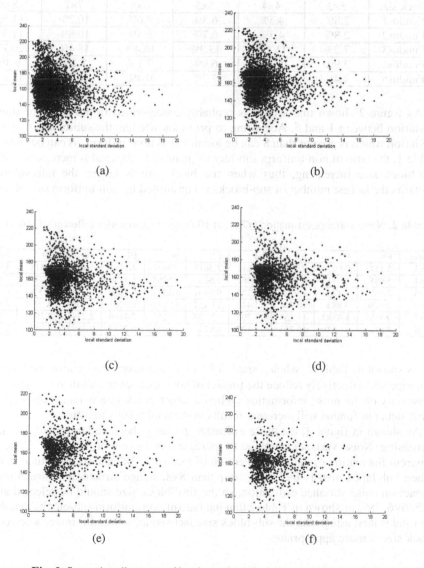

(a)   (b)

(c)   (d)

(e)   (f)

Fig. 2. Scattering diagrams of local standard deviations and local mean

Generally retinal blood vessels occupy most areas of the fundus image, the local standard deviation of sub-blocks which contain blood vessels and other image details is larger and responses in the right area in figure 2. So the number of sub-blocks is reducing with the increase of sub-block size, the number of sub-blocks which contains structural information turns large.

Table 1. Number of non-uniform blocks percentage of the total number of blocks in 5 fundus image

Block size	3×3	4×4	5×5	6×6	7×7	8×8
Fundus1	2.8%	4.5%	6.3%	8.6%	10.2%	11.5%
Fundus2	2.8%	4.3%	6.7%	8.4%	10.8%	11.5%
Fundus3	7.2%	11.4%	13.3%	16.4%	18.4%	20.7%
Fundus4	3.9%	5.4%	6.9%	7.1%	8.2%	9.4%
Fundus5	4.4%	6.8%	7.7%	9.4%	10.2%	12.2%

As figure 2 shown that scatter is probably concentrated in the range of standard deviation between 1 and 5. According to priori knowledge, the sub-block of standard deviation which is greater than 8 can be identified as non-uniform, as can be seen from Table 1, the ratio of non-uniform sub-blocks' number to the total is increasing with the sub-block size increasing, thus when the block size is larger, the interval which contains the largest number of sub-blocks is constituted by non-uniform sub-blocks.

Table 2. Noise variance estimation (σ_n) of 10 fundus pictures with different sub-block size

block	F1	F2	F3	F4	F5	F6	F7	F8	F9	F10
3×3	5.1543	7.0378	7.2789	8.3611	7.3611	7.9834	4.1494	5.1111	3.511	5.9931
4×4	5.985	8.0115	8.7926	9.8087	9.663	8.6265	5.4836	6.8748	4.2366	6.6614
5×5	7.3654	9.1213	9.9174	11.287	9.6164	10.377	6.2048	6.9333	5.0063	6.7814
6×6	7.1786	9.885	10.721	11.442	11.451	11.581	6.6201	6.9638	4.8971	7.1274
7×7	9.8182	8.6887	11.134	11.792	12.174	12.251	7.4149	8.2383	5.203	7.5953
8×8	8.8243	11.536	10.705	11.107	10.24	11.613	6.2337	7.3692	4.8849	7.1556

As shown in Table2, when a smaller block size is taken, non-uniform information in image will effectively reduce the impact of noise estimation, while it will reduce the sensitivity of the noise information ;When a larger block size is taken, the impact of some detail in fundus will increase, results in the final results to differ.

As shown in figure 3, the noise estimates gradually become larger with block size increasing. Noise variance continue to increase in selection of 3×3, 4×4. There is an apparent fluctuation of values in selection of 5×5, 6×6, 7×7 and its amplitude is small. When sub-block size is 8×8 or greater than 8×8, image structural information will impact on noise variance estimation, so the final block size should be selected among 5×5, 6×6, 7×7, as shown in Table 2 that the ratio of non-uniform sub-blocks' number to the total is increasing with the sub-block size increasing, so Fundus image selected 5×5 block size is more appropriate.

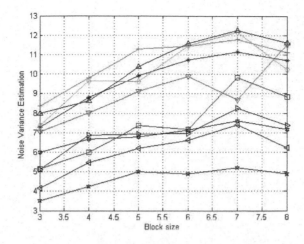

Fig. 3. Noise variance estimation of 10 fundus pictures with different sub-block size

4   Conclusion

Image noise variance is an important indicator which is reflected in the image processing system of performance. In this paper, author took fundus image to verify effectiveness of LMLSD based on Gaussian wave extraction and the impact of this algorithm of sub-block size. Experimental results show that in fundus images using 5×5 sub-block size estimates obtained higher accuracy. The result of this study provides an accaurate sub-block size for fundus image noise estimation,and reduces the blindness of the process of fundus image noise estimation.

References

1. Liang, Y., Chang, H.: Research on FRIT of Color Fundus Image Based on Prior Knowledge. In: e-Business and Information System Security (EBISS 2010), pp. 1–4 (2010)
2. Olsen, S.I.: Noise Variance Estimation in Image. GVGIP:Graphic Models and Image Processing 55(4), 319–323 (1993)
3. Tsur, D., Ullman, J.D., Abitboul, S.C.: Query Flocks:A Generalization of Association-rule Mining. In: ACM-SIGMOD Conference on Management of Data, Seattle, Washington, pp. 1–12 (1998)
4. Donoho, D.L., Johnstone, I.M.: Ideal Spatial Adaptation via Wavelet Shringkage. Biometrika 81, 425–455 (1994)
5. Gao, B.C.: An operational method for estimating signal to noise ratios from data acquired with imaging spectrometers. Remote Sensing of Enviroment 43, 23–33 (1993)

The Application of Particle Filter Algorithm in Multi-target Tracking

Jiaomin Liu[1], Junying Meng[1,2], Juan Wang[1], and Ming Han[1]

[1] College of Information Science and Engineering, Yanshan University,
Qinhuangdao, China
[2] Department of Computer Science, Shijiazhuang University,
Shijiazhuang, China
daishan74@126.com

Abstract. Particle filter is a probability estimation method based on Bayesian framework and it has unique advantage to describe the target tracking non-linear and non-Gaussian. In this paper, firstly, analyses the particle degeneracy and sample impoverishment in particle filter multi-target tracking algorithm, and secondly, it applies Markov Chain Monte Carlo (MCMC) method to improve re-sampling process and enhance performance of particle filter algorithm. Finally, the performance of the proposed method is certificated by experiment that tracking multiple targets of similar appearance and complex motion. The results show the efficacy of the proposed method in multi-target tracking.

Keywords: Particle filter, Multi-target tracking, Sequential important sampling, MCMC.

1 Introduction

In recent years, many single-target video sequence tracking system are successfully developed one after another, but the multi-target tracking system is still a challenging project, especially the tracking of multiple targets of similar appearance and in complex motion. At present, the classic algorithms that widely applied. NNF algorithm is proposed by Singer, etc. and it needs less computation and not relies on clutter distribution model, good for easy project implementation. However, it only uses the measurement in statistical sense that closest to the predicted position of tracked target as a candidate measurement. Therefore, NNF algorithm often has tracking errors and tracking lost occurred [1]. JPDA algorithm is proposed by Bar-Shalom, etc., which is regarded as one of most effective algorithms for solution of multi-target data association under intensive measurement, and its tracking success rate is relatively high under all circumstances. With JPDAF algorithm, the search of associated solution is actually a seeking of combination issue, and the computation amount of searching process is in exponential growth trend as quantity of targets and measurement grown, so it is difficult for such algorithm to be used widely in practical project [2]. Reid proposed MHT algorithm of multi-target tracking based on "All Neighbor" optimal filter proposed by Singer and concept of confirmation matrix proposed by Bar-Shalom. However, such algorithm will have computation amount rise rapidly as target number and observed quantity increased, so its application is limited in practical project [3]. Particle filtering (PF) [4] technique is a Optimal Regression Bayesian filtering algorithm based on MCMC simulation, which is not limited by linear error and high Gaussian noise hypothesis and is applicable for non-linear non-Gaussian model.

2   Multi-target Tracking Technique and Bayesian Method

Many scientific research need to estimate the system state changes over time. Generally, two model should be built when we analyses a dynamic system, one is the state transition model, describe the state change over time, the other is the observation model, which is a model of state measure value. Bayesian filter can provide a strict universal framework to solve this problem.

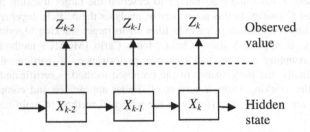

Fig. 1. Dynamic state space model

To solve dynamic state system using Bayesian method, people hope to build state posterior probability density function based on all the information that could be obtained. Since it embodies all available information, this function can be considered is the complete solution of this estimation problem. In principle, the optimal estimation of the state can be got from probability density function, when get a measured value, can also get an estimate. Therefore, Bayesian recursive filter is a very good choice. When a new measured value comes, we deal with it in order and don't have to save all of the data. So, the filter consists of two steps: prediction and fix.

3   Particle Filter Multi-target Tracking Technique

3.1   Particle Filter Theory

Particle filter is a probability estimation method based on Bayesian framework and it is very suitable to describe the target tracking uncertainty. Particle filter approach provides a flexible framework, and many traditional vision tracking methods can make a great robustness enhancement through slight model modification and embedment into the particle filter frame work alone. Moreover, particle filter approach has unique advantage in handling non-linear non-Gaussian multi-modal cases, and it is more suitable for target tracking, compared with traditional Bayesian filter method, and outstands from the various algorithms and becomes the mainstream in the field of vision tracking algorithm [5]. The essence is to realize Bayesian filter in non-parametric Monte Carlo simulation method.

3.2 Sequential Important Sampling

The core idea of particle filter algorithm is to use weighting of a series of random samples and posterior probability density required by expression, to get the estimated state value. When the sample number is very large, such probability estimation will be equal to posterior probability density. Assume Ns indicate the particle number, then $\{X_k^i, i = 1,..., N_s\}$ means a support point set, and its corresponding weight is $\{w_k^i, i = 1,..., N_s\}$, and normalized weight is $\sum_{i=1}^{N_s} w_k^i = 1$, then $\{X_k^i, w_k^i\}_{i=1}^{N_s}$ indicates the random particle set describing posterior density. Thereupon, posterior probability density at the time k can use discrete weight sum that is approximate to:

$$p(X_k|Z_{1:k}) \approx \sum_{i=1}^{N_s} w_k^i \delta(X_k - X_k^i) \tag{1}$$

In which, the weight w_k^i can be sampled and selected from important density function $q(X_k|Z_{1:k})$ in sequential important sampling method.

If the sample X_k^i can be obtained from important density $q(X_k|Z_{1:k})$, then the weight of the i'th particle can be defined as:

$$w_k^i \propto \frac{p(X_k^i|Z_{1:k})}{q(X_k^i|Z_{1:k})} \tag{2}$$

If the important density function can be decomposed as follows:

$$q(X_k|Z_{1:k}) = q(X_k|X_{k-1}, Z_{1:k})q(X_{k-1}|Z_{1:k-1}) \tag{3}$$

Then the posterior probability density can be expressed as:

$$
\begin{aligned}
p(X_k|Z_{1:k}) &= \frac{p(Z_k|X_k)p(X_k|Z_{1:k-1})}{p(Z_k|Z_{1:k-1})} \\
&= \frac{p(Z_k|X_k)p(X_k|X_{k-1},Z_{1:k-1})p(X_{k-1}|Z_{1:k-1})}{p(Z_k|Z_{1:k-1})} \\
&= \frac{p(Z_k|X_k)p(X_k|X_{k-1})}{p(Z_k|Z_{1:k-1})} p(X_{k-1}|Z_{1:k-1}) \\
&\propto p(Z_k|X_k)p(X_k|X_{k-1})p(X_{k-1}|Z_{1:k-1})
\end{aligned}
\tag{4}
$$

And updated formula for weights is obtained therefore:

$$
\begin{aligned}
w_k^i &\propto \frac{p(Z_k|X_k^i)p(X_k^i|X_{k-1}^i)p(X_{k-1}^i|Z_{1:k-1})}{q(X_k^i|X_{k-1}^i, Z_{1:k})q(X_{k-1}^i|Z_{1:k-1})} \\
&= w_{k-1}^i \frac{p(Z_k|X_k^i)p(X_k^i|X_{k-1}^i)}{q(X_k^i|X_{k-1}^i, Z_{1:k})}
\end{aligned}
\tag{5}
$$

Weights can be normalized as:

$$\tilde{w}_k^i = \frac{w_k^i}{\sum_{i=1}^{N_s} w_k^i} \tag{6}$$

If $q(X_k|X_{k-1},Z_{1:k}) = q(X_k|X_{k-1},Z_k)$ is achieved, namely the important density function only depends on X_{k-1} and Z_k, then only storage sample X_k^i but not X_{k-1}^i and the past observation $Z_{1:k-1}$ is needed, therefore computation storage can be greatly reduced. At this time the weight is revised as:

$$w_k^i \propto w_{k-1}^i \frac{p(Z_k|X_k^i)p(X_k^i|X_{k-1}^i)}{q(X_k^i|X_{k-1}^i,Z_k)} \tag{7}$$

Thus, the posterior probability density at the time K can use discrete weight sum that approximate to:

$$p(X_k|Z_{1:k}) \approx \sum_{i=1}^{N_s} \tilde{w}_k^i \delta(X_k - X_k^i) \tag{8}$$

Therefore, particle filter algorithm is to obtain samples mainly from important density function, and get corresponding weight in iteration as successive arrival of measured values, and finally represent the posterior probability density in the form of weight sum, and get the estimated state value.

For multi-target tracking system, N (quantity) particles are involved in initial particle set $S_0 = \left(s_0^n, \frac{1}{N}\right)_{n=1,...,N}$, in which each element $s_0^{n,i}$ from $i = 1,...,M$ is obtained from independent $p(X_0^i)$ sampling. The particle set at the time t-1 is assumed as $S_{t-1} = \left(s_{t-1}^n, p_{t-1}^n\right)_{n=1,...,N}$, in which $\sum_{n=1}^{N} p_{t-1}^n = 1$. Each particle is a vector of dimension $\sum_{i=1}^{M} n_x^i$, and $s_{t-1}^{n,i}$ represents the i'th element in s_{t-1}^n, and n_x^i represents the state vector dimension of the i'th target.

Each iteration in particle filter algorithm is divided into two steps: prediction and weight updating. Prediction means sampling from proposed density function F_t^i, and proposed density function is consistent with the target motion model; weight updating is to make the weight at the time t-1 multiplied by the observation likelihood.

$$\tilde{s}_t^n = \begin{pmatrix} F_t^i\left(s_{t-1}^{n,1}, v_t^{n,M}\right) \\ \vdots \\ F_t^M\left(s_{t-1}^{n,M}, v_t^{n,M}\right) \end{pmatrix} \tag{9}$$

For the likelihood calculation of the nth particle, the observed value $\tilde{s}_t^n, n = 1,...,N$ can be expressed as:

$$p\left(Z_t = \left(z_t^1, ..., z_t^{m_t}\right) \middle| X_t = \tilde{s}_t^n\right) = \prod_{j=1}^{m_t} p\left(z_t^j \middle| \tilde{s}_t^n\right)$$

$$\propto \prod_{j=1}^{m_t}\left[\frac{q_t^0}{V} + \sum_{i=1}^{M} l_t^i\left(z_t^j; \tilde{s}_t^{n,i}\right)q_t^i\right]$$
(10)

In which, $l_t^i\left(z_t^j; \tilde{s}_t^{n,i}\right) = p\left(z_t^j \middle| K_t^j = i, X_t^i = \tilde{s}_t^{n,i}\right)$, $q_t^i, i = 1, ..., M^t$ means the probability of the j'th observed value from the i'th target, and M^t means the quantity of target at the time t.

3.3 Particle Degeneracy and Re-sampling

The basic problem to be solved in sequential importance sampling algorithm is particle degeneracy, after a few or multiple recursion, the weights of most particles become very small and only a few particles have a relatively large weights.

Re-sampling technique is used herein to solve particle degeneracy, namely removing the particles of small weight and reproducing those of large weights. Detailed process is as follows.

After systematic observation, the first step is to recalculate and confirm the weight ranges of the particles. The realistic particles will be granted relatively large weights and those deviating from reality will be given relatively small ones.

The second step is re-sampling process, in which the particles of large weights will derive much more "offspring" particles and those of small weights will correspondingly derive less ones, moreover, the weights of "offspring" particles will be re-set.

The third step is system state transition process, in which the state of each particle at the time t will be predicted through adding a random amount of particles.

The forth step is system observation process at time t, similar to the first step, the final representation of target state will be obtained through weighting of a numbers of particles.

3.4 Strategy for Sampling Improvement

Particle filter re-sampling inhibits the weight degeneracy, but also introduces other problems. At first, the particles are no longer independent, reducing the opportunity of parallel computing because of continuous re-definition of new particle set; second, the particles of relatively large weights will be chosen for many time, weakening the particle diversity, and the sample particles contain many duplicate points, when the system noise is small, the said will be obvious, and after several iteration, all particles will converge to a point, and this is known as particle depletion.

Particle depletion resulted from re-sampling process makes the number of particles expressing PDF state too small and therefore inadequate, while unlimited increase of particle number is not realistic.

MCMC method is introduced to generate samples from target distribution through constructing Markov chain, which has a good convergence effect. In each process of iteration of sequential important sampling, the particles can move to different places by combining with MCMC, so that particle depletion is avoided, and furthermore, Markov chain can push the particles to the places closer to PDF state and make the sample distribution more reasonable. There are many MCMC methods put into application, and MetropolisHasting method is adopted herein.

Specific re-sampling process is as follows:

According to the samples uniformly distributed in the range (0,1), thresholds $u \sim U(0,1)$ are obtained; Sampling $x_t^{*(i)}$ as per distribution probability $p(x_t \mid x_{t-1}^{(i)})$, i.e.

$x_t^{*(i)} \sim p(x_t \mid x_{t-1}^{(i)})$; Accept $x_t^{*(i)}$, if $u < \min[1, \frac{p(y_t \mid x_t^{*(i)})}{p(y_t \mid \tilde{x}_t^{(i)})}]$; otherwise, drop $x_t^{*(i)}$, make $x_t^{(i)} = \tilde{x}_t^{(i)}$.

4 Conclusion

The paper introduces the particle filter such practical estimation problem-solving method into the field of vision tracking, constructing the tracking framework based on particle filter and combining with characteristics of targets at all levels, to manufacture trackers of good performance that have "multimodal" tracking features and be able to improve robustness. In specific implementation, re-sampling in MCMC method will be applied to solve particle degeneracy and sample impoverishment in particle filtering visual multi-target tracking algorithm. MCMC method can effetely enhance the performance of particle filter algorithm and reduce the computational complexity.

References

1. Song, T.L., Lee, D.G., Ryu, J.: A probabilistic nearest neighbor filter algorithm for tracking in a clutter environment. Signal Processing 85(10), 2044–2053 (2005)
2. Oussalah, M., De Schutter, J.: Hybrid fuzzy probabilistic data association filter and joint probabilistic data association filter. Information Sciences 142(1-4), 195–226 (2002)
3. Herbland, A., Lasserre, R., Lemetayer, P., Clementy, J., Gosse, P.: Malignant hypertension (MHT): a systematic therapeutic approach through blockade of renin angiotensin system. American Journal of Hypertension 17(5), S155 (2004)
4. Nummiaro, K., Koller-Meier, E., Van Gool, L.: An adaptive color-based particle filier. Image and Vision Computing 21(1), 99–110 (2003)
5. Shao, X., Huang, B., Lee, J.M.: Constrained Bayesian state estimation – A comparative study and a new particle filter based approach. Journal of Process Control 20(2), 143–157 (2010)
6. Milstein, A., Sanchez, J.N., Williamson, E.T.: Robust global localization using clustered particle filtering. In: Proceedings of the National Conference on Artificial Intelligence, Edmonton, pp. 581–586 (2002)
7. Doucet, A., Gordon, N.J., Krishnamurthy, V.: Particle filters for state estimation of jump Markov linear systems. IEEE Transactions on Signal Processing 49, 613–624 (2001)
8. Collins, R.T.: Mean shift Blob Tracking through Scale Space. In: Proc.of the IEEE CS Conference on Computer Vision and Pattern Recognition, pp. 234–240 (2003)

Environmental Sensor Networks:
A Review of Critical Issues

Jinxin He[1,2], Jonathan Li[3], and Haowen Yan[3]

[1] College of Earth Sciences, Jilin University, Changchun 130061 P.R. China
[2] College of Geo-Exploration Science and Technology, Jilin University,
Changchun 130026 P.R. China
[3] Department of Geography and Environmental Management, University of Waterloo,
Waterloo N2L3G1 Canada

Abstract. "Environmental Sensor Networks (ESNs)" are a subset of sensor
networks which are specifically tuned to an environmental application. During
the last decade, for many applications the aim is to move towards Wireless
Sensor Networks (WSNs), as cables are often impractical, obtrusive and can
disturb the environment being monitored. Hart et al. reviewed over 50
representative examples of ESNs before 2006. However, in the recent years,
there have been some new trends in ESNs. Thus in this paper, the goal is to
present a comprehensive review of the recent literatures of ESNs since 2006.
Following a top-down approach, we give an overview of various aspects of
ESNs, and then review the literatures on several typical applications of ESNs.
Finally, we draw our conclusions and propose the future directions of ESNs.

Keywords: Environmental Sensor Networks, Wireless Sensor Networks,
Geosensor Networks, Sensor Web, Spatial Sensor Web, Geospatial Sensor Web.

1 Introduction

"Environmental Sensor Networks (ESNs)" are a subset of sensor networks which are
specifically tuned to an environmental application [Hart et al., 2006]. A sensor network
is composed of a large number of tiny sensor nodes and possibly a few powerful control
nodes (also called sink nodes), which are densely deployed either inside the
phenomenon or very close to it. ESNs can be wired together, and there are a number of
very significant projects (particularly those underwater) that use this method [Hart et
al., 2006]. During the last decade, the advances in Micro-Electro-Mechanical Systems
(MEMS) technology, wireless communications, and digital electronics have enabled
the development of low-cost, low-power, multifunctional sensor nodes that are small in
size and communicate untethered in short distances [Akyildiz et al., 2002]. Hence, for
many applications the aim is to move towards Wireless Sensor Networks (WSNs), as
cables are often impractical, obtrusive and can disturb the environment being
monitored [Hart et al., 2006]. At present, ESNs have received many attentions due to
their wide applications in water quality monitoring, air quality monitoring, and some
others related to environment and earth system sciences. In the recent years, there have

D. Jin and S. Lin (Eds.): Advances in MSEC Vol. 2, AISC 129, pp. 425–429.
springerlink.com   © Springer-Verlag Berlin Heidelberg 2011

been some new trends in ESNs. For instance, ESNs have been combined with GIS (Geographic Information Systems), GPS (Global Positioning Systems) and Remote Sensing technologies more frequently; Geosensor Networks, Spatial Sensor Web, Geospatial Sensor Web, and some other concepts related to Sensor Network and Sensor Web have been proposed. Thus in this review paper, the goal is to present a comprehensive review of the recent literatures since 2006. Following a top-down approach, we give an overview of various aspects of ESNs, and then review the literatures on some typical applications of ESNs.

2   What Are ESNs?

ESNs are typically arrays of devices containing sensors and interconnected using a radio network. ESNs vary in their scale and function, and Hart et al. (2006) divided ESNs into four categories: large scale single function networks; localized multifunction sensor networks; biosensor networks; heterogeneous sensor networks [Hart et al., 2006]. In general, typical ESNs comprise an array of sensor nodes and a communication system which allows their data to reach a server. [Hart et al., 2006].

2.1   Sensor Nodes

In ESNs, sensor nodes typically have a set of design goals including sensor integration, data quality, size, cost, robustness and power management [Hart et al., 2006]. A sensor node is usually composed of one or a few sensing components, which are able to sense conditions (e.g. temperature, humidity and pressure) from its immediate surroundings, and a processing and communication component, which is able to carry out simple computation on raw data and communicate with its neighbor nodes in short distances. The control nodes may further process the data collected from sensor nodes, disseminate control commands to sensor nodes, and connect the network to a traditional wired network sensor nodes are usually densely deployed on a large scale and communicate with each other through wireless links. The sensor nodes can store data, make decisions about what data to pass on (e.g. local area summary) and even make decisions about when and what to sense (when conditions are appropriate). The mobility of sensor nodes or base stations may be high and require location systems [Hart et al., 2006].

2.2   Communication System

Communication system is an essential part of an ESN. Most systems relay their data via radio links or wired networks. They use radio systems, satellite phones, mobile phones or short text messaging [Hart et al., 2006], much research has been done to develop low-power, robust, and secure communication protocols between sensor nodes. Since each node has a very limited reliable communication range (10–100 m), sending of messages in a sensor network is performed in a multi-hop way relaying the messages between sensor nodes until they arrive at their destination. Therefore, such as 802.15.4, ZigBee and so on, protocols need to focus on how to route messages from a node to a destination using the least amount of energy and a robust topology [Nittel, 2009].

2.3 Data Management

The traditional WSNs can't visualize and analysis the sensed data directly; hence, some researchers combined WSNs with GIS (Geographic Information Systems), GPS (Global Positioning Systems), and Remote Sensing, and proposed a few new concepts in recent years. For example, Nittel called GeoSensor Networks (GSNs) are specialized applications of WSNs technology in geographic space that detect, monitor, and track environmental phenomena and processes [Nittel, 2009]. In addition, although sensor networks have to date been deployed for a wide variety of applications, the communication links among these sensor networks have typically been lacking [Liang, 2007]. Thus, Liang proposed Spatial Sensor Web (SSW), which is a revolutionary concept for achieving collaborative, coherent, consistent and consolidated sensor data collection, fusion and distribution [Liang, 2007]. Similarly, Di defined the Geospatial Sensor Web (GSW) as the Sensor Web that performs Earth Observations (EO) and envisioned that major new EO sensors in the future will be web-ready [Di, 2007]. In the ESNs' infrastructure including the above mentioned concepts, RDBMS (Relational DataBase Management Systems), such as Oracle, SQL Server, MySQL, etc, and XML (eXtensible Markup Languages) are usually used to store all kinds of sensed data.

3 Applications

3.1 Water Quality Monitoring

As water quality perturbations related to escalating human population growth and industry pressures continue to increase in coastal and inland areas, effective water quality monitoring has become critical for water resource management programs [Glasgow, 2004]. Both remote monitoring and in situ monitoring are two most popular approaches to monitoring water quality. As to the former, Satellite remote sensing, aerial remote sensing and other remote sensing technologies have been applied to water quality monitoring for many years. However, during the last decade, ESNs have become one of the most useful approaches to monitoring water quality. For instance, Quinn et al. (2010) used a number of state-of-the-art sensor technologies that have been deployed to obtain water and salinity mass balances for a 60,000 ha tract of seasonally managed wetlands in the San Joaquin River Basin of California. These sensor technologies are being combined with more traditional environmental monitoring techniques to support Real-Time Salinity Management (RTSM) in the River Basin [Quinn et al., 2010]. Additionally, wireless communication, data management, etc. are also important components to Capella et al.'s water quality monitoring system [Capella et al., 2010].

3.2 Air Quality Monitoring

Air pollution monitoring is considered as a very complex task but nevertheless it is very important. Traditionally data loggers were used to collect data periodically and this was very time consuming and quite expensive. The use of ESNs can make air pollution monitoring less complex and more instantaneous readings can be obtained [Khedo et al., 2010]. Ma et al. (2008) presented a distributed infrastructure based on

WSNs and Grid Computing technology for air pollution monitoring and mining, which aims to develop low-cost and ubiquitous sensor networks to collect real-time, large scale and comprehensive environmental data from road traffic emissions for air pollution monitoring in urban environment [Ma et al., 2008]. In addition, Jung et al. (2008) used two systems to monitor air quality: sensor network control system and air pollution monitoring system. The control system supports the operators which control sensor network such as sampling interval change and network status check. The operators are useful for keeping the good status of data transmission. The air pollution monitoring system supports sensor data abstraction and air pollution prevention models for understanding the pollution level and area. The models are used for providing alarm message and safety guideline for people in pollution area [Jung et al., 2008]. Khedo et al. (2010) designed and implemented an air pollution monitoring system that used an Air Quality Index to categorise the various levels of air pollution. It also associates meaningful and very intuitive colors to the different categories, thus the state of air pollution can be communicated to the user very easily [Khedo et al., 2010].

4   Conclusions and Future Directions

ESNs can offer a powerful combination of distributed sensing capacity, real-time data visualization and analysis, and integration with adjacent networks and remote sensing data streams. These advances have become a reality as a combined result of the continuing miniaturization of electronics, the availability of large data storage and computational capacity, and the pervasive connectivity of the Internet. Nevertheless, at present, ESNs are still in the infant stage. On one hand, ESNs' development requires the development of information and communication technology, such as the power supplies of sensor nodes, the standards of wireless communication protocols, safety and security problems, and so on are most important factors. On the other hand, ESNs should be applied into more and more applications related to environment and earth system sciences in the future. In particular, the sensed data from ESNs or WSNs should be fused with video sensed data, audio sensed data and remote sensed data to a higher level.

Acknowledgements. This work was supported by the Interdisciplinary Foundation for Front Science of Jilin University (Grant No.: 200903028).

References

1. Akyildiz, I.F., Su, W., Sankarasubramaniam, Y., et al.: Wireless Sensor Networks: a survey. Computer Networks 38, 393–422 (2002)
2. Capella, J., Bonastre, A., Ors, R., et al.: A Wireless Sensor Network approach for distributed in-line chemical analysis of water. Talanta 80(5), 1789–1798 (2010)
3. Di, L.: Geospatial sensor web and self-adaptive Earth predictive systems (SEPS). In: Proceedings of the Earth Science Technology Office (ESTO)/Advanced Information System Technology (AIST) Sensor Web Principal Investigator (PI) Conference, pp. 1–4. NASA press, San Diego (2007)

4. Glasgow, H., Burkholder, J., Reed, R., et al.: Real-time remote monitoring of water quality: a review of current applications, and advancements in sensor, telemetry, and computing technologies. Journal of Experimental Marine Biology and Ecology 300, 409–448 (2004)
5. Hart, J.K., Martinez, K.: Environmental Sensor Networks: A revolution in the earth system science. Earth-Science Reviews 78, 177–191 (2006)
6. Khedo, K.K., Perseedoss, R., Mungur, A.: A wireless Sensor Network Air Pollution Monitoring System. International Journal of Wireless & Mobile Networks 2(2), 31–45 (2010)
7. Liang, S.: An interoperable and scalable gisservice architecture for the world-wide sensor web. Ph.D. thesis. York University. Toronto (2007)
8. Ma, Y., Richards, M., Ghaenem, M., et al.: Air Pollution Monitoring and Mining Based on Sensor Grid in London. Sensors 8, 3601–3623 (2008)
9. Nittel, S.: A survey of Geosensor Networks: Advances in Dynamic Environmental Monitoring. Sensors 9, 1–15 (2009)
10. Quinn, N., Ortega, R., Rahilly, P., et al.: Use of environmental sensors and sensor networks to develop water and salinity budgets for seasonal wetland real-time water quality management. Environmental Modelling & Software 25(9), 1045–1058 (2010)

Research on Service-Oriented Geospatial Information Sharing Mechanism and Technical Architecture

Zhigang Li[1], Tian Liang[1], and Wunian Yang[2]

[1] Chengdu University of Technology, College of Administrative Science,
Chengdu, China, 610059
[2] Chengdu University of Technology, College of Earth Sciences,
Chengdu, China, 610059
{cdlglzg,qinaidemama0909}@163.com

Abstract. SOA is providing conceptual design pattern for service-oriented distributed systems, and Web service is one of standards-based implementation of SOA technologies, deployed on the Web as an object or component. Survey the whole paper, get two points. First of all, it compares the spatial data oriented sharing with the spatial information service oriented sharing on technology and feature, and analysis of the service-oriented spatial information sharing mechanism and the implementation of ideas; Secondly, it frames a service-based geospatial information sharing platform, expounding the key technology and organizing way of spatial information based on Web Service. Offer solutions for GIS data integration and spatial information sharing that platform-crossed and sector-crossed.

Keywords: GIS, SOA, Geospatial Information, Sharing.

1 Introduction

The application of GIS has been deep into many aspects like government decision-making, social security, urban planning, land management, traffic management, environmental protection, disaster relief, intelligent district management and so on. However, these systems lack of integrating with each other on resource and information, the phenomenon of information isolated, then resulting in most of the systems take little exertion. After the development of GIS into the Web GIS phase, Geospatial spatial information service engender, it gone through three stages of geography information distributing, the geography information browsing and inquiring, geography information Web Service. Especially, The Web Service approach to service consumers with a proper, open, security and configuration according to need condition which is assembled used[10].

SOA (Service Oriented Architecture) is a kind of software engineering architecture that applies with mutative condition. Import the concept of SOA into geospatial information sharing, namely, converts data-sharing into service-sharing. It quests for an on-demand-service, business-agility, low-cost GIS service model, in the end, actualize multi-source and heterogeneous distributed sharing in geographic information.

D. Jin and S. Lin (Eds.): Advances in MSEC Vol. 2, AISC 129, pp. 431–436.
springerlink.com   © Springer-Verlag Berlin Heidelberg 2011

2   The Service-Oriented Spatial Information Sharing Mechanism and Realization Measure

According to different granularity between systems, the spatial information sharing can be divided into data-level sharing and application-level sharing. Data-level sharing refers to system shares the data, that low-granularity sharing, for extraction and delivery of data which is high-volume, pure database-level, and be independent of transaction control logic. Application-level sharing refers to provide high degree, transaction control logic correlated data-shift function and general-service sharing function to business information systems, with high-granularity sharing, such as geo-coding service, buffer analysis service, street map service etc[5]. Now, contrast spatial data oriented sharing with spatial information service oriented sharing on the major technical criterion [3].

Transferring interface: spatial data oriented sharing uses for the data interface, but spatial information service oriented sharing using standardization service protocol [9];

Updating way: the former makes use of periodic updates, the latter takes dynamic, continuous updates;

Sharing platform: spatial data oriented sharing in the way of data format conversion, spatial information service oriented sharing takes different granularity service;

Requirement of operator: the former needs for professional and experienced staff, the latter requires the professionals work with the public together;

Security: the one prefers data security protocol or data encryption, the other bases on contract to provide content, manner and quality of service;

Flexibility: that data search is passive, when this is capable of register and visibility;

Integration and interoperability: the former is difficult for interoperability, the latter carries CORBA, EJB and DCOM through normative protocol to get interoperated.

Based on the two types of information sharing above, the spatial information sharing of kernel mechanism should include the following [5]:

(1) Establish an open geospatial Information resource directory with the core of metadata, to integrate and manage distributed data, and achieve the shift and sharing on data-level [9].

(2) Web Service specifies the discovery, configuration, alternation, assemblage, movement and management of service, realizes the sharing between operation logic and commercial logic [10].

(3) To combine metadata with Web Service, build and implement a variety of application systems which based on data platform and service platform, lastly, achieve the application-level sharing. Ground on the above mechanism, it takes the following steps:

Firstly, around the network environment, according to metadata standardization, get the metadata information registered that is from each functional department and city-based GIS data, form a metadata service center, make different types of customers can find the related data so much as required data, where the platform provides the service interface.

Secondly, deploy Web Service groupware in each department to achieve standardized Transaction of communal heterogeneous data and packaging of operation logic and commercial logic, the specific Web Service development follows the OpenGIS framework of OGC Web Services specifications;

Thirdly, some publics query the metadata through metadata search system that Web Service-based on information portal website;

Finally, different jurisdiction users transfer different departments (nodes) Web Service processing data along with different metadata description, complete the download or resource interoperability in order to share information intervenient data-level and application-level.

3   Architecture of Service-Oriented Geospatial Information Sharing Platform

Adoption of SOA and Web Service technology, with construction principle of data distributed and independent, set up the database into a multi-source, heterogeneous distributed database, also a spatial information sharing platform based on B / S pattern. The structure of three layers including Web Server layer, GIS application server layer and Data layer ,which installed in different servers[1,3]. thereof, the data layer management of spatial data; spatial data services running on GIS application server layer, posted spatial data to GIS server videlicet; Web server transfer GIS application server, it is the client for the GIS application server. Web Server layers through the operation of these services to achieve functions by calling GIS application server, in which needs a browser [8] (Fig. 1).

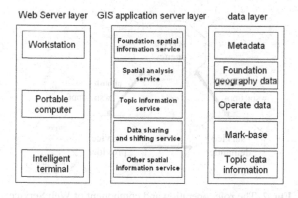

Fig. 1. SOA-based geospatial information sharing framework

Depend on the framework, each professional sector maintains respective spatial geographic information database. Spatial database register and release on the "Data Center" directory server, "GIS Application Server" supply "Data Service" sub-database coordination. SOA follows the concept of "software as a service, service that is", encapsulate a variety of geospatial information resource into granularity suitable, reusable, maintainable, interoperable, cross-platform, programming language independent software component by Web Service technology, follow the consecution of "Data-Services-Application" to organize. Achieve an open department data directory by metadata, compound and expand service, and utilize software resources scattered in the network in the form of Web Service, thereby, arrive at function sharing, create a loosely coupled, extensible, manageable, self-service environment ultimately. Here, we regard Web Service as an object or component on the Web. The main features of structure:

1) Cross-platform and multi-format integration: balance Web Service and traditional integration method then receive a conclusion, the former solve the problem of multi-format GIS data integration under multi-platform; actualize direct accessing and sharing between heterogeneous data.

2) External sharing information interface: structure use Web Service technology to integrate multiple heterogeneous databases, packing sorts of heterogeneous data that shared into Web Service and putting out it. Both it has external patulous function, affording sharing service towards other systems with Web Service interfaces [10].

The basic idea of SOA can be summed up into service; all the resources may be used to express the service in this framework. In this structure, there are three kinds of alternating roles: service provider, service agency (service registry center) and service demander (user). Interaction involve publish, inquire and bind operations. Service provider defined description of the Web Service, and publishes it to service or service agency; service demander searches description from a local intermediary or service agency with a find operation, then ties service providers and description to achieve interaction with invoking Web Service. It is logical structure for service provider and service demander; as a result it can be expressed in two features [3, 6] (Fig. 2).

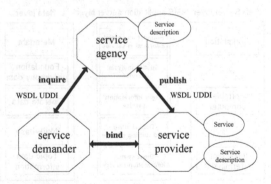

Fig. 2. The role, operation and component of Web Service

Three basic operations consist of publish, inquire and bind act reciprocity. Service provider publishes service to service agency [6]. And service demander inquires service in needs towards service agency, with that bind them.

Service provider: bring services specification software entity;

Service demander: software entity that transfer the service provider. Traditionally, it referred to as "client." Service demander can be End-User applications or other services;

Service agency: it can send request of service to one or more service providers;

Visibility, the structure of Web services is a natural medium act on message exchange between users and providers. Web Service is a sort of idiographic manner to implement SOA, also a more canonical conception, along with introduce into XML, for bringing better interoperability between users and providers.

4   The Key Technologies to Service-Oriented Spatial Information Sharing Platform

. NET, J2EE of next-generation Internet technologies grows, it supplies GIS a new exploitive and running environment. XML-based GIS data model in line with the current direction of geospatial information sharing, adapt to Web-based data search and data mining, then packaged the convention into standardized Web service, so as to be called by different GIS. And XML WebGIS betake of XML technology, actualize SOA-based geospatial data sharing and issuance [1].

SOA provides a conceptual design model for the service-based distributed system. Web service is a new Web application, in a position to publish, search and calling by Web, with a self-contained, self-described and modular features. Web Service can react to a simple customer request, also can complete a complex business flow, it represents implementation of SOA, which is most popular.

Spatial information system is footstone of SOA Web service, at the present time; it has relatively mature methods and techniques, including:

1) Web service is a standard-based technology of performing SOA, which defines how application interoperability on the Web. Client can call the Web Service by WSDL document in any language or on any platform. Client generates a SOAP request message in the way of WSDL describing document [2]. The SOAP request is send to Web server, and then Web server transmit these requests to Web Service processor. Processor parses incoming SOAP request, calling Web Service, and return an SOAP response. After Web server get the SOAP response, send back to client by HTTP.

2) Transfer messages based on XML and SOAP, using XML and SOAP make spatial information integration and interoperability becomes relatively simple. XML provides a cross-platform data encoding and organization, where SOAP is established above, and defined simple cross-platform exchange method. SOAP protocol bound on HTTP, can carry on the long-distance procedure call by multi-language and cross operating system, caused the language and the platform independency [4].

3) Make use of UDDI as service information of Unification description, which is a kind of standardized criterion of web-based, distributional and providing information registration center for Web. It contains a group of visiting protocol executable standardization.

4) Because Web Service take simple and understandable Web protocol as module surface to describe and cooperate with criterion, It has shielded the difference of different software platform completely, regardless of CORBA, DCOM or EJB may carry mutual manipulation by standardized protocol, carry out the environment integrative. Moreover, Web Service works on the code level, once after Web service disposes, other application program might discover and transfer the serve directly, and might exchange data with other software. Ultimately, it forms a interactive application system.

5   Conclusions

Conclusion of this paper: first of all, Web Service is a new notion carries out SOA of spatial geographic information services, possess of unified interface, supports standardized data format of XML, sustains integration manners of data-driven and API-driven simultaneity; Secondly, Web Service actualizes data and function interoperability between systems by building XML-based normative correspondence protocol, that can solves the integration and sharing of GIS data on cross-platform as well as cross-departmental effectively. Thirdly, SOA-based geospatial information sharing wipes off the problem of anti-synchronized, reconstruction and inefficiency in traditional geospatial data sharing, indeed, it achieves multi-source, heterogeneous distributed sharing of geospatial information. Fourthly, SOA-based geospatial information sharing holds the principles of data distribution and independence, not only has concentrated data center, but also allows data to distribute on any node in the network.

References

[1] Yi, M., He, Z.: Research of SOA-based geospatial information sharing. Mapping and Geospatial Information (12), 102–104 (2009)
[2] Huang, Y., Cheng, J.: Key technologies of geography information sharing service in Shenzhen. Geography Information World (8), 24–28 (2009)
[3] Li, D., Huang, J., Shao, Z.: Design and implement of Service-Oriented digital city Sharing platform frame. Journal of Wuhan University (Computer Science Version) (9), 881–885 (2008)
[4] Chen, Y., Cui, T.: Structure of SOA-based spatial information service. Geography Information World (12), 49–52 (2008)
[5] Ding, H., Chen, J., Yu, J.: Study in SOA-based digital city information sharing method. Computer Project and Design (20), 4632–4635 (2009)
[6] Lu, Z., Zhao, Y., Chen, R.: Discussion of SOA-based spatial information resources conformity and service pattern. Computer and the Digital Project (9), 125–127 (2009)
[7] Brown, P.C.: Implementing SOA: Total Architecture in Practice. Pearson Education,Inc., Publishing as Addison-Wesley, U.S (2008)
[8] Peng, M., Fan, W.: Service-oriented government geographic information sharing platform's design and implementation. Geospatial Information (6), 59–61 (2009)
[9] Feng, M., Liu, S.-G., Euliss Jr, N.H., Yin, F.: Distributed Geospatial Model Sharing Based on Open Interoperability Standards. Journal of Remote Sensing 6, 1067–1071 (2009)
[10] Deng, H., Wu, F., Wang, Z.: Research on spatial information service that Web service-based. Computer Engineering and Design (23), 4450–4452 (2006)

The Research and Practice on Computer Graphics Extended Education

Zhefu Yu and Huibiao Lu

Transportation equipment and Logistics Engineering College,
Dalian Maritime University, Dalian, 116026, China
yuzf629@sina.com

Abstract. For the problems existing in combination teaching of engineering drawing and computer graphics. A extension teaching method was proposed according to the features of engineering specialty. This teaching method takes the fact into account that the engineering specialty have many drawing operations after engineering drawing course. The teaching content of computer graphics can be extended and be optimized. For students, the learning of computer graphics will be throughout the university studying. Computer graphics and professional practice will be integrated more closely.

Keywords: Engineering drawing, Computer graphics, Extended teaching.

1 Introduction

Since the 1980s, China began to promote the use of CG technology. To date, CG technology has been applied in many industries universally. Mastering CG is the community's needs. Suzuki, the Tokyo University professor said: "CG is the embodiment of modern graphics, CAD is not only a drawing tool and a means for creativity and thinking"[1]. For engineering students in college, mastering CAD technology and having product development expertise with CAD are essential. After learning CAD technology, engineering students can learn other professional software by analogy. For example: students can learn finite element analysis software more quickly.

2 Teaching in Computer Graphics

In the universities, CG has been included in engineering drawing course for a long time. most schools relying on a popular applied software in the teaching CG process, such as: autoCAD. Engineering drawing course has been developed into a combination of "Engineering Drawing" and "computer graphics". The teaching content and teaching methods has undergone a qualitative change. The combined model bridges the gap between traditional engineering drawings and modern CAD technology [2]. "Computer graphics" teaching in various colleges and universities, "Engineering Drawing" course what percentage of the share. For this, the author

D. Jin and S. Lin (Eds.): Advances in MSEC Vol. 2, AISC 129, pp. 437–440.
springerlink.com   © Springer-Verlag Berlin Heidelberg 2011

carried out some research. Can be seen from Table 1, different types of colleges and universities "computer graphics" is different from the allocation of teaching hours, even if the same types of colleges and universities "computer graphics" is different from the allocation of teaching hours. For the small hours of the universities, teachers have the responsibility through innovative teaching methods so that students have more limited hours of content.

Table 1. The current situation of computer graphics teaching hours in the whole arrangement of engineering drawing course of various colleges and universities

University	Specialty	Total hours	Hours of computer graphics
Dalian University of Technology	Civil Engineering	84 hours	28 hours
Shandong University of Technology	Mechanical Engineering	112 hours	10 hours, Additional 16 exercise hours
Guangdong University of Technology	Civil Engineering	72 hours	15 hours
Lanzhou Institute of Technology	Civil Engineering	78 hours	Additional 24 hours
Yangzhou University, School of Mechanical Engineering	Mechanical Engineering	90 hours	Independent courses 30 hours
Chung-Kai Institute of Agricultural Technology	Mechanical Engineering	80 hours	Additional 12 hours 20 exercise hours
School of Mechanical Engineering, Xinjiang University	Mechanical Engineering	112 hours	Additional 16 hours
North China Institute of Water Conservancy and Hydropower	Civil Engineering	126 hours	23 hours, 22 exercise hours
Zhejiang Textile Garment Institute	Mechanical Engineering	120 hours	30 hours

3   Computer Graphics Teaching in DMU

Dalian Maritime University, Department of Mechanical basis is mainly responsible for graphics teaching job of engine management specialty in past. In recent years, the University added some professions, such as: mechanical engineering, civil engineering. These professions' students need grasp CAD to higher extent than those of engine management specialty. To the newly created civil engineering, for example, the computer graphics part occupies 10 hours in of the total 80 hours of engineering drawing. the main teaching content is two-dimensional graphics program, draw, modify, skills, writing text, dimensioning, creating block and other basic elements. computer graphics content also has been infiltrated in the other teaching hours. After several rounds of teaching practice, we have achieved good teaching results. But from the feedback situation, we found the following problems:

1) the teaching hours are too concentrated and teaching cycles is too short, to reinforce the knowledge. Teaching results are difficult to achieve desired level.

2) limited teaching hours leads to limited content. The content includes 2-D graphics rendering , editing, layers, dimensioning ,drawing skills, the block and attributes. The depth is not enough.

3) teaching resources are limited, such as our computer drawing room just meets the teaching hours need, but cannot cover after-school job.

4   Implementation of the Extended Teaching

Extension education means extension of time and content in teaching CAD, which includes three aspects.

4.1   Extending to After-School

We provided a series of detailed operating instructions for students to complete the CAD homework independently and to stimulate students' interest in learning CAD.

4.2   Extending to the Follow-Up Courses

Taking into account the fact that there are many course design in the follow-up curriculum in engineering disciplines, such as: for civil engineering, there are surveying practice, concrete structures design, steel structures design, basis design and construction courses following up engineering drawing course. They are all involved in CG drawing work. After investigating follow-up curriculum of every engineering disciplines. we cooperated with the teacher of profession. CG teaching was coordinated with the various professional courses. This cooperation ensure that students must complete a certain number of CG work in four years. It resulted in a improvement of students' CG skills. CG teaching Infiltrate into the various professional courses, so that "computer graphics" and a professional practice link up more tightly. The extension of time makes the extension of CG teaching content. We add some new content, such as: 3D modeling and secondary development, on the provisions of the syllabus. We implement this part through the online teaching page built by our department, multimedia courseware were released on the web to help students overcome difficulties encountered in these course. It's also a CAD forum. We published material and resources about CAD, such as: the autolisp programming, which can automatically add steel head hook at the ends of a line. This can improve students' interest in learning and self-learning ability.

4.3   Extending to the Extra-Curricular

We encourage students to participate in technology innovation activities, such as: national CAD/CAM technology competition. the innovation activities will improve their overall ability to use knowledge.

5   The Purpose of Extended Teaching

Implementation of the teaching extension is based on engineering characteristics and focus on practical results. It enable students to master the basic theory and basic skills of computer graphics, and to have the ability developing graphics software, students can really put it into practice and can lay a solid foundation for the future.

References

1. Giesecke, F.E.: Engineering Graphics, 8th edn., pp. 1–2. Higher Education Press, Beijing (2003)
2. Zhou, X., Luo, K., Wei, Y.: The Teaching Practice of Computer Graphics. Journal of Guangdong University of Technology (Social Science Edition) 8, 87–89 (2008)

Study of the Border Port Logistics Equilibrium Based-On DIS/HLA under Comprehensive Transportation System

Dou Zhi-wu and Li Yan-feng

School of Business, Yunnan University of Finance and Economics,
650221 Kunming, Yunnan
zhiwudou@163.com

Abstract. In order to solve the port logistics demand inadequacy and supply enough coexist, logistics resource shortage and superfluous with logistics service problems, taking the Yunnan border port as an example, researching the modeling, analysis, simulation, optimization and equilibrium methods of the border port logistics system under the comprehensive transportation system. The distributed interactive simulation (DIS) framework based-on high level architecture (HLA) was suggested at first, and the system model of the border port logistics system was established using system model language (SysML), then, the system model was transform to simulation federation using HLA federate development tool. Simulation experimental results show that the method is feasible and effective, and important significance to reveal operational law of the border port logistics system.

Keywords: Border port, Logistics equilibrium, High level architecture, Distributed interactive simulation.

1 Introduce

In recent years, with the rapid development of China's economy and global economic integration, the demand to logistics continues to grow. According to statistics, in 2010, the national import and export value was 297276 billion dollars, increasing 34.7%, the exports value was 157793 billion dollars, increasing 31.3%, and import 139483 billion dollars, growing 38.7%. Predicting to 2015, China import and export value will exceed 400000 billion dollars. Only the Yunnan Province port import and export value was 468 billion dollars, the volume of import and export freight was 7.89 million tons, the rate is 39.3% compared to the same period, country port 24. At the same time, the phenomenon of Customs delaying, load and unload delaying is serious with port logistics every day. Investigating its reason, on one hand, port hinterland economic development caused the imbalance of the regional logistics demand and structure; on the other hand, in order to capture the port hinterland blind and excessive investment out of practice lead to market disequilibrium situation. Border port logistics imbalance becomes the stone in the way of the national economic development so the paper aims to suggest a method to reveal the operational law of border port logistics system.

D. Jin and S. Lin (Eds.): Advances in MSEC Vol. 2, AISC 129, pp. 441–446.
springerlink.com   © Springer-Verlag Berlin Heidelberg 2011

2   Border Port Logistics Research

In currently, logistics transportation system (highway, railway, port, aviation) modeling study is comparing mature. The popular methods are operations research models [1, 2] and system simulation model based on queuing theory and inventory theory [3]. In view of the complexity and uncertainty, multiple Agent model, Petri model and discrete event simulation model [4] are more popular.

Research on equilibrium is focused on the relationship between upstream and downstream supply chain and various transportation modes selection [5], and more strategy discussed, and the research on cohesion balanced is rare in both domestic and foreign. Port linking is interested field in recent years [6], distributed Agent simulation model begins to use in supply chain and logistics coordination problem [7,8], While has not been studied at the cohesion of border port.

The High Level Architecture (HLA) [9] was developed by the U.S. Defense Modeling and Simulation Office (DMSO) to provide simulation interoperability and reusability across all types of simulations and was adopted as an IEEE standard. The Runtime Infrastructure (RTI) is an implementation of the HLA standard that provides the platform for distributed simulations [6].

3   Border Port Logistics Equilibrium System and Its Distributed Interactive Simulation System

3.1   Border Port Logistics Equilibrium System

Fig.1 shows a simple structure of the border port logistics system. In the system there are many kinds of traffics, work joins, logistics stations and port operation areas. The system includes 3 subsystem: Transport network and its Connecting portion subsystem; Transport production and operation subsystem; Transport organization and Coordination; According to the distributed and complexity characteristics, the conventional methods hard to reach the goal, the paper studies using SysML to build

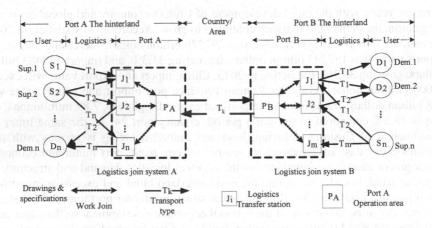

Fig. 1. The simple structure of the border port logistics system

the system model each entity, using HLA method to build the overall simulation framework, and using HLA simulation federate development tool to integrate the system model into federation object model. The simulation overall framework shows as Fig.2.

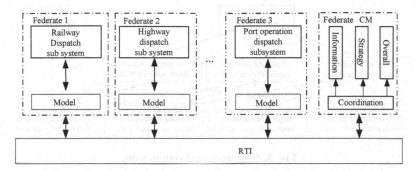

Fig. 2. The simulation overall framework

3.2 Simulation Modules Function

The system mainly completes 4 modules function:

(1) RTI service
In addition to execute 6 management functions (federation management, declaration management, object management, data distributed management, time management, ownership management), RTI service module need to manage computing resources from different organizations and the availability of these computing resources that may change during the execution of the simulation.

(2) Network information
Network information takes charge of the simulation resource monitoring, simulation estate evaluation, load balancing, federation transferring and code transferring. From the beginning of the first federation establishing, the network information module begins to monitor the whole simulation processes and the estate of every federate. The information is the base of the simulation decision.

(3) Federation module
The function of the module is to establish federation, registry federate instance, start-up federate process and delete federate exited federation.

(4) Network basis establishment
The network basis establishment is the physics basement of simulation running.

4 Border Port Steel Stockyard System SysML Models and Experiment Result Analysis

4.1 SysML Models

In this paper taking a border port steel stockyard as an example, using SysML to build system demand model, system structure model and system behavior model, and then use the development tool to convert it to a federate object model.

System requirement model: Used to describe the function and the performance conditions system must realize and reach. Steel stockyard system demand diagram as shown in fig. 3.

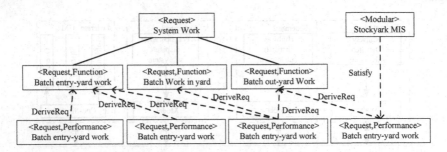

Fig. 3. Steel stockyard system work

System structure model: It describes and analyzes the system structure using the module diagram, the internal module diagram, and system parameters. Here only to give the system module diagram, as shown in fig. 4.

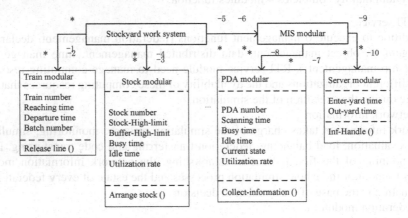

Fig. 4. The system module diagram

System behavior model: The system behavior model mainly refers to the system activity diagram, sequence diagram and operation diagram. By the limited of the Length, does not give.

Through the above system modeling the simulation federation object classes and attributes can clearly be obtained, interaction classes and interaction parameters can be defined, thus the Federation Object Model (Federation Object Model, FOM) can be developed by FOM construction tools.

4.2   Experiment Results Analysis

Tab.1 shows the experiment data.

Table 1. The experiment data

index	value	index	value
The number of stock	1000	loading and unloading line number	12
Stock height (m)	1.8	Load speed of factory bridge (m/min)	35.5
buffer height (m)	0.3	factory bridge no-load speed (m/min)	52
factory bridge	8	The factory bridge capacity (piece/day)	600
PDA	16	vehicle load speed (m/min)	45.4
steel transport vehicles	8	vehicle no-load speed (m/min)	63
worker	50		

The admission operations is scheduled at begin of every day, then the other operation. The aim of the simulation is to analyze the utilization rate of the operation resource and the active time matching degree.

The rules of parameters: Train arrives one time every morning, the batch number of loading plate obey the NORM (32, 8) normal distribution; each batch of steel plate number obey the NORM (40, 5) normal distribution; the thickness of the steel plate obey UNIF (10, 40) uniform distribution, unit mm; each batch steel plate mean store time obey UNIF (27, 33) uniform distribution, unit day; PAD scanning a sheet need time to UNIF (2, 4) uniform distribution, seconds; PAD playing steel in-and-out yard information time respectively to UNIF (12, 18) and UNIF (4, 8) uniform distribution, seconds. Distributed simulation experiment was performed on 365 days, every day at 8 hours. The simulation results are shown in tab. 2.

Table 2. The simulation experiment results

index	results	Index	results
Train mean staying time	3.04	Batch mean out-yard time (min/batch)	15.24
Line mean utilization rate %	57.24	Out-yard Mean scanning time (s/batch)	55.11
Workers mean utilization rate %	52.07	Into-yard Mean processing time(s/batch)	54.74
The highest utilization rate %	98.50	Batch mean into-yard time (min/batch)	40.05
Stock mean utilization rate %	63.23	Into-yard Mean scanning time(s/batch)	32.02
Vehicle mean utilization rate %	32.12	Out-yard Mean processing time(s/batch)	8.21
Bridge mean utilization rate %	60.54		

Tab.1 shows that: The mean time of train staying in yard is shorter conducive to the yard supply; the highest utilization rate of stock shows yard can ensure the operation, the low average utilization rate reflects yard operation normal. While the other utilization rate slightly lower, indicating to other resources not sufficient, there are areas for improvement. The admission time and information processing time is less which has small effect on storage yard operation.

446   Z.-w. Dou and Y.-f. Li

5   Conclusion

Through the construction of distributed interactive simulation of border port logistics system under comprehensive traffic system, border port logistics system operation rules can macroscopically be understanding, easy to port logistics equilibrium management; and the resource coordination and influence relationship can microcosmically be understanding, benefit to seek for the logistics balance method; thereby realizing the equilibrium operation of the border port logistics system, and optimizing the utilization of resources.

Acknowledgement. Natural Science Foundation of Yunnan Province (2009ZC087M).

References

[1] Li, J., Zhu, Y.-L., Shen, H.: Three stage logistics network Location-Routing Problem Modeling and Algorithm Research. Control and Decision 25(8), 1195–1206 (2010)
[2] Ma, C.-W.: Multimodal segments divided transport carrier selection Based on Agent. Journal of Harbin Institute of Technology 39(12), 1989–1992 (2007)
[3] Jin, C., Zhong, Q.-Y., Wu, L.-W.: Distributed Simulation Architecture for Coordination of Production Scheduling in Supply Chain. In: International Conference on Management Science and Engineering (ICMSE 2008), Harbin, China, August 8–10 (2008)
[4] Chun, J., Lu, Z., Peng, G.: Multimodal transport system on container terminals for coordination of resource allocation optimization. Journal of System Simulation 21(3), 900–908 (2009)
[5] Defense Modeling and Simulation Office (DMSO), The High Level Architecture Homepage (1997), http://www.dmso.mil/projects/hla/
[6] Dou, Z.-W., Li, H.-W.: Research on self-adaptived dynamic grid-based method in data distributed management. Systems Engineering-Theory & Practice 28(12), 128–132 (2008)
[7] Dou, Z.-W., Deng, G.-S., Mao, H.-J.: Dynamic grid-based method research in data distributed management. Systems Engineering-Theory & Practice 27(1), 137–142 (2007)
[8] Dou, Z.: The Application of DIS/HLA in the Petrochemical Industry Product System. In: 2010 the 3rd International Conference on Computation Intelligence and Industrial Application, vol. 12, pp. 24–27 (2010)
[9] Zhou, Y., Dai, J.-W.: HLA simulation programming design, pp. 4–13. Publishing house of electronics industry, Beijing (2002)

The Construction of the Program Design Online Community

XueFeng Jiang and JunRui Liu

College of Computer Science, Northwestern Polytechnical University, Xi'an, China

Abstract. Be aimed at the bad teaching effect of program design course caused by the shortage of teaching resources, the program design online community has been build. The community includes the online classroom, theoretical examination, experimental analysis, basic exercises, skills upgrading, discussing and exchanging, and so on. Through the online teaching resources, theoretical follow-up assessment, tasks management and automatic review system, this community can help the students to inquiry learning deficiency and remedy, guarantee the students' practice intensity and quality, and cultivate the students' innovative thinking and develop skills. The practice results in my school show that the course community is useful for the students' learning and the teachers' teaching, and it is an auxiliary teaching platform with promotional value.

Keywords: course community, online classroom, classroom test, comprehensive exam, skills upgrading, BBS.

1 Introduction

Program design is a public computer basic course after the computer foundation course, and it is the most professional required courses in China's colleges and universities. However, in many schools, the teaching effect is not very satisfactory, the students can't accurately grasp the program design language, and they have poor language ability. Investigate the reason, it is because of the relative shortage of teaching resources, the students can not be timely guided in the learning procedure, and their practice intensity and their practice quality can not be guaranteed, then they can't digest and consolidate the content of the classroom.

In view of this situation, this paper proposed the establishment of the online community of the program design course, and the author hopes that the online community can help the students to learn and urge the students to learn, so that the learning effect can be improved.

2 The Architecture of the Online Community

After the analysis of program design course characteristics and the present situation of the teaching activities, the author thinks the course community should at least have the following functions:

D. Jin and S. Lin (Eds.): Advances in MSEC Vol. 2, AISC 129, pp. 447–452.
springerlink.com   © Springer-Verlag Berlin Heidelberg 2011

(1) The leak detection of theory classroom function. The class community should provide the students to find out the knowledge with bad learning effect in the classroom and learn again, so that the teaching effect can be ensured.

(2) Urging and guiding the students to practice. Because the program design course is a very practical course, the practice intensity and quality had decided the course study effect in the very great degree.

Therefore, the course community should be able to help the teachers to urge students to practice and guide the students when they encountered difficulties in the practical process.

(3) Cultivating the students' innovative thinking and enhancing their ability. Although the program design is a language class, the programming language is just a tool, mastering the basic language knowledge and using the tool to solve complex problems are the ultimate training target. Therefore, the course community should have to cultivate the students' thinking, stimulate their innovative ability.

According to the function demand, the author has researched and designed the course online community. The architecture of the course community is shown in figure 1.

Seen from Figure 1, the course community includes mainly the following five major functional modules:

(1) Theory studying. The module is mainly used to check theoretical classroom leakage, fill a vacancy, and consolidate the effect of theoretical study. The module includes three function modules: the online classroom, classroom test and Comprehensive exam. Online classroom provides the students with the teacher's classroom video, PPT, electronic materials and common questions etc. The students can access such resources according to their own theoretical study condition. Classroom test provides the timely test function for each knowledge unit, and this module requires the students full timely the test and submit timely the test result after the teaching of each knowledge unit is filled, so that the teachers can master each unit of knowledge learning situation of each student, and adjust the content arrangement based on these case. Comprehensive exam helps the students to review and consolidate the course knowledge after the class is over. This module provides multiple sets of simulation examination questions and can review automatically papers, and limit the exam time, so that the students can carry on the same test training as the final test.

(2) Practice training. The practice training module is set for the programming curriculum practice. As a result of most domestic colleges and universities teaching resources are not very rich, the curriculum practice intensity and quality cannot be guaranteed.。 The module can make up the deficiency through the software function to a certain extent. The actual analysis includes the experiment analytical video and the experiment analytical webpage, and the students can view relevant information to get help when they encountered difficulties in the experiment. Basic training is oriented to the basic application of the language, this module requires the students to fulfill the corresponding experiment task in a certain time and submit the results to receive the machine automatically review, so that the teachers can understand all students' experiment situation while the student can know his own experiment situation. The skills upgrading module hope that the students can innovate based on the skilled use of

language, it provides a lot of curriculum design topic and the task tips for the students to choose, the students can choose the appropriate subject according to their actual situation and present the course design work documentation and code for teachers to check.

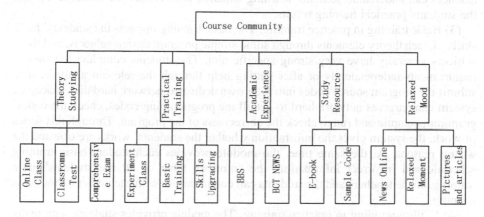

Fig. 1. The architecture of the course community

(3) Academic experience. This module provides the platform which the teachers and the students can discuss technical problems.

(4) Study resources. This module has collected much helpful resources for the students' learning, such as electronic books, other related teaching video, development paradigm and software development etc.

(5) Relaxed mood. The relaxing module provides the users the computer industry related news, jokes, humor video, the essay, and lets the students relax in the learning space, and open up their own eyes.

3   The Main Function Module of the Online Community

The following describes the main function modules of the course community in detail.

(1) Classroom test in theory learning. The classroom test module provides the test to the corresponding theoretical knowledge unit, the questions forms include the choice questions, the filling in the blanks and the judging. When the students complete the test within the prescribed time and submit the results, the system will write the students' answers into the text files and save in the network hard disk space. So that the teachers can get the students' answers files and the analysis document automatically generated by the system, and they can understand and master the learning condition of the students.

(2) Comprehensive exam in theory study. The comprehensive exam hopes that the students can adapt to the actual exam and became familiar with the exam environment, so the comprehensive exam requires the students to finish the test in time and submit. In the comprehensive exam, all theoretical knowledge are included, the questions forms include the choice questions, the filling in the blanks, the judging, the program

correction, the program reading and the program completing. Compared to the classroom test, the comprehensive exam is more difficulty and more comprehensive, and it mainly examines the knowledge comprehensive application. The students can perform the comprehensive exam repeatedly according to their own situation, and the teachers can understand students' learning situation and remedy the shortage through the students' practical training results.

(3) Basic training in practice training. The basic training inspects the students' basic skills of each theory elements through some simple program design subjects, and these subjects generally have very strong specific aim. The students complete the subject requirements independently or after getting help through the relevant analysis, then submit the program source codes into their own dedicated network hard disk space, the system can retrieves network hard to get all the programming codes, check the codes' grammar, compile and run to check the correctness of the program. Through this series of work, the system gives the information whether the students' works are true and the wrong reasons. At the same time, the module provides each student basic training schedule and incorrect information, the summary table of all the students' basic training, so the teachers and the students can communicate with each other on the basis training.

(4) Skills upgrading in practice training. The module provides students with many curriculum design topics with different difficulty coefficient, and presents these questions targeted tips. The students choose their own interest and competent subject to develop according to their own situation. Development process must follow the software engineering requirements. After development is completed, the students need to submit various software documentations of the curriculum design works, the curriculum design source code, the executable program and the screenshots of the running process according to the requirements of the system. After the teachers login network disk, they can access the relevant information of the students' curriculum design works, and guide and score. Also, the teachers can control all students of the course according to the document analysis and the summary tables generated by the system.

4   The Results

The program design course community has set up and use three years, the effect is shown as following.

Firstly, from the perspective of students, the influence of the course community includes:

(1) The students grasp the theory knowledge more firmly. Course community includes online courses, classroom test and comprehensive training, so that students can make up their own study insufficient in theory whenever and wherever possible, and check their theory learning situations, then consolidate the theory learning effect.

(2) The students' programming ability is strengthened, and their programming ideas ingrained. With the help of practice training in basic training of the course community, each student in my school completed 100 experimental subjects in a semester, and the lines of their programming code increases from the past hundred lines and even a few lines to thousands of lines, in a course semester all students completed 300,000

experiment topics, and each topic has be given a correct or error analysis information, which are unable to imagine in traditional experimental models. In addition, since all of the test questions require the students programming in the actual programming environment for the preparation and validating, then can be submitted to score, so the students are no longer programming outline on the paper, the students' basic skills and programming skills are enhanced.

(3) The students dare to challenge the difficult and large software projects. After the basic practice with highly difficult in basic training, the students' language ability has been greatly improved. The skills upgrading in practice training makes students dare to innovate based on the language basic skills.

(4) The learning interest of the students is stronger. The course community gives the students multi-mode and multi-angle study platform, the students can get help and guidance according to their own actual condition in any time, then they can adjust their own learning, all these increased the students' learning interest. The click rate of my school program design course community reached tens of thousands of people daily, and this is also a phenomenon of strong learning interest.

Secondly, describes from the perspective of teachers:

(1) The teachers' workload is reduced. The course community makes the teachers need no longer to fulfill the document management work and the practice reviews work, and the course community shares the teachers' guidance in largely.

(2) The students' situation can be master more clearly. The course community makes the teachers can understand each student learning case, and can control the course global situation, which is not able to do in the past because the teachers' energy is limited.

5   Conclusion

The course community has rich functions. It provides the students with a variety of learning method, ensures the practice intensity and the practice quality, and supervises and guides the students' learning, so it can consolidates the students' learning effects and improves the students' skills and innovative ability. At the same time, the course community reduces the teachers' burden. Therefore, the course community is a worthy popularizing teaching auxiliary platform.

References

1. Shunli, Z.: The reform and practice of the program design course teaching. Chinese Adult Education 5, 137–138 (2010)
2. Luo, X.: The research of the students' occupation ability cultivation in program design. Chinese Adult Education 9, 163–164 (2010)
3. Lu, E.: The design and application of C # on-line training and learning system. Laboratory Research and Exploration 8, 81–83 (2009)
4. Han, J.: The design and implementation of the programming online evaluating auxiliary teaching system. Journal of Inner Mongolia Normal University (Natural Science Edition) 9, 473–476 (2010)

5. Chen, G.: The C + + program design teaching mode and method. Computer education 11, 135–137 (2011)
6. Xu, Y.: The design of the computer examination and automatic marking system. Open Education Research 6, 80–83 (2005)

The Reform of the Many-Branched Setting in Software Development Technology Course

JunRui Liu and XueFeng Jiang

College of Computer Science, Northwestern Polytechnical University, Xi'an, China

Abstract. Be aimed at the situation that the software development technology curriculum has too many contents, is unable to meet the needs of various professional and personalized requirements, and has poor development situation, the many-branched setting in software development technology course is put forward. After the analysis of the professional requirements, the strategy of the many-branched setting suggests that the software development technology course could be divided into four branched courses, and each course focuses on a technology field of software development. After the software development technology course is set into four branches, each branched course establishes one's own system, its content is more pertinence, and can always grasp the related technology development trends, so the strategy of the many-branched setting can enhance the teaching effect of the course.

Keywords: the many-branched setting, Windows advanced programming, web development, embedded development technology, numerical calculation and algorithm design.

1 Introduction

Software development technology course is also known as software technology or software foundation, it is a computer public course of science and engineering professional afterwards computer foundation and program design course. Based on program design course, software development technology focuses on cultivating students' idea of software development and software development technology, and its teaching range is very large in many science and Engineering School. In recent years, along with the computer application and the relevant professional development, the new computer technology continue to emerge, the professional requirements of the computer application also emerge in an endless stream, so the students' computer application training target is also advancing with the times.

However, the development pace of software development technology curriculum is relatively slow, and this situation makes the course unable to meet the computer application training requirements of the most students, so the students which study the course have become less. In view of this situation, this paper puts forward the many-branched settings of software development technology course, and uses multiple branching forms to meet computer applications requirements in many areas, while fully absorbs the latest computer technology. So that software development technology course can be promoted into a new stage of development.

D. Jin and S. Lin (Eds.): Advances in MSEC Vol. 2, AISC 129, pp. 453–456.
springerlink.com

454   J. Liu and X. Jiang

2 The Many-Branched Setting Strategy in Software Development Technology Course

At present, software development technology course takes the software development as the main clues, its content includes software engineering, database technology, VB or VC visual technology, network programming, and multimedia technology, etc. In some schools, OS and data structure are also placed in this course. The students of this course study all course contents, or according to their own professional demand, the course contents are adjusted slightly. Because the course has too many contents to introduce each part in detail, to take each professional personalized teaching, and to accommodate more new technology, this paper proposes the multi-branched setting in this course to solve these problems.

The many-branched strategy setting of software development technology course thinks that the course is divided into the following four directions:

(1) Windows advanced programming. This branched curriculum is set for general engineering major, such as mechanical and electrical engineering, active professional, and meets the development needs of the centralized MIS system or C / S software. The contents of the course include software engineering, interface programming, multimedia programming, database programming technology and network communication technology. In the course teaching, the case teaching method is used, and kinds of technology used in windows advanced programming. At the same time, the curriculum offers VB and VC to choose according to the different needs of the different majors.

(2) Network development. The direction is set mainly for the majors which have the needs of web site development and network programming needs. Its contents include software engineering, database technology and web development technology such as ASP.net. The course uses a dynamic website development to link all technology knowledge, and hopes the students to understand the related technology of the network development in the case teaching.

(3) Numerical calculation and algorithm design. This branched course meets the needs of engineering computation and algorithm design in some majors, such as aerospace, aviation, navigation and so on. The course content includes the data structure, algorithm design and algorithm analysis and scientific computing. Its teaching is mainly the complex data organization and processing mode, common numerical calculation tools and their using methods in different development environment, common scientific computing libraries and their using methods.

(4) Embedded development technology. The branched course meets the needs of embedded development in electronic, automation and other majors. Its contents include software engineering, embedded system and development environment, embedded development technology, and explain software development technology in the embedded system.

3   The Advantage of the Many-Branched Setting

After software development technology course are divided into many branched course, the curriculum system is huge, and its management and implementation is more difficulty than the original, but the advantage of the many-branched setting can not be ignored.

The advantage of the many-branched setting in software development technology course is described as following:

(1) Many branches meet different professional personalized demand. Because the course is divided into many branches, and each branched course has its own characteristics and can thorough presentation of a technology, can enhance the effect of teaching.

(2) Many branches include the more comprehensive technology. Due to the multiple branches setting in the course, each branch can orientate within its own key technology and all contents can be described in detail, and will not cause the whole course content.

(3) The many-branched setting can grasp the technical developments timely. Due to the multiple branches setting in the course, each branch can orientate within its own key technology, so it can update the course contents in real-time while will not affect other branched course and will not cause the big change of the whole course.

(4) Many-branched setting make the teachers could do their best. Previous curriculum puts many kinds of technology into a course, its teaching needs the teachers understanding of the each technique in the curriculum task, and it's a very difficult task for the teachers, especially in the times which the computer software and new technology emerge in an endless stream, even many older teachers can't fulfill the teaching tasks. After many branched curriculum is set, each teacher can only focus on their own expertise in branched direction, and he can optimize the teaching process.

(5) Many-branched setting enables the students to learn clearly and rightly. After the course is divided into many branches, the students which take the course learn only their needs and their interest in a field of technology, and their learning is trenchant primary and secondary, and can get the better learning effect.

(6) Many-branched setting conforms to the curriculum development trend and the national requirements of the talents training. After the course is divided into many branches, every major can found the branch which suits their own needs, and the curriculum has the better development prospect. At the same time, with many-branched settings, each student learned all need technology, masters these techniques greatly, and has the practice ability of the related technology, which is in line with the state on the training demand.

4   The Effect of the Many-Branched Setting

The many-branched setting strategy of software development technology has been in operation for five years in our school, and the branched courses increase from the two branches MIS and numerical calculation to four branches such as Windows advanced programming, numerical calculation and algorithm design, network development and embedded development technology, its implementation effect is good.

(1) The elective courses range expands exponentially. Our software development technology course was toke by only a dozen majors in the past. Now, there is 33 majors whose students study the software development technology course, namely all non-computer majors have selected the different branched courses.

(2) The teaching effect is greatly improved. Due to the course is set in many branches, each teacher can choose different branch according to his specialty, while the students only need to learn their interest and necessary techniques, the teaching contents is more targeted and more advanced, so the learning effect is greatly improve, and the students' ability of software development is enhanced. Even some students' curriculum design works participated in various software design contest and gained good ranking directly.

5   Conclusion

The many-branched setting of the software development technology course puts forward a kind of new curriculum development strategy when the curriculum development is poor. The many-branched setting strategy not only meets the demand of different majors, but also can improve the teaching effect and always grasps the software technology dynamic development, so the course has the forward-looking and exuberant vitality. In conclusion, the author thinks that the many-branched setting of the software development technology course is in line with the course's development trend, and is useful for the teaching both the learning. At the same time, the strategy also puts forward a develop proposal for some other curriculum which situation is not good.

References

1. Tian, J., Peng, X.: The thoughts and suggestions of the cross curriculum in the students' fostering. Optical Technique 11, 305–306 (2007)
2. Wang, G.: The reform of the college computer public computer curriculum. Journal of Jilin Institute of Technology 3, 59–60 (2001)
3. Tian, B.: The significance of the separate teaching mode. Chinese Society of Education 9, 32–35 (2010)
4. Chen, Q., et al.: The social needs investigation report of the computer basic course teaching contents and curriculum system in non-computer majors. Computer Education 16, 104–108 (2009)
5. Zhong, D., et al.: The exploration and study of the university public computer course teaching reform. Journal of Chongqing Academy of Arts and Sciences (Natural Science Edition) 2, 70–72 (2007)

Visual Information Interactive Design on Web Interface

Peng Shi

School of Mechanical and Vehicular Engineering,
Beijing Institute of Technology,
Beijing, China
Spanish2001@163.com

Abstract. Facing the new era, social economy and culture is experiencing rapid and profound changes. The influx of a large number of information has made our original values, and ways of thinking and the knowledge structure are facing with various new challenges. Network is the most effective window to obtain information, and a lot of irrelevant information like noises disturb our attention and make a lot of useful information submerged in large amounts of useless information, which produce certain difficulties for visitors to gain useful information. Web interface design belongs to one kind of human-computer interactive design. This conforms to the general man-machine interface design principle. At the same time, according to the particularity of website, visual information interactive design has its own characteristic and principle. Through interpretation of the web interface design and visual arts, this paper try to discover the principle and design method of visual information interactive design in a web page----Organization and arrangement of elements on interface according to aesthetic and man-machine interactive principles. Guide the visitor's sight according to the intention flow of design to make the audience clear and quickly accept web information.

Keywords: web design, interface design, visual interactive guide.

1 Introduction

Web interface is a media to transmit and exchange information between computer and people, including information input and output. Ideal design of web interface have advantage of simple to operate, aesthetic ,well decorate and have function of guide users to find useful information. Hence, users sense pleasures and generate strong interest in web page, this is an important factor to improve the traffic on website.

Web interface design is a crossover fields which include: computer science and psychology, the art of design, cognitive psychology and ergonomics. Web interface design is a creative activity, combination of technical and artistic, perceptual and rational.

A web interface generally contain a large amount of information and pictures, but most pages are ignored the importance of visual information interactive design. A large stack of Information, make the overall visual impression of page multifarious, disorder, affect man-machine interactive efficiency on pages. For web design,

D. Jin and S. Lin (Eds.): Advances in MSEC Vol. 2, AISC 129, pp. 457–462.

efficient communication of information is by various factors: Text and graphics, image, logo, the color and so on. Only through accurate definition for the page, can ensure the right design direction, attracting the viewer's attention, convey information clearly and accurately, seek for unify of function and form on web interface.

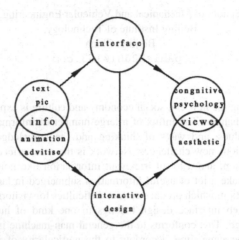

2   Visual Guidance of Web Interface

Visual guidance is a kind of "space movement" is a process when viewer's Points of focus follow with each element in the space of web pages along certain trajectory. Through the visual guidance and the rule of visual focus movement, according to the public's psychological and physiological characteristics on Visual. Organize all kinds of visual information for example text, graphics, images, logos, colors, etc, according to certain aesthetic principles. Guide the visual focus of the reader to flow according to the design intention. Thus information of web pages transmitted to audience will be clear, smooth, fast. Elaborate visual guidance design, make site unique, increase the goodwill of readers to the site, and improve website hits.

Rational layout is essential to visual information interactive. Studies show when browse the Web, people's attention usually fall perpendicularly on center of the computer screen, and then walk around. Web designer should consider arrange important information or viewer's visual spots in the best area, make the whole web design theme clear at a glance. In the interface, information placed in different regions, will cause the different attention levels by the visitors, which gives different mental feeling. Web designer can reasonable arrange proper place with proper information or decoration, according to the need of expression and the focus of page. In visual guidance design of web interface, we must design the order for browser and make clear what the user will see first and what will see next.

Overall planning—Visual psychology thought that the whole is not equal to the sum of parts, and the character of the whole determines the nature and significance of the parts. [1]

In a web interface design, each region relative to the whole page is a component. The entire page visual effect is made from these parts, if between those parts visual layout is not clear with each other, Space management will in chaos, unable to distinguish between primary and secondary area and will inevitably affect the efficiency to communicate information and weaken the overall Visual effect of the page. Based on the way of segmentation and arrangement to web information space, Make sub--information arranged regular and logical, and make sure the importance between those sub-information in order.

And it needs to consider the characteristics of visual effect from whole and sub-sectors, so that each sub- sectors both have its own characteristics and not separate from requirements of harmonious and overall visual perception.

Highlight theme—Grasp the site's theme precisely. Communicate the theme to the viewer precisely, according to the visual psychology regularity. Only the theme is understood and accepted by people timely, people can satisfy with the efficiency and practical need. [2] At the same time properly strengthened the artistic and aesthetic feeling will help topic outstanding.

Unified form and content—Content is the sum of all the internal elements which consist of the design, the form is the external manifestations of content and internal structure of various elements. In web interface design, we should maintain the consistency between the form and the content. What elements on page express the function effective, choose effective visual communication means and style of decoration design and abandon irrelevant elements. All those problem are essential to research.

Arrangement of visual focus—By highlighting the main parts of information, Enhance the strength of information identification and guidance for information demand. General website interface contains numerous areas, People are likely to feel chaos on Visual perception. Therefore, the grasp of visual focus is particularly important. Through sub-sectors arrangement and location can adjust the visual focus on page, divide the page regularly. Based on arrange the sequence of important level of information intelligently, formed somewhat stressed visual flow for user. Control the order and process according to the order and importance of information, so that enhance the comprehension of information.

Area segmentation—Through various approaches can segment visual area into sectors, such as linear, rectangular color, text, advertisements. When segmentation, Have to control size and rhythm of the information area gap between sectors. when the gap broad, the rhythm slows down, the visual flow appears to stretch: And too much increases the gap, will loss the contact between sectors which can not respond with each other, the visual flow looks weak. [3] when the gap narrow , reinforce the rhythm powerfully, the visual flow appears to compact: And too much decreases the gap, will Cause your visitors visual fatigue information confusion and subject unclear.

Visual scopes Optimization—Arrange Pause of visual flow in the most concentrated area, this should be the best Visual region for audience. For Different region, the level of visitor's attention is different, psychological feeling is different also. Top area of the page usually gives viewer feeling of light, floating, positive visual experience, bottom area of the page gives viewer impression of heavy, depressed, limiting, short spacing and stable. left area of the page gives viewer impression of light, free, stretch, full of vitality; right area of the page gives viewer impression of depression, constraints, but serious. According to human visual experience, the sequence of priority visual area is upper left, upper right, lower left, lower right in sequence. Page layout should also take into account what function of visual elements of each component will play in page. Place each element in the best position, So that the effect of each constituent element on the page highlighted to greatest extent.

3   Unique Feature and Visual Artistic Influence

Web design also requires a certain artistic influence. chaotic, mediocrity, short of aesthetic web pages will not be attract visitors, The visual design aesthetic principles also applies to web interface design ,those principles include Order, harmony, diversification and so on. Use of these rules, attracting the attention of the reader effectively.

Unique Interface and new graphics design will be artistic and creativity, the layout is more reasonable which will attract visitors attention, Wonderful page can excited visitors when visit website and increase the click rate. Web graphic artist and development group will need to use all means and most advanced technology to artistic treat and decorate the pages, so the page will be with more artistic, charming, more attractive, more sense of visual impact. Ensure that each page can maintain a strong individuality and also belong to the style of entire page. take advantage of a lot of visual elements such as color, graphics, symbols, animation, etc. not only understand how to use the most advanced software and technical means, but also have to know when and where use or abandon these elements, because the efficient transmission of useful information to the audience is the most important and critical. [4]

A web interface of integral style and integral atmosphere expression should match with content of web site, integrate visual elements----logo, navigator, layout, color, fonts, pictures, banner, etc, and strengthen integral visual effect. On the web interface, make use of unique creative style, through the way of strong visual impact and spatial visual imagination to guide visitors, This is extremely effective means to enhance visual communication on web interface.

4   Coordination of Visual Elements

In graphic design, visual communication depends on the visual elements and means of elements combination. Web interface design takes advantage of basic visual elements in normal graphic design to achieve the purpose of beautification and information communication. These visual elements include text, image, color and so on, Web design should meet the basic requirements of the visual aesthetics rules, at the same time to meet the needs of the public demand from mental to visual; this requires us to consider co-ordination of all elements.

Text—Text as the primary means of information transmission is an essential element of the page, play a major role of information describing and explaining. So the page text will occupy a large area, the performance of text is good or bad will affect the quality of the page directly. During The page layout and design of the text, because the computer provides us with a wide selection of fonts, which tend to diversification. Text in visual communication page as one of the image elements, beside convey information, it also has the function of conveying emotions.

The main function of text is information communication. In order to achieve the effectiveness of the communication, we must consider the overall effect of text edition. The main intention of text edition is have a good grasp of The font style, font size, spacing, between line etc. To give people a clear visual impression, avoid complicated and messy on page, decrease the unnecessary decoration.

When layout and design of text on Page, the emphases is lay on Display mode which should be subordinated to the requirements of the nature and characteristics of content. Its style should consistent with the characteristics content expressed, not to separated from each other, Rather than conflict with each other.

Logo, graphics image—Graphic possess of incomparable advantage in visual expression, compare with text in many respects, amount of information conveyed by Pictures is larger than which conveyed by text. Pictures can also achieve the goal to make user clear at a glance. Through way of combine pictures with text at the same time, greatly increase the way it provides information. The use of image and picture can achieve effect of stressing, when use it appropriately, it will produce strong visual impact. But if too much used, it will like a uniform noise, fail to emphasize the theme.

Graphic images can make communication of information more direct, Credible, enrich amount of information page provided, and also greatly beautify the network page. Location, size, quantity, form, direction of place is directly related to the effect of visual communication on page.

Video, animation—Web design graphic art also has a dynamic aspect: flashing text, changing colors and Flash animation. During the creation of works of Web graphics art design, In order to further attract visitors to view site, Make the site information, notes and documents more clear, rich, Can use video, animation, etc. can achieve good results. Video and animation design allows overall effect of the page not only natural, clear, simple but also novel and unique.

Web Design involves various means the video, banner, animation and other visual elements., Video can provide accurate information and instructions, More vivid, convenient, accurate and Highly persuasive in description. Flash animation is a common form on web pages, Animation has a strong visual impact and visual guidance effect, add animation to a static Web page Can achieve good visual effects of association of activity and inertia.

Color design—Color is rich of emotion and meaning in people's life. Appropriate use of color can produce strong visual effects, when users browse the web, first impression is the color of the page, and its take as the direct factor impact the user's interest. Therefore, the use of color can produce strong visual effects; make the style of page more vivid and clear. Color design should consider: choose the color of the entire page, set a main color, Reflect the theme of the site can help determine the design style. [5] Selection of contrast level between the background color and text color not too violent, not lead to over visual exciting, otherwise, will produced read fatigue, Harmonious colors make reading easy and enjoyable. [6]

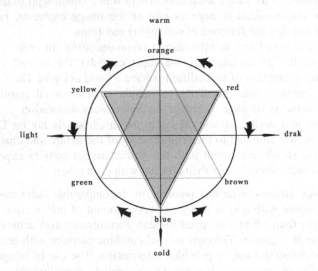

5   Conclusion

Web page visual interactive design is a kind of comprehensive design, it involves wide the range, include consumer psychology, visual art design, ergonomics, science and technology etc. In this article, discussion and analysis is only from the perspective of visual communication, design and technology not only to meet the needs of the users, more important is to create a pleasant visual environment, So they feel a Wholehearted amusement and resonance.

References

1. Zhang, F.: Art and Design Web Interface Design (2002)
2. Hou, H., Qian, Y.: Guide of Web Interface Design and Creativity (2009)
3. Lewis, P.: Principles of graphic design (2005)
4. Gao, B.: Design and Analysis of Web Visual Information Transmission
5. Preece, J., Rogers, Y., Sharp, H.: Interaction Design - beyond Human-Computer Interaction (2003)
6. Cao, T.: Nature and Characteristics of Digital Art (2007)

A Fast Direct Calculation Inference Algorithm

HaiYang Chen[1], XiaoGuang Gao[2], and Jian Xu[1]

[1] College of Electronics and Information, Xi'an Polytechnic University,
Xi'an 710048, China
[2] School of Electronics and Information, Northwestern Polytechnical University,
Xi'an 710129, China
chy_00@163.com, xggao@nwpu.edu.cn, xu9010@sina.com

Abstract. At present, among the inference algorithms on Dynamic Bayesian Networks (DBNs), the advantages of the Direct Calculation Inference (DCI) algorithm is that it needn't perform the complex graphic transformation, its calculation formula is simple, and easy to program, however, the disadvantage is that the inference efficiency is quite low when there are many time slices. In this paper, after analyzing the algorithm complexity, we managed to find the crucial steps for decreasing the algorithm complexity, proposed a Fast Direct Calculation Inference (FDCI) algorithm based on optimizing the calculation means. It is proved by the simulation experiments that the inference results of the FDCI algorithm and the DCI algorithm are equal, but the inference efficiency of the proposed algorithm is much higher.

Keywords: dynamic Bayesian networks, inference, soft evidences, complexity.

1   Introduction

Bayesian networks are a tool that can apply the probability statistics to the complex domain to perform the uncertain inference and the data analysis. Bayesian networks is the extension of Bayes method, is the combination of artificial intelligence, probability theory, graph theory and decision analysis, and is one of the most efficient theoretical models in the areas of uncertain knowledge and probabilistic reasoning. It visually expresses joint probability distributions of a set of variables in the graphic form, and the calculation complexity of probabilistic reasoning has been reduced greatly by using the conditional independence assumption, it provides a better solution to the complex uncertain inference problem, and is widely applied to the areas of medical diagnosis[1], industry[2], finance[3], computer systems[4], agriculture[5], ecology[6], and military[7]. Bayesian networks are divided into static Bayesian networks and DBNs. At present, the study about the inference algorithm of static Bayesian networks is very mature, its application areas are very wide[8-11]. The study about DBNs has recently become the focus. Most of the inference algorithms on the DBNs are based on the complex graphic transformation, such as the interface algorithm[12], during its computation process, first perform graphic transformation, then perform forwards and backwards recursive operation, the programming is very complex and cumbersome. Only a few inference algorithms don't need graphic transformation, such as the forwards-backwards algorithm[12,13]

D. Jin and S. Lin (Eds.): Advances in MSEC Vol. 2, AISC 129, pp. 463–469.
springerlink.com   © Springer-Verlag Berlin Heidelberg 2011

and the DCI algorithm[14], their reasoning formulas are intuitive, easy to understand and the programming is simple. The forwards-backwards algorithm can only handle DBNs of Hidden Markov Models, but the DCI algorithm can deal with DBNs with more complex structures, however, its disadvantage is that inference efficiency is quite low when there are more time slices, which influences its application, so a FDCI algorithm is proposed. On the base of analyzing its complexity, find the core step which affects the complexity of the algorithm, then optimize its calculation means to reduce the calculation amount.

2   Types of Evidence

DBNs are the extension of static Bayesian networks, the process of the change of its variable states can be viewed as a series of snapshots, each of which describes the state of the world at a particular time. Each snapshot, called time slice in this paper, contains a set of random variables, some of which are observable and some are not. On any time slice, the states of observable variables are known, these observation values are called evidence. If the obtained evidence can exactly show which state the variable is in, it is called hard evidence, otherwise, is called soft evidence.

3   The DCI Algorithm and Complexity Analysis

3.1   The DCI Algorithm

As the time goes by, a static Bayesian networks with n hidden nodes and m observed nodes can develop into a DBNs with T time slices. Let $X_1^t, X_2^t, ..., X_n^t$ donate the hidden nodes and $Y_1^t, Y_2^t, ..., Y_m^t$ be the observed nodes of Bayesian networks at time slice t. Let x_i^t donate a certain state of the ith hidden node X_i^t, and y_j^i be the observation value of the jth observed node Y_j^i at time slice t.

Under the condition of hard evidences, literature [14] gives the DCI algorithm by Bayes formula and the independence assumption of Bayesian networks:

$$P(x_1^1, ..., x_n^1, ..., x_1^T, ..., x_n^T \mid y_1^1, ..., y_m^1, ..., y_1^T, ..., y_m^T)$$
$$= \frac{P(x_1^1, ..., x_n^1, ..., x_1^T, ..., x_n^T, y_1^1, ..., y_m^1, ..., y_1^T, ..., y_m^T)}{\sum_{x_1^1, ..., x_n^1, ..., x_1^T, ..., x_n^T} P(x_1^1, ..., x_n^1, ..., x_1^T, ..., x_n^T, y_1^1, ..., y_m^1, ..., y_1^T, ..., y_m^T)}$$
$$= \frac{\prod_{i,j} P(x_j^i \mid Pa(X_j^i)) \prod_{k,l} P(y_l^k \mid Pa(Y_l^k))}{\sum_{x_1^1, ..., x_n^1, ..., x_1^T, ..., x_n^T} \prod_{i,j} P(x_j^i \mid Pa(X_j^i)) \prod_{k,l} P(y_l^k \mid Pa(Y_l^k))} \qquad (1)$$

where $Pa(X_j^i)$ is the parents of X_j^i in graph, $Pa(Y_l^k)$ represents the parents of Y_l^k, let $x_1^1, ..., x_n^1, ..., x_1^T, ..., x_n^T$ be a combined state of the hidden variable, and $i, k = 1, 2, ..., T$, $j = 1, 2, ..., n$, $l = 1, 2, ..., m$.

If the obtained observation evidences are not hard evidences, but soft ones, that is, some observation variables all have certain probabilities in their more than one state, then Equation (1) is modified as

$$
\begin{aligned}
&P(x_1^1,\ldots,x_n^1,\ldots,x_1^T,\ldots,x_n^T \mid y_{lo}^1,\ldots,y_{mo}^1,\ldots,y_{lo}^T,\ldots,y_{mo}^T) \\[4pt]
&= \frac{P(x_{1:n}^1,\ldots,x_{1:n}^T,y_{1:mo}^1,\ldots,y_{1:mo}^T)}{\displaystyle\sum_{x_{1:n}^1,\ldots,x_{1:n}^T} P(x_{1:n}^1,\ldots,x_{1:n}^T,y_{1:mo}^1,\ldots,y_{1:mo}^T)} \\[4pt]
&= \frac{\displaystyle\sum_{y_{1ms}^1,\ldots,y_{1ms}^T}\prod_{i,j} P(x_j^i \mid Pa(X_j^i))\prod_{k,l}\left[P(Y_l^k = y_{ls}^k \mid Pa(Y_l^k))P(Y_l^k = y_{ls}^k)\right]}{\displaystyle\sum_{x_{1:n}^1,\ldots,x_{1:n}^T}\sum_{y_{1ms}^1,\ldots,y_{1ms}^T}\prod_{i,j} P(x_j^i \mid Pa(X_j^i))\prod_{k,l}\left[P(Y_l^k = y_{ls}^k \mid Pa(Y_l^k))P(Y_l^k = y_{ls}^k)\right]}
\end{aligned}
\tag{2}
$$

where $y_{1:ms}^1 = \{y_{1s}^1,y_{2s}^1,\ldots,y_{ms}^1\}$, $y_{1:mo}^k = \{y_{1o}^k,y_{2o}^k,\ldots,y_{mo}^k\}$, $x_{1:n}^i = \{x_1^i,x_2^i,\ldots,x_n^i\}$, y_{lo}^k represents the state of the lth observation variable Y_l^k at time k, $P(Y_l^k = y_{ls}^k)$ means the probabilities to which the observation value belongs to its sth state y_{ls}^k.

The advantages of this algorithm are that it needn't perform complex graphic transformation, reasoning formulas are intuitive, easy to understand and can handle DBNs with more complex structures.

3.2   Complexity Analysis

On a Discrete Dynamic Bayesian Networks (DDBNs) with T time slices, there are n hidden nodes and m observed nodes on each time slice, the number of the node states is no more than N, the combination states of the hidden nodes have N^{nT} kinds. If obtained evidences are hard evidences, the complexity of Equation (1) is $O\big((m+n)TN^{nT}\big)$. If they are soft evidences, the combination states of all the nodes on T time slices are $N^{(m+n)T}$ at most, we need do computation $(2m+n)TN^{mT} - 1$ times for the joint probability distribution $P(x_{1:n}^1,\ldots,x_{1:n}^T,y_{1:mo}^1,\ldots,y_{1:mo}^T)$, $(2m+n)TN^{(m+n)T} - N^{nT}$ times for the numerator in Equation (2), $N^{nT} - 1$ times for the denominator in Equation (2), so the complexity of Equation (2) is $O\big((2m+n)TN^{(m+n)T}\big)$.

From the above complexity analysis, it's easy to see that if the obtained evidences are hard evidences, the complexity of the FDCI algorithm has the exponential relationship with the product of the number of the hidden nodes on the single time slice and the number of the time slices. If the obtained evidences are the soft ones, the complexity has the exponential relationship with the product of the number of the nodes on the single time slice and the number of the time slices, so to the same networks, the complexity of handling the soft evidences is higher than that of the hard ones.

4   The FDCI Algorithm

The DCI algorithm for handling hard evidences is using chain rule of probability and the condition independence, factorize the joint probability into a series of products of condition probabilities, however, to the soft evidences, add the process of expectation summation. By the complexity analysis of the DCI algorithm, it's obvious that the complexity of Equation (2) is decided by the calculation amount of the numerator, if the amount of computing numerator can be reduced, the complexity of this algorithm will be reduced, too.

4.1   Deduction

By the above analysis, the complexity of DCI algorithm is mainly decided by the amount of computing the numerator, the calculation means for the numerator can be optimized, that is, the order of numerator summation and the product operation in Equation (2) needs to be changed, which means, in the Equation (2), first perform the product operation, then summation, now we change the order like this, first perform summation, then product operation, but the results are equal, then we obtain

$$P(x_1^1,\ldots,x_n^1,\ldots,x_1^T,\ldots,x_n^T \mid y_{1o}^1,\ldots,y_{mo}^1,\ldots,y_{1o}^T,\ldots,y_{mo}^T)$$

$$= \frac{P(x_{1:n}^1,\ldots,x_{1:n}^T,y_{1:mo}^1,\ldots,y_{1:mo}^T)}{\displaystyle\sum_{x_{1:n}^1,x_{1:n}^2,\ldots,x_{1:n}^T} P(x_{1:n}^1,\ldots,x_{1:n}^T,y_{1:mo}^1,\ldots,y_{1:mo}^T)}$$

$$= \frac{\displaystyle\prod_{i,j}P(x_j^i \mid Pa(X_j^i))\prod_{k,l}\left[\sum_{p=1}^{s_l}P(Y_l^k = y_{lp}^k \mid Pa(Y_l^k))P(Y_l^k = y_{lp}^k)\right]}{\displaystyle\sum_{x_{1:n}^1,\ldots,x_{1:n}^T}\prod_{i,j}P(x_j^i \mid Pa(X_j^i))\prod_{k,l}\left[\sum_{p=1}^{s_l}P(Y_l^k = y_{lp}^k \mid Pa(Y_l^k))P(Y_l^k = y_{lp}^k)\right]}$$

$$(3)$$

where s_l means that the lth observation variable has s_l states on the single time slice, and $P(Y_l^k = y_{lp}^k)$ represents the membership degree to which the lth observation value belongs to its pth state at the time slice k.

4.2   Complexity Analysis

Assume that the given DDBNs are the same as the ones in section 2.2. When the observation evidences are soft evidences, perform computation for $(2Nm+n)T-1$ times for the joint probability distribution $P(x_{1:n}^1,\ldots,x_{1:n}^T,y_{1:mo}^1,\ldots,y_{1:mo}^T)$, $((2Nm+n)T-1)N^{nT}$ times for the numerator in Equation (3), and perform addition operation for $N^{nT}-1$ times for the denominator in Equation (3), and computing posterior probabilities of all the combined states of the hidden variables need do division operation for N^{nT} times, so the complexity of the Equation (3) is $O((2Nm+n)TN^{nT})$.

By the comparison of complexities in Equation (2) and (3), you can see that the exponent has changed from $(m+n)T$ into nT, the calculation amount has been reduced because of $n+m>n$. When $m+n\gg n$, the calculation amount will be reduced more greatly. It's obvious that the proposed FDCI algorithm is more efficient and has more practical value than the DCI algorithm.

5 Experimental Results and Analysis

The DDBNs as shown in Fig. 1, there is only a hidden node and four observed nodes on each time slice, where the hidden node X has four states, they are $\{1,2,3,4\}$. The observed node Y_1 has three states $\{y_{11}, y_{12}, y_{13}\}$, Y_2 has two states $\{y_{21}, y_{22}\}$, Y_3 has three states $\{y_{31}, y_{32}, y_{33}\}$, Y_4 has three states $\{y_{41}, y_{42}, y_{43}\}$. The parameters of Fig. 1 are given in Table 1, the prior probabilities are $P(X=1,2,3,4)=\{0.08,0.3,0.4,0.22\}$, table 2 for the observation data.

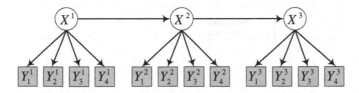

Fig. 1. The DDBN model

Table 1. Parameters of Fig. 1

	Transfer Probabilities					Conditional Probabilities			
	$X=1$	$X=2$	$X=3$	$X=4$		$P(Y_1|X)$ (y_{11},y_{12},y_{13})	$P(Y_2|X)$ (y_{21},y_{22})	$P(Y_3|X)$ (y_{31},y_{32},y_{33})	$P(Y_4|X)$ (y_{41},y_{42},y_{43})
$X=1$	0.93	0.05	0.02	0.0	$X=1$	0.0,0.05,0.95	0.95,0.05	0.95,0.05,0.0	0.98,0.02,0.0
$X=2$	0.05	0.7	0.2	0.05	$X=2$	0.15,0.35,0.5	0.2,0.8	0.15,0.7,0.15	0.05,0.85,0.1
$X=3$	0.02	0.2	0.68	0.1	$X=3$	0.4,0.35,0.25	0.25,0.75	0.1,0.5,0.4	0.0,0.6,0.4
$X=4$	0.0	0.1	0.15	0.75	$X=4$	0.1,0.3,0.6	0.02,0.98	0.05,0.25,0.7	0.0,0.2,0.8

Table 2. Observation data

Time slice	Y_1	Y_2	Y_3	Y_4
1	0.0,0.25,0.75	0.1,0.9	0.15,0.85,0.0	0.0,0.75,0.25
2	0.0,0.25,0.75	0.15,0.85	0.2,0.8,0.0	0.0,0.75,0.25
3	0.0,0.2,0.8	0.1,0.9	0.15,0.85,0.0	0.0,0.8,0.2
4	0.0,0.15,0.85	0.2,0.8	0.2,0.8,0.0	0.0,0.85,0.15

By the observation data in Table 2, the comparison of simulation results of two inference algorithms mentioned above are as shown in Table 3. Fig. 2 is for the comparison of the elapsed time of the two algorithms when there are different time slices.

Table 3. The comparison of the inference results by the DCI algorithm and the FDCI algorithm

Time slice	The results of the DCI algorithm	The result of the FDCI algorithm
1	0.00316937,81.3908,15.0738,3.53218	0.00316937,81.3908,15.0738,3.53218
2	0.0017254,89.7287,8.6035,1.66603	0.0017254,89.7287,8.6035,1.66603
3	0.00111371,91.3493,7.37063,1.27898	0.00111371,91.3493,7.37063,1.27898
4	0.03277,87.2356,10.6436,2.08802	0.03277,87.2356,10.6436,2.08802

By comparison of the two columns of data in Table 3, it's easy to see that the inference results of two algorithms are the same on each time slice, although their reasoning mechanisms are different, the results are equal.

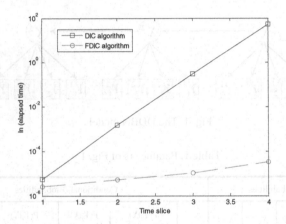

Fig. 2. The comparison of elapsed time run by Matlab programs of the DCI algorithm and the FDCI algorithm

The ordinate in Fig. 2 is for taking logarithm of the elapsed time, from the curve, we know that the logarithm of the elapsed time of the two algorithms has linear relationship with the number of the time slices, the slope of the DCI algorithm is larger, however, the FDCI algorithm's is smaller, so using the FDCI algorithm can reduce the elapsed time significantly, and improve the efficiency of the inference.

6   Conclusion

Among many theory problems about DBNs, how to reduce the complexity of the algorithm and improve calculation efficiency is always the focus of the Bayesian networks research. In this paper, on the base of optimizing the calculation means of the DCI algorithm, a FDCI algorithm is proposed, by comparing with the DCI algorithm, its complexity is reduced remarkably, so it has more practical value, and it's proved by the simulation experiments that this inference algorithm is very efficient.

References

1. Ucar, T., Karahoca, D., Karahoca, A.: Predicting the Existence of Mycobacterium Tuberculosis Infection by Bayesian Networks and Rough Sets. In: 15th National Biomedical Engineering Meeting, pp. 1–4. IEEE Press, Antalya (2010)
2. Garcia, J.I., Gomez Morales, R.A., Miyagi, P.E.: Supervisory System for Hybrid Productive Systems Based on Bayesian Networks and OO-DPT Nets. In: 13th IEEE International Conference on Emerging Technologies and Factory Automation, pp. 1108–1111. IEEE Press, Hamburg (2008)
3. Wang, S.C., Shao, R.Q., Leng, C.P.: Dynamic Bayesian Network Model for the Enterprise Financial Risk Warning. In: 2009 International Conference on Electronic Commerce and Business Intelligence, pp. 431–434. IEEE Press, Washington (2009)
4. Abdi, M.K., Lounis, H., Sahraoui, H.: Predicting Change Impact in Object-Oriented Applications with Bayesian Networks. In: 33rd Annual IEEE International Computer Software and Application Coference, pp. 1: 239–1: 243. IEEE Press, Washington (2009)
5. Bacon, P.J., Cain, J.D., Howard, D.C.: Belief Network Models of Land Manager Decisions and Land Use Change. Journal of Environmental Management 65(1), 1–23 (2002)
6. Zhang, J., Gong, H.L., Li, X.J., Ross, M.: Assessment of Nitrogen Loading in Miyun Reservoir Beijing Using Bayesian decision Network. In: 3rd International Conference on Bioinformatics and Biomedical Engineering, pp. 1–4. IEEE Press, Beijing (2009)
7. Chen, H.Y., Gao, X.G., Fan, H.: Inference Algorithm of Variable Structure DDBNs and Multi-Target Recognition. Acta Aeronautica et Astronautica Sinica 31(11), 2222–2227 (2010) (in Chinese)
8. Jensen, F.V., Nielsen, T.D.: Bayesian Networks and Decision Graphs. Springer, New York (2001)
9. Neapolitan, R.E.: Learning Bayesian Networks. Pearson Prentice Hall Upper Saddle River, New York (2003)
10. Zhang, L.W., Guo, H.P.: Introduction to Bayesian Networks. Science Press, Beijing (2006) (in Chinese)
11. Li, H.J., Ma, D.W., Liu, X.: Bayesian Networks in Fault Detection Equipment. National Defense Industry Press, Beijing (2009) (in Chinese)
12. Kevin, P.M.: Dynamic Bayesian Networks: Representation, Inference and Learning. PhD thesis, University of California, Berkeley (2002)
13. Russell, S., Norvig, P.: Artificial Intelligence: A Modern Approach. Prentice Hall Upper Saddle River, New Jersey (2003)
14. Shi, J.G., Gao, X.G.: Direct Calculation Inference Algorithm for Discrete Dynamic Bayesian Network. Systems Engineering and Electronics 27(9), 1626–1630 (2005) (in Chinese)

Disassembly Sequence Planning Based on Improved Genetic Algorithm

JiaZhao Chen, YuXiang Zhang, and HaiTao Liao

Xi'an Research Inst. of Hi-Tech,
Xi'an, Shaanxi Province, P.R.C.
Jazhch@sina.com

Abstract. This paper researches on the optimization method of disassembly sequences based on genetic algorithms (GA). In order to find the best disassembly sequence, the paper gives encoding mode, fitness function, and crossover and mutation method. By adopting adaptive method to change crossover rate and mutation rate, genetic algorithm is improved to ensure its convergence and to accelerate its searching speed of the best solution. The simulation result proves that the improved genetic algorithm is effective when it is used to optimize disassembly sequence.

Keywords: genetic algorithm, adaptive method, sequence planning.

1 Realization of GA in Disassembly Sequence Planning

Assembly sequence panning is a class of combinatorial optimization problems [1], and the conventional graph search algorithms are difficult to be effective. Being able to get optimized solution without need to traverse the whole searching space, genetic algorithm (GA for short) can quite well solve the combinatorial explosion problem in sequence planning. Because detachable parts must be installed, this paper adopts genetic algorithm to optimize product's disassembly sequence.

The basic principle of genetic algorithm is starting from a population standing for a feasible solution set, by carrying out genetic operations, to produce a population standing for a new solution set which is more adaptable for environment than the original generation. The best individual in the last generation can serve as the problem's optimized solution. The key problems of genetic algorithm are encoding, fitness function, and genetic operation.

1.1 Encoding

This paper adopts real number encoding mode [3], namely, using part number to encode directly, thus dispensing us from encoding and decoding. A three-place gene code set is used to express the part information in an assembly, including part number, detachment tool, and removal direction. Gene code set is expressed as:

$$\text{Part}_i = \{\text{Number}_i + \text{Tool}_i + \text{Direction}_i\} . \tag{1}$$

D. Jin and S. Lin (Eds.): Advances in MSEC Vol. 2, AISC 129, pp. 471–476.

Where, $Number_i$, $Tool_i$ and $Direction_i$ stand for the part number, detaching tool and removal direction of part i. In a gene set, the code places except for part number are used to store disassembly information. For a certain part, its detachment information is certain and doesn't participate in genetic operation, so the disassembly sequence can be described by the chromosome formed by part number.

1.2   Setting-Up of Fitness Function

The fitness function is set up by using disassembly time composed of three parts, i.e., basic disassembly time, orientation time and tool transform time, which can comprehensively express the information involved in part disassembly process. Suppose detaching part i in disassembly sequence S_j needs a time of $dt(i, S_j)$, then,

$$dt(i, S_j) = w_1 bt(i, S_j) + w_2 ot(i, S_j) + w_3 tt(i, S_j) \quad (i \geqslant 2). \tag{2}$$

Where, $bt(i, S_j)$, $ot(i, S_j)$ and $tt(i, S_j)$ stand for basic disassembly time, orientation time and tool transform time, and w_1, w_2, w_3 are weights. The total disassembly time of the first i parts is:

$$DT(i, S_j) = \omega_1 \sum_{j=1}^{i} bt(i, S_j) + \omega_2 \sum_{j=1}^{i} ot(i, S_j) + \omega_3 \sum_{j=1}^{i} tt(i, S_j) \quad (i \geq 2). \tag{3}$$

Define a fitness function as follows:

$$Fitness = \frac{C}{[a + DT(i, S_j)]} \quad (1 \leq i \leq n, 1 \leq j \leq m). \tag{4}$$

1.3   Genetic Operation

Selection Operation. This paper adopts the optimal saving selection, i.e., carries out selection operation by using roulette at first, and then completely copies the individual with best fitness value in the current population into next generation. This selection mode ensures the search result gotten at the end of the genetic algorithm is the individual with best fitness value appeared in all generations.

Crossover Operation. This paper adopts precedence preservative crossover (PPX for short) [5] to realize crossover operation. PPX uses a pair of sequences with same length as the parent to decide the crossover and offspring production mode. To carry out crossover, first empty the two child individuals, select the value in the place of child chromosome according to the value in the place of crossover operator, and this part is selected to be removed from the parent generation. Repeat this process until all parts are selected and a child sequence is produced. For example, for two parent disassembly sequences of a given population G(n): S(n, 1)=(4, 6, 7, 2, 1, 9, 8, 0, 5, 3) and S(n, 2)=(6, 8, 2, 3, 0, 9, 7, 5, 4, 1), define two crossover operators PPX1: 2122211212 and PPX2: 2211121221. Offspring from crossover operation are: 6, 4, 8, 2, 3, 7, 1, 0, 5, 4, and 6, 8, 4, 7, 2, 3, 1, 9, 5, 4.

Mutation Operation. This paper first select a disassembly sequence to mutate by a certain mutation rate, and then randomly select a certain place in the sequence, and keep the part of the sequence before the place unchanged and reproduce a feasible disassembly sequence for the part after the place, and combine the two parts together to form a new feasible disassembly sequence. For example, for a disassembly sequence S(n, i)=(4, 6, 7, 2, 1, 9, 8, 0, 5, 3), select place 5, and a child sequence from mutation is S(n, i)=(4, 6, 3, 2, 1, 9, 8, 0, 5, 7).

2   Adaptive Genetic Algorithm and Its Improvement

2.1   Adaptive Genetic Algorithm

Genetic algorithm cannot ensure global convergence in some situations. To solve this problem, Srinivas proposed adaptive genetic algorithm (AGA for short) [4], where crossover rate p_c and mutation rate p_m changes with change of fitness value. When fitness values of individuals tend to be unanimous or local optimal, p_c and p_m rise up, or when fitness values are scattered, p_c and p_m drop down. AGA ensures GA's convergence and accelerates searching speed of optimal solution while keeps population's diversity. p_c and p_m are adjusted as follows [2]:

$$P_c = \begin{cases} \dfrac{k_1(f_{max} - f')}{f_{max} - f_{avg}}, f' \geq f_{avg} \\ k_2, f' < f_{avg} \end{cases}$$

$$P_m = \begin{cases} \dfrac{k_3(f_{max} - f)}{f_{max} - f_{avg}}, f \geq f_{avg} \\ k_4, f < f_{avg} \end{cases}$$

(5)

Where, f_{max} and f_{avg} are the maximum and average fitness value of every generation; f' and f are the larger fitness of the two crossover individuals and the fitness value of the mutation individual. k_1, k_2, k_3, and k_4 take values in (0, 1).

2.2   Improved Adaptive Genetic Algorithm

From equation (5), the nearer the fitness value approaches the maximum, the less the crossover rate and mutation rate are, and when the fitness equals the maximum, crossover rate and mutation rate equal zero. This is suitable for the later stage of the population's evolution, but isn't advantageous to the first stage, because the better individuals are almost unchanged in the first stage and the best individual is not necessarily the global optimal solution [6]. To solve the problem, the crossover rate and mutation rate of the individual with maximum fitness can be set nonzero to increase the crossover rate and mutation rate of the excellent individuals, thus making the evolution avoid reaching a stalemate. Accordingly, p_c and p_m are adjusted as follows:

$$P_c = \begin{cases} P_{c1} - \dfrac{(P_{c1} - P_{c2})(f_{max} - f')}{f_{max} - f_{avg}}, f' \geq f_{avg} \\ P_{c1}, f' < f_{avg} \end{cases}$$

$$P_m = \begin{cases} P_{m1} - \dfrac{(P_{m1} - P_{m2})(f_{max} - f)}{f_{max} - f_{avg}}, f \geq f_{avg} \\ P_{m1}, f < f_{avg} \end{cases}$$

(6)

3   An Example of Disassembly Sequence Planning

3.1   Planning Example

Taking a throttle valve as the example, this paper uses genetic algorithm to optimize the disassembly sequence. Fig. 1 is the part composition diagram, and Table 1 lists the parts and their detachment tools, removal direction, and basic detachment time.

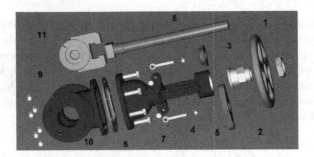

Fig. 1. The explosive diagram of throttle valve

There are 40 feasible disassembly sequences of the throttle valve (omitted). Taking them as the original population, this paper uses genetic algorithm to optimize so as to get the optimal disassembly plan.

3.2   Parameter Setting

The algorithm program in this paper is carried out by Visual C++2003.net and the computer configuration is Core2 （TM） Q8200 processor with a frequency of 2.33GHZ and a memory of 3.5G.

Table 1. Basic disassembly information of the throttle valve parts

No.	Name	Quantity	Detachment tool	Removal direction	Basic detachment time
1	stem nut	1	spanner	+Z	2
2	wheel	1	none	+Z	1
3	fixed nut	1	spanner	+Z	2
4	valve cover	1	none	+Z	0
5	gland	1	none	+Y	1
6	fastener2	6	spanner	+Z	12
7	Fastener1	2	spanner	+Z	8
8	valve stem	1	none	-Z	2
9	valve seat	1	none	-Z	1
10	washer	1	none	-Z	1
11	dam	1	hammer	+Y	2

The paper takes the number of running generation as termination condition. The parameters of fitness function are set as follows: $w_1=0.25$, $w_2=2$, $w_3=3$, $C=100$, $a=17$. From calculated data and literature [5, 6], when population volume is 40 and running generation number is 200, if p_c=0.60, p_m=0.40, genetic algorithm can get good result, and the planning efficiency is highest. When adaptive genetic algorithm is adopted, if p_{c1}=0.9, p_{c2}=0.5, p_{m1}=0.4, p_{m2}=0.06, the best optimized result can be gotten.

3.3 Planning Result

Fig. 2 shows the optimization results of the disassembly sequence of the throttle valve of genetic algorithm and adaptive genetic algorithm. From the figure, it can be seen that when genetic algorithm runs 139 generations, the fitness value of the population reaches the maximum 2.70, and the corresponding disassembly sequence is 1, 6, 2, 3, 7, 9, 10, 8, 4, 5, 11, while the improved genetic algorithm can gain the same optimal solution when it runs 92 generations.

Run genetic algorithm before and after improvement 50 times respectively and the result is listed in Table 2. It can be seen that genetic algorithm before and after improvement all can find the optimal solution, but the improved genetic algorithm can get faster, and the relative time consumption is shorter. This indicates the improved genetic algorithm can accelerate the convergence speed and reduce the relative time consumption.

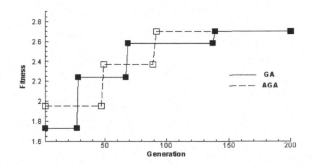

Fig. 2. Comparison of genetic algorithm before and after improvement

Table 2. Comparison of genetic algorithm before and after improvement

Algorithm	Times of the optimal solution being found	Average generations of the optimal solution being found	Average time (s) of running 200 generations	Relevant time consumption (s)
GA	48	163	3.479	2.835
AGA	46	117	3.813	2.231

4   Conclusion

This paper indicates that genetic algorithm is an effective means to optimize disassembly sequence. Genetic algorithm is easy to convergence too early and convergences slowly near the optimal solution. To counter these disadvantages, adaptive method can accelerate the searching speed of genetic algorithm.

References

1. Dai, G.: Research on the techniques of modeling and sequence planning for digital pre-assembly. PhD thesis, Nanjing university of science and technology (2007)
2. Zhang, M., Wang, S.: Improved genetic algorithm of adaptive real range search. Journal of Xi'an Jiaotong University 36, 226–256 (2002)
3. Yu, Y., Liu, Y., Yan, G.: Encoding theory and application of genetic algorithm. Computer Engineering and Applications 3, 86–89 (2006)
4. Sheng, Q.: Study on digital assembly sequence planning system. Master thesis, Nanjing university of science and technology (2006)
5. Han, J.: Research on disassembly sequence planning based on genetic algorithms. Master thesis, Huanzhong university of science and technology (2007)
6. Zhou, K., Li, D., Pan, Y.: Complex product assembly sequence planning based on GSAA. Mechanical Science and Technology 3, 277–280 (2006)

A Method to Raise Test Efficiency of SRM Gas-Tightness

JiaZhao Chen, WenTao Yu, and Liang Qi

Xi'an Research Inst. of Hi-Tech,
Xi'an, Shaanxi Province, P.R.C.
Jazhch@sina.com

Abstract. Taking a solid rocket motor (SRM for short) with 12-front-winged cylindrical charge as example, this paper uses software Fluent to calculate the variation of temperature during gas filling and pressure stabilization in gas tightness test of SRM. The result shows that gas tightness test efficiency can be raised by shortening the time of pressure stabilization.

Keywords: solid rocket motor, gas tightness test, efficiency, Fluent.

1 Introduction

Gas tightness test is an important examination item of solid rocket motor (SRM for short). SRM must keep good gas-tightness to ensure reliable ignition of its main charge as well as its normal and stable work.

Usually, the pressure difference method is used to check SRM's gas-tightness, that is, fill the motor with gas in a certain pressure, and measure the pressure of the inner cavity of SRM after holding the pressure for a period of time. If the pressure difference between the two times is within a certain range, the tightness is believed up to standard, otherwise it is not qualified. This method is simple, easy to be operated and low in cost, but the test time is so long as to influence test efficiency seriously. Shortening the pressure stabilization time is an important way to shorten the whole test time so as to significantly raise test efficiency. To eliminate the influence of temperature rise during gas filling on the pressure, it needs to stabilize the pressure before holding the pressure. Therefore, the gas tightness test usually includes three periods, i.e., gas filling, pressure stabilizing and pressure holding. By using numerical calculation method, this paper analyzes temperature variation during gas filling and pressure stabilizing to seek probability to shorten the time of pressure stabilization.

2 Modeling and Simplifying

Generally, SRM is composed of combustion chamber with charge, nozzle and ignition device, and combustion chamber is composed of shell, heat insulation layer and grain. For some SRMs, a part of nozzle and ignition device sticks into the combustion chamber. Moreover, there is a block cover at the nozzle throat, thus making a closed cavity formed inside the combustion chamber. The gas is charged into this cavity from a filling port in the top cover of the ignition device when gas tightness test.

D. Jin and S. Lin (Eds.): Advances in MSEC Vol. 2, AISC 129, pp. 477–482.

2.1   Geometrical Model Analysis and Simplification

The real structure of SRM is complex, and the grain shape is varied. This paper takes a SRM with 12 front- winged cylindrical grain as example to calculate. The SRM has a diameter of 1.4m, combustion chamber length of 1.666m, inner cavity diameter of 0.348m and wingspan of 0.45m. For easy to calculate, simplification is done when modeling.

(1) Ignore the volume of inner cavity occupied by ignition device as well as the shape variation of the charge column at the joints and the front and rear enclosure.

(2) Neglect the influence of the submerged section of the nozzle on the volume of the inner cavity, and simplify the cavity into a simple winged cylindrical container.

(3) Set a gas filling inlet in the middle of the front enclosure.

A geometrical model of SRM is set up by using software Gambit and the model is divided into tetrahedral grids by using program TGrid [3]. The inner cavity is divided into 17020 grids, as shown in Fig. 1.

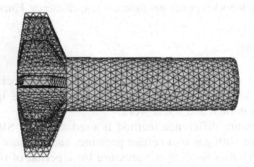

Fig. 1. Inner cavity model and its grid division of SRM

2.2   Gas Charging Model and Simplification

The gas flow and pressure variation are very complex in real gas filling process, so there is hypothesis as follows for easy to calculate.

(1) The gas charged is ideal gas which accords with the state equation of ideal gas.

(2) Neglect the influence of such outside factors as gravity during gas filling.

(3) Only influence of gas filling on the pressure and temperature of the gas in the inner cavity of SRM is considered instead of the influence on the stress and strain of the charge column.

Turbulent flow model RNG k-ε, an improved model based on the standard k-ε model, is taken as the gas flow model during gas charging in the inner cavity of SRM. Because the gas flow is very complex and there is such non-uniform turbulent flow as vortex, RNG k-ε model coincides with the real flow and has high precision when it is used to calculate the flow with large velocity gradient [1]. The heat-transfer model of gas filling is the same as that of pressure stabilization, and the wall is set as temperature boundary condition. After the calculation of gas filling is finished, change pressure inlet boundary condition to wall boundary condition.

2.3   Heat-Transfer Model and Its Simplification

During gas-tightness test of SRM, there is heat exchange between the gas in the inner cavity and outside of SRM, including convection between the gas in the cavity and the charge column, and between the air outside SRM and the shell, as well as heat conduction between the charge column and the heat insulation layer, and between the heat insulation layer and the shell. During test, the thermal radiation is little and can be omitted. The heat exchange between outside and the gas in the cavity can be divided into two parts:

(1) Heat transfer through the enclosure and the block cover in the nozzle;
(2) Heat transfer through the multi-layer cylindrical wall [4] formed by the cylindrical shell of the combustion chamber, heat insulation layer and grain.

Calculation of the thermal resistance of the two parts shows that the cylindrical thermal resistance R2 is much smaller than the thermal resistance R1 of the block cover and enclosure [2]. So it can be believed that heat transfer of the cylindrical section is a major part in whole heat transfer, and the heat transfer through the front and rear enclosure and the block cover can be treated according to the same method as the cylindrical section. For easy to calculate, neglect the influence of the convection of the external environment. During calculation, the wall temperature is believed to be constant 293K as the same as that of the environment from beginning to end.

3   Calculation Result and Its Analysis

In boundary condition, set inlet total pressure as 800000Pa, and total temperature as 293K. In fluid materials, set material attribute as nitrogen (N2), the temperature 293K, and the conductivity as 0.59. Set X Velocity as 0 and temperature as 293K. Build three monitoring points X=0.5, 1.0 and 1.5, using Fluent to calculate the temperature in the cavity of SRM during gas filling and pressure stabilization.

3.1   Calculation Result of Gas Filling

In fact, the main task of pressure stabilization is to eliminate the influence of temperature rise during gas filling on pressure. Fig. 2 shows the temperature variation at X=1.5 during gas filling, and those at the other two monitoring points are basically the same.

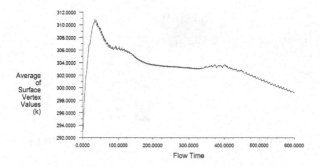

Fig. 2. Temperature variation graph at X=1.5

From the graph, it can be seen that the temperature during gas filling goes up rapidly at first and then falls down slowly, and the top temperature is near 311K. That is because the gas filling speed is high and the temperature is low at the beginning, and the heat produced is more than that transferred, but later the filling speed drops down with the rise of the pressure and the temperature goes up, so the heat produced reduces while the heat transferred increases. At last the heat produced is less than that transferred. That makes the temperature falls down. At the end of gas filling, the temperature drops to 301K.

3.2   Calculation Result of Pressure Stabilization

Fig. 3 shows the temperature distribution contour in the cavity of SRM at 900s' and 1800s' pressure stabilizing (including 600s' gas filling). Fig. 4~6 are the temperature graph at X=0.5, 1.0 and 1.5 during pressure stabilizing (after 600s' gas filling).

It can be seen from the figures:

(1) The temperature variation all over the cavity of SRM is almost the same, and the three points for monitor can largely show the temperature variation tendency in the cavity of SRM.

(2) The temperature variation graphs at the three monitoring points show that the temperature drops fast in the first period because of the high gas temperature in the cavity of SRM, and the falling speed gets slower and slower with the temperature close to the environment temperature. It can be seen that the falling speed gets slow after 10 minutes' stabilization, and it gets very slow after it reaches 294K, but it will take a very long time to reach real balance with the environment. Because little temperature difference results in little pressure difference which has little influence on the result of gas tightness test, it can be believed that the temperature between the inside and outside the SRM has reached balance and has no need to continue pressure stabilization when the temperature drops to some extent. At this time, the pressure stabilization period can be considered over and the pressure holding period starts.

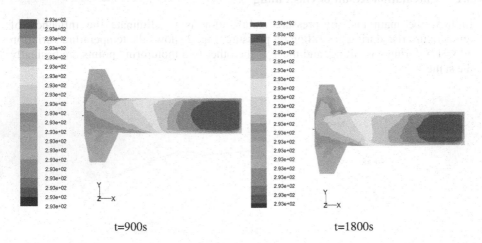

t=900s   t=1800s

Fig. 3. Temperature distribution contour

Temperature variation graph, it can be seen that it will drop temperature slowly in later period, and at last be established as balance. The pressure stabilization where the temperature difference has little difference. Generally, the accuracy of the pressure gauge used to get experiment of ±1K and its definition tendency is 1k/2s. corresponds to relative tendency of ±2K. Therefore, the pressure difference caused by the temperature difference for distinguished leak pressure is stabilized for ±4k. the pressure difference experiment factor, when the SRM is 10.000s. when the stabilized leak pressure... So the temperature data is less for the leak pressure can be identify the pressure stabilization.

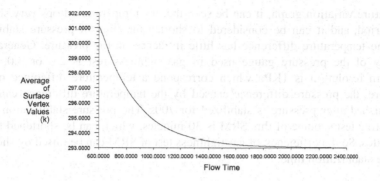

Fig. 4. Temperature variation graph at X=0.5m

References

1. Tian, Z., wang, J., Li, x.: The DZY-I Used in My. Institute of engineering research of calculation of theory and test Evaluation, Test (1985) (in Chinese)
2. wang, J., Li, x.: Jet engine control temperature control of the series awards of the Journal of Weapons Equipment, 7, 113—115 (in Chinese)
3. wu, x.: ... numerical mmm. HJB-12. 11—17 (1987) (in Chinese)
4. Dai, x.: ... Laser of Technology 8 (1987)
5. liu, x., x.i.x.: Aerospace Engineering Press, 500 (in Chinese)

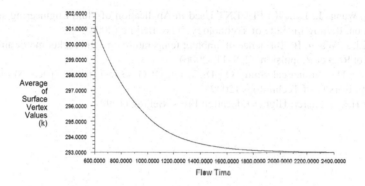

Fig. 5. Temperature variation graph at X=1.0m

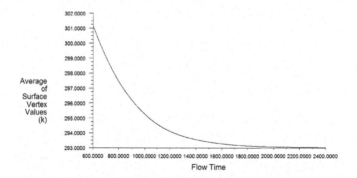

Fig. 6. Temperature variation graph at X=1.5m

4 Conclusion

The pressure stabilization is the process to make the temperature of inner cavity of SRM to drop down and reach balance with the outside environment. From the temperature variation graph, it can be seen that the temperature drops very slowly in later period, and it can be considered to shorten the time of pressure stabilization when the temperature difference has little influence on the pressure. Generally, the accuracy of the pressure gauge used in gas tightness test is 2.0 or 3.0, and its minimum resolution is 1KPa which correspond a temperature difference of 0.8K. Therefore, the pressure difference caused by the temperature difference can not be distinguished after pressure is stabilized for 700s. The pressure stabilization time in the existing test scheme of this SRM is 30 minutes, which can be shortened to about 15 minutes. So the efficiency of gas tightness test of SRM can be raised by shortening pressure stabilization time.

References

1. Han, Z., Wang, J., Lan, X.: FLUENT Used in Application of Fluid Engineering simulation calculation. Beijing Institute of Technology Press, Beijing (2004)
2. Ai, C., Li, J., Wang, B.: Influence of ambient temperature on solid rocket motor airtight test. Journal of Rocket Propulsion 32, 8–11 (2006)
3. Wang, Z.: The Numerical Study On Diffusion Of Gase in Limited Space. Master thesis, Dalian University of Technology (2009)
4. Dai, G.: Heat Transfer. Higher Education Press, Beijing (1999)

Fast NLMS Algorithm with Orthogonal Correction Factors Designed for Adaptive Transversal Equalizers

Huxiong Li

Department of Automatic Control Northwestern Polytechnical University
Xi'an 710072, China
lhxiong@126.com

Abstract. The generalized affine projection algorithm called normalized least mean square with orthogonal correction factors (NLMS-OCF) attempts to accelerate the convergence rate of the normalized least mean square algorithm by adapting weights based on the past input vectors. The NLMS-OCF provides complete flexibility in choosing the past input vectors. A fast version of NLMS-OCF is then derived for the adaptive transversal equalizers based on the input signal characteristic. Simulation results that compare the NLMS-OCF and fast version of NLMS-OCF are presented.

Keywords: Affine projection algorithm, NLMS with orthogonal correction factors, fast affine projection algorithm, equalization.

1 Introduction

The affine projection (AP) algorithm discovered by Ozeki and Umeda [1] applies updates to the weights in a direction that is orthogonal to the most recent input vectors. This speeds up the convergence of the algorithm over that of the normalized least mean square (NLMS) algorithm. Reference [2], based on the direction vector, gives a definition for the AP algorithm and presents the desirable decorrelation properties. Reference [3] introduces a space–time decision feedback equalizer whose coefficients are recursively renewed on the basis of the modified AP algorithm that employs a hyperplane projection scheme to reduce computational complexity without any degradation in its convergence characteristic and tracking capability. Reference [4] presents a fast affine projection algorithm for acoustic equalization which shows the expected tradeoff between convergence performance and computational complexity. Reference [5] proposes low-complexity reduced-rank filters based on finite impulse response filters with adaptive interpolators and develops NLMS and AP adaptive algorithms for the adaptive transversal equalizers.

Reference[6], based on the idea that, the best improvement in weights occurs while the successive input vectors are orthogonal to each other, proposes the NLMS with orthogonal correction factors (NLMS-OCF). These two algorithms—AP and NLMS-OCF—which are independently developed as a result of various interpretations and different perspectives, can be viewed as the same algorithm that updates the estimated weights on the basis of multiple input signal vectors. And the NLMS-OCF provides complete flexibility in choosing the past input vectors. This flexibility provides improved

D. Jin and S. Lin (Eds.): Advances in MSEC Vol. 2, AISC 129, pp. 483–488.
springerlink.com   © Springer-Verlag Berlin Heidelberg 2011

convergence rate over the AP algorithm. The only disadvantage of the NLMS-OCF algorithm is its higher complexity [6]. In this paper, a fast version of NLMS-OCF is proposed to ameliorate this problem for the adaptive transversal equalizers.

2   Nlms-ocf for Adaptive Transversal Equalizer

We use the equalizer problem framework as established in [7]. The resulting equalization problem is depicted in Fig. 1. The channel is assumed to be ideal, i.e., it causes no intersymbol interference. Also, the equalizer is fixed to operate in the training mode—it has exact knowledge of what is transmitted. The above simplification was employed merely to make the current analysis tractable. The input signals are subject to several assumptions as follows. Both two input signals— x_n and ε_n— are assumed to be complex random processes that are zero-mean, wide-sense- stationary, mean-ergodic, and proper. And these two input signals— x_n and ε_n—are mutually independent.

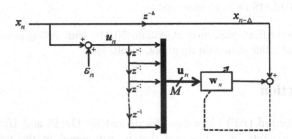

Fig. 1. Adaptive transversal equalization problem

The transversal equalizer structure is determined by two parameters: the number of input taps N and the desired signal delay Δ with respect to the most recent input sample $u_n \triangleq x_n + \varepsilon_n$. The input process is converted into input vectors, via a tapped delay line, and is defined as

$$\mathbf{u}_n = \begin{bmatrix} u_n & u_{n-1} & \cdots & u_{n-N+1} \end{bmatrix}^T \qquad (1)$$

where $(\bullet)^T$ is the transpose operator. The delay Δ for the desired signal must be chosen to be less than N.

The weight vector $\mathbf{w}_n \in \mathbb{C}^N$ is adapted using the NLMS-OCF algorithm with step-size $\mu \in (0,2)$. The number of the orthogonal correction factors is chosen as $M < N$ [6]. The corresponding estimation error signal is written as

$$e_n = x^{n-\Delta} - \mathbf{u}_n^T \mathbf{w}_n \qquad (2)$$

The input vector $\mathbf{u}_n$, which is subject to the l_2 norm, is computed as

$$\left\|\mathbf{u}_n\right\|^2 = \left\|\mathbf{u}_{n-1}\right\|^2 + u_n^2 - u_{n-N}^2 \tag{3}$$

And the μ_0 is defined as

$$\mu_0 = \frac{\mu e_n}{\left\|\mathbf{u}_n\right\|^2} \tag{4}$$

The variable μ is known as the step-size. The new estimate for the weights is implemented as follows

$$\mathbf{w}_n^1 = \mathbf{w}_n + \mu_0 \mathbf{u}_n \tag{5}$$

And the $\mathbf{u}_n^0$ is set as

$$\mathbf{u}_n^0 = \mathbf{u}_n \tag{6}$$

For $k = 1, 2, \cdots, M$, repeat steps from (7) to (10). And $\mathbf{u}_n^k$ is defined as

$$\mathbf{u}_n^k = \mathbf{u}_{n-k} - \sum_{i=0}^{k-1} \frac{\mathbf{u}_{n-k}^T \mathbf{u}_n^i}{\left\|\mathbf{u}_n^i\right\|^2} \mathbf{u}_n^i \tag{7}$$

The corresponding estimation error signal e_n^k is given as

$$e_n^k = x^{n-\Delta-k} - \mathbf{u}_{n-k}^T \mathbf{w}_n^k \tag{8}$$

The μ_k is defined as

$$\mu_k = \begin{cases} \dfrac{\mu e_n^k}{\left\|\mathbf{u}_n^k\right\|^2} & if \ \left\|\mathbf{u}_n^k\right\|^2 \neq 0 \\[2mm] 0 & otherwise \end{cases} \tag{9}$$

The new estimate for the weights is iterated as follows

$$\mathbf{w}_n^{k+1} = \mathbf{w}_n^k + \mu_k \mathbf{u}_n^k \tag{10}$$

At last, we set

$$\mathbf{w}_{n+1} = \mathbf{w}_n^{M+1} \tag{11}$$

Repeat the steps from (2) to (11) for each n, and this constitutes the NLMS-OCF algorithm. The total number of computations needed per NLMS-OCF iteration is [6]

$$2N + 4 + \sum_{k=1}^{M} \left[3N + 2 + (2N+1)k\right] \tag{12}$$

$$= NM^2 + 4NM + M^2/2 + 2N + 5M/2 + 4$$

Summarizing (12), the NLMS-OCF algorithm has a computational complexity of $O\left(NM^2\right)$.

3   Fast Nlms-ocf for Adaptive Transversal Equalizer

The iterated direction of the NLMS-OCF algorithm $\mathbf{u}_n^k, 0 \leq k \leq M$ are orthogonal each other [6]. Under the assumption that these two input signals— x_n and ε_n — are complex random processes that are zero-mean, wide-sense-stationary, mean-ergodic, and proper, we can obtain the result

$$E\left(\mathbf{u}_{n-k}^T \mathbf{u}_n\right) = 0, k = 1, 2, \cdots, M \tag{13}$$

Based on (6), (7) and (13), the following result can be obtained

$$E\left(\mathbf{u}_{n-1}^T \mathbf{u}_n^0 \mathbf{u}_n^0\right) = 0 \tag{14}$$

So $n \to \infty$, based on (7) and (14), we can consider that

$$\mathbf{u}_n^1 \approx \mathbf{u}_{n-1} \tag{15}$$

The same idea as (15), we can obtain the following results

$$\mathbf{u}_n^k \approx \mathbf{u}_{n-k}, k = 1, 2, \cdots, M \tag{16}$$

The estimate for the weights and error signal are defined as $\mathbf{w}_n = \mathbf{w}_n^0 = \mathbf{w}_{n-1}^{M+1}$ and $e_n = e_n^0$, respectively. Therefore, a fast version of NLMS-OCF algorithm is obtained and summarized as the following:

$$\left\|\mathbf{u}_n\right\|^2 = \left\|\mathbf{u}_{n-1}\right\|^2 + u_n^2 - u_{n-N}^2 \tag{17}$$

For $j = 0, 1, 2, \cdots, M$, repeat steps from (18) to (20).

$$e_n^j = x^{n-\Delta-j} - \mathbf{u}_{n-j}^T \mathbf{w}_n^j \tag{18}$$

$$\mu_j = \frac{\mu e_n^j}{\left\|\mathbf{u}_{n-j}\right\|^2} \tag{19}$$

$$\mathbf{w}_n^{j+1} = \mathbf{w}_n^j + \mu_j \mathbf{u}_{n-j} \tag{20}$$

At last, we set

$$\mathbf{w}_{n+1}^0 = \mathbf{w}_n^{M+1} \tag{21}$$

Repeat the steps from (17) to (21) for each n, and this constitutes the fast version of NLMS-OCF algorithm. The total number of computations needed per fast NLMS-OCF iteration is

$$2 + \sum_{j=0}^{M}(2N+2) = 2(NM + N + M + 2) \tag{22}$$

From (22), it can be concluded that computational complexity of the fast version NLMS-OCF algorithm is $O(NM)$ for the adaptive transversal equalizers.

4   Comparison with Simulation Results

In this section, we compare the mean square error (MSE) learning curves from simulations between the NLMS-OCF and fast NLMS-OCF. The initial estimate for the weights is zero. The parameter N is selected to be 32 and the delay Δ is chosen to be 16. Both the amplitudes of the real and imaginary parts of the input signals x_n are set to be one.

Case 1. The noise ε_n is assumed to be absent. We use the number of the orthogonal correction factors $M=8$ and the step-size $\mu=0.3$. The steady-state simulations are given by averaging steady-state iterations from 2000 to 5000. The MSE behavior predicted by the NLMS-OCF is shown in Fig. 2, together with the result predicted by the fast NLMS-OCF. We observe that the predicted MSE results by the two algorithms are much closer each other.

Case 2. Both the amplitudes of the real and imaginary parts of the noise ε_n are assumed to be equal 10^{-4}. We use the number of the orthogonal correction factors $M=8$ and the step-size $\mu=0.3$, which is the same as case 1. The steady-state simulations are given by averaging steady-state iterations from 1000 to 4000. The MSE behavior predicted by the NLMS-OCF and fast NLMS-OCF are shown in Fig. 3. It can be concluded that the fast NLMS-OCF reduces computational complexity without any degradation in its convergence characteristic compared with the NLMS-OCF.

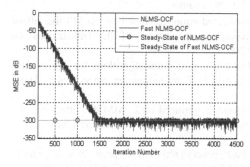

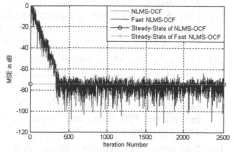

Fig. 2. Learning curves of NLMS-OCF and fast NLMS-OCF

Fig. 3. Learning curves of NLMS-OCF and fast NLMS-OCF

5   Conclusion

Under the assumption that the input signals are the complex random processes that are zero-mean, wide-sense-stationary, mean-ergodic, and proper, a fast version of the NLMS-OCF is proposed for the adaptive transversal equalizers in this paper. This algorithm allows for flexibility in the choice of the input vectors used for adaptation. The fast NLMS-OCF reduces computational complexity without any degradation in its convergence characteristic compared with the NLMS-OCF. In some situation, the simulation results show that the convergence rate of the fast version of the NLMS-OCF is a slightly better over the NLMS-OCF. We think the reason is that the iterated direction of the NLMS-OCF algorithm is $\mathbf{u}_n^k$, which is different from the direction $\mathbf{u}_{n-k}$ that causes the estimation error signal. However, the iterated direction of the fast version of the NLMS-OCF is $\mathbf{u}_{n-k}$, which direction also causes the estimation error signal. The price paid for the improvement comes from the increased memory requirements. But it is not serious drawbacks since DSPs have a large amount of on-chip memory and a high throughput in present day.

Acknowledgement. This work is supported by foundation of Wenzhou science and technology bureau (G20090104). The authors are grateful for the anonymous reviewers who made constructive comments.

References

[1] Ozeki, K., Umeda, T.: An adaptive filtering algorithm using an orthogonal projection to an affine subspace and its properties. Electronics and Communication in Japan 67-A(5), 19–27 (1984)
[2] Rupp, M.: A family of adaptive filter algorithms with decorrelating properties. IEEE Transactions on Signal Processing 46(3), 771–775 (1998)
[3] Lee, W.C.: Space–time adaptive decision-directed equalizer based on NLMS-like affine projection algorithm using iterative hyperplane projection. IEEE Transactions on Vehicular Technology 56(5), 2790–2797 (2007)
[4] Bouchard, M.: Multichannel affine and fast affine projection algorithms for active noise control and acoustic equalization systems. IEEE Transactions on Speech and Audio Processing 11(1), 54–60 (2003)
[5] Lamare, R.C.D., Sampaio-Neto, R.: Adaptive reduced-rank MMSE filtering with interpolated FIR filters and adaptive interpolators. IEEE Signal Processing Letters 12(3), 177–180 (2005)
[6] Sankaran, S.G. (Louis) Beex, A.A.: Fast generalized affine projection algorithm. International Journal of Adaptive Control and Signal Processing 14(6), 623–641 (2000)
[7] Ikuma, T. (Louis) Beex, A.A., Zeidle, J.R.: Non-wiener mean weight behavior of LMS transversal equalizers with sinusoidal interference. IEEE Transactions on Signal Processing 56(9), 4521–4525 (2008)

An Improved Kerberos Intra-domain Authentication Protocol Based-On Certificateless Public-Key Cryptography

Wang Juan[1], Cao Man-cheng[1], and Fang Yuan-kang[1,2]

[1] Chizhou College, Chizhou, Anhui
[2] Information Science and Technology School Nanjing University of Aeronautics and Astronautics Nanjing China
Fyk80@163.com

Abstract. The paper sums up some improvements in Kerberos intra-domain authentication protocol included in many domestic and foreign literatures. By analyzing the limitations of those improvement schemes, an improvement in Kerberos intra-domain authentication protocol based on certificateless public-key thought is proposed. The analysis shows that the improvement proposal can overcome some defects in the original Kerberos intra-domain authentication protocol, such as the key escrow problem and network intermediaries attack, etc. Moreover, the improvement also meets the demand of security proposed by key agreement protocol, which has a certain security and perspective of application in the process of network identity authentication.

Keywords: Kerberos, Intra-domain Authentication, Security Attributes, Certificateless Public key cryptography.

1   Introduction

Kerberos[1], which has intra-domain and inter-domain authentication modes, is a widely-used identity authentication protocol based on the trusted third party, developed firstly by the Project   of Athena in Massachusetts Institute of Technology (MIT). The nucleus of Kerberos is the authentication center – Key Distribution Center (KDC), which consists of authentication server AS and Ticket Granting Server (TGS) . The basic principal is as follows: If a user wants to access some application server, it must get its identity authentication in KDC and obtain the ticket to visit the application server, which provides direct service for the user with the ticket.

The method adopted by traditional Kerberos intra-domain authentication protocol is symmetric data encryption standard DES, in which there exists such limitations as clock synchronization being difficult, password guessing attack, complicated storage and management of keys, not providing digital signature, and undeniable mechanism[2], which lead to poorer internet protocol security.

Due to the limitations of traditional Kerberos intra-domain authentication protocol, many literatures at home and abroad have improved it by adopting asymmetric (public key) encryption system RSA, called for short, Kerberos RSA protocol[3-6], which, to

D. Jin and S. Lin (Eds.): Advances in MSEC Vol. 2, AISC 129, pp. 489–496.
springerlink.com

some extent, overcomes those limitations existing in traditional Kerberos. However, RSA is not so perfect, because it has such weaknesses as slower encryption/decryption speed, and lower execution efficiency[7]. If public key system is adopted to encrypt and decipher in the process of transmitting data, the authentication efficiency must be affected.

In order to make up the defects of poorer security of symmetric key system and lower execution efficiency of public key system, literature [8-9] has improved Kerberos protocol by mixing symmetric data encryption system DES with asymmetric (public key) encryption system RSA, called for short, mixed-system Kerberos intra-domain authentication protocol. The improvement, to some extent, has eased the limitations in traditional Kerberos and Kerberos RSA by combining higher execution efficiency of symmetric encryption system with higher security of public key encryption system. Moreover, in order to prevent the possible internal attack existing in mixed-system Kerberos intra-domain authentication protocol, literature [9] has proposed a safer Diffie-Hellman key exchange protocol – public keys in mixed encryption scheme should adopt Diffie-Hellman key exchange protocol.

But after analyzing the improved Kerberos intra-domain authentication protocol in literature [9], the researcher found there are still drawbacks in it. That's because, after clients (C) and application server (S) passed identity authentication, the two sides' exchanging of parameters used in producing consultation session keys, is still transmitted with Kc.s encryption generated by Kerberos. Thus, Kerbers may still possibly intercept and capture the parameters exchanged by both sides, and then gain session keys by impersonating as C and consulting with S. Therefore, such session keys are not safe.

In sum, on the basis of a certificateless public key Cryptography system[10], the paper has proposed Kerberos intra-domain authentication key agreement scheme, which may solve the above problems efficiently,and can meet the demands of the present known key agreement protocol's security attributes.

2   Related Pre-knowledge

2.1   Key Agreement Protocol's Basic Security Attributes

Literature [11-12] lists several security attributes needed to investigate while making a security analysis of most protocols at present.

a) Key Hidden Authentication（KHA）. Each user of the protocol believes that only the protocol's participants know session keys, which cannot be obtained by attackers. Providing key agreement protocol identified by keys hidden can resist man-in-the-middle attack.

b) Known Session Key Security（KKS）. Even if some previous session key is exposed or obtained initiatively by attackers, the attackers cannot get access to any other session keys.

c) Forward Security（FS）. Long-term private key exposure of a protocol participant cannot affect the security of his previous session keys.

d) Resist Key Compromise Impersonation Attack(KCI). If A's long-term private key is breached, the attacker may disguise as A, but he cannot disguise as any other entity in the name of A.

e) Unknown Key Shared Security (UKS). If the shared session key of both transmitter and receiver is K, the attacker cannot enforce the session key shared by both sides as K'.

f) Keys' Uncontrollability (KU). All participants of the protocol cannot control the output of session keys, which is called session keys' uncontrollability of the protocol.

2.2 Linear Diffie-Hellman Problem

Set G1, G2 respectively for a q order group, q is a large prime number, G1 is an additive group; G2 is a multiplicative group; P is a generator of G1. Discrete logarithm problem in G1 and G2 is an intractable problem. If the map ê : G1×G2→G2 satisfies the following properties, this map is called an admissible bilinear map. [13]

1) Bilinearity: Given arbitrary P,Q∈G1 and arbitrary a, b∈Zq*, then the equation ê(aP, bQ)=ê(P,Q)ab can be established.

2) Non-degeneracy: If P, Q∈G1 exists, the inequality ê(P,Q)≠1 can be established.

3) Calculability: For any P, Q∈G1, there is an effective algorithm to calculate ê(P,Q).

Difficult problems related to cryptography calculation:

(1) Calculating discrete logarithm problem (DLP): Given P, Q, assume that Q=n P(n∈Zq*) exists, find n.

(2) Calculating Diffie-Hellman problem (CDH): Given P, aP, bP, among which, a, b ∈ Zq*, calculate abP.

(3) Calculating Bilinear Diffie-Hellman problem (BDH): Given P, aP, bP, cP, among which, a, b, c∈Zq*, calculate ê(P,P)abc

2.3 Certificateless Public-Key Cryptography Principles and Processes

At the Asian Cryptography Meeting in 2003, some experts like Al-Riyami proposed the thought of certificateless public-key cryptography, which, still on the foundation of linear Diffie-Hellman problem, is a cryptosystem based on public key infrastructure (PKI) and identity characteristics[13-14]. The cryptosystem is equipped with a key generation center (KGC), whose primary role is to create a partial private key for users. Following that, the users can obtain their long-term private keys by combining one of their random secret values with the partial private key produced by KGC; and gain their public keys by combining the secret value with the system public key of KGC. That is to say, in the certificateless public-key system, the user's private keys are produced through his own calculation with the participation of KGC. Thus, the user's private keys are only known by the user himself, which solves the key escrow problem in the public key system. The process of a certificateless Public-key cryptosystem is consisted of the following four steps:

(1) System Initialization: G1, G2 are the two cyclic groups with order for q on an elliptic curve. The map ê : G1×G2→G2 is a bilinear map. Choose a one-way cryptographic hash function

H1 : {0,1}*→G1 ; H2 : {0,1}n×G2→Zq* (n stands for plaintext length). A random number s∈Zq*, generated by KGC and saved as a system master key, together with P∈G1, a generator of G1, can create KGC system public key Ppub=sP. Then the system parameters params={G1,G2,ê,q,P, Ppub,H1 } can be disclosed.

(2) Extract Partial Private Key: The user A provides his identity information IDA to KGC. After KGC verifies A's identity, QA=H1(IDA) and partial private key DA=sQA can be extracted. Then, through a secret security channel, QA, DA can be transmitted to the user A, who may verify the authenticity of DA by means of the equation ê(DA,P)=ê(QA,Ppub).

(3) Choose A Secret Value: The user A chooses randomly a value xA∈G1 as his own secret value.

(4) Generate Private Key and Public Key: At the client side, after the user A inputs the partial private key created by KGC and his own secret value xA, A's long-term private key SA =xADA=xAs QA and public key PA= xAPpub= xAsP can be generated.

3 The Improvment of Kerberos Intra-domain Authentication Protocol Based-On Certificateless Public-Key Cryptography

3.1 Certificateless Intra-domain Authentication Key Agreement Protocol Adopted by This Paper

The protocol includes three consultation entities,A key generation center and both parties of intra-domain communication A and B. A and B must use the shared key obtained through consultation to start a secure session. KGC's public parameters are {G1,G2,ê,q,P,H1 } and each parameter description is the same as above. KGC generates randomly a system master key s∈Zq* and calculating system public key Pkgc=sP.

According to the theory of Certificateless Public-key Cryptography in section 1.3, clients A and B must submit respectively their identity information IDA and IDB to KGC, which will return the results to them after it calculates both parties' partial private keys DA=sQA=sH1(IDA) and DB=sQB=sH1(IDB). Then the combination of the secret value xA,xB∈Zq*, chosen randomly and separately by A and B, with the partial private keys DA and DB, returned by KGC, will generate their own long-term private keys: SA=xADA and SB=xBDB, public keys: PA=xAP and PB=xBP, and temporary session keys:

S'A=DA+xAQA=(s+xA)QA

S'B=DB+xBQB=(s+xB)QB

The processes to obtain the shared session keys through consultation are as follows:

(1) After A and B choose randomly and separately the secret number r1,r2∈Zq*, calculate TA=r1QA and TB=r2QB.

(2) A must transmit <IDA,TA,PA, MACKA(IDA,TA,PA)> to B, and meanwhile B must transmit <IDB,TB,PB, MACKB(IDB,TB,PB)> to A. Moreover, they must verify the integrity of their messages, among which MACkx is the user X's message authentication code used to guarantee the data's integrity.

A→B:IDA, TA, PA, MACKA(IDA, TA, PA)

B→A:IDB, TB, PB, MACKB(IDB, TB, PB)

If:

MACKA(IDA, TA, PA)= MACKB(IDB, TB, PB)

MACKA(IDB, TB, PB)= MACKB(IDB, TB, PB)

The equations illustrate that the message exchange process is not subject to malicious attacks, which may ensure to generate a shared session key through negotiation. If the verification results do not match, key negotiation must be carried on again.

(3) A and B calculate respectively KA and KB :

$KA=\hat{e}(S'A,P)r1\cdot \hat{e}(TB,\qquad Pkgc+PB)\qquad =\hat{e}(S'A,P)r1\cdot \hat{e}(TB,\qquad Pkgc+PB)$
$=\hat{e}((s+xA)QA,P)r1\cdot \hat{e}(r2QB,(s+xB)P)=\hat{e}(QA,P)r1(s+xA)\ \hat{e}(QB,P)r2(s2+xB)$;

$KB=\hat{e}(S'B,P)r2\cdot \hat{e}(TA,\qquad Pkgc+PA)\qquad =\hat{e}(S'B,P)r2\cdot \hat{e}(TA,\qquad Pkgc+PA)$
$=\hat{e}((s+xB)QB,P)r2\cdot \hat{e}(r1QA,(s+xA)P)=\hat{e}(QB,P)r2(s+xB)\ \hat{e}(QA,P)r1(s+xA)$

This time, the equation K= KΛ= KB may be verified, and so K is the shared session key obtained through negotiation.

3.2   Improved Kerberos Intra-domain Authentication Protocol

In the new improved scheme, the function of KGC (Key Generati on Center) in certificateless public key system is integrated into that of KDC (Key Distribution Center) in Kerberos, where the registered user's public key is stored. Shown as Figure 1, if the registered client visits intra-domain application server, KDC will at first generate the system master key s∈Zq* and the system public key Pkdc=sP. Then the registered client's partial private key DC and the intra-domain application server's partial private key Ds will be generated by KDC, too. After that, C (client) and S (server) choose their own secret values c and s, and generate, through calculation, their respective public key, private key and temporary key PC/RC/ TC and Ps/Rs/Ts. The newly improved Kerberos intra-domain authentication protocol may be described with formula's symbolization as follows: E and D represent Encryption Algorithm and Decryption Algorithm respectively; KPx and KRx indicate X's public key and private key respectively; Authenticatorx,y signifies that x is the authentication ticket to access y; and r stands for a random number extracted to prevent replay attacks.

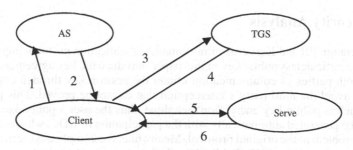

1 to request ticket license; 2 ticket license; 3 to request service ticket; 4 service ticket;
5 to request service; 6 to offer two-way authentication

Fig. 1. Kerberos Intra-domain Authentication Processes

Step 1: From AS, C obtains the ticket TGT to visit TGS.

1) C->AS: DKRc(IDc) ,IDtgs,R1

Client (C) sends a request message to get access to TGS from AS, and the message includes C's signing messages and a random number R1 used to illustrate that the access request toward AS is new.

2) AS->C: Ptgs, TGT

Once AS verifies C's identity, it will transmit to C TGS' public key and the ticket to access TGS: TGT= EKPtgs(IDc, ADc, Pc,Lifetimes1).

Step 2: From TGS, C obtains the service ticket (ST) to access application server.

3) C->TGS:TGT,Authenticatorc,tgs

Client (C) shows TGT to TGS, and submits the authentication ticket Authenticatorc,tgs = EKPtgs (IDc,IDs,R2) to TGS,

4) TGS->C:EPc(Ps, ADs), ST

TGS transmits to C the generated ticket ST= EKPs(IDc, ADc,IDs,Pc,Lifetimes2,R2) and S' public key and address encrypted with C's public key.

Step 3: C and S accomplish the two-way identity authentication.

5) C->S: ST, Authenticatorc,s

Holding the service ticket (ST) issued by TGS, C submits to S the authentication ticket Authenticatorc,s =EKPs(IDc,IDs,R3), used to get access to the application server.

6) S->C:EPc(R3)

S returns R3 to C.

The above processes mainly accomplish the two-way identity authentication between C and S. According to the certificateless key agreement protocol proposed by this paper, C and S may obtain the shared session key through a key negotiation. The processes are as follows:

7) C->S:EKRc(Tc=r1Qc)

8) S->C:EKRs(Ts=r2Qs)

C calculates the session key Kc=ê(S'c,P)c • ê(Ts, Pkdc+Ps) =ê(S'c,P)c • ê(Ts, Pkdc+Ps) =ê((s+xc)Qc,P)c • ê(sQs,(s+xs)P)=ê(Qc,P)c(s+xc) ê(Qs,P)s(s+xs)) ;

S calculates the session key Ks=ê(S's,P)s • ê(Tc, Pkdc+Pc) =ê(S's,P)s • ê(Tc, Pkdc+Pc) =ê((s+xs)Qs,P)s • ê(cQc,(s+xc)P)=ê(Qs,P)s(s+xs) ê(Qc,P)c(s+xc)

K=Kc=Ks is the final session key obtained through negotiation.

4   Security Analysis

Both literature [9] and Kerberos intra-domain authentication protocol improved on the basis of certificateless public key in this paper introduce the key agreement protocol, in which both parties of communication obtain the session key through consultation in order to avoid the third party's interception that cannot be proved. This paper adopts certificateless public-key encryption technique, thus the user's public key and private key can be generated automatically with the participation of KDC, which solves the key escrow problem in the original protocol. Meanwhile, the complicated verification of the public key system in PKI (Public Key Infrastructure) can also be omitted, which enhances the system's operating efficiency. Moreover, Kerberos' server only needs to save all users' public keys. Therefore, even if the server is breached, attackers can only obtain the users' names and their public keys. Without obtaining the users' private keys, the attackers cannot get the system service.

As for the security of the key agreement protocol, this scheme completely meets the demands proposed in Section 1.1.

a) Key Hidden Authentication (KHA). Supposed an attacker X communicates with S in the name of C, it may choose a random number x and transmit $Tx = xQc$, Pc to S, and therefore obtains Ts and Ps. However, X cannot gain C's temporary private Tc. Besides, to work out s from $Ts = sQs$ is equal to solving a discrete logarithm problem (DLP). Therefore, the protocol can provide the function of key hidden authentication.

b) Known Session Key Security (KKS). While executing each key agreement protocol, both participants C and S may reselect a random number as the secret value and obtain a new session key through consultation. Therefore, if a session key is let out, it cannot influence the conversations before or after the session key.

c) Forward Security (FS). Presumed an attacker obtains C's long-term private key Sc, he cannot work out Xc through $Sc = xcDc$ due to a discrete logarithm problem (DLP). Thus, he cannot obtain C's short-term key S'c. Moreover, the attacker does not know the temporary secret random number r1 chosen by C, so the session key K cannot be influenced. Therefore, the exposure of C's private key cannot lead to the reveal of its session key. Even if KDC's primary secret key s is exposed, and an attacker can work out their partial private keys, but because KDC does not know both Client and Server's private keys, the attacker cannot obtain Client's long-term and temporary keys. Likewise, he cannot work out the session key. For that reason, the protocol has the attribute of forward security.

d) Resist Key Compromise Impersonation Attack (KCI). Presumed an attacker X knows the application server S' private key $Rs = xs \cdot Ds$, and if he wants to personate client C to communicate with S, X must accurately figure out $K = Kc = Ks$. $Kc = ê(S'c,P)c \cdot ê(Ts, Pkgc+Ps)$. Not knowing the secret value c, X cannot exactly calculate Kc. Similarly, not knowing the short-term key Ts and the secret value s, X cannot calculate Ks, either. Hence, the protocol has the ability to resist key compromise impersonation attack.

e) Unknown Key Shared Security (UKS). Supposed attacker A enforces C and S to share the session key K', but it is impossible for them to share a session key because both C and S' identities are unauthenticated and there is no consultation between them. After the key agreement is reached, C and S need to confirm the message integrity to verify the validity of the session key. Therefore, the agreement has the unknown key shared security.

f) Keys' Uncontrollability (KU). In the agreement, the parameter values required to generate the session key , such as <IDc, Tc, Pc>, <IDs,Ts,Ps> are provided by the parties involved in the agreement. Namely, the session key is generated through C and S' joint consultations, in which one party is not controlled by the other. Furthermore, either party cannot pre-determine the session key value. Therefore, the agreement has keys' uncontrollability.

5 Summary

This paper has improved Kerberos intra-domain authentication protocol on the basis of certificateless public-key cryptography thought. The analysis shows that the improvement proposal can solve more effectively some problems existing in the original Kerberos intra-domain authentication protocol, such as the shared session key escrow problem, the third party's interception of secret message that cannot be proved, etc. Moreover, the improvement scheme is more practical because it meets the security demand proposed by the key agreement protocol in literature [11-12]. However, the scheme also has its own defect – to increase the amount of calculation of all communication parties. With the deep and further study of the protocol, the corresponding optimization measures will be adopted to satisfy people's higher requirements for security and practicality of the identity authentication technique in the complicated network environment.

Acknowledgment. The authors would like to thank the anonymous referees for the useful suggestions for improving this paper. This project is supported by Anhui Natural Science Foundation of (KJ2011B108) and the National High Tech Research and Development Plan of China under Grant No.2009AA010307.

References

[1] Steiner, J.G., Neuman, B.C., Schiller, J.I.: Kerberos:An Authentication Service for Open Network Systems. In: USENIX Conference Proceedings, pp. 191–202 (February 1988)

[2] Bellovin, S.M., Merritt, M.: Limitations of the Kerberos Protocol. In: Winter 1991, USENIX Conference Proceedings, pp. 253–267. USENIX Asociation (1991)

[3] Ganesan, R.: Yaksha: augmenting Kerberos with the public key cryptography. In: Proceedings of the Internet Society Symposium on Network and Distributed System Security, pp. 132–143. IEEE Computer Society Press (1995)

[4] Liu, K.-L., Qing, S.-H., Meng, Y.: An Improved Way on Kerberos Protocol Based on Public-Key Algorithms. Journal of Software (6), 872–877 (2001)

[5] Mo, Y., Zhang, Y.-Q., Li, X.: Study of the Attacks on Kerberos Protocol and Countermeasures. Computer Engineering 31(10), 66–69 (2005)

[6] Tian, J.-F., Bi, Z.-M., Zhang, J.: An Improved Way on Kerberos Protocol Based on Public-Key Algorithms. Microelectronics & Computer 25(9), 161–164 (2008)

[7] Zhou, T., Wang, J.-Y., Li, M.-J., Li, Z.-J.: Analysis and Comparison of the Kerberos Protocol's Versions. Computer Science 36(2), 119–128 (2009)

[8] Tang, W.-D., Li, W.-M., Zhou, Y.-Q.: Improving Kerberos protocol with ElGamal algorithm. Computer Engineering and Design 27(11), 2063–2065 (2006)

[9] Hu, Y., Wang, S.-L.: Research on Kerberos identity authentication protocol based on hybrid system. Journal of Computer Applications, 1659–1661 (June 2009)

[10] Al-Riyami, S.S., Paterson, K.: Certificateless Public Key Cryptography. In: Laih, C.-S. (ed.) ASIACRYPT 2003. LNCS, vol. 2894, pp. 452–473. Springer, Heidelberg (2003)

[11] Blake Wilson, S., Johnson, D., Menezes, A.: Key agreemen tp rotocols and their security analysis. In: Proc. of the 6 th IMA Inter2national Conference on Cryptography and Coding, p. 30245. Springer, Heidelberg (1997)

[12] Liu, W.-H., Xu, C.-X.: Certificateless two-party key agreement scheme without bilinear pairing. Application Research of Computers 27(11), 4287–4292 (2010)

[13] Lippold, G., Boyd, C., Gonzalez Nieto, J.: Strongly Secure Certificateless Key Agreement. In: Shacham, H., Waters, B. (eds.) Pairing 2009. LNCS, vol. 5671, pp. 206–230. Springer, Heidelberg (2009)

[14] Zhang, L., Zhang, F.-T.: A Method to Construct a Class of Certificateless Signature Schemes. Chinese Journal of Computer 32(5), 940–945 (2009)

On Evaluating English On-Line Study Websites in China

Hanxiong Zhu, Ruiting Xie, and Yi Xie

School of Foreign Languages,Wuhan University of Technology, 430071 Wuhan,
Hubei Province, China
zhuhx@whut.edu.cn

Abstract. The paper conducts a comparative study of English on-line study
websites in China. First it looks into the basic requirements and disciplines for
English study websites and selects four such websites to examine their
characteristics under the integrated index. Analytical Hierarchy Process (AHP) is
applied to set up a website evaluation system. Based on the research findings, the
paper proposes measures for improving the quality of English study websites in
China.

Keywords: English Study Website, Analytical Hierarchy Process, Website
Evaluation System.

1   Introduction

English learning is very important in China while it is also challenging for many
Chinese learners. With the development of Information Technology, studying online
becomes increasingly popular. However, there are certain problems in many
newly-emerged English study websites. Therefore, it is necessary to set up an
evaluation system for these websites so as to help English learners to make good use of
them and to help the websites' self-improvement and optimization.

On-line teaching websites have been studied by western scholars for some years. As
early as 1983, Desmond Keegan studied on-line teaching system.[1] There are already
a large number of research results and literature on common websites such as the
researches about commercial websites and service websites, but very few on English
on-line study websites.[2]

In China, scholars' study mainly focuses on the applications, designs and
developments, but few on website evaluations and managements.[3] The research of
e-learning on English study website includes the development of English on-line
teaching, the on-line curriculum and the comprehensive evaluation of the website.
Since the information era arrives, the studies of English language teaching assisted by
network have been largely developed and produced certain results. Wu Heping(2000),
after studying the innovative establishment of English teaching environment with the
assistance of network, proposed that on-line teaching websites can provide language
environment and psychological environment for language teaching thus have great
potential for growth.[4] As for the studies of English on-line curriculum, scholars focus
on some specific websites with the design and development of the courses. Some also
study the development and evaluation of English on-line study websites based on their

D. Jin and S. Lin (Eds.): Advances in MSEC Vol. 2, AISC 129, pp. 497–501.

distinguishing characteristics. For example, Han Yongbo(2003) , based on the interactive e-learning of English, made the researches on e-learning, feedbacks of immediate information and applications of multimedia.[5]Through literature review, it is clear that the study on an evaluation system for the websites is rare. This paper aims to look into this area and establish an evaluation system for English on-line study websites based on the user's experience. In doing so it hopes to provide guidance to English learners in China in making use of the websites and give suggestions to the websites for self-improvement.

2   Elements and Characteristics of English On-Line Study Website

The evaluation of websites can be many faceted, but five basic elements are required to ensure the quality of English on-line study websites, which are (1) Adaptability for users. A comprehensive English study website should be adapted to most users of different English levels. In other words, the convenience and effectiveness of learning is available to learners of all levels; (2) Catering to users' demand. The websites can provide abundant resources to meet users' personal requirements; (3) Convenience in e-learning. Users can have e-learning anywhere and at anytime on the website, and study on line as long as they can; (4) Interactiveness. The website can interact with users by providing them with exercises and tests, such as on-line self examining exercises, mock tests, posting questions and so on; (5) Interface and unique features. English on-line study websites should present friendly interfaces, logical layouts and unique features.

The above five basic elements serve to benchmark the quality of English on-line study websites and can form the basis for the establishment of website evaluation standards.

Apart from the five basic elements, other factors such as Matthew Effect and Path Dependence and Lock-up also play key roles in the website evaluation system. According to Matthew Effect, when the websites in their early stage were well managed, they could attract a large number of users. Even when they lose their advantages when new and better websites emerge, they still retain the same number of users. This is likely to cause confusion in the evaluation of the websites. On the other hand, some users keep on using the first website they happened to find without any evaluations or comparisons with others websites. Just like Matthew Effect, this Path Dependence and Lock-up Phenomenon also can a negative impact in the selection of an appropriate website, and may hinder the development of a promising on-line study website. Therefore, setting up a scientific evaluation system is imperative as it benefits both users and the websites.

3   Selection of English On-Line Study Websites

Through Google Search Engine in simplified Chinese version, this paper investigates online the Chinese English study websites. By typing "On-line English study" in the Google search bar, it appeared 60 links in the first six pages, which included 8 pieces of news, 7 pieces of advertisement and 45 pieces of website information. In order to select the most popular English study websites, this paper selects 8 websites according to the

ALEXA international click rate ranking, and then chooses the first 4 which have higher click rates than others for further study. They are: Dict, Putclub, Ebigear and Tingroom. (Statistics as shown in Table 1, by July 28th, 2011)

Table 1. First 8 websites according to the ALEXA international click rate ranking

ALEXA ranking	ALEXA ranking in 3 months	IP visits in a day	links
1.Dict	1602	2475000	www.dict.cn
2.Putclub	16399	264000	www.putclub.com
3.Ebigear	18878	237000	www.ebigear.com
4.Tingroom	22801	186000	www.tingroom.com
5.Englishbaby	45580	130200	zh.englishbaby.com
6.Wwenglish	41177	126000	www.wwenglish.com
7.24en	54094	96300	www.24en.com
8.gyii	1494638	2700	www.gyii.com

4   Evaluation of English On-Line Study Website Based on AHP Model

Due to the complexity of different websites, it is difficult to make direct comparisons and ranking among them. It is more feasible to find and select median index for evaluation and ranking. After determining the importance index in the four chosen websites, this paper finds the weight of each by integrating the data and applies AHP to make a comprehensive ranking of these popular English study websites.

4.1   Website Evaluation System Based on AHP Model

This evaluation method is an application of AHP. In this evaluation system, median indexes include adaptability for users, catering for user demand, convenience in e-learning, on-line interactiveness and interface & characteristics of website.(as shown in Figure 1)

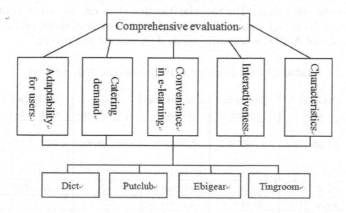

Fig. 1. Evaluation system of English on-line study websites

4.2   Index Weighing

According to the different contributions the factors of criterion layer makes to the evaluation objective, the paper establishes a judgment matrix of intermediate indexes aimed at the target layer. Through the hierarchical single order of the matrix which includes: the adaptability for users, satisfaction to the demands, the convenience of e-learning, on-line interactivity, and the interface and characteristics of website, then the relative weight vector of criterion layer can be worked out: which is $W_u = (0.103092784, 0.309278351, 0.515463918, 0.051546392, 0.020618557)^T$; and through the consistency check, the random consistency ratio is: C.R=0<0.1, which achieves a high consistency on all of the five elements. With the same measurement, the paper works out the weight vectors of each hierarchical single level of goal layer, which are:

1) Adaptability for users
$W_1 = (0.071428571, 0.285714286 , 0.428571429, 0.214285714)^T$ C.R=0<0.1
2) Catering for user demands
$W_2 = (0.090909091, 0.181818182 , 0.363636364, 0.363636364)^T$ C.R=0<0.1
3) Convenience in e-learning
$W_3 = (0.058823529, 0.352941176 , 0.352941176, 0.235294118)^T$ C.R=0<0.1
4) On-line interactiveness
$W_4 = (0.125, 0.5, 0.25, 0.125)^T$ C.R=0<0.1
5) Interface and characteristics of website
$W_5 = (0.083333333, 0.333333333, 0.5, 0.083333333)^T$ C.R=0<0.1

It has been demonstrated that all of the random consistency ratios of goal layer are less than 0.1, which means this model passes the consistency check.

4.3   Comprehensive Evaluation

According to the statistics in Table 2, the second and third elements draw more attention from users than other three intermediate indexes. It suggests that the on-line study website developers should make more efforts on the two aspects so that the website can be widely appreciated by the public. Meanwhile, it is interesting to find that the Dict has the highest click rate in the ALEXA ranking, but it is less appreciated by users compared to the other three. The fact clarifies that click rate or visitor volume can not reveal the overall characteristics of websites but one aspect of them; on the other hand, it is approved that AHP is a scientific method on website evaluation.

Table 2. General ranking of English on-line study websites

	Adaptability for users	Catering for demands	Convenience in e-learning	On-line interactiveness	Interface and characteristics	
	0.103092784	0.30927835	0.51546392	0.051546392	0.020618557	Hierarchy General Ranking
Dict	0.071428571	0.09090909	0.05882353	0.125	0.083333333	0.0739629
Putclub	0.285714286	0.18181818	0.35294118	0.5	0.333333333	0.300262
Ebigear	0.428571429	0.36363636	0.35294118	0.25	0.5	0.3617718
Tingroom	0.214285714	0.36363636	0.23529412	0.125	0.083333333	0.2640033

5   Conclusion

Instead of using traditional web evaluation method, this paper evaluates the English on-line study websites through AHP. By discussing the importance among the five intermediate indexes, it provides a better website evaluation system for the web developers, as well as for the users to select appropriate study websites. According to the statistics, the paper concludes that the satisfaction to the user demands and the convenience of e-learning, on the one hand, have vital importance on a website's comprehensive evaluation; on the other hand, they are the focuses and challenges in the building of websites. Due to the limited size of sample websites and other factors, the newly established evaluation system requires further research and verification. It is hoped that the present study will arouse more interest among researchers and the English study websites in China will develop in a productive way and serve users' needs more readily.

References

1. Li, Y.: Characteristics and Design of Websites for Learning Chinese as a Second Language. Gansu Science and Technology 26, 25–37 (2010)
2. Xuan, W.H.: Studies on Evaluation Theories of Websites for Learning Chinese as a Second Language and its Application. Postgraduate Thesis. Zhongshan University (2009)
3. Jiang, C.X., Feng, X.Q.: Studies on Websites for Special Subject Learning. Distance Education in China 4, 45–72 (2006)
4. Wu, H.P.: English Teacher and English Teaching in Internet Era. Research on Audio-visual Education Program 8, 41–44 (2000)
5. Han, Y.B.: Analyses of Interactive Language Teaching in On-line English Learning. Educational Information Technology 7, 34–36 (2003)

Comparison of Three Motor Imagery EEG Signal Processing Methods

Dan Xiao

Jiangxi Blue Sky University, Nanchang, P.R. China, 330098
blackhuman@gmail.com

Abstract. Feature extraction and classification of EEG signals is core issues on EEG-based brain computer interface (BCI). Motor imagery EEG signals can be difficult to classification because EEG sensor signals are mixtures of effective signals and noise, which has low signal-to-noise ratio. So signal processing methods should be used to improve classification performance. In this paper, three methods were used to process motor imagery EEG data respectively, and the Fisher class separability criterion was used to extract features. Finally, classification of Motor Imagery EEG evoked by a sequence of randomly mixed left and right image stimulations was performed by multilayer back-propagation neural networks (BPNN). The results showed that using of the three methods significantly improved classification accuracy of Motor Imagery EEG, and SOBI method had done a best job in this situation.

Keywords: Motor Imagery EEG, Second-order Blind Identification (SOBI), Phase Synchronization measure, Energy entropy.

1 Introduction

Brain computer interface is an area of research that has recently received much attention. A brain computer interface is a way of using an individual's thought processes to control computer or electromechanical hardware without using overt muscle activities [1, 2]. This type of system has the potential to provide a new form of communication and control options for individuals paralyzed from high-level spinal cord injury, severe neuromuscular disorders, or amyotrophic lateral sclerosis (ALS). The methods and utility of a BCI are currently being investigated in the field of rehabilitation [3] and neurorobotics [4, 5].

A kind of Analysis brain–computer interface system based on analysis of EEG. Various signal processing methods have been used in EEG analysis for best classification accuracy, such as based on event-related (de-)synchronization (ERD/ERS)[6], based on different rhythms of EEG [6], based on wavelet transform [7], based on wavelet entropy [8] and have achieved certain results.

In this paper, three methods were used to process motor imagery EEG data respectively, and the Fisher class separability criterion was used to extract features. Finally, classification of Motor Imagery EEG evoked by a sequence of randomly mixed left and right image stimulations was performed by multilayer back-propagation neural networks.

D. Jin and S. Lin (Eds.): Advances in MSEC Vol. 2, AISC 129, pp. 503–508.
springerlink.com   © Springer-Verlag Berlin Heidelberg 2011

2   Data Description

The dataset of BCI competition 2003 is used in this investigation. This dataset was provided by Graz University of Technology. The subject sat in a relaxing chair with armrests. The task was to perform imagery left hand or right hand movements according to a cue. The order of cues was random. The experiment consists of several runs (>= 6)) with 40 trials each after each; after trial begin, the first 2s were quite, at t=2s an acoustic stimulus indicated the beginning of the trial, and a cross "+" is displayed; then from t=3s an arrow to the left or right was displayed for 1s; at the same time the subject was asked to imagine a left hand or right hand, respectively, until the cross disappeared at t=7s (Figure. 1). Each of the 2 cues was displayed 10 times within each run in a randomized order.

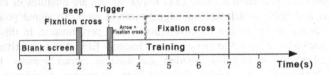

Fig. 1. Timing of the paradigm

3   Method

3.1   Data Preprocessing

The description data above was sampled with 250 Hz, it was filtered between 1 and 50Hz with Notchfilter on. To enhance the difference between these two tasks and reduce the effect of artifacts, common average reference (CAR) was used to re-reference it. Figure 2 shows the autoregressive (AR) model power spectrum over a trial of each task before and after CAR on electrodes C3 and C4. The features of motor imagery EEG signals appear in the 8-30Hz (see figure2). So, C3 and C4 channel is band pass filtered at 8-30Hz and CAR were applied in order to sharp the feature for motor imagery.

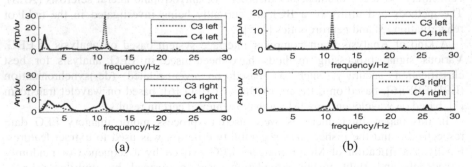

(a)   (b)

Fig. 2. (a) AR power spectrum before CAR (b) AR power spectrum after CAR

3.2   Second-Order Blind Identification

SOBI decomposes n-channel continuous EEG data into n components, each of which corresponds to a recovered putative source that contributes to the scalp EEG signal. Let x(t) represent n-dimensional vectors which correspond to the n continuous time series from the n EEG channels. Then $x_i(t)$ corresponds to the ith EEG channel. Each of the $x_i(t)$ can be assumed to be an instantaneous linear mixture of n unknown components or sources $s_i(t)$, via an unknown n × n mixing matrix A. SOBI uses the EEG measurement x(t) and nothing else to generate an unmixing matrix W that approximates A^{-1}.

SOBI exploits the time coherence of the source signals to decompose the mixture of sources. SOBI finds W through an iterative process that minimizing the sum-squared cross-correlations between one component at time t and another component at time t + s, across a set of time delays. Because such cross-correlations are sensitive to the temporal characteristics within the time series, temporal information contained in the continuous EEG data affects the results of source separation.

3.3   Energy Entropy

The original EEG signal was time domain signal and the signal energy distribution was scattered. The signal features was buried away in the noise. In order to find the features, the EEG signal was subjected to time–frequency (TF) analysis. The goal of TF analysis was to give a description of the signal energy as a function of time and frequency. A good TF representation aids to extract features correctly and easily.

In this study, the TF distribution (TFD) was constructed from short-term Fourier transform (STFT), which used function Spectrogram in Matlab toolbox.

Short-term Fourier transform was a time-frequency analysis, which could analyze nonstationary time-varying signals on frequency domain and time domain at the same time, so the dynamic change of signal energy with time could be observed. Energy Entropy defined on the basis of this could characterize signal complexity with the changes in time, and also many of the characteristics in frequency domain, which had a good time-frequency local capabilities.

3.4   Phase Synchronization Measure

There are many different methods to measure synchronization between two time signals. The most commonly used for analyzing EEG signals is the classical coherence Cohij(f) and termed the Phase Locking Value (PLV). In this work, we examine PLV measure that quantifies interaction between EEG signals and has a more direct interpretation.

This phase value can be calculated by using either the Hilbert transform or by convolution with a complex Gabor wavelet; as there is not much difference between those two possibilities when applied to EEG data. Hilbert transform is defined as

$$\tilde{x}_i(t) = \frac{1}{\pi} PV \int_{-\infty}^{\infty} \frac{x_i(\tau)}{t-\tau} d\tau \tag{1}$$

PV denotes that the integral is taken in the sense of Cauchy principal value, and the instantaneous phase is calculated as follows:

$$\Phi_i(t) = \arctan \frac{\tilde{x}_i(t)}{x_i(t)} \tag{2}$$

For discrete signals, the phase locking value is calculated as follows:

$$PLV = \left| \frac{1}{N} \sum_{n=1}^{N} e^{i\Delta\Phi} \right| \tag{3}$$

A PLV is equal to the average length of all unit vectors $e^{i\Delta\Phi}$ in one window. When the phase difference is constant (phase synchronization), all phase difference vectors will be aligned resulting in a PLV equal to 1. If the phase differences are randomly distributed over $[0, 2\pi]$. The vector sum, and thus the PLV, will be 0.

3.5   Fisher Distance

The Fisher class separability criterion [31] was used preparatory to extract features. The Fisher distance of two classes was calculated as

$$F = \frac{(\mu_1 - \mu_2)^2}{\sigma_1^2 + \sigma_2^2} \tag{4}$$

where μ was equalizing value and σ was variance.

3.6   Back-Propagation Neural Networks

Multilayer back-propagation neural networks were trained to classify motor imagery EEG into two stimulus classes: left and right motor imagery. The traingdx learning algorithm have been used to train the network, which uses gradient descend with momentum and variable learning rate in batch learning mode. Gradient descend with momentum can avoid a shallow local minimum and a variable learning rate can make the learning as fast as possible while maintaining stability. The batch learning was used to update the network weights after all training data was presented.

For each of the three subjects, a series of neural networks were trained to 99.99% classification accuracy using 50% of the trials (training set). Subsequently, the trained network was used to classify the remaining 50% of trials (testing set). This two-step process was repeated 60 times with different training sets to obtain the average for the data from each subject.

4   Results and Disscussion

The dataset provided by Graz University of Technology was used in this investigation, which included three subjects: K3b, L1b and K6b. The data of each subject included two types of motor imagery EEG- left hand movement and right hand movement. The numbers of sample of K3b's two types of motor imagery were 36, 37. That of K6b's were 20, 26, and L1b's were 20, 21. Half of the samples (36 K3b's, 23 L1b's, 20 K6b's) were randomly selected as training set, the other as testing set. Two electrodes - C3 and C4 were adopted in this experiment, from the EEG signal of each electrode 100 characteristic points were selected respectively, altogether 200 features. For each of the three subjects, a series of neural networks were trained to 99.99% classification accuracy using the training set. Subsequently, the trained network was used to classify the testing set. This two-step process was repeated 60 times with different training sets to obtain the average for the data from each subject.

To examine whether the three methods offered any advantage in motor imagery EEG classification, we compared classification accuracy using the three methods processing signals respectively to using the sensor signals. The average classification accuracy achieved recorded in table 1, which showed that, average classification accuracy of the sensor signals, the SOBI processing signals, the energy entropy processing signals, the Phase Synchronization measure processing signals respectively achieved 72.6%, 90.6%, 84.4%, 86.2%.

Table 1. Classification accuracy on different signal processing methods

	Classification accuracy of sensor signals	Classification accuracy of SOBI	Classification accuracy of energy entropy	Classification accuracy of Phase Synchronization measure
K3b	70.2%	88.9%	91.6%	93.0%
K6b	70%	90.9%	80.4	85.4%
L1b	77.5%	91.9%	81.2%	80.2%
Average	72.6%	90.6%	84.4%	86.2%

The ability to classify motor imagery EEG signals has important theoretical and practical implications for both basic and applied research, which could be used to generate better input and feedback signals for brain computer interfaces in clinical applications.

The results of the studies related with the EEG signals classification indicated that all of the methods used for feature extraction have different performances and no unique robust feature has been found. Therefore, the EEG signals classification was considered as a typical problem of classification with diverse features. The present study dealt with five-group classification problem, which is the assignment of segments to one of five predetermined groups.

The results presented in table 1 showed that using of the three methods significantly improved classification accuracy of Motor Imagery EEG, and SOBI method had done a best job in this situation. But, it has been to mention that classification accuracy is depended on not only signal processing method but also feature extracting method and classification method. So we believe that the current discrimination performance may be further improved because a rather crude fisher distance was used to extract features. Some useful feature may be lost.

In future work, we plan to further improve motor imagery EEG classification performance by other feature extracting method and try few trials even single-trial classification.

Acknowledgements. This work was supported by IT Project of Jiangxi Office of Education [GJJ10274]. The authors are grateful for the anonymous reviewers who made constructive comments.

References

1. Nicolelis, M., Dimitrov, D., Carmena, J., et al.: Chronic, multisite, multielectrode recordings in macaque monkeys. Proc. Natl. Acad. Sci. USA 100, 11041–11046 (2003)
2. Wolpaw, J.R., McFarland, D.J., Vaughan, T., et al.: The Wadsworth Center brain–computer interface (BCI) research and development program. IEEE Trans. Neural Syst. Rehabil. Eng. 11, 204–207 (2003)
3. Pfurtscheller, G., Muller, G., Pfurtscheller, J., et al.: Thought'–control of functional electrical stimulation to restore hand grasp in a patient with tetraplegia. Neurosci. Lett. 351, 33–36 (2003)
4. Kositsky, M., Karniel, A., Alford, S., et al.: Dynamical dimension of a hybrid neurorobotic system. IEEE Trans. Neural Syst. Rehabil. Eng. 11, 155–159 (2003)
5. Moore, M.: Real-world applications for brain–computer interface technology. IEEE Trans. Neural Syst. Rehabil. Eng. 11, 162–165 (2003)
6. Pfurtscheller, G., Stancak, A., Neuper, C.: Event-related synchronization (ERS) in the alpha band—an electrophysiological correlate of cortical idling: a review. Int. J. Psychophysiol. 24, 39–46 (1996)
7. Rosso, O.A., Martin, M.T., Figliola, A., Keller, K., Plastino, A.: EEG analysis using wavelet-based information tools. Journal of Neuroscience Methods 153, 163–182 (2006)
8. Rosso, O., Blanco, S., Yordanova, J., Kolev, V., Figliola, A., Schürmann, M., Basar, E.: Wavelet Entropy: a new tool for analysis of short time brain electrical signals. J. Neurosci. Meth. 105, 65–75 (2001)

Multi-agents Simulation on Unconventional Emergencies Evolution Mechanism in Public Health

Qing Yang* and Fan Yang

Management School, Wuhan University of Technology,
430070 Wuhan, P.R. China
yangq@whut.edu.cn

Abstract. Based on the cellular automaton principle and multi-agents theory of complex systems, this essay studied the public health unconventional emergencies generation and evolution mechanism, established evolution model and carried out simulation of the public health unconventional emergencies evolution mechanism, and finally took SARS emergency for an example. Research results showed that the evolution of the public health emergency often promots other linkage emergencies, the damage of linkage system is larger than that of promotion system, and the damage is uncontrolled except for controlling the promotion system effectively, just like isolation measures or inject vaccine for individual of the promotion system so as to prevent promotion system from producing linkage hazards.

Keywords: Public Health Unconventional Emergencies, Evolution Mechanism, Cellular Automaton, Multi-Agents Simulation, Promotion System, Linkage System.

1 Introduction

H1N1 influenza, SARS and other emergencies (hereafter referred to as emergencies) put forward new requirements for the public health emergencies management. In order to conduct emergencies management scientifically and effectively, we should study and figure out the evolution mechanism of emergencies. According to the system theory, emergencies' generation-evolution mechanism can be concluded as a kind of interaction among three basic elements composed of man, substance and environment [1], and its evolution process can be divided into five periods which includes incubation period, outbreak period, development period, recession period and death period [2].

In this essay, we take the advantage of the multi-Agents simulation technology as the development platform [3] and complex system critical theory [4] as the basis to construct the cellular automata model [5] of the evolution process. The model

---

* Author introduction: Qing Yang (1962-), doctor of management, professor, doctoral supervisor, research direction: crisis management complex system, venture capital and high technology industrialization, etc.; Fan Yang (1980-), doctoral candidate, research direction: crisis management complex system.

simulates the generation and evolution process of the emergencies, in order to observe how individuals lead to group phenomena and to analyze the simulation results of SARS.

2   Evolution Model

2.1   Evolution Mechanism

In the study of the causes of emergencies, scholars often regard superficial induction factors as the fundamental cause, and overlook the critical state and internal energy accumulation in a complex system when the stability of complex system undergoes damage. Emergency is a state which derivates from some kind of imbalance in complex system. The essential reason is that energy accumulates to a certain extent which throws the system in a critical state, an incentive to break the balance. Such critical state always occurs spontaneously in imbalanced system, called as the self-organized critical state. The critical state bridges the gap between the rules and disorder in concept, as a medium state. This state is not static, not lapsable casually but approximate to an extremely unstable balance, just like sand is always on the edge of the upheaval [6].

Public health emergencies inner mechanism coincides with that above indeed. The inner cause of its evolution is due to final explosion of virus infection caused by energy accumulation progress along with the diffusion of virus among people and this explosion without subsequent timely and proper control will lead to other explosions as to generate successive linkage explosions of emergencies. In this essay, the model is on the basis of secondary linkage emergencie which means an explosion of the public health emergency triggered another associated emergency.

2.2   Cellular Automata Mechanism

This essay applies cellular automata theory to construct the evolution model of public health emergency, while the cellular automata model is a kind of dynamic system in which time and space are discrete. The principle of the model is that every cell in limited discrete state scattered in Lattice Grid, following the same reaction rules, update synchronously according to a certain local rules. A large number of cells bring about the evolution of a dynamic system through simple interaction [7]. Different from general dynamics model, cellular automata is not strictly defined by physical equation or function, but composed by a structure of rules determined by a series of models.

2.3   Model Design

This essay constructs the model based on the linkage evolution mechanism and cellular automata above, of which there are two kinds of cells named A and B, constitute the independent subsystem respectively. The promotion system for A, while the linkage system for B, these two subsystems have the same grid space structure and the same size. There exists cells interaction between subsystems and two subsystems internal. Inside the subsystem, the interaction activities occur among adjacent cells in coordinate frame, that is to say, one cell can interact with four adjacent cells. Interactions between subsystems could occur if the two cells, A_i and B_i, at the same position but different coordinate grid space, A_i in explosion state will trigger B_i from stable state to latent state or explosion state. The evolution of the B_i is caused by A_i or by of the same kind adjacent B_j.

Evolution control is implemented by the real-time control and pre-definition for state parameters of A_i and B_i, respective internal transfer operators in A or B, and transfer operators of A_i to B_i.

Refer to figure 1, it shows the demonstration of the evolution structure of A_i to B_i, in which C_{ii} represent the unilateral interactions between them.

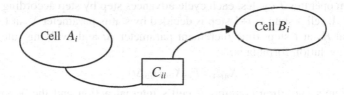

Fig. 1. Cells interaction model

3   Construction of Multi-agents Simulation Model

The model is one application of swarm2.2, a multi-agents platform. The concept of agent was proposed by American scholar-Minsky at the earliest, in Intelligence Society[8], which means one kind of entity have the abilities of self-adaptation and self-government, to recognize and simulate intelligent actions[9]. The format definition is below:

Agent:= $\{S_m, Ag_i\}$

S_m represents the internal state of Agent; Ag_i represents the function or external interaction[10].

This simulation model establishes promotion system A consisting of M cells, the linkage system B consisting of M_2 cells. Every cell is in its respective grid unit. The grid space of them have the same numbers of grids-M units (M=world X × world Y, world X and world Y both mean the verge of space). But M_2 is less then M, the number M_2 depends on the density (ρ) of B system. And ρ is an input value with range from 0 to 1. Each cell owns a coordinate position (xPos, yPos). After that, it can make cell A_i and B_i to transfer energy between them or in them, and to simulate the process of explosion of the emergency. Cells have three distinct states which consist of stable state, latent state and explosion state. The initial state of system A is except for one random cell in latent state, all the other cells in stable state. All cells of system B are in stable state and are dispersed in grid unit randomly.

The simulation process as shown in Figure 2:

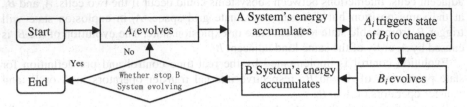

Fig. 2. Simulation flow chart

System operates T cycles, each cycle advances step by step according to its transfer operator. A_i cell's state at $t+1$ step is decided by status parameter X_{it} at t step, adjacent total input H_{it} at t step, time increment parameter Δt and evolving rule F_i of cell A_i. Transfer evolution formula is:

$$X_{i(t+1)} = F_i(X_{it}, H_{it}, \Delta t) \qquad (1)$$

B system's evolution contains B_i cell's internal action and the action of A_i to B_i. The former transfer evolution formula is the same with formula (1). Latter evolution pattern is that a single A_i evolves to the extent of explosion which may trigger B_i to change state. If once B_i is aroused to evolve, energy $E(A_i)$ of A_i will be changed into zero, and this B_i could be given some energy from it to change it's state or to enhance its internal energy.

B_i cell's state at $t+1$ step is decided by status parameter X_{2it} at t step, adjacent input H_{it} at t step, time increment parameter Δt and evolving rule F_{2i} of cell B_i. Transfer evolution formula is:

$$X_{2i(t+1)} = F_{2i}(X_{2i\,t}, H_{2i\,t}, \Delta t) \qquad (2)$$

4   Simulation Analysis of SARS Case

Combined with the characteristics of SARS, the simulation makes energy transfer from system A to B or in the internal of them to simulate the process of SARS explosion. System A represents SARS itself and the system B represents medical scare emergency triggered by SARS.

The results of simulation test are graphs for three observation measured respectively by latent value as the cells in the latent state in current cycle, the explosion value as total amount of cells in explosion state in current cycle and total destructive value, which means the accumulation of energy in current cycle. Moreover, two curves respectively signify each situation of system of A and B in observation graph.

Known from SARS, it's linkage emergency has a higher density in space. Because its impact is considerable wide range, the initial influence degree is high as well, the linkage system performs weak defensive ability, even nothing at all, while the transfer action of SARS self to linkage system is so strong and linkage system evolution ability is strong too, so, input value of density ρ of B space is set to 0.8, the defensive ability z of B against A is equal to 0.1, the rate c of action transfer from A to B is set to 0.8 and B's gene length parameter $L_{(B)}$ is set to 25. The result of simulation as shown in figure 3:

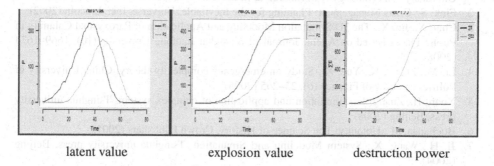

latent value   explosion value   destruction power

Fig. 3. Results of SARS linkage evolution simulation

In figure 3, the latent value figure shows that latent value of SARS is over that of linkage system before 42 cycle and maximum of both can reach higher, while SARS reach max 225, a comparative higher amount which indicates that late harm is more serious. Seen from explosion value, linkage system explode slightly later in curve than SARS and restrained by model, the space density of linkage system is limited to comparative higher 0.8, lower than the density of SARS system (equal to 1). Therefore, the explosion value of linkage has been below the SARS system. However, if the situation reversed, linkage system will show more power of explosion. Destruction power graph clearly shows the destructive force of the two systems, the linkage system is obviously more harmful than that from SARS system itself.

The results of simulation approach to real world, that is to say the model can reflect the destruction of SARS and linkage destruction from SARS. The linkage destruction would lead to a larger damage without timely and proper control.

5   Conclusion

The evolution of the public health emergency often promots other linkage emergencies, which explode more swiftly and follow at their motivation systems heel. Furthermore, higher destructive power will be brought up since the complexity of situation, high density and weak defensive ability of linkage systems. Therefore, in order to control the linkage destruction effectively, we must carry out effective and swift isolation measures or inject vaccine for individual of the promotion system so as to prevent promotion system from producing linkage hazards.

Acknowledgments. This research is supported by National Natural Science Foundation of China (Grant No. 91024020).

References

1. Ma, Y., Wang, C.: Analysis of the Causes of Important Sporadic Events. Journal of Wuhan University of Technology (Information & Management Engineering) (9), 92–95 (2006)
2. Fink, S.: Crisis Management: Planning for the Inevitable. Universe, Inc., Lincoln (2002)
3. Chai, X., Jin, X.: The 3D Simulation Modeling and Application of Paroxysmal Calamity in Public Places Based on Agent. Journal of Shanghai Jiaotong University (10), 1669–1674 (2008)
4. Li, M., Zhang, K., Yue, X.: Study on emergency complexity theory. China University of Politics and Law of Finance (6), 23–265 (2005)
5. Xuan, H., Zhang, F.: Simulation and application of complex system. Tsinghua university press, Beijing (2008)
6. Buchanan, M.: Ubiquity: The Science of History. Crown Publishers (2001)
7. Ji, H., Wang, X.: System Modeling and Simulation. Tsinghua university press, Beijing (2004)
8. Minsky, M.L.: Society of Mind. Touchstone Press (1988)
9. Yang, Q., Shi, Y., Wang, Z.: Multi-Agent Research on Immunology-based Emergency Preplan. In: Proceedings of 2010 International Conference on e-Education, e-Business, e-Management and e-Learning (IC4E 2010), pp. 407–410. IEEE Computer Society (2010)
10. Khalil, K.M., et al.: Multi-Agent Crisis Response Systems - Design Requirements and Analysis of Current Systems. Working Paper (2009)

The Application of Knowledge Management Based on Ontology in College Education Management

Xuejun He

Zhejiang Economic & Trade Polytechnic

Abstract. The main purpose of this paper is the study of the experimental method of knowledge management in university. We put forward a method based on learning and its components, set up a group of memory of science, technology and management knowledge capital university life. Our constitution is of particular interest to a library of resources for the study and research. The deposit of the dominant knowledge management files, including: training course, mat, e-books, etc. On the basis of the production process to provide an online learning and actor start to knowledge management. Effective file use an ontology classification index. The ontology can create for search and sailing on the capitalization of knowledge.

Keywords: Knowledge management, e-learning, information and communications technology.

1 Introduction

The knowledge management (km) is a wide range of concept, put forward a kind of range of strategic and application practice in identifying, create, organize, representative, storage, sharing, communication, search, analysis and improve its insights and experience. Such knowledge and experience, including knowledge reflected in personal or embedded in the organizational process or practice [1]. These strategies of KM are commonly by methods and software tools to help capture and knowledge organization, and the resource, material and technology to achieve strategic objectives [2].

2 Backgrounds

The importance of the work, and demonstrates the in universities and colleges kilometers of KM some examples, such as student registration process, the distribution of sales budget. The method of this paper in education, the extent of it kilometers also presents a road map, the next countermeasures. The main purpose of this work is put forward Suggestions for practical solutions to improve the reform decisions knowledge sharing and belongs to the university institutions. Finally, we emphasize actually a surprising enthusiasm in virtual learning environment, and its goal is to provide education content, administrative information, online documents etc.

D. Jin and S. Lin (Eds.): Advances in MSEC Vol. 2, AISC 129, pp. 515–520.
springerlink.com   © Springer-Verlag Berlin Heidelberg 2011

3   Aspects of Knowledge Management

The concept of KM generally raises the problem of the exact meaning of knowledge. In this section we try to demystify this concept in a general framework. We then present some aspects of KM that are unique to academic environments of interest.

3.1   The Concept of Knowledge

3.1.1   Data, Information Knowledge
This definition of knowledge is a important issue, many debates in epistemology. In the field of knowledge management in several explanation, visions of, this concept is available in the literature [9], [10], [11]. Part of our work, we use a kind of practical and operation visual difference between data, make information and knowledge according to hierarchical model.

In this model, the data can be considered a primitive element completely out of context. Information is a data set in context. Information is not knowledge, but can become so if it is to understand and be personal.

More specifically, and to extend what knowledge, some authors like [9], [10] a crucial difference make dominant and recessive knowledge. In this view, tacit knowledge, not formally or specific knowledge can be integrated in a person's head, according to their experience and rooted in action, in the process, in a specific context. The dominant knowledge is the knowledge transfer and compiled a formal, the system of language (documents, information system, and so on). Such knowledge is not static, dynamic process and changes their rich persistently.

3.1.2   Individual and Collective Knowledge
Knowledge can be individual or collective. Personal knowledge is a part of the implicit or explicit knowledge of a role of organization. Hold As a representative of the collective knowledge organization of the accumulation of knowledge stored in rules, procedures, program code, guide and sharing problem solving activities and the interaction between the models of organization agent. In the collective knowledge, some very easily transmissible (the dominant knowledge), others are more difficult to capture, clarify and spread the recessive knowledge within the organization (), the collective knowledge, by establishing personal knowledge flow and feed cycle comes from within or outside the organization.

3.2   Ontology

Ontology is a formal theory used to explicit knowledge. The primary objective of ontology is to model knowledge. Indeed, they provide definitions of concepts and terms used to describe a domain, logical and semantic relations between concepts and terms and the constraints of their use. Practical descriptions on ontology have shown their importance in several respects:

Ontology involves the factorization of knowledge. Like the oriented object approach, knowledge is not repeated in each instance of a concept.

Ontology provides a unifying framework to reduce, eliminate ambiguities and conceptual and terminological confusion.

Ontology can significantly increase the performance of search engines. Through the semantics provided, ontology can address problems such as noise and silence of the traditional search engines.

Ontology can support the sharing and reuse of knowledge. Indeed, if a group of researchers want to create or extend ontology in a particular field, it can reuse existing ontology and extending them.

Ontology implements mechanisms of deductive reasoning, automatic classification, information retrieval, and ensure interoperability between systems.

In the next sections, we will focus our presentation on all of these aspects in the tourism domain. We will particularly explain the importance of KM in the field of tourism; specify the architecture of a KMS and the procedure to acquire and to represent ontology in that domain.

4 Characteristics of Km in Higher Education

A university is its main traditional organization by a group of individuals, structure, order and drive resources (material, people, and money) to achieve a common goal. But it is very specific; its characteristic is through the three major factors:

The nature of the various powers-a university organization in view of the different actors has special needs. It can be around five major types of actor (students, teachers, and management personnel, technical personnel and partners). These actors expectations depends on many factors, it is difficult to imagine. The teacher and students, for example, create or search general information and knowledge. The purpose of this paper is to achieve the basic search documents (books, journal articles, paper, etc.), and meet the need to create a course, write research papers or file to complete a course, etc;

All kinds of resources, in addition to regular resources, a university organization provides some resources, different types of actors. Our list includes: administration documents, course, training, and training mat directory, the library catalog, eBook, electronic journals, patent, the report of the commission, the study object, etc. All these fund resources for every customer's concept of the formation and management.

5 Knowledge Capitalization Methodology

The phase of capitalizing knowledge uses different methodologies from the field of engineering knowledge. Among these, we can cite: the Common ADS methodology, MKSM and REX. These methodologies have been put in place to handle the whole process of learning, from the gathering of knowledge to the development of a complete system.

In the case of University System (US) considered, we opted for a more natural option, incremental and participatory, which involves different actors in the creation of knowledge. It is actually a solution based on e-learning.

For that we used the educational platform Model. In this section, we present mainly the capitalization of the knowledge for the administration staff. Thus, the staff takes courses on the platform, answered questionnaires, file documents, make annotations to documents filed in response to individual or collaborative assignments.

5.1 The Choice of an Institution

The work that we propose concerns the establishment of a resource center in a university for training and research. We chose the National School of Commerce and Management (ENCG) as the first institution of the University of Agade IBN ZOHR on which our study focused view of our membership in that institution.

The ENCG for 2010 includes about 900 students, 45 teachers and 15 people with administrative, technical, and service, as well as library staff. Several types of training are provided, we can mention: the diplomas ENCG, the Masters, continuing education and doctoral training. This is a learning environment where actors change with different production requirements, research and exploitation of knowledge. Knowledge generally used are distributed, heterogeneous and difficult to find, with the exception of library books and student papers. Some existing databases in each department, however, allow locating some resources.

5.2 The Choice of Learning Content

Two modules were followed by administrators as learners and creators of knowledge. The first module is to introduce first the importance of using Information Technology in various daily tasks of these actors, and also to deepen their basic knowledge in computing. This module includes courses on the use of the Internet, and also the Microsoft office tools (Word, Excel, FrontPage, and PowerPoint). The second module focused on techniques for knowledge management. A part related to the production and archiving of documents have also been the subject of a collaborative work.

5.3 Description of Content

The Model platform (Figure 1) is the main entry point to the knowledge of the particular organization, and also to information found on the web. In the platform developed different types of educational resources have been developed to meet various objectives initially set. The various tools such as forums, chats, assignments and multiple choices, helped to extract most of tacit and / or explicit knowledge. They constitute a first draft to the memory MUS (Memory of the University System).

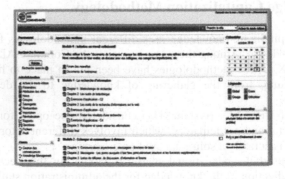

Fig. 1. Model Learning Portal

6   Ontology and Architecture

Knowledge of capture by the above methods provides resources devoted to ontology knowledge representation. The ontology is used to index files and resources, so as to promote their search and the concept of navigation relationship. This section provides some elements of our development domain ontology.

6.1   Indexing Resources

In this case, it can create complex queries, for example to find files according to their author (name, function, etc.). Finally, another level of semantic is through the semantic relationship development work items (, connecting to, is a part, is a partner, interaction for the premise, complete, and so on). These types of relationship are used to show the user research, or perform a automatic extension of research. For example, if a student is seeking a course, he usually happy to know that the necessary conditions of course need to modify, for case study and the class.

6.2   The Ontology

In the case of the MUS project, we used the editor Protégé2000. The model protégé2000 offers a graphical environment for the design of ontology (Fig. 2). It helps to define the classes and hierarchies with multiple inheritance, attributes, etc. We take this work to illustrate some examples of implementation from the ontology "Ontologie_US_ENCG" (Figure 2).

Fig. 2. Part of ontology "Ontologie_US_ENCG"

7   Conclusion

The research in the field of design has become interested in kilometers of organization memory use ontology. Our research suggests that is outstanding the fact is: not only in educational knowledge organization memory (courses, etc.), and in different knowledge in the context of use the university system, administrative, education or whether technology.

References

[1] Wiki. Knowledge Management,
    http://en.wikipedia.org/wiki/Knowledge_management
    (consultated at 2010)

[2] Megder, E., Cherkaoui, C., Mammass, D.: Impact du travai colaboratif sur la gestin des connaissances. In: er Congrès International sur les Technologies Numériques de l'Information et de Communication Educative - Expériences et Perspectives' Marrakech, May 2-5 (2007)

[3] Nonaka, I.: A dynamic theory of organizational knowledge creation. Organisation Science 5(1), 14–37 (1994)

[4] Grunstein, M., Barthès, J.: AN Industrial View of the Process of Capitalizing Knowledge (1996)

[5] Ermine, J.L.: "La gestion des connaissances, un levier de l'intelligence économique", "De l'intelligence économique à l'économie de la connaissance". Economica, 51–68 (2003)

[6] Mikulecká, J., Mikulecky, P.: University Knowledge Management. Issues and Prospects,
    http://eric.univ-lyon2.fr/~pkdd2000/Download/WS5_12.pdf
    (consulted at 2010)

[7] Petrides, A.L., That, R.N.: Knowledge Management in Education. Research report. Institute for the Study of Knowledge Management in Education,
    http://iskme.path.net/kmeducation.pdf (consulted at 2010)

[8] Perry, M.: Knowledge Management as a Mechanism for Large-Scale Technological and Organizational Change Management (E-learning and ERP) in Israeli Universities -The Second ILAIS Israel Association for Information Systems. Bar-Ilan University (2007)

[9] Polanyi, M.: The tacit dimension. Routledge & Kegan Paul Ltd., London (1974)

[10] Nonaka, I.: A dynamic theory of organizational knowledge creation. Organisation Science 5(1), 14–37 (1994)

[11] Lille: Misions de l'université, consulté en Février (2009),
    http://www.univ-lille3.fr/fr/

Development and Application of the Virtual Intelligent Digital Camera Teaching System

YangNa Su

Department of Education, HanShan Normal University, China

Abstract. Aiming at the existing problems in the experiment teaching of Photography and the lack of intelligence and professionalization in current virtual camera, according to the taxonomy of learning outcomes theory and operant conditioning and reinforcement, applying Flash+ActionScript technology, a teaching system of virtual camera is designed and developed, which is featured with strong interaction, professionalization and intelligence in evaluation. The system introduces the principle of camera and the process of installation and shooting. With the example of M-module, it focuses on the process of virtual shooting and the realization of professional and evaluation functions such as focusing, zooming, exposure and depth of focus. And the effects in teaching application is discussed.

Keywords: photography, virtual camera, intelligence, zooming, exposure.

1   Problems Identification

The application of virtual reality in education and teaching is the hotspot in the study of educational technology. And the development of virtual device has great educational and economical effects[1] · Photography is a major course with strong operation and practicalness. Traditional experimental teaching mainly relies on demonstration of material object. However, professional experimental instruments are too expensive for school to provide each student with a device. Experiment can only carried out in groups. And novice students' non-standard operation would easily damage the devices. What's more, the flying upgrade speed of electronic instruments makes new ones become outdated in a short time. Camera in reality is mechanical. And the existing teaching materials are mainly in the form of video. Video can be perceived directly and vividly, but its one-way transmission is in lack of "two-way interaction" which is an important function in instruction process. The limited number of courseware and animation can only simulate some simple functions of common camera but not able to offer intelligent evaluation which is the special function of professional camera. Some instructional disks are mostly played in steamline. Learners can only watch without participation.

Based on the experiment need of Photography instruction, a virtual digital cameral is developed with strong interactive functions featured with professionalization and intelligence. The camera can simulate professional camera with different effects under different modes. It can also handle parameters such as shutter and aperture intelligently to adjust exposure degree and depth of field and, make effective analysis and reasonable evaluation on shooting results.

D. Jin and S. Lin (Eds.): Advances in MSEC Vol. 2, AISC 129, pp. 521–527.

2   Theoretical Basis for the Development of Virtual Camera

2.1   The Theory of Classifying Learning Outcomes

Based on Gagne's *category of learning outcome*, the learning of traffic regulations belongs to knowledge skill and motion skill. Gagne thought, actual motion presentation can be used in the learning of motion skill. When presenting motions, instructors should reinforce which external clues need to be paid attention to in order to control motion reaction. Although these clues are important to the learning of motion skill, their effect is limited because motion skill not only controlled by external stimulates but also by internal stimulates i.e. muscle motion clue. This internal stimulate from muscle motion is more significant for skills to reach a higher level. And the only pass to gain and make use of this internal control is to practise in person. The development of virtual camera provide students with favorable experimental environment for practicing their stills with camera, enabling students to participate in practical operation in specific timing and even when devices are in shortage.

2.2   The Theory of Operant Conditioned Reflex

Skinner's theory of operant conditioned reflex believes that all behaviors are composed of reflexes and any simulate-reflex unit should be seen as reflex. Operant conditioned reflex is related to two common principles[4]:

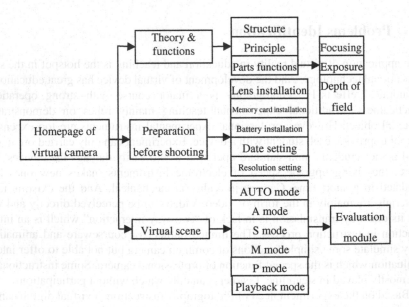

Fig. 1. The structure of modules of the virtual camera system

① Any reflex followed by reinforcement (reward) stimulus will tend to repeat itself.
② Any stimulus that can enhance operant reflex rate is reinforce stimulus. The powerful intelligence function of virtual camera not only can offer guideline and help to learners before operation but also can make reasonable analysis and evaluation on the outcome of operation, i.e. the correct outcome of each shooting operation of the camera.

3   The Framework and Functions of Virtual Camera

3.1   System Function Introduction

Mainly based on the need of the experimental course Photography, according to the primary functions of FUJI camera FinePix S5100 / FinePix S5500, the study uses Photoshop to optimize picture materials, and applies Flash+ActionScript to construct interactive intelligent virtual animation. The system not only simulates the structure and working principle of the camera but also the whole process from installation of the camera to virtual shooting and playback.

3.2   Structure and Functions of Modules

Virtual digital camera system is composed of the following three modules, shown as Figure1.

3.2.1   Principle and Function Module

The structure and principle of camera is the imaging principle of convex glass. Suppose object distance is u, image distance is v and focal distance is f. According to the formula of convex glass imaging principle $1/u+1/v=1/f$, when mouse drafts the object, the position and size of the virtual image will change and, the position of camera body and lens can also be dragged. The execution method is to apply ActionScript collision detection function hitTest() and the drag and drop function startDrag(), stopDrag() to realize the functions of dragging, auto-calculation of object distance, image distance and amplify times.

3.2.2   The Module of Preparation before Shooting

Before using camera for shooting, students should install the camera correctly and set the system in order to be prepared. Only when it's prepared the experimental shooting operation can be carried out. This module, mostly applying drag and match pattern, simulates the preparation work before shooting, including "installation of lens, installation of memory card, installation of battery, date setting and resolution setting". Each module has intelligent feedback function.

3.2.3   The Core Module of Virtual Scene

This module mainly simulates five different shooting modes of professional camera, i.e. Auto mode, P mode, A mode, S mode, M mode and playback mode. In order to identify different shooting effects under different modes, this module simulates all kinds of shooting effects with typical parameter and extreme value, thus it's better for students to understand the concept and effect of exposure, depth of field and motion trail, and for them to master professional shooting skills.

4   Key Technologies in the Development of Virtual Camera

4.1   Focusing Function

Focusing is the primary task before shooting and is the operation that is easy to be ignored by students in practice. Only correct focusing leads to clear image of shooting target and the success or failure of focusing decides the image quality of shooting directly. There are auto focusing and manual focusing modes, the principle of whichever mode is to blur the photo before focusing. Focusing is realized by controlling the transparency of the film's _alpha.

4.2   Zooming Function

Zooming is the most basic operation when finding a view and composing a picture. And composition a picture is also the key to the success of the work. The process of zooming can be realized by increasing the angle of view or pulling in the lens. The effect can be simulated by reducing or amplying picture. And T/W button is used to control attributes _xscale and _yscale of the film.

4.3   Exposure Effect

Exposure is the most important of all shooting factors. In order to enhance students' understanding about the concept of exposure and to master proper exposure operation and parameter setting, the system simulates three different exposure effects under different combination of aperture coefficient value and shutter speed parameter: insufficient exposure, over exposure and normal exposure. The principle is to add to the picture a movie clip which changes gradually from white to black, and then control the exposure by controlling the alpha transparency of the movie clip. The switch of the states of exposure compensation button is realized by setting the valuable Flag. Flag=0 means the button is not pressed and Flag=1 means the button is pressed.

4.4   Depth of Field Effect

Depth of field is one of the nouns in professional photography technology. In order to reach higher simulation for students to operate camera in professional level, the system simulate the effect of depth of field. When the shutter speed value is 500 and the aperture coefficient value is 5.6, the near depth of field possible to be shot is shown as Fig. 2-a. When shutter speed value is 30 and aperture coefficient value is 22, the far depth of field possible to be shot is shown as 2-b. The principle of depth of field is to use Photoshop to blur part of the picture, and then generate a movie clip in frame animation, and finally go over the frames according to the aperture coefficient value.

4.5   Motion Trail Effect

Motion trail effect is one of the special technology effects of professional photography. In order to consummate the simulated professional functions of camera, the system simulates motion trail effect. As shown in Fig. 3-a, when shutter speed value is 250 and aperture coefficient value is 8, the motion object of simulated shooting effect is positioned accurately. As shown in Fig. 3-b, when shutter speed value is 60 and

Fig. 2-a. Simulated near depth of filed

Fig. 2-b. Simulated far depth of field

Fig. 3-a. Simulated shooting effect of motion accurately positioned

Fig. 3-b. Simulated shooting effect object of motion object trail

aperture coefficient value is 16, the motion object generate trail. The principle of motion trail is to use movie clip to save motion object and use the codes in blank movie clip to control the generation of motion trail. After freezing the motion object, capture the motion trails between two frame and reduce its transparency to 30%, then 20% and finally 10%, thus the motion trail effect from deep to superficial will be generated.

5   Realization Result and Evaluation

5.1   The Experimental Process

To simulate shooting experiment, M mode is taken as an example. With the combination of different shooting modes, focusing modes, shutter speed values and aperture coefficient values, the shooting effects are shown as the following Fig.s from 4-a to 4-f. Fig. 4-a and 4-b show the shooting effects of no focusing and inaccurate

Fig. 4-a. Wrong mode and AF   **Fig. 4-b.** Shutter 125 aperture 22   **Fig. 4-c.** Shutter 1000 aperture 8

Fig. 4-d. Shutter 8 aperture 22   **Fig. 4-e.** Shutter 125 aperture 8   **Fig. 4-f.** Shutter 125 aperture 16

Table 1. M mode experimental parameters and shooting effects

Times	Mode	AF	Focusing	Shutter	Aperture	Exposure
1	Auto	S-AF	Unfocusing	125	11	Normal
2	M	MF	Unsuccessful	125	22	Insufficient
3	M	MF	Successful	1000	8	Insufficient
4	M	C-AF	Successful	8	22	Over
5	M	MF	Successful	125	8	Normal
6	M	MF	Successful	125	16	Normal

Table 2. Results in Evaluation Module

Your shooting results are as follows:				Total score: 64	
	Mode	AF	Focusing	Exposure	Times
Correct	42	52	35	Normal	25
Error	18	8	25	Over	15
				Insufficient	20

focusing. Fig. 4-c, 4-d and 4-f show the exposure effects of insufficient exposure, over exposure and normal exposure. Fig. 4-b and 4-3 show different depths of field. The experimental parameters and text feedback are collectively shown in table 1.

5.2   Experimental Results and Evaluation

The evaluation module of the virtual camera is the feature and innovation of this system. The function is to carry out automatic evaluation on students' operation results. When students are operating virtual settings and shooting, the system will test automatically the correctness of the setting and operation of each step. It can test the correctness of shooting mode, AF focusing mode and focusing result. It can also judge exposure as normal, over or insufficient. Finally in the result module it can add up the times of correctness and error in each step's setting and operation. Take the parameters in 4-a to 4-f under M mode as example, in outdoor environment with same lighting, when shutter speed and aperture coefficient are combined with different parameters, the system will automatically carry out statistic evaluation on each shooting effect, and the operation score will be calculated according to the proportion of correctness and error. The statistic result is shown as Table 2.

6   Conclusion

The shooting effect of virtual camera is almost closed to real camera, but the function of sharing, network and feedback & evaluation of virtual camera are the advantages that real camera can not compare. The evaluation on shooting results enables students to know about the correctness of their operations, so that they can adjust next shooting operation according to the evaluation and feedback. In this way, their correct operation will be enforced, wrong operation will be avoided or reduced, feedback will inspire students to think and summarize during shooting process. This will stimulate learners' activeness in a virtuous cycle and promote the advancement of shooting skills. It is accord with the two common principles of Skinner's operant conditioned reflex theory.

The intelligent virtual camera provides students with a new way of learning. Its development and application will have transformative influence over experimental instruction. In the system of virtual camera, students can complete experimental projects with self-plan and self-control and evaluate the experiment results. Besides, the system of virtual camera can also act as the role of an instructional teacher, who can create lifelike experimental scenes to add to students' interest in learning and can also supply abundant information resource and benign technology support service. Analyzing from the angle of a teacher, the virtual camera makes experimental instruction more directly perceived and interactive, which reduces greatly teachers' workload of experimental instruction, thus promote the advancement of experimental instruction efficiency. The features of network and sharing of virtual camera realize one student one INSTRUMENT of virtual experimental operation, which reduces the economical cost of investment in experiment instruments and realizes the teaching mode of resource sharing across time and space. The practice of instructional application shows that virtual digital camera has significant instructional value in advancing the informationization level of instruction, consummating the training environment of experimental instruction, optimizing instruction process, facilitating learning effectiveness and increasing students' learning interest, hands-on ability and innovation ability. The application and popularization of virtual experiment in instruction will have unlimited development prospects.

References

[1] Jones, M.: Vanishing Point: Spatial Composition and the Virtual Camera Animation 1746-8477 2(3), 225–243 (2007)
[2] Klein, G., Murray, D.W.: Simulating Low-Cost Cameras for Augmented Reality Compositing. IEEE Transactions on Visualization and Computer Graphics 1077-2626 16(3), 369–380 (2010)
[3] Kong, S.-H.: Comparison of two- and three-dimensional camera systems in laparoscopic performance: a novel 3D system with one camera. Surgical Endoscopy 0930-2794 24(5), 1132–1143 (2010)
[4] Li, S., Xu, C.: Efficient lookup table based camera pose estimation for augmented reality. Computer Animation and Virtual Worlds 1546-4261 22(1), 47–58 (2011)
[5] Han, M.: A perspective factorization method for Euclidean reconstruction with uncalibrated cameras. Computer Animation and Virtual Worlds 1546-4261 13(4), 211–223 (2002)

Improvement Project on Collecting and Integrating Video Resources

Weichang Feng[1] and Xiaomeng Chen[2]

[1] School of Computer and Communication Engineering, Weifang University
fweichang@163.com
[2] School of Mathematics and Information Science, Weifang University
Weifang 261061, China
wfxycxm@126.com

Abstract. E-Yuan multimedia system is developed for the rich audio and video resource on the Internet and on its server side, it can automatically search and integration of network video and audio resources, and send to the client side for the user in real-time broadcast TV viewing, full use of remote control operation, Simply it's a very easy to use multimedia system. This article introduces its infrastructure, main technical ideas and you can also see some details about server side and client side. At the same time, the improvement on how to collect and integrate video resources is comprehensively elaborated.

Keywords: Linux, Embedded system, Web 2.0, Server, Client, Video search.

1 Introduction

E-Yuan multimedia system is developed for the rich audio and video resource on the Internet and on its server side, it can automatically search and integration of network video and audio resources, and send to the client side for the user in real-time broadcast TV viewing, full use of remote control operation, Simply it's a very easy to use multimedia system. System is divided into server side and client side. On the server side, for different network multimedia resources, we developed different search branch, through the unified data format, every parts of the system could use and generate data for others and finally generate the data for client side. Server's framework is extensible, if there is a new network media resource, we just dynamic increase a appropriate search branch for it, and no change in the overall framework. On the server side, the content for the client side is filtered to ensure that it's health and effectiveness. On the client side, beside the network resource browsing, also supports browsing the most mainstream formats of video, audio and pictures in the removable storage media.

2 Main Technical Characteristic of Server

By developing different search branch for different network multimedia resource, server architecture achieve a good scalability.

D. Jin and S. Lin (Eds.): Advances in MSEC Vol. 2, AISC 129, pp. 529–534.
springerlink.com   © Springer-Verlag Berlin Heidelberg 2011

For example, the search branch for YouTube, it does the following work. First, it get the latest multimedia information through the RSS protocol and then transformed it into the unified XML format, and put it into the data storage center for the entire system.

The unified XML multimedia data format helps every parts of sever to share data, do the collection and analysis automatically, and make the entire system work efficiently and orderly, also it's a good foundation for further server expansion.

3   Client Side Introduction

E-Yuan client side based on embedded systems , because it wants to minimize the costs.

Embedded system is undoubtedly the most popular areas of IT applications, especially with the intelligent household electrical appliance. As we are usually familiar with mobile phones, electronic dictionaries, video phone, MP3 players, digital video (DV), high-definition television (HDTV), game consoles, smart toys are typical of embedded systems.

E-Yuan client side separates into two parts (Hardware part and software part).

3.1   Client Hardware Structure

Client hardware includes: motherboard, power supply, memory, hard drive, infrared remote control and the system enclosure.

The major part of system hardware is an industrial motherboards. Usually an Industrial motherboard designs for an fixed use and not easy to expend its function. With the market requirements of the board become more complex, it requires industrial motherboard has a strong expansion capabilities to meet different customer needs.

This system picks up J7F2WE1G5D embedded high-performance motherboard.

3.2   Client Software Structure

Client software includes the embedded Linux operating system and its applications. Linux operating system is the most successful applications in the embedded field. After the hardware design is complete, we need software to achieve every kind of functions. The value-added of embedded system largely depends on the level of the embedded software application. That is the intelligence level of products is determined by the software.

Software system includes the base subsystem, multimedia subsystem, and human-computer interaction subsystems. Functional description of each subsystem are as follows:

Base subsystem: Contains the basic library and some base functions. Other subsystems depends on this subsystem.

Multimedia Subsystem: The core subsystem, including all media-related input and output functions.

HCI subsystem: allows the users to quickly and easyly interact with the system.

System data flow control principle shown in Fig. 1. (MMS means Multimedia Subsystem)

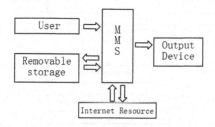

Fig. 1. System data flow control principle

4 Improvement Project on Collecting and Integrating Video Resources

The design project of collecting and integrating audio and video resources automatically is changed. Instead of using RSS subscription functionality originally, we designed a special video vertical search engines to realize the site directional search. Moreover, we designed a web video retrieving central database to achieve the purpose of integrating high quality web video resource. Thus more rich and the broadcast program source can be provided for the embedded online play equipment E-sourcebox which is developed independently.

The improved project of collecting and integrating video resources can be divided into four modules: video collecting module, video data updating module, the terminal interface module and client player interface supporting moduleA. Adjusting instruction architecture according to the cpu on the motherboard.

4.1 Video Collecting Module

The main functions of the video collecting module are as fellow : developing and improving the customization Nutch worms, capturing video website data, preprocessing the website data, judging the effectiveness of resource and saving the valid data into website video searching centre database, as shown in Fig. 2.

4.2 Video Data Updating Module

The video data updating module implements the function of data update timely which is the important indicator of the precise online playing on the client side.

Although the Nutch worms limited the search scope, but in order to guarantee the data accuracy of the database centre, the capture interval is designed short, such as a day or even shorter time to update a database to ensure the correctness of client play resources data.

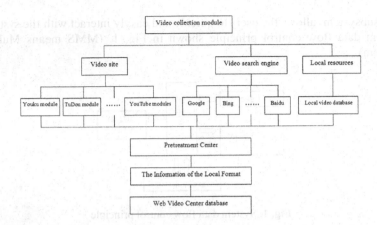

Fig. 2. Web video collection module

The video data updating module can carry out independently to specific video website. It can update and maintain regularly according to the column setting style of different website and generate the XML file used by the web pages and terminals, to ensure the synchronism and accuracy of display list of client data. As this module has heavy workload, so it needs more manpower. The follow chart of the video data updating module is shown in Fig. 3.

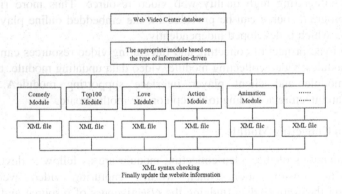

Fig. 3. Web video data update module

4.3   The Terminal Interface Module

According to the peculiarity that the E-sourcebox uses the remote operating, the client interface adopts the tree list to display the programme source information so as to the location and search of remote. Its functions are shown in Fig. 4.

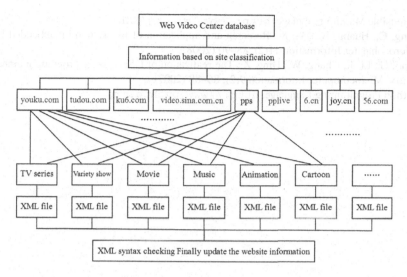

Fig. 4. Composition of the client front interface

4.4   The Client Player Interface Supports Module

Some famous video sites such as Youku and Tudou adopt customized player to broadcast video program. It is very easy to download and install a video player on the computer, and each video website supports download and installation. But it is not so easy to install a video player designed by a web site in the embedded multimedia device. The system has been achieved to supply interface in the E-sourcebox client to play Youku, ppstream, Google and many other video resources by using their own custom player.

5   Conclusion

Innovation of this article: According to the high speed of WEB2.0 technology increasing, E-Yuan multimedia system will have more features and supply the data user need more quickly and accurately. With the rapid growth of the Internet and the rapid increasing quantity of bandwidth user, the internet real-tine media industry should have a bright feature and I believe this system has a wide marketing prospect.

References

1. O'Reilly, J.C., Alessandro, R., Greg, K.H.: Linux Device Drivers, 3rd edn. Oreilly & Associates Inc., San Francisco (2005)
2. Lombardo, J.: Embedded Linux. SAMS Technical Publishing, Indianapolis (2001)
3. O'Reilly, D., Bovet, P., Marco, C.: Understanding the Linux Kernel. Oreilly & Associates Inc., San Francisco (2007)
4. My Media System, http://mymediasystem.org/

5. Extensible Markup Language, http://www.w3.org/XML/
6. Teng, C., Huang, B., Ma, X.: Research and application of file system in embedded Linux. Microcomputer Information 11(2), 88–90 (2008)
7. Xiao, G., Li, J., Chang, W., Ding, Z.: The realization of USB driving program in embedded Linux. Microcomputer Information 9(2), 86–88 (2007)
8. Python Programming Language, http://www.python.org

JND Model Study in Image Watermarking

Wei Li, Cheng Yang, Chen Li, and Qiu Yang

Information Engineering School, Communication University of China, Beijing, China
{liwei601,lichen200808,yangqiu1005}@126.com,
cafeeyang@gmail.com

Abstract. The Just Noticeable Distortion (JND) model is used to determine the optimum strength for watermark embedding and to provide an imperceptible and robust watermarking scheme .This paper gives detailed analyses and comparisons on existing models, and then finds that these models do not take full advantage of the stratified perception characteristic of human vision, therefore the ability in looking for a balance of the watermark perceptibility and robustness is limited. Base on these analyses, this paper proposes to constitute a JND model based on the content decomposition and stratified perception characteristic in DWT domain, in order that the contradiction between the watermark perceptibility and robustness can be better balanced.

Keywords: HVS, JND model, imperceptibility, prime sketch, content watermarking.

1 Introduction

Digital watermarking achieves the digital content copyright protection and content authentication by embedding watermark information into the original digital works. Effective digital watermarking must have three basic characteristics-imperceptibility, robustness and larger watermark capacity. But there is a contradiction among the three properties, which matters with the characteristic of original digital content and the region of the watermark embedded etc. Therefore ,on condition that watermark is imperceptible ,watermarking algorithm should select the appropriate region to embed watermark information by the highest possible bit rate and the highest possible intensity ,while the watermark is imperceptible according to the characteristic of application [1, 2].

JND model based on HVS (Human Visual System) provides the visual threshold depending on the characteristic of human vision and image itself [3].it represents the maximum image distortion that human eye cannot detect ,which is generally a comprehensive reflection of the image frequency sensitivity , luminance masking, contrast masking and other characteristics of HVS .Provided that watermarking algorithm wants to hide information as much as possible but not to bring about visual quality distortion ,it should employ different threshold in different intensity region according to JND model. Therefore, through the research and application of JND model, watermarking algorithm can avoid damaging image visual quality to the maximum.

D. Jin and S. Lin (Eds.): Advances in MSEC Vol. 2, AISC 129, pp. 535–543.
springerlink.com   © Springer-Verlag Berlin Heidelberg 2011

By studying existing JND models, we find that the current models are still mainly concentrated on the choice of the temporal-spatial domain,frequency domain and simple pixel characteristic, and have few associations with the characteristic of image content, therefore robustness is difficult to meet the needs of practical application, while the ability in coordination with the relationship of robustness, perceptibility and watermark capacity is not satisfactory .In recent years, some new theories and methods of image analysis and graphics and other related areas, particularly the latest research of primal sketch theory, give us a great inspiration to start from the bottom problem of image processing - the content decomposition of image ,to explore a more efficient JND model based on HVS and image content characteristic.

2   General Framework of JND Model Based on HVS

Through the observation of the human visual phenomenon, and the research of visual physiology, visual psychology and other interrelated subject, people discover various perception characteristics of human vision, and apply them into digital watermarking scheme to determine the watermark embedded region and embedded strength. Currently the researchers have proposed lots of JND models based on the characteristics of the HVS, such as frequency sensitivity, luminance masking and contrast masking, etc. Utilizing these characteristics reasonably has great significance for enhancing watermark imperceptibility and robustness [4, 5].

HVS processes visual information using a multi-channel mechanism [6], which decomposes the input image into different perception component. Different perception channels deal with the specific perception component, each channel has a corresponding perception threshold , if a stimulation is lower than the threshold of perception channel, the human eye will not perceive the impact of the stimulation , while we must consider the interaction of the multiple stimulations , which will lead to variation of perception threshold . Perception threshold makes human eye cannot detect image distortion below the threshold, while masking phenomenon makes the perception threshold enhanced, and tolerate more undetected image distortion . Therefore, in the process of JND modeling, we must consider the impact of various perception components on threshold, and allocate the energy of watermark signal reasonably to maximize the watermark embedding strength, and meanwhile, make the embedded watermark information does not affect the visual quality of original image.

Fig.1 shows a general framework of JND modeling, which considers the visual characteristics of the human vision synthetically, and makes the objective calculation more in line with the subjective evaluation. Based on current understanding of HVS, its complexity makes us cannot constitute a complete perception model, consequently, we must continue to explore the characteristics of HVS and the visual perception process for an accurate JND model.

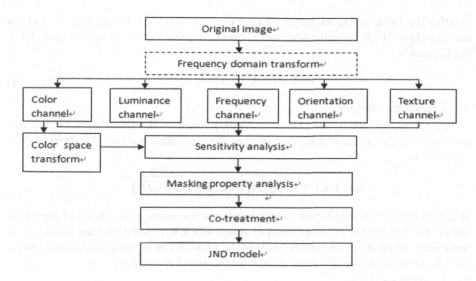

Fig. 1. General framework of JND model constitution based on HVS

3   JND Model Comparison in Digital Watermarking System

3.1   JND Model in Spatial Domain

Generally, the existing JND models have two categories, one is spatial JND model which is also called pixel-wise JND model [7, 8]. Although spatial JND model provides a more direct view of JND map of the original image, it does not exploit the HVS comprehensively. Another one is subband JND model which is determined in transform domain.

3.2   JND Model in DCT Domain

Transform domain algorithms modify the transform domain coefficients for watermark embedding, which have great advantage in imperceptibility and robustness. A large number of image and video coding standards perform in the DCT domain, therefore JND modeling causes a highly interest of the researches in DCT domain. A better JND model is proposed by Ahumada[9], which gives the frequency sensitivity threshold for each DCT coefficient by incorporating CSF. Waston et al[10]consider the luminance masking and contrast masking effect to the base threshold, then give a more accurate JND model. Wei et al [11] improve Waston JND model by proposing a new frequency sensitivity, luminance masking, contrast masking, and a new calculation method.   However, the computational complexity of the model is high, and the practical application effect is worse than Waston model, therefore, this paper will introduce the Waston JND model mainly

1)   Frequency sensitivity

Watson model defines a sensitivity table t(i,j), which indicates the minimum variation detected by the DCT coefficients in the absence of masking noise, the smaller the value is, the higher the frequency sensitivity of human eyes is. Q(i,j)is the standard JPEG quantization table, then frequency sensitivity of each DCT coefficient is:

$$t(i, j) = Q(i, j)/2 \tag{1}$$

2) Luminance masking

Waston model adjusts t(i,j) for pixel block k based on the DC (that is, the average luminance of the image block), then obtains the luminance masking threshold of each coefficient.

$$t(i, j, k) = t(i, j) \times [F(0,0,k)/\overline{F(0,0)}]^{\alpha} \tag{2}$$

Where F (0,0) is DC coefficient ,which represents the average luminance of the whole image. F (0,0,k) is the DC coefficient of image block k, α is the relevant parameter of luminance, Waston recommends 0.649 of the value. It can be seen that relatively larger modifications are not easily aware in high bright area of an image.

3) Contrast masking

Contrast masking refers to the reduction in the perception of one image component by the presence of another, contrast masking threshold s(i,j,k) will be:

$$s(i, j, k) = Max\left\{ t(i, j, k), |F(i, j, k)|^{w(i,j)} t(i, j, k)^{1-w(i,j)} \right\} \tag{3}$$

4) Waston JND model

The quantization error e (i,j,k) in the process of JPEG compression is the difference between the DCT coefficient of original image and Compressed coefficient. The DCT coefficient is divided by quantization matrix, and then rounded to the nearest integer value, then the value is multiplied by the quantization matrix , the final result is Compressed coefficient.

$$e(i, j, k) = F(i, j, k) - U(i, j, k)Q(i, j) \tag{4}$$

$$U(i, j, k) = Round[F(i, j, k)/Q(i, j)] \tag{5}$$

Then quantization error is divided by the respective contrast masking threshold, then the resulting perceived distance d (i,,j,k) of each coefficient is the JND threshold.

$$JND_{Watson} = d(i, j, k) = \frac{e(i, j, k)}{s(i, j, k)} \tag{6}$$

3.3   JND Model in DWT Domain

DWT is a signal analysis method of multi-scale and multi-resolution, which has the ability to characterize local signal in the time /frequency domain, and its good characteristic of time-frequency decomposition is more in line with the HVS. Therefore, the watermarking algorithm based on JND model in DWT domain can effectively improve the imperceptibility and robustness of watermarking, which is the leading research field of digital watermarking scheme.

In 1992, Lewis et al [12] propose a JND model in DWT domain which take into account the frequency sensitivity, luminance masking, texture masking, and give the maximum variation that each DWT coefficient can tolerate. Barni et al [13] explore the HVS further, construct a JND model which is more consistent with the perceptibility of HVS. Zheng et al[14,15]propose a JND model based on texture measure. This model is based on approximate wavelet sub-band so that JND threshold is more robust, but this method does not take into account the characteristics of HVS sufficiently, therefore the imperceptibility needs be improved. So this paper will introduce the Barni JND model mainly.

1) Frequency sensitivity

The human eye is less sensitive to the noise in different direction and subband of the medium-high frequency, especially for 45° of high frequency. Firstly, the original image is decomposed through DWT in L levels, and s = HL, LH, HH represents the orientation of vertical, horizontal and diagonal, then

$$frequency\ (l,s) = \begin{cases} \sqrt{2} & if\ (s = HH\ ) \\ 1 & otherwise \end{cases} \times \begin{cases} 1.00 & if\ (l = 0) \\ 0.32 & if\ (l = 1) \\ 0.16 & if\ (l = 2) \\ 0.10 & if\ (l = 3) \end{cases}. \tag{7}$$

2) Luminance masking

$$L(l, x, y) = 1 + L2(i, j, k). \tag{8}$$

$$L1(l, x, y) = \frac{1}{256} I^{3,l} (1 + \left\lfloor \frac{x}{2^{3-l}} \right\rfloor, 1 + \left\lfloor \frac{y}{2^{3-l}} \right\rfloor). \tag{9}$$

Taking into account that the human eye is less sensitive to noise in the dark and bright areas, Barni propose:

$$L2(l, x, y) = \begin{cases} 1 - L1(l, i, j) & if L1(l, i, j) < 0.5 \\ L1(l, i, j) & otherwise \end{cases}. \tag{10}$$

3) Texture masking

The human eye is less sensitive to noise in highly textured region, but more sensitive near the edges.

$$T(l, x, y) = \sum_{k=1}^{3-l} \frac{1}{16^k} \sum_{s}^{HH,HL,LH} \sum_{i=0}^{1} \sum_{j=0}^{1} (I^{k+l,s}(i + x/2^k, j + y/2^k))^2$$

$$\cdot var(I^{3,LL}(1 + j + x/2^{3-l}, 1 + i + y/2^{3-l}))_{\substack{x=0,1 \\ y=0,1}}. \tag{11}$$

4) Barni JND model

Based on these considerations above, Barni JND threshold of each DWT coefficient is:

$$JND_l^s(x, y) = 0.5 \times frequency(l, s) \times L(l, x, y) \times T(l, x, y)^{0.2} \qquad (12)$$

We believe that there are still some deficiencies of the Barni JND model. For one thing, the calculation of luminance masking depends on the high frequency coefficients completely, which considers the luminance masking between the highest level and other levels, it will be more accurate to consider luminance masking between adjacent levels; For another, the calculation of texture masking ignores the energy of high frequency coefficient, but the high frequency coefficients in low-level reflect the image texture characteristics more accurately; while the model dose not fully consider the stratified perception characteristic of human vision.

4   Experiment and Analysis

In this paper, we do simulation experiments on Waston JND model in DCT domain and Barni JND model in DWT domain, and obtain JND threshold respectively. Afterwards we perform a global energy adjustment of the original image based on JND thresholds, eventually evaluate performance of the two models on the premise of imperceptibility. We use matlab programming for simulation experiments, while the original image (Fig.2) is 256 grayscale image Lena(resolution is 512×512).

1) The statistical analysis of JND threshold in DCT and DWT domain

Fig.3, Fig.4 is the image of frequency domain through DCT and DWT. We can see that Fig.3 exhibits highly noticeable blocking effects, which will produce relatively larger errors in the process of image reconstruction. The DWT has good characteristic of temporal-spatial and direction, and the process of decomposition of original image through DWT is very similar to the characteristic of the human eye perception.

Fig. 2. Original image   **Fig. 3.** 8×8 block DCT transform image

Fig. 4. DWT transform image in four levels

Table 1. The transform coefficient and JND threshold in DCT, DWT domain

location	Coefficient maximum	Coefficient minimum	JND threshold maximum	JND threshold minimum
DWT domain				
cH1	0.351	-0.2941	0.6767	0.0161
cD1	0.1784	-0.1686	0.9565	0.228
cV1	0.4745	-0.4569	0.6764	0.0161
cH2	0.8902	-0.7225	0.3768	0.009
cD2	0.4422	-0.5029	0.5329	0.0127
cV2	1.1245	-0.8873	0.3768	0.009
cH3	1.4377	-1.6162	0.3281	0.0078
cD3	0.824	-1.052	0.4639	0.0111
cV3	2.3441	-1.5583	0.3281	0.0078
cH4	2.2696	-2.5576	0.357	0.0085
cD4	1.4826	-2.3309	0.5048	0.012
cV4	3.5324	-3.2123	0.357	0.0085
cA4	12.9799	2.4336	0.357	0.0085
DCT domain				
Original image	6.7853	-1.7017	0.4555	-0.4554

Waston JND model divides the original into 8 × 8 block, and each block is transformed into its DCT, and then calculates the perception threshold of each Coefficient according to the Waston model. Barni JND model decomposes the original image into a series of sub-band signal in different spatial resolution, frequency and direction by using four levels Haar wavelet transform, and then calculates the perception threshold of wavelet coefficients according to the Barni model. From the Table 1, we can see that JND thresholds in DWT domain are

2)   The evaluation of JND model's performance in DCT and DWT domain

Fig.5, Fig.6 is the energy-adjusted image in DCT and DWT domain. The human eye cannot detect the difference between the original image and Fig.5 or Fig.6 from the subjective visual perception, while peak signal to noise ratio (PSNR) of the energy-adjusted image in DWT Domain is significantly lower than the DCT domain's, which shows that the DWT coefficients can tolerate larger distortion without affecting the visual quality of image, therefore the watermarking algorithm in DWT domain can better balance the contradiction between robustness and perceptibility.

Fig. 5. Energy-adjusted image in DCTdomain (PSNR=51.8261)

Fig. 6. Energy-adjusted image in DWT domain (PSNR=37.9210)

5   Conclusion

Based on the above discussions, we can see that the existing JND models in the DWT domain do not fully take into account the stratified perception characteristic of human vision, therefore we consider to constitute a JND model based on the content decomposition and stratified perception characteristic. We know that the image consists of texture and profile based on the prime sketch theory [16,17,18],which calls the image portion with distinguishable components as sketchable , and the portion without distinguishable components is said to be non- sketchable. Therefore, we consider to combine the prime sketch theory with the multi-level wavelet decomposition for the JND modeling. First, we decompose the image into texture region and profile region according to the prime sketch, and then analyze the sensitivity, masking effect and other perception characteristics of texture and profile, eventually constitute the JND model for content watermark embedded in DWT domain. We believe that the watermarking algorithm based on this model can better balance the contradiction between the robustness and perceptibility.

Acknowledgement. The work on this paper was supported by National Nature Science Foundation of China (60902061), and the 3rd phase of 211 Project of Communication University of China.

References

1. Cox, I.J., Miller, M.L.: The first 50 years of electronic watermarking. EURASIP Journal on Applied Signal Processing 1 (2002)
2. Fang, W., Chen, K.: A Wavelet Watermarking Based on HVS and Watermarking Capacity Analysis. J. International Conference on Multimedia Information Networking and Security (2009)
3. Bouchakour, M., Jeannic, G., Autrusseau, F.: JND mask adaptation for wavelet domain watermarking. In: IEEE International Conference on Multimedia and Expo. (2008)
4. Jung Yong, J., Kang Ho, K., Ro Yong, M.: Novel Watermark Embedding Technique Based on Human Visual System. In: Proc. SPIE (2001)
5. Qi, H., Zheng, D., Zhao, J.: Human visual system based adaptive digital image watermarking. Signal Processing 88(1) (2008)
6. Ninassia, A., Le Meur, O., Le Callet, P., Barba, D.: On the performance of human visual system based image quality assessment metric using wavelet domain. In: Proc. SPIE, vol. 6806, p. 680610 (2008)
7. Girod, B.: Psychovisual aspects of image communication. Signal Processing (1992)
8. Chou, C.-H., Li, Y.-C.: A perceptual tuned subband image coder based on the Measure of just-noticeable-distortion profile. IEEE Trans. Circuits Syst. Video Technol. 5(6), 467–476 (1995)
9. Ahumada, A.J., Peterson, H.A.: Luminance-mode based DCT quantization for color image compression. In: Proceedings of the Human Vision, Visual Processing, and Digital Display III, pp. 365–374. SPIE Press, San Jose (1992)
10. Watson, A.B.: DCTune:A technique for visual optimization of DCT quantization matrices for individual images. In: A. in Soc. Inf. Display Dig. Tech. Papers XXIV, 946–949 (1993)
11. Wei, Z., Ngan, K.N.: Spatial Just Noticeable Distortion Profile for Image in DCT Domain. In: IEEE Int. Conf., Multimedia and Expo. (2008)
12. Lewis, A.S., Konwles, G.: Image compression using the 2-D wavelet transform. IEEE Trans. Image Processing (1992)
13. Barni, M., Bartolini, F., Piva, A.: ImProved Wavelet-Based watermarking through Pixel-wise masking. IEEE Transactions on Image Processing (2001)
14. Zhu, L., Zhao, L.: A Digital Watermarking Scheme Based On Texture Measures in Wavelet Domain. IEEE Image and Signal Processing (2009)
15. Zheng, X., Zheng, X.: A Multiwavelet Based Digital Watermarking Algorithm Using Texture Measures. In: Proceedings of the 2010 International Conference on Wavelet Analysis and Pattern Recognition, Qingdao (2010)
16. Marr, D.: Vision. W.H. Freeman and Company (1982)
17. Guo, C.E., Zhu, S.C., Wu, Y.N.: A mathematical theory of primal sketch and sketchability. In: Proc. Int'l Conf. on Computer Vision (ICCV) (2003)
18. Li, Z., Gao, R., Guo, C., Dong, J.: A new image compression method based on primal sketch model. In: IEEE International Conference on Multimedia and Expo. (2007)

Illegal Node Detection of Wireless Mesh Network Based on Node Reputation Value

Huai Yang

School of Information Engineering, Chongqing City Management College, Chongqing
401331, China
cgyyhtx@163.com

Abstract. Multi-hop nature of Mesh wireless network has stronger communication then traditional wireless communication, leading to larger security challenge. Based on multi-hop characteristics of wireless mesh network, current wireless intrusion detection mechanisms were introduced and their advantages and disadvantages were compared. The single node intrusion detection mechanism was selected and a intrusion detection schema to detect illegal mode effectively based on node reputation was presented. Wireless network intrusion simulation was performed with NS2, which also verifies effectiveness of the schema.

Keywords: WMN, multi-hop, intrusion detection, illegal node.

1 Introduction

Wireless mesh network is a new type of multi-hop wireless network [1], it can be seen as a special WLAN. In addition to lower low mobility, wireless mesh is essentially an Ad hoc network. In general sense, wireless mesh network is a static wireless network consisted of routers and terminals connected by wireless links, which is wireless edition of Internet [2]. The topology is shown in Fig. 1.

From Fig. 1 we can see that mesh network mainly consists of two parts of backbone network and terminal users. Backbone network is consisted of wireless routers, which has lower mobility and bridge users and Internet. Terminal user network is mainly constituted of various wireless terminals. As each node AP in the network can transceiver signal, each node can directly communicate with one or more peer nodes [3]. Therefore, when the nearest AP congested by too much traffic, data can automatically route to an adjacent node with less traffic for transmission. In this way, packet continuously route to next adjacent node for transmission based on network circus till it reach the destination. The access way is multi-hop, which has stronger communication function there traditional wireless network.

As to the multi-hop characteristics, mesh wireless network is faced with larger security challenges. As we known, wireless communication is vulnerable to passive attacks and active attacks [4]. These security risks will be further amplified in multi-hop mesh network. Each user in mesh wireless LAN connects to AP, so it will help management by administrators. However, as mesh network is a multi-hop network, all security management is concentrated on one end wireless gateway. It will delay

D. Jin and S. Lin (Eds.): Advances in MSEC Vol. 2, AISC 129, pp. 545–550.
springerlink.com   © Springer-Verlag Berlin Heidelberg 2011

detection and response of attacks, which will benefit attackers. As wireless router is near or far away from Internet AP, the node far from Internet AP may access small bandwidth. Reasonably designed protocol is important to ensure fairness among nodes, which also bring new challenges to protection of fairness [2]. Routers in wired network always are properly protected, so it is not so convenient to attack on routers in wired network. Wireless routers are always deployed outdoors, so they can not be properly protected physically. It is likely to lead to attacks on wireless routers, such as attackers forge as legal nodes and publish error route information and cause network congestion by frequently publishing information.

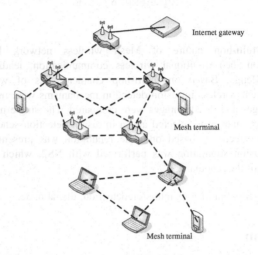

Fig. 1. Mesh network topology

Now there are mainly three types of wireless network intrusion detection systems, namely single node IDS, distributed collaborative IDS and mobile agent IDS [5]. The structure if single node IDS is robust. When some nodes in the network can not run IDS, the effect of them on other nodes is small. The watchdog mechanism is single node IDS [6]. But the detection efficiency of single node IDS is not high and it has large power consuming. Mobile agent is program with some intelligence, which can autonomously operate and provide corresponding services. Albers et al have presented intrusion detection framework using mobile agent techniques, but not design generation and management of mobile agent in detail. Collaborative IDS intrusion detection and response can be achieved only by nodes, but need information interaction with adjacent nodes. Distributed IDS effectively overcome shortcomings that lower detection efficiency and large power consumption of single node IDS. It is more mature compared with mobile agent IDS. Sun B. et al brought out a non-overlapping mobile Ad hoc intrusion detection framework and a kind of clustering algorithm for gateway node. Lu J. D. et al proposed a intrusion detection mechanism based on adjacent node monitoring. They all belong to distributed IDS. The networking manner of wireless mesh network determines its security vulnerability has distributed characteristics, so distributed IDS is more suitable for wireless network. The paper uses collaborative IDS.

2 Intrusion Detection of Mesh Based on Node Reputation Mechanism

The paper presented an improved reputation node mechanism based on collaborative IDS theory to perform intrusion detection of wireless mesh network. NS2 is object-oriented and discrete event driven network environment simulator, which can simulate whole network environment. The paper simulated wireless mesh network with NS2 and set a bad node. In case of packet transmission in network, the node tries to intrude into network and interferes normal transmission of packets. If adjacent node finds abnormal packet transmission through monitoring, it determines bad node with reputation node mechanism and delete its information from route, then re-select wireless link and restore normal transmission of data. Finally, simulation data is analyzed to verify effectiveness and accuracy of reputation node mechanism.

2.1 Wireless Environment Simulation

Fig. 2 shows the simulated mesh wireless network scene with NS2. There are 7 wireless nodes in the scene, where node 0 and node 1 is source_node and target_ node respectively. At 1.0s of scene, node 0 starts sending broadcast message and induces response from other adjacent nodes. Then node 0 starts sending TCP packets to node 1. with broadcast messages, we can know that source node 0 has two adjacent nodes, namely node 2 and node 4.

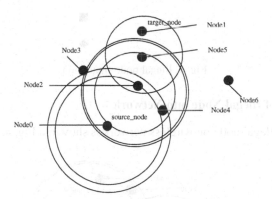

Fig. 2. Wireless mesh simulation

If node 0selects node 2 as relay node, the node 2 can not directly forward packet to node 1 after receiving TCP packets. So it selects its adjacent node 5 as relay node and finally forward packet to node 1 through node 5, the total hops is hop=3.

If node 0 select node 4 as its relay node, node 4 needs to select adjacent node 2 as relay node, the total hops of wireless link is hop=4.

Wireless routing protocol determines minimum hops as principle for wireless link selection. Based on the principle, node 0 send TCP packet to node 1 at 1.0s and select the first wireless link. The experiment simulation result shown in Fig. 2 also confirms the principle.

2.2   Illegal Node Simulation

There are many intrusion ways on wireless network, the most common way of which is node intrusion, namely attacker forge a legal node intrude network with an illegal node, leading to network congestion or even paralysis with the following program section:

```
$ ns_node-config-ifqLen 0
Set node_(6) [$ ns_node]
$ node_(6) random-motion()
$ ns_initial_node_pos & node_(6) 20
```

Re-set node 6 in the wireless scene, where $ ns_node-config-ifqLen 0 indicates that the queue length of node 6 is set 0. It means when other node send packets to node 6, it can not record packets but only produce packet loss, as shown in Fig. 3.

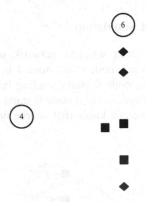

Fig. 3. Illegal node simulation

2.3   Intrusion of Illegal Node into Network

The intrusion of illegal node on wireless network is shown in Fig. 4.

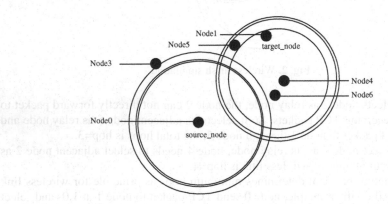

Fig. 4. Illegal node intrusion

At 1s of simulation scene, node 1 sends packets to node 1 via node 2 and node 5. At 72.4s of scene time as wireless nodes moves, node 0 needs to refer to node 6 as relay node to complete data transmission. But the ifqLen queue length of node 8 has been set 0, node 6 will discard all packets from node 0. Then target node 1 can not receive packet, which is an invasion.

2.4 Intrusion Detection

In the paper, node reputation mechanism is used to perform intrusion detection. The whole detection mechanism consists of three components. Firstly after node send packet to adjacent node, the node will monitor packet-receiving node and timer start timing. If the packet-receiving node is not target node, when the node detects packet-receiving node and did not forward packets, compute node reputation. If the reputation decrease below predetermined value, delete route information of the node from table and re-select a link.

Parameters of reputation node mechanism include ID_Value, $\triangle$Value, $\triangle$Time, ID_Time, $\triangle$I and M. ID_Value is the node reputation value of monitored node; $\triangle$Value is increase or decrease value of reputation at single detection; $\triangle$Time is the monitoring time interval; ID_Time is initial monitoring time and M is gain coefficient.

The computation formula of node reputation is ID_Value=ID_Value$\pm\triangle$i$\cdot$M$\cdot\triangle$Value. If monitored node completed packet forwarding, the formula takes addition operation. If the node has not completed packet forwarding, the formula takes subtraction operation.

Sometimes, packet can not be sent for network congestion. If the event is classified as an intrusion, it is a mis-determination. The problem has not been solved in [7]. Therefore, to minimize times of mis-determination, the parameter of $\triangle$Time is set. When the monitor node starts timing, count once each $\triangle$Time except for the node send packets out. The algorithm can be described as following:

```
Open Monitor Function
Begin monitor if node be used
Set Δi 1
if Message is not transport then
  Start time
  if time>ΔTime then
    set time 0
    start time
    Δi+
  end if
end if
```

$\triangle$Time interval divides monitoring time into small pieces and computes total times. If node forwards packet in time, the value of $\triangle$i is always 1. Otherwise, the value of $\triangle$i increases corresponding to time and $\triangle$Time.

3   Conclusion

Based on structure of WMN and characteristics of it, the paper introduced existing wireless intrusion detection mechanisms and compared there advantages and disadvantages. The optimal detection manner was selected. Simulation results were arrived with NS2 and a specific intrusion detection schema was presented. Compared with program in [7], it can further decrease false determination rate. As WMN is a new type network with promising future, the intrusion detection of WMN should be further strengthened.

References

1. Zhu, D.-Q.: Interview for Wireless MESH Network. Data Communications 2, 14–18 (2007)
2. Guo, C., Hong, P.-L., Xue, K.-P.: An interference-aware multi-path routing protocol in IEEE 802.11 wireless mesh networks. Journal of the Graduate School of the Chinese Academy of Sciences 27, 809–817 (2010)
3. Heberlein, L.T., Dias, G.V., Levitt, K.N.: A network seurity monitor. In: Proceedings of IEEE Symposium on Research in Security and Privacy, pp. 296–304. IEEE Computer Society press, Los Alamitos (1990)
4. Zhao, Y.-F., Yang, G.: Research of Cross-Layer Security Architecture in Wireless Mesh Networks. Jiangsu Communication 23, 5–10 (2007)
5. Huan, J., Ju, S.-G.: Survey on intrusion detection in mobile Ad Hoc networks. Computer Engineering and Design 28, 3066–3069 (2007)
6. Marti, S., Giuli, T.J., Lai, K.: Mitigating routing misbehavior in mobile ad hoc networks. In: Proceedings of the 6th Annual International Conference on Mobile Computing and Networking, pp. 255–265. ACM Press, New York (2000)
7. Wang, Z.-Z., Guan, Y., Lu, J.-D.: Routing Security Mechanism Based on Neighbor Nodes Monitoring and Detecting for MANET. Computer Engineering 33, 148–151 (2007)

Utilizing Free Software to Improve the Teaching and Learning of Computer Major Courses

XianBo He[1], BaoLin Li[1], MingDong Li[1], and YouJun Chen[2]

[1] Computer College of China West Normal University, Nanchong, Sichuan 637002, China
[2] Math and Information College of West Normal University, Nanchong, Sichuan 637002, China

Abstract. It is an important means to help students deeply comprehend the related theory and improve their programming ability to introduce some free software to the teaching and learning of computer major courses in colleges and universities. The research and learning on source code of free software such as Linux, eCos, GNU tools is very helpful to the learning and teaching of courses such as the Operating System, Data Structure, programming languages, Embedded Software Development and Compiling Principle. Meanwhile, the course designs and projects development based on these free software can largely improve students' design ability on complicate software using programming language such as C language, C++ language and assembly language.

Keywords: Free software, programming ability, Linux operating system.

1 Introduction

How to train applying talents who have good programming skill and programming abilities is an important task of computer major education in colleges and universities. How to deeply and whole train students' programming ability by combining the theory of related courses with practice application in the limited teaching time and poor experiment environment is the common problem confronted in computer majors of colleges and universities.

The contents of the computer major courses such as operating system, data structure and compiling principle are very abstract. Combining these abstract contents with concrete implementation schema enable students deeply comprehend those abstract theories. In learning some programming languages such as C, C++, assembly language, combining those basic syntax with their applications in complicate free software is able to better comprehend their spirits and improve students' programming ability with these programming languages.

In addition, to effectively utilize development environment, customizing some developing tools or the whole development environment is necessary. So, learning how to organize an integrated software development environment is necessary.

Some large mature free software such as Linux, eCos and GNU tools illustrate their respectively frontier and mature technique implementation schema. Meanwhile, in those free software, the advantages and characteristics of programming languages are fully demonstrated, and the concepts of software engineering is applied.

D. Jin and S. Lin (Eds.): Advances in MSEC Vol. 2, AISC 129, pp. 551–556.

Analyzing and researching on these free software and combining them with practice teaching and application schemas, are effectively means to improve the teaching and students' software design abilities of computer majors.

2   Free Software in Common Use

Free software is a matter of the users' freedom to run, copy, distribute, study, change and improve the software. It generally agrees to obey GPL(General Public License),and it means that the program's users have the following essential freedoms:

1) The freedom to run the program for any purpose.

2) The freedom to study and modify the program, and change it to make it do what you wish.

3) The freedom to copy the program so you can help your neighbor.

4) The freedom to improve the program, and release your improvements to the public, so that the whole community benefits.

Access to the source code is a precondition for the second freedom, so freedom software is an open source software. Just because of its free and transparency, free software and open source projects enjoy great support of governments and software and hardware enterprises all over the world. By the software developers' dedication and efforts, free software have become the experimental plots and birthplaces of many new technologies in software development field and mature free software have become a strong rival of corresponding commercial software.

Among open source software, Linux, eCos, GNU series tools (such as gcc and gdb) are helpful to computer major course teaching.

2.1   Linux

Linux is a multi-user and multitask open source operating system with high security policy. Its kernel has characteristics such as powerful functions, steady performance, supporting multi-processor and rich application software. The coding and design of Linux kernel embodies mature theory and excellent programming skill of complicate computer software. Owing to experts and programming lovers' efforts and unselfish devotion, Linux also have become the experimental plots of many new techniques and new device drivers in operating system field, and learning its source code is one of the best ways to improve programmers' software design ability.

The Linux kernel version have improved from initial version 0.01 with simper function to version 2.6.x with real-time kernel characteristic (current newest version 2.6.37), and its official website address from that its kernel source code may be downloaded is http://www.kernel.org.

2.2   eCos

eCos(embedded Configuration operating system) is an open source real-time operating system developed by Cygnus company. It is one of the fastest growing real-time operating systems (RTOS) in the embedded marketplace and the most widely adopted open source RTOS. It has the characteristics such as novel design concepts, unique and concise kernel component configuration interface. Its kernel is almost all implemented according to object-oriented programming principles and thinking, and its programming language is C++.

2.3   GNU Software Development Tools

GNU Open source software development tools includes compilers of programming languages in common use (such as gcc), debuggers (such as gdb) and the analysis and handling tools on object files(such as objdump, nm, strip and objcopy).

2.4   make Tool and makefile Script Source Program

Make is the automatic organization and management tool of GNU open source software project. make tool organizes the whole procedure from function module configuration to compilation and linkage by makefile script file. A makefile file is composed of some make rule descript script sentence that mainly describes the dependence relation between the target and source code program file and the necessary operations to build the target. To compile a source code program, if a source code file is changed, all source code file and library depending it should certainly be recompiled and upgraded. So, being able to auto-recognize and re-compile objects by time stamp is the fundamental function of make tool and make script file.

3   The Schema of Introducing Free Software to Teaching and Learning

3.1   Introducing Linux and eCos to the Operating System Course

The Operating System is a fundamental core course of computer major. Owing to its contents are very abstract, the teaching and learning is very difficult. To its process management and schedule, memory management, device driver and file system, the corresponding implementations in Linux kernel are very suitable for learning reference; especially by studying and research on the virtual file system and the unique device file system implementation mechanism of Linux kernel, students may actually "touch" concrete application and renovation of operating system theory. Meanwhile, it can cultivate Linux programming talents by introducing students to improve and rebuild Linux kernel to meet the need of the actual specific application field and programming the device drivers based on Linux. In addition, by tracing the Linux kernel history of various versions, students may learn the improvement development history of the theory and technique of operating system. Owing to the newest version of Linux kernel absorb the most mature and frontier technique, to introduce the newest technique of Linux kernel to teaching and learning can keep the teaching content of Operating System   course   along the times.

eCos implemented a real-time operating system kernel with object-oriented concepts. The process control block, process schedule, semaphore and timer are all implemented by the class and object in C++ programming language. In one hand, learning eCos source code can enable students to deeply comprehend design ideas of the real-time operating system, it can be very helpful to train students' ability of designing the complicate software such as operating system with the object-oriented thinking and concepts.

3.2   Introducing Linux and eCos to Data Structure Course

Data structure is also a fundamental core course of computer major. To this course, the application instances of the basic data structure, which is its important contents, such as linear list, stack, queue, tree and hash table can be find in many place of Linux and eCos source code. For example, in the implementation of Linux kernel and eCos kernel, almost all kinds of process queues are organized by double directed linkage queue data structure; all local variables of a process are allocated in memory stack; the family relations of all processes are described by binary tree or triple trees data structure; in order to speed up data searching, a lot of hash tables and hash functions are skillfully applied too. In addition, the special location of the operating system requires the implementations of all functions as efficient as impossible. Learning those algorithms is very helpful for students to design efficient algorithms for special functions.

To the teaching materials of Data Structure whose contents are described in C language, introducing Linux kernel source code is suggested. But to that whose contents are described in C++ language , the eCos source code is undoubtedly the best choice, and due to eCos is an real-time operating system, students can learn how to balance the memory space and time consumption of a function algorithm from the source code implementation of eCos kernel.

3.3   Introducing the Linux Source Code and Makefile Script Source Program to Related Courses on the Embedded Software Development

Now, embedded systems has become one of the   widest computer application fields, and various improved embedded Linux versions based on standard Linux also become very important component of an embedded software development platform. It is an embedded software programmer's necessary skill how to improve the real-time performance of the standard Linux and how to program the driver of a new hardware device.

Meanwhile, reconfiguration, clipability and expandability are the important characteristics of an embedded software development environment. So, it may enable students deeply comprehend how to construct and customize an integrated development environment of an embedded system based on specific application.

3.4   Introducing Open Source Software Code to Other Courses of Computer Majors

Linux kernel is programmed in assemble and C language, and its code design present excellent programmers' intelligence and success experience. So it can be helpful to cultivate student's good programming habit and improve their programming skill of C language and assemble introducing Linux kernel source code to the teaching on C language and assemble. And introducing eCos source code to the course of C++ programming can improve students' ability on design complicate software in object-oriented concepts using C++ . In addition, it can deepen the comprehension on how to design compiler and debugger. analizing the implementation of GNU tools such as gcc and gdb in source level.

4 How to Studying Open Source Software Code

The source code scale of some open source software such as Linux, eCos and GNU tools are very large, the following are some studying suggestion.

4.1 Establishing Corresponding Interesting Teams and Organizations

Establishing various interesting teams or organizations among students and teachers is an important means to learn and research open source embedded software. In the initial learning stage, the team or organization members should mainly consist of related course teachers and excellent students. Then, by offering some specific Lectures on the learning of embedded open source software, establishing source code learning and communication website on the embedded open source software and developing some project based on correspondingly open source software, it can train some core members who can abstract more students to join.

4.2 Carefully Selecting Introducing Contents of Open Source Code According to Teaching Courses

The source code scale of some open source software such as Linux, eCos and GNU tools are very large and the referred technique points are more. It is very important to introduce suitable source code into the teaching according teaching contents and students' knowledge fundament. The selection rules that introducing source code enable students comprehend course contents easily and be helpful to improve students' programming ability. Meanwhile, the introduction contents should be dynamically modulated according to students' efficiency feedback.

4.3 Considering on the Whole, Implementing Step by Step

To different grades of the undergraduate and post graduate in computer major, correspondingly the introduced source code contents should be suitable for learned objects. The basic introducing principle should obey these rules: learning them from the easy to the difficult and complicated, spreading them over a whole area from one point and studying them step by step. In grade one of undergraduate, the learning emphasis should be good programming style, programming custom and the language syntax criticality and programming skills. Then in grade two and three, combining with the major courses such as computer operation system and data structure, the analysis and research on complicate programming software module is considered to be introduced and the learning emphasis should be put on the combination of theory and its application practice; in this phase, conducting some excellent students to work out some complicated course designs is probable. Finally, in the grade four of undergraduate and the phase of postgraduate, the main work is to train their ability to independently develop complicated application software by improve and rebuild corresponding open source software such as improving and customizing Linux kernels for specific embedded system application, rebuild embedded integration software development environment and adding graphic operating interface to gdb; in this phase, the emphasis is on how to cultivate the specialized talents.

5   Conclusion

How to improve students' programming ability is the criticality of computer major education, and it is also the bottleneck in most colleges and universities. It can implement good combination of abstract theory and their application and enable students comprehend teaching contents better by Introducing some free software such as Linux, eCos and GNU tools into the teaching and learning of computer major courses such as Operating System, Data Structure, Embedded System, C/C++ Programming, Assembly programming and Compilation Principle. Meanwhile, it can improve students' ability developing complicate software in large degree by means of conducting students to participate in software developing course design and application project based on corresponding free software.

Acknowledgement. Supported by Application of infrastructure projects of office of science and technology in Sichuan province (project NO: 010JY0151), scientific research fund of Sichuan Provincial Education Department (project NO:08ZA015), teaching reform research project fund of China West Normal University (project NO: JGXM0945) and teaching reform research project fund of Sichuan Provincial Education Department (project NO: P09262).

References

1. He, X., et al.: The embedded software development technique fundamentals. Tsinghua university press, Beijing (2011)
2. The Free Software Definition,
   http://www.gnu.org/philosophy/free-sw.html

The Combined Homotopy Methods
for Optimization Problem
in Non-convex Constraints Region

Yunfeng Gao

College of Arts and Science, Jilin Agricultural Science and Technology University,
Jilin 132109, Jilin Province, China
304790969@qq.com

Abstract. In this paper, we study the homotopy method of the optimization problem in non-convex constraints region and introduce the constructive method about quasi-normal. We prove that the chosen mappings on constrained grads are positive independent and the constrained quasi-normal cone satisfies the quasi-normal cone condition. As a by-product of our analysis, we adopt the predictor-corrector procedure and make up the computation procedure with the Matlab language and have got a good digital result.

Keywords: Non-convex constrains region, Homotopy method, the quasi-normal cone condition, Numerical examples.

1 Introduction

As a globally convergent method, more and more attention has been paid to the homotopy method. In paper [1], the combined homotopy method was produced for solving functions min-problems in non-convex region; prove the whole convergence under the condition "normal cone condition". In this paper, we give the constructive method of the quasi-normal cone condition in non-convex region which is formed by partial reverse convex constrains, show that this method is feasible and efficient.

We consider the following programming problem:

$$(NP) \quad \begin{cases} \min f(x), \\ s.t.\, g_i(x) \le 0, i = 1,2,\cdots,m \end{cases} \tag{1}$$

Where $x \in R^n$, f, g_i are twice continuously differentiable functions.

Let $\Omega = \{x \in R^n \mid g_i(x) \le 0, i = 1,\cdots,m\}$,

$\Omega^0 = \{x \in R^n \mid g_i(x) < 0, i = 1,\cdots,m\}$,

$\partial\Omega = \Omega \setminus \Omega^0$, $I(x) = \{i \in \{1,\cdots,m\} \mid g_i(x) = 0\}$,

$\Omega_i^0 = \{x \in R^n \mid g_i(x) < 0\}$, $g(x) = (g_1(x),\cdots,g_m(x))^T$,

$\nabla g(x) = (\nabla g_1(x),\cdots,\nabla g_m(x))$.

D. Jin and S. Lin (Eds.): Advances in MSEC Vol. 2, AISC 129, pp. 557–561.
springerlink.com © Springer-Verlag Berlin Heidelberg 2011

2   Some Definitions and Properties

Definition 2.1[2]. If the smooth map $\eta_i : R^n \to R^n (i \in \{1,2,\cdots,m\})$ satisfies :
$\forall x \in \partial \Omega$, if

$$\sum_{i \in I(x)} (y_i \nabla g_i(x) + \alpha_i \eta_i(x)) = 0, \quad y_i \geq 0, \quad \alpha_i \geq 0,$$

we have $y_i = 0, \alpha_i = 0, (i \in I(x))$. Then, we call $\eta(x) = (\eta_1(x), \cdots, \eta_m(x))$ is the positive independent mapping about $\nabla g(x)$.

Definion 2.2[2]. If there exists a positive irrelative smooth map $\eta_i(x)$ about $\nabla g(x)$ and satisfies

$$\forall x \in \partial \Omega, \left\{ x + \sum_{i \in I(x)} \alpha_i \eta_i(x) \mid \alpha_i \geq 0, i \in I(x) \right\} \cap \Omega^0 = \Phi.$$

Then Ω about $\eta(x)$ satisfy quasi-norm cone condition.

We need to construct quasi-norm cone for solving min-function problem in non-convex feasible region Ω [3]. Subsequently, we present the conditions of quasi-norm cone for a class of non-convex constrains region.

3   Main Results

In Non-convex region:

$$\Omega = \{x \in R^n \mid g_0(x) \leq 0, g_1(x) \leq 0, g_2(x) \leq 0\}.$$

Where $g_0(x) = \dfrac{1}{2}x^T x - a_0^T x + b_0 \leq 0$; $g_1(x) = -\dfrac{1}{2}x^T x + a_1^T x - b_1 \leq 0$;

$$g_2(x) = -\frac{1}{2}x^T x + a_2^T x - b_2 \leq 0. \quad a_0, a_1, a_2 \in R^n, b_0, b_1, b_2 \in R.$$

Ω satisfies : $a_0^T a_0 - 2b_0 > a_1^T a_1 - 2b_1 > a_2^T a_2 - 2b_2 > 0$;

$\{x \mid g_1(x) = 0, g_2(x) = 0\} \neq \Phi$, $a_2^T a_2 - a_1^T a_2 + b_1 - b_2 \leq 0$;

$$a_1 + \sqrt{a_1^T a_1 - 2b_1} \cdot \frac{a_1 - a_0}{\|a_1 - a_0\|} \in \Omega_0^C, \quad a_1 - \sqrt{a_1^T a_1 - 2b_1} \cdot \frac{a_1 - a_0}{\|a_1 - a_0\|} \in \Omega_0^0,$$

$$a_1 \neq a_0 \ ; \ a_1 + \sqrt{a_1^T a_1 - 2b_1} \cdot \frac{a_1 - a_2}{\|a_1 - a_2\|} \in \Omega_0^C, \quad a_1 \neq a_2 \ ;$$

$\forall x \in \{x \mid g_2(x) = 0\}$, we have $g_0(x) < 0$.

We give the positive independent mapping:

$$\eta_0(x) = \nabla g_0(x) = x - a_0;$$

$$\eta_1(x) = a_1 + \sqrt{a_1^T a_1 - 2b_1} \cdot \frac{a_1 - a_0}{\|a_1 - a_0\|} - x;$$

$$\eta_2(x) = a_1 + \sqrt{a_1^T a_1 - 2b_1} \cdot \frac{a_1 - a_2}{\|a_1 - a_2\|} - x \tag{2}$$

Theorem 3.1. Suppose the mapping $\{\eta_i \mid i = 0,1,2\}$ be given by(2), then $\{\nabla g_i(x) \mid i = 0,1,2\}$, i.e. $\forall x \in \partial\Omega$, if

$$\sum_{i \in I(x)} (\alpha_i \nabla g_i(x) + \beta_i \eta_i(x)) = 0, \alpha_i, \beta_i \geq 0,$$

then $\alpha_i = \beta_i = 0$.

Proof: When $I(x) = \{2\}$, we have $g_2(x) = 0$, $g_1(x) < 0$,

$$(\nabla g_2(x))^T \eta_2(x) = (a_2 - x)^T (a_1 + \sqrt{a_1^T a_1 - 2b_1} \cdot \frac{a_1 - a_2}{\|a_1 - a_2\|} - x)$$

$$= \left(\frac{1}{2} x^T x - a_2^T x + \frac{1}{2} x^T x - a_1^T x\right) + \frac{a_1 - a_2}{\|a_1 - a_2\|}(a_1^T a_2 - a_2^T a_2 - b_1 + b_2) + a_1^T a_2$$

$$> -b_2 - b_1 + a_1^T a_2 \geq a_2^T a_2 + b_1 - b_2 - b_2 - b_1 = a_2^T a_2 - 2b_2 > 0,$$

So $\eta_2(x)$ is the positive independent mapping about $\nabla g_2(x)$.

Theorem 3.2. Suppose the mapping $\{\eta_i(x) \mid i \in M\}$ be given by(2), then the feasible set Ω satisfies the qusai-normal cone condition about $\{\eta_i(x)\}$ .i.e. $\forall x \in \partial\Omega$, we have

$$\{x + \sum_{i \in I(x)} \alpha_i \eta_i \mid \alpha_i \geq 0, \sum_{i \in I(x)} \alpha_i > 0\} \cap \Omega^0 = \Phi$$.

Proof: When $I(x) = \{1,2\}$, $g_1(x) = 0$,

we know

$$g_1\left(a_1 + \sqrt{a_1^T a_1 - 2b_1} \cdot \frac{a_1 - a_2}{\|a_1 - a_2\|}\right) = 0.$$

For Ω_1^C if convex set, we have

$$x + \alpha\eta_2 = x + \alpha\left(a_1 + \sqrt{a_1^T a_1 - 2b_1} \cdot \frac{a_1 - a_2}{\|a_1 - a_2\|} - x\right)$$

$$= (1-\alpha)x + \alpha\left(a_1 + \sqrt{a_1^T a_1 - 2b_1} \cdot \frac{a_1 - a_2}{\|a_1 - a_2\|} \right) \in \Omega_1^C \subset \Omega^C,$$

$\forall 0 < \alpha < 1.$

As $\alpha \geq 1$, $x + \alpha\eta_2(x) \in H^+ = \{x \mid a_1^T x + b_0 - b_1 > 0\}$, so

$$\{x + \alpha\eta_2(x) \mid \alpha \geq 0\} \cap \Omega^0 = \Phi.$$

As $0 < \alpha < 1$, $x + \alpha\eta_1(x) \in \Omega_1^C$, $\alpha \geq 1$, $x + \alpha\eta_1(x) \in H^+$, we have

$$\{x + \alpha\eta_1(x) \mid \alpha \geq 0\} \cap \Omega^0 = \Phi.$$

Because Ω_1^C and H^+ are all convex set, so

$$\{x + \alpha\eta_1(x) + \beta\eta_2(x) \mid \alpha, \beta \geq 0\} \cap \Omega^0 = \Phi.$$

The K-K-T equation of (1):

$$\begin{cases} \nabla f(x) + \nabla g(x) = 0, \\ Y g(x) = 0, \\ g(x) \leq 0, y \geq 0. \end{cases} \tag{3}$$

We construct homotopy equation under cottle constrains condition:

$$H(\omega, \omega^{(0)}, t) = \begin{pmatrix} (1-t)(\nabla f(x) + \nabla g(x)y + t\eta(x)y^2) + t(x - x^{(0)}) \\ Y g(x) - t Y^{(0)} g(x^{(0)}) \end{pmatrix} = 0,$$

we can easily prove the existence and convergence of a smooth homotopy path from almost any interior initial point to a solution of the KKT system of (1).

4   Algorithm and Example

Algorithm 4.1(Euler-Newton method):

Step 0: Give an initial point $(\omega^{(0)}, \mu_0) \in \Omega^0(t) \times R^m \times \{1\}$, an initial step length $b_0 > 0$, three small positive numbers $\varepsilon_1, \varepsilon_2, \varepsilon_3, k = 0$;

Step 1 : Compute the direction :

(1.1) Compute a unit tangent vector $\xi^{(k)} \in R^{n+m+1}$ of $DH(\omega^{(k)}, \mu_k) = 0$.

(1.2) Determine the direction $\eta^{(k)}$,

If the sign of the determinant $\begin{vmatrix} DH(\omega^{(k)}, \mu_k) \\ \xi^{(k)} \end{vmatrix}$ is $(-1)^{m+1}$, let $\eta^{(k)} = \xi^{(k)}$, or

$\eta^{(k)} = -\xi^{(k)}$, go to step 2 ;

Step 2 :

(2.1) $(\overline{\omega}^{(k)}, \overline{\mu}_k) = (\omega^{(k)}, \mu_k) + b_k \eta^{(k)}$,

(2.2) 若 $(\overline{\omega}^{(k)}, \overline{\mu}_k) \notin \Omega^0(t) \times R^m \times (0,1], b_k = b_k / 2$, go to (2.1) , or go to step 3 ;

Step 3 : Compute a corrector point :

(3.1) $(\omega^{(k+1)}, \mu_{k+1}) = (\overline{\omega}^{(k)}, \overline{\mu}_k) - (DH(\overline{\omega}^{(k)}, \overline{\mu}_k))^+ H(\overline{\omega}^{(k)}, \overline{\mu}_k)$,

(3.2) If $(\omega^{(k+1)}, \mu_{k+1}) \notin \Omega^0(t) \times R^m \times (0,1], b_k = b_k / 2$, go to (2.1) ,

(3.3) If $\left\| H(\omega^{(k+1)}, \mu_{k+1}) \right\| \le \varepsilon_1, b_{k+1} = 4b_k$, go to 4 ;

(3.4) If $\left\| H(\omega^{(k+1)}, \mu_{k+1}) \right\| \in (\varepsilon_1, \varepsilon_2), b_{k+1} = b_k$, go to 4 ;

(3.5) If $\left\| H(\omega^{(k+1)}, \mu_{k+1}) \right\| \ge \varepsilon_2, b_{k+1} = b_k / 2, k = k+1$, go to 2 ;

Step 4 : If $t \le \varepsilon_3$,then stop, output $(\omega^{(k+1)}, \mu_{k+1})$; or $k = k+1$, go to step1.

Example 4.1. min $f_2(x) = x_1^2 + (x_2 + 2)^2 + x_3^2$

$$x_1^2 + x_2^2 + x_3^2 - 9 \le 0$$
$$\text{s.t } - x_1^2 - (x_2 - 1)^2 - x_3^2 + 4.41 \le 0$$
$$- x_1^2 - (x_2 + 1)^2 - x_3^2 + 2.25 \le 0$$

Table 1. Numerical solution for example 4.1

$x^{(0)}$	x^*	y^*	$f_2(x^*)$
(0.1, -2.8, 0)	(0.0000, -2.5000, 0)	(0.0000, 0.0000, 0.6667)	0.2500
(-0.1, -2.8, 0)	(0.0000, -2.5000, 0)	(0.0000, 0.0000, 0.6667)	0.2500

References

[1] Feng, G.C., Lin, Z.H., Yu, B.: Existence of Interior Pathway to a Karush-Kucher-Tucker Point of a Nonconvex Programming Problem. Nonlinear Analysis, Thero, Methods & Applications 32(6), 761–768 (1998)
[2] Liu, Q.-H., Feng, G.-C., Yu, B.: The Homotpy Interior-point Method for Non-linear Programming under Quasi Normal Cone Condition. Acta Mathematicae Applicate Sinica 26(2), 372–377 (2003)
[3] Liu, Q.-H.: A Combined Homotopy Inter-point for Solving Non-convex Programming: [Ph D Thesis]. Jilin University, Changchun (1996)

A Novel Method for Dead-Time Compensation
of the Inverter Using SVPWM

YiMin Gong, XiaoJiao Chen, and Ying Huang

State Key Laboratory of Automobile Dynamic Simulation, ChangChun, 130012, China
Gongym@jlu.edu.cn, chn0907@163.com

Abstract. The Dead-time effect of the three phases bridge inverter is analyzed in this paper. A Dead-time compensation strategy is presented for a permanent-magnet synchronous motor drive taking zero-current clamp and parasitic capacitance effects into account. It improves the Dead-time effect, with practicality and little calculation .The validity of theory analysis and this method is proved by the experiment results, the method is applied to the controlling of Air conditioner motor.

Keywords: Three-level inverter, Dead time, Compensation, PWM.

1 Introduction

Three-phase PWM voltage-source inverters have gained increasing popularity in industrial motor driving systems recently. There has a voltage loss that cannot be neglected between the reference voltage and the actual output voltage due to the dead time effect [1],[2]. The distortion of voltage leads to serious problems such as current waveform distortion and torque oscillation of motor driving system [5]. Many authors have made research in dead-time effect and advanced several strategies to compensate it. Generally, two methods are proposed to improve the performance. One compensation strategy is based on average error of voltage [3], [4], [7]. It is easy to actualize, but not always exact enough. The other method is time compensation [6]. This method can achieve high accuracy, but detection of current direction in dead-time compensation is important [8], [9], [10]. The misjudgement of zero-current point will lead to error compensation [1]. This paper analyses a method of trapezoidal time SVPWM dead-time compensation. The experiments on Renesas R8C34E controlling chip before and after compensation achieve a satisfactory effect.

2 Analysis of Dead-Time Effect

There are two directions of a-phase current, one is positive current, and the other is negative. Fig.1 shows the channel flow ·of the a-phase current for the positive direction ($i_a>0$). During on-period T_{on}, the a-phase current i_a will flow through the switching device s_a^+. Otherwise, it flows through diode D_a^- during both the off-period T_{off} and dead-time period T_d. Therefore, for the off-period of s_a^+, the output voltage of a-phase inverter v_{AN} is equal to the voltage during dead-time period. Fig. 2 shows

D. Jin and S. Lin (Eds.): Advances in MSEC Vol. 2, AISC 129, pp. 563–568.
springerlink.com

the relationship between the ideal and actual inverter output voltages for $i_a > 0$. If we taking into account the dead-time, the actual voltage becomes. Moreover, because of the switching device needs turn-on time and turn-off time, the actual voltage becomes. Besides, there are some other corrections for the actual voltage, the on – state voltages of switching device and of the diode, Finally, we have the actual voltage. These undesirable time and voltage components are integrated into the dead-time compensation time (DTCT) can be represented as follows:

$$T_c = T_d + t_{on} - t_{off} + \frac{V_{on}}{V_{dc}} T_S \qquad (1)$$

Where T_d, t_{on}, t_{off}, V_{on}, V_{dc} and T_S are dead-time, turn-on time, turn-off time, average on-state voltage, dc-link voltage, and switching period, respectively, i.e., Average on-state voltage is given as follows:

$$V_{on} = \begin{cases} \dfrac{T_{on}}{T_{off}} v_S + \dfrac{T_{off}}{T_S} v_D & \text{for } i_a > 0 \\[3ex] \dfrac{T_{off}}{T_S} v_S + \dfrac{T_{on}}{T_S} v_D & \text{for } i_a < 0 \end{cases} \qquad (2)$$

Where v_D, v_S, T_{on} and T_{off} are on-state voltages of diodes and switching devices, on-period, and off-period of the a-phase leg, respectively. In consequence of that the on-period T_{on} for $i_a > 0$ is almost identical to off-period T_{off} for $i_a < 0$, it should be noted that the average on-state voltage for positive current direction is almost the same as the negative. From a small analysis, the difference between ideal and practical inverter output voltages becomes $\Delta v_{AN} (= v_{AN}^{ideal} - v_{AN}^{dtc})$.

Similarly, the difference between ideal and practical inverter output voltages for negative a-phase current ($i_a < 0$) can be analyzed.

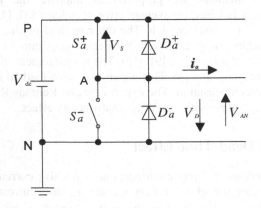

Fig. 1. Channel flow of a-phase stator current ($i_a > 0$)

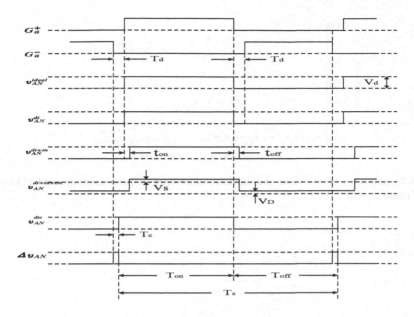

Fig. 2. Relationship between a-phase ideal and practical inverter output voltage ($\dot{i}_a > 0$)

3   Dead-Time Compensation Time

It is seen that parasitic capacitance have effect on rise time in [1]. Fig. 3 shows the characteristics of the inverter output voltage for instantaneous values of phase current during the period of rise and fall. It is note worthy that fall time exponentially increases as low magnitude of phase current. fall time depends on the magnitude of positive current.

In our research, we combine the dependence of turn-off time on the magnitude phase current into DTCT quantitatively. Fig. 4 shows how to adjust DTCT. Mark 1 shows when fall time is t_{tr0}, the output voltage of the inverter at $i_a = 0.8A$. In mark 2, when fall time increases, the magnitude of phase current decreases, mark 3 shows the compensated quantity because of the shade area. We can transform the shade area in trace 3 into a rectangle, shown it the mark 4.The equivalent formula is :

$$t_{etr} = \frac{1}{2}(t_{tr} - t_{tr0})$$ (3)

t_{etr} is the Equivalent transition time and the practical transition time is t_{tr} . Mark 5 illustrates t_{etr} is incorporated into DTCT, i.e. DTCT is adjusted as :

$$T_c = T_{cn} - t_{etr}$$ (4)

T_{cn} is a nominal value of DTCT, the characteristic of Equivalent transition time of phase current is as explained in Figure 5.It is utilized for adjusting DTCT.

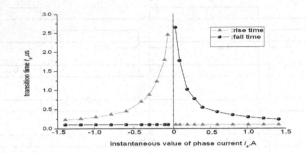

Fig. 3. Characteristics of rise time and fall time of inverter output voltage for instantaneous value of phase current

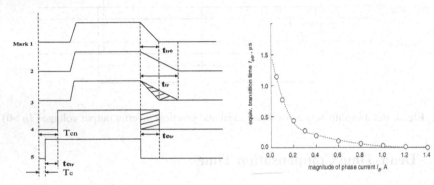

Fig. 4. Adjustment of DTCT for positive current (for fall time)

Fig. 5. Equivalent transition time for magnitude of phase current

4   Verifications

Fig. 6 is the block diagram of permanent magnet synchronous motor vector control system using the method of dead-time compensation above. It uses Renesas R8C34E chip to implement the vector control algorithm and the dead-time compensation algorithm. The parameters of permanent magnet synchronous motor are shown in Table 1.

Table 1. Specification of Tested PMSM

Rated power	P_n	70W
Rated torque	τ_n	0.4275N·m
Maximum rated speed	Nn	3500r/min
Armature resistance	R	53Ohm
Armature inductance	Ld	0.1739H
Armature inductance	Lq	0.1936H
EMF constant	Ke	0.1425V·s/rad
Number of pole pairs	P	4

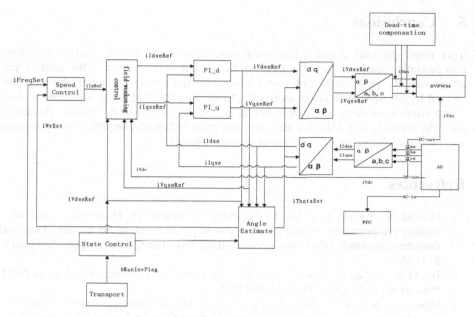

Fig. 6. Vector controlled PMSM drive with proposed dead-time compensation

Fig. 7 shows the results of experiments before compensation,after square-wave compensation and after trapezoidal compensation in the same conditions but different rotate speeds.The trapezoidal dead-time compensation can improve the current waveform and the output of the system more effective than the square compensation.

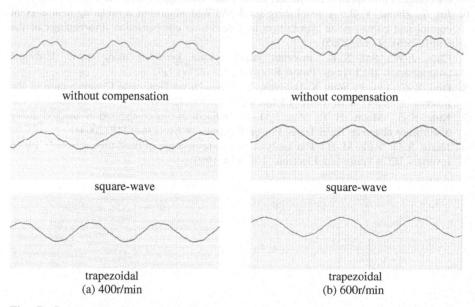

Fig. 7. Current waveform at 400r/600r rotate speed with square-wave compensation and trapezoidal compensation

5   Conclusions

This paper presents a method of trapezoidal time-based PWM inverter dead-time compensation used in vector control of permanent servo system. This method can ameliorate the performance of motor output by improving the wave abnormalities caused by the dead-time effect more effectively than the traditional square-wave time compensation. Besides, it is simple to realize and easy-to-modular. Finally, the author uses the Renesas R8C34E chip to verify the proposed dead-time effect compensation and obtain a relatively satisfactory compensation result.

References

1. Urasaki, N., Senjyu, T., Kinjo, T., Funabashi, T., Sekine, H.: Dead-time compensation strategy for permanent magnet synchronous motor drive taking zerocurrent clamp and parasitic capacitance effects into account. IEEE Proc.-Electr. Power Appl. 152(4), 845–853 (2005)
2. Hu, Q.-B., Lü, Z.-Y.: A novel method for dead-time comepensation based on SVPWM. Proceedings of the CSEE 25(3), 13–16 (2005)
3. Urasaki, N., Senjyu, T., Uezato, K., Funabashi, T.: Online dead-time compensation method for voltage source inverter fed motor drives. In: Proc. Conf. on Applied Power Electronics, Anaheim, CA, pp. 122–127 (2004)
4. Choi, J.-W., Sul, S.-K.: A New compensation Strategy Reducing VoltageKurrent Distortion in PWM VSI Systems Operating with Low Output Voltages. IEEE Transactions on Industry Applications 31(5), 1001–1008 (1995)
5. Lee, J.-S., Takeshita, T., Matsui, N.: Stator-flux-oriented sensorless induction motor drive for optimum low-speed performance. IEEE Trans. Ind. Appl. 33(5), 1170–1176 (1997)
6. Jin, S., Zhong, Y.-R.: A novel three-level SVPWM algorithm considering neutral-point control and narrow-pulse elimination and dead-time compensation. Proceedings of the CSEE 25(6), 60–66 (2005)
7. Choi, J.-W., Sul, S.-K.: Inverter output voltage synthesis using novel dead-time compensation. IEEE Trans. Power Electron., 221–227 (1999)
8. Dou, R.-Z., Liu, J., Wen, X.-H., Hua, Y.: Research on Dead-time Compensation of the Inverter using SVPWM. 'Power Eletronics 38(6), 59 6 (2004)
9. Kim, H.-S., Moon, H.-T., Youn, M.-J.: Onlinedead-timecompensation method using disturbance observer. IEEE Trans. Power. Electron. 18(6), 1136–1345 (2003)
10. Jeong, S.-G., Park, M.-H.: The analysis and compensation of dead-time effects in PWM inverters. IEEE Trans. Ind. Electron., 108–114 (1999)

The Application of Wavelet in Digital Image Watermarking Pretreatment

Chen Li, Cheng Yang, and Wei Li

Information Engineering School, Communication University of China, Dingfu East Road 1,
Beijing, China
{lichen100808,liwei601}@126.com, cafeeyang@gmail.com

Abstract. With the development of digital image watermarking, digital image watermarking pretreatment technique is paid more and more attention. Digital image watermarking pretreatment is to remove the redundant parts and retain the effective parts which content watermark message. This paper establishes a digital image watermarking model based on wavelet. Digital image watermarking pretreatment and the application of wavelet transform in the digital image pretreatment watermarking are mainly researched in the model. Different methods of digital image watermarking pretreatment are compared which based on the robustness and the invisibility. MATLAB simulate results prove that wavelet is good at digital image watermarking pretreatment and biorthogonal wavelet is better in wavelet bases at the same time.

Keywords: Digital image watermarking model, Pretreatment, Wavelet transform.

1   Introduction

In this paper, through a practical case study, the application of RE to the shape matching of a car is explained in detail. The principles and methods are not only applicable to the autobody panels, but also to any other industry.

In the recent years, cultural industries and media industries have been developed rapidly. The transmitted extensive and using of media content easily make it be paid more and more attention in the security needs of copyright protection, piracy tracking and monitoring of content security. However, the traditional measures of protection, such as media content encryption and authorization etc., couldn't effectively meet these needs because of its pre-control features. At the same time, the digital watermarking technology has become more and more important because it has the natural advantage. It bounds media content with copyright, fingerprints, the content identification information and so on closely, and determines the media content's copyright, source of pirate and legality through information detection and validation.

In the digital image watermarking, not only the robustness of watermarking, but also the human visual system (HVS) should be considered. Because the wavelet transform's space-frequency characteristics are similarities with the HVS's some characteristics[1] the wavelet transform has become widely use in digital watermarking.

D. Jin and S. Lin (Eds.): Advances in MSEC Vol. 2, AISC 129, pp. 569–576.

This paper establishes a digital image watermarking model based on wavelet, and analyzes the features of wavelet transform in digital image watermarking pretreatment which based on the robustness and the invisibility. The mainly feature of the model is that it combining digital watermarking with wavelet closely, using the good properties of wavelet, applying wavelet transform into pretreatment of watermark message image and cover image.

Traditional digital image watermarking pretreatment methods are chaotic sequence [2], Arnold transform [3] and so on. These techniques only change the watermark image's form, couldn't extract the effective watermark message for embedding. This paper proposes that the effective watermark message should be extracted in the pretreatment. In this paper, wavelet transform is compared with other methods which used in extracting digital watermarking message in the theory, compression evaluation and so on, combined with HVS to discuss the wavelet's selection bases in digital image watermarking pretreatment.

2   Digital Image Watermarking Model

Fig.1 shows a digital image watermarking model based on wavelet transform. Figure 1 shows that we should first pre-treat the cover image and the watermark message image before embedding. Pretreatment will reduce the redundancy in the watermark image. Watermark image embedding in the cover image is to embed the important message that representing the watermark message image, but too many redundancies contained in the gray image, which though the wavelet transform, quantization and encoding after the watermark message image pretreatment to achieve image compression. Then watermark message image does encoding, such as spread spectrum encoding and error correcting encoding, in the open channel to improve watermark message's accuracy and security in the public interference channel. The last is subliminal channel encoding, researching from the visual perception model, the masking threshold and the information content measured, to pre-treat for the subliminal channel, which in order to implement the watermark message embedding.

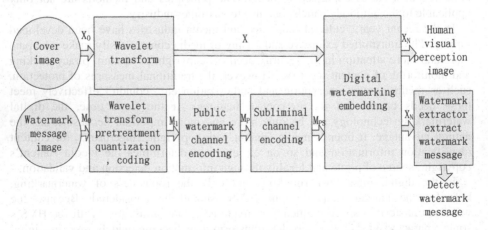

Fig. 1. Digital image watermarking model based on wavelet transform

Cover image pretreatment is to ensure embedding enough watermark messages which have achieved the request of robustness and invisibility. A suitable embedding method which is benefit of the watermark message embedding in the cover image should be chosen after the pretreatment. The cover image that contained watermark message transmits through the media, and arrives at the receiver. The receiver judges the quality of the watermarking embedding through the users' visual perception effect and the watermark message image's quality which extracted form cover image.

3   The Selection of Different Method in Watermark Image Pretreatment

Binary watermark image and grayscale watermark image are commonly used as the digital watermark image. Binary watermark image is used more frequently than grayscale watermark image, but grayscale image has great advantages in the amount of information and identify performance [3]. Grayscale image contains more watermark message than binary image. But because of this reason, the spread application of grayscale image is limited [4].It is called for pre-treating to grayscale image to keep useful message and remove redundant message. It is called image compression.

The methods which used in grayscale watermark image compression are VQ compression [5,6,7], DFT transform [8], DCT transform [9,10,11], DWT transform [12,13] and so on.

VQ compression is an effective loss-compression technique whose characteristics are high compression ratio and simple decoding. VQ compression treats watermark message image as a string of data which contained m data and divided those data into n segments to make up x groups. Each group selects a data vector to represent those data. Since VQ compression can compress and extract image's features effectively, its robustness is good. Because using password in the compression for using individual parts to replace a lot of messages, it will cause some additional distortion which resulted in bad quality in image extracting. At the same time, it will form a code book which more complex and longer operating time when compression.

DFT transform is a powerful tool in linear system analysis. Slowly change wave describes the low-frequency message, and rapidly change wave describes the high-frequency message. Because of the natural correlation of image, low-frequency message can describe the mainly message of the image and high-frequency message can describe the detail of the image. Most high-frequency messages are in a small magnitude range, which good for image compression and can be used in the digital image watermarking pretreatment. But in the DFT transform, once the window function been determined, the corresponding window has been determined too. If you want to get a good low-frequency message, you should take a long window, but if you want to get a good high-frequency message, you should take a short window. It will lead to some contradictions.

DCT transform is a good orthogonal transform which good at compression [14]. DCT transform can concentrate much message energy into a small number of coefficients and then achieve compression though quantization. Because the global nature of the DCT transform, each data in the space will affect every pixel in the image. In order to limit the global effect of DCT transform, blocking effect [15] will emergences inevitable. Blocking effect is due to coarse coding at low-bit quantization and the adjacent blocks' sampling of DCT coefficients falls in different quantitative, which will affect the human's visual perception.

Wavelet transform can remove blocking effect in DCT transform though wavelet's soft-threshold [12,13], In the DCT transform, high-frequency components which contained image details and edge messages are often ignored. Different form DCT transform, wavelet transform usually ignores the invalid wavelet coefficients in image, and retains effective wavelet coefficients. The features of effective wavelet coefficients are spread distribution and few numbers, which can improve the compression evaluation, the robustness and the invisibility. It is suitable for application in the digital watermark message pretreatment.

4   The Selection of Wavelet Bases in Digital Image Watermarking Pretreatment

Wavelet usually can be divided into continuous wavelet, discrete wavelet, biorthogonal wavelet and multiwavelet, etc. Different wavelet bases have different features, which needed to use special wavelet bases in special application. The aim of wavelet in transform in digital image watermarking pretreatment is compression. The commonly wavelet bases used in compression is haar wavelet, multiwavelet, biorthogonal wavelet and so on.

Haar wavelet is one of the most common wavelet bases. Compared with db wavelet, sym wavelet, morlet wavelet and so on, haar wavelet has excellent performance. In the compactly supported orthogonal wavelets, only haar wavelet has symmetry. The wavelet of symmetry can reduce the distortion in the signal reconstruction to get the larger compression. But haar wavelet's localization performance is bad, which would get blocking effects [16] in the encoding algorithm. It will affect the human's visual perception, so rarely used in practice.

Multiwavelet is a promotion of single-wavelet. It has features such as orthogonal, symmetry, short support set and high vanishing moments, etc. The characteristics of two-scale equation are impossible to single-wavelet. But when multiwavelet in fact of cutting in large area, the smooth region will be destroyed and the loss of message is more than rich-texture region. So watermark image with multiwavelet is weak in cropping, rotating, shifting and so on. The robustness is bad too. At the same time, the algorithm of multiwavelet is difficult and high complexity [17].

Biorthogonal wavelet is good at image processing and been liked by researchers because it taking orthogonal, compactness and symmetry into account. Biorthogonal wavelet is including two scaling function and two wavelet function - analysis wavelet and synthesis wavelet. The positions of analysis wavelet and synthesis can be interchange. The robustness and invisibility of biorthogonal wavelet is good. The calculating algorithm is simple and calculating speed is fast.

Combined with robustness and invisibility in image processing, it is known that haar wavelet's visual perception is bad, multiwavelet's robustness is bad and its computation is large, biorthogonal wavelet's robustness and invisibility are relatively good. So it is more suitable to use biorthogonal wavelet as wavelet base.

5   Simulation Results and Analysis

The images which pre-treated by VQ, DFT, DCT and DWT are shown in Fig.2:

Fig. 2. The images that are pre-treated by different methods

Table 1. PSNR, SNR, MSE, CQ in different methods

Performance evaluation in different methods	VQ	DFT	DCT	DWT
PSNR	4.9431	22.6085	25.3495	37.9747
SNR	3.3037	17.0944	19.8354	32.4607
MSE	3.622	5.0054	8.0029	10.2039
CQ	57.5360	82.5824	123.5878	151.5097

It is can be known from Fig.2 that the watermarking pretreatment with DWT is the best in VQ, DFT, DCT and DWT based on human's visual invisibility. In the table 1, PSNR is peak signal to noise ratio between the original image and the pre-treated image, SNR is signal to noise ratio between the original image and the pre-treated image, MSE is mean square error between the original image and the pre-treated image, CQ is correlation quality between the original image and the pre-treated image. It is can be known from table 1 that the performance of VQ, DFT, DCT and DWT is ascending order in PSNR, SNR, MSE and CQ. The wavelet transform's performance is the best. Combined with Fig.2 and Tab.1, the performance of digital watermark message pretreatment with wavelet transform is relatively good.

The images after pre-treating by different wavelet bases are shown in Fig.3:

Fig. 3. The images after pre-treated by different wavelet bases

Table 2. Compression rate, PSNR, SNR, MSE, CQ in different wavelet bases pretreatment

Performance evaluation in different wavelet bases pretreatment	Haar wavelet	Multiwavelet	Biorthogonal 3.9	Biorthogonal 7.9
Compression rate	66.1606	66.1610	66.1611	66.1617
PSNR	37.9747	38.1938	38.2796	38.2972
SNR	32.4607	32.6798	32.7155	32.8231
MSE	9.6056	9.3040	9.7020	10.2039
CQ	151.3854	151.1317	151.3858	151.5097

It is can be known from Fig.3 that the differences among these types of wavelet bases are slightly in human's visual perception. In the Tab.2, PSNR is peak signal to noise ratio between the original image and the pre-treated image, SNR is signal to noise ratio between the original image and the pre-treated image, MSE is mean square error between the original image and the pre-treated image, CQ is correlation quality between the original image and the pre-treated image. It is can be known from table 2 that the differences among these types of wavelet bases aren't great in PSNR, SNR, MSE and CQ, but using biorthogonal wavelet transform will better in these wavelet bases.

6   Conclusion

This paper establishes a digital image watermarking model based on wavelet transform, and analyzes the digital image watermarking pretreatment in this model. Based on the evaluation of robustness and invisibility, the theory and experiment evidence that wavelet transform's effect is better than others methods. In the different wavelet bases, biorthogonal wavelet's performance is the best in digital image watermarking pretreatment. Digital image watermarking pretreatment is the early stage during the watermarking embedding. It would be better to embed watermark message image after pretreatment. At the same time, due to the advantage of wavelet transform, the idea of wavelet transform can be extended to the other processes of digital watermarking. Through the theory analysis and experiment evidence, a better watermarking algorithm will be found.

Acknowledgments. The work on this paper was supported by National Nature Science Foundation of China (60902061), and the 3rd phase of 211 Project of Communication University of China.

References

1. Boggess, A., Narcowich, F.J.: Wavelet and fourier analysis based, pp. 183–239. Electronics Industry Press, Beijing (2002)
2. Tian, L., Zhang, J.-S.: A chaotic digital watermarking method based on the private key and wavelet transform. Journal of the China Railway Society 26, 68–71 (2004)
3. Niu, X.-M., Lu, Z.-M., Sun, S.-H.: Digital watermarking of still image with gray-scale digital watermarking. IEEE Trans. Consumer Electronics 46, 137–145 (2000)
4. Wei, J.: The digital watermarking algorithm based on biorthogonal wavelet transform. Master degree's thesis of Heibei University of Technology (2006)
5. Sun, S.-H., Lu, Z.-M.: Principles and applications of vector quantization, pp. 1–93. Science Pres, Beijing (2002)
6. Zhou, J., Zhou, Y.-H.: Wavelet transform and VQ compression in image. Journal of Shanghai Jiaotong University 31, 133–136 (1997)
7. Yan, H.: Research of wavelet transform in image and VQ compression. Maser degree's thesis of University of Electronic Science and Technology (2001)
8. Braudaway, G.W.: Protecting Publicly-available Images with An Invisible Image Watermark. In: International Conference on Processing, vol. 1, pp. 524–527 (1997)

9. Zhou, P.-Y., Shen, L., Tian, X.-L., Xia, S.-W.: The new watermarking algorithm based on wavelet-SVD. Computer Science 05 (2010)
10. Sun, S.-H., Lu, Z.-M., Niu, X.-M.: Digital watermarking technology and applications. Science Press, Beijing (2004)
11. Huang, J.W., Shi, Y.Q.: Embedding Color Watermarks in Color Images. In: The 2001 IEEE International Symposium on Circuits and Systems (ISCAS 2001), vol. 5, pp. 239–242 (2001)
12. Rangsanseri, Y., Dachasilaruk, S.: Filter evaluatiaon in image debocking by wavelet thresholding. In: TENCON 1999 Proc. IEEE Region 10 Conf., pp. 613–616 (1999)
13. Wu, S., Yan, H., Tan, Z.: An efficient wavelet-based deblocking algorithm for highly compressed images. IEEE Trans. Circuits Syst. Video Technol. 11, 1193–1198 (2001)
14. Zhang, Z.-P., Liu, G.-Z.: Video image compression research based on wavelet. Electronic Journal, 883–888 (2002)
15. Chen, L.: The research of blocking effects removing based on wavelet. Academic degree's thesis of Shanghai University (2007)
16. Liu, J.-F., Huang, D.-R., Hu, J.-Q.: Orthogonal wavelet in the digital watermarking. Electronics and Information Technology Journal 04 (2003)
17. Tang, X.-N.: Digital image watermarking technique research based on the characteristics of BOM multiwavelet. P.H.D thesis of Jilin University (2009)

The Design of Clothing Sales Management Platform

Tian-Min Cheng

School of Economics and Management, Zhongyuan University of Technology,
41 Zhongyuan Road, 450007 Zhengzhou City, Henan Province P.R. China

Abstract. The clothing enterprise must decrease the stock to rapid feedback for the customer requirement. In this paper, the design and implement of clothing sale platform is presented, it have three child systems: league business system, league query system and shop website. The C/S structure is used in the League business system and League query system. The B/S structure is used in shop website system. This system can cut down the time spent on information transmission and feedback, supply the manages the important information, improve their response speed to the market and promote the enterprise competitive ability.

Keywords: clothing enterprise, sales management, shop website, business analysis.

1 Introduction

At present market is undergoing deep fundamental changes. Product life cycles shorten, the variety of product increases, the batch size decreases, and the user pay attention to the product delivery, price and quality, pursue after personalized clothing. Clothing enterprises not only reduce inventory, but also respond quickly to customer needs and expand sales; they should integrate the upstream fabric manufacturers and downstream sales outlets, efficiently manage enterprise logistics, information flow and capital flow[1].

As the fashionable and seasonal products, once clothing missed the selling period, and its value will be a sharp decline. The slightest error will lead into the overstock, which can quickly destroy even the entire enterprise. The information construction will be benefit to solve the problem, and sales information construction is most urgent, which can understand the changing market supply and demand quickly, and guide product development and production, reduce the overstock[2].

Web technology bring a new approach for collecting sales and inventory data of the clothing store outlet business, which can greatly reduces data collection costs and improve data collection accuracy and timeliness, greatly improves the efficiency and correctness of the decision-making.

2 System Design Goals

Rationalizing the enterprise management system. All the data such as sales data, customer information, league information, is collected and stored in computer, realizing sales cashier automatically, it is accurate, complete, objective, can reflect the economic activity dynamically.

Making decision intuitively and scientifically. The system will provide statistical analysis reporting about product sales information, which supports scientific decision-making. It also provides various types of curves and graphs, which are more intuitive, more accurate for the leader.

Strengthening the logistics management. With the help of web technology, the sale data can be exchanged smoothly among the corporate headquarters, its leagues, warehouses, and shops; we can monitor the goods distribution, locate the goods quickly, make the inventory control, use the warehouse reasonably, and reduce the overstock.

3   The Work Flow of Clothing Sales

The typical work flow of clothing sales is shown in figure 1[3].

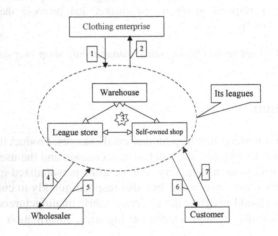

Fig. 1. The work flow of clothing sales

1. When the Clothing enterprise delivers the goods to its league, the league warrant, increase in inventory.

2. When the goods have quality problems or other issue, the league returns to the goods, publishes warehouse receipt, decrease in inventory.

3. The goods can be allocated between self-owned shop and its league store, requisition slip will be published. The shipper decrease in inventory, at the same time the receiving store increase in inventory.

4. Self-owned shop and the league store can carry out wholesale business, when the store sells a batch of clothing, it publish sale order as achievement.

5. When the wholesalers return the defective goods, the receiving store increase in inventory.

6/7. Retail Business happens between store and end-user, including purchase and return.

4   The Data Flow Analysis

Data flow diagram is shown in figure 2[4].

All the leagues abstracted into two categories. One owns sub-leagues and stores, called leagues A; the other only owns the stores, called leagues B;

Because of its larger scale, leagues A will be equipped with a database server. The subordinate leagues store the sales data to the database of leagues A through the data exchange program. The store owned by leagues A can upload its sales data to the database of leagues A through the internet web interface.

Leagues B must not be fit out with database server, only with a proxy database server, provided by the third-party, all the sore of leagues B upload its sales data to the proxy server. The store owned by leagues B can upload its sales data to the database of leagues A through the internet web interface. All the league database servers or the proxy servers require to report its sales data, which is released with web service.

The business leadership obtains their required information through the dynamic calls to various published web service.

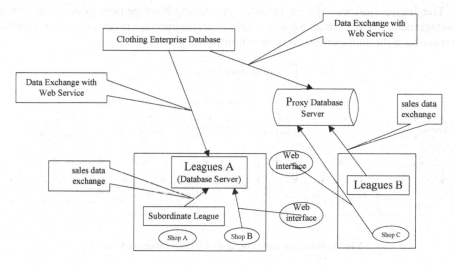

Fig. 2. Data flow diagram

5   The System Framework Design

Clothing Sales Management Platform includes a shop website (called shop system), a league business system and a query system.

The shop website allows the staff in the shore to input sales data immediately; every store is assigned an account with the help of authority management. The website can also query the goods code, store inventory and sales, other brother store inventory. It is shown in figure 3.

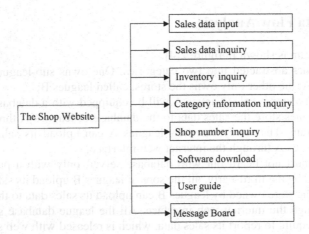

Fig. 3. Structure and function of shop website

The league business system includes purchasing management system, allocation management, return management, sales management, inventory management, communication management. It is shown in figure 4.

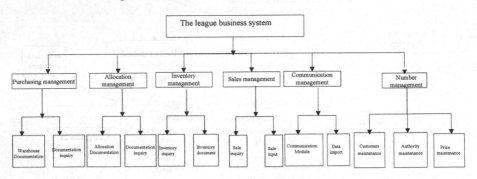

Fig. 4. Structure and function of league business system

The query system provides the inquiries about sales, inventory and price. It is shown in figure 5.

System communication module will realize the data share between the league business system and the stores website, the league business can obtain the retail information from the shop system, and the shop should access the new information about inventory and category. The league business export released data according to a certain format, which will be delivered to web database server using the HTTP protocol. When the stores website receive data upload instructions, it calls the corresponding program to export data, it can be transported to the league business system through Internet[5].

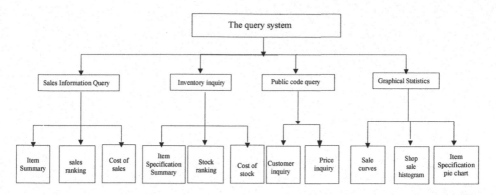

Fig. 5. Structure and function of query system

The league business system and the query system use client/server structure. The shop website is developed using web page with browser/server structure. System communication module uses soap and web service technologies.

6   Conclusion

The research on application of web database technology is becoming increasingly extensive with the rapid development of Internet Technology. The dynamic exchange of resource and information on the internet comes true. The platform can achieve sales control perfectly, help enterprise, league, customer, its sub-organization and the staff can communication, query, track, plan.

References

1. Wu, X.-J.: Analysis of textile and apparel rapid response system. Shanghai Textile Science & Technology 34(5), 6–9 (2006) (in Chinese)
2. Wang, L.-Y., Mao, L.: Causes and avoiding strategies for over stocking in textile and garment enterprises. Shanghai Textile Science & Technology 37(9), 49–51 (2009) (in Chinese)
3. Liu, Y.-M., Wang, Z.-H., Wang, Y.: Intelligent IMA for Garment Industry Based on Product Life Cycle. Journal of Donghua University, Natural Science 30(2), 34–39 (2004) (in Chinese)
4. Zhang, L., Zhang, W.-X.: Research on Sales Opportunity Management System Oriented to Textile Enterprise. Journal of Wuhan University of Technology 32(13), 156–160 (2010) (in Chinese)
5. Wang, J.-H.: The design and implement of a garment distribution sale system. Master Dissertation Southeast University (2004) (in Chinese)

Study of the Model of Agile Supply Chain Management

Tian-Min Cheng

School of Economics and Management, Zhongyuan University of Technology,
41 Zhongyuan Road, 450007 Zhengzhou City, Henan Province P.R. China

Abstract. The agile supply chain management has become the research focus at home and abroad, the paper makes the frame of the agile supply chain management system, and puts forward the model based on coordinated decision-making centers and workflow management technology, by which the core the enterprise and its suppliers can share the information with each other, and plan the resource of the whole supply chain neatly and reasonably, have enough strength to face challenge in quick response to the customers' need.

Keywords: agile supply chain management, coordinated decision-making centers, workflow management technology.

1 Introduction

Economic globalization makes an increasingly competitive international market, and enterprises are facing severe problems of survival and development. The past competition between enterprises has gone, replaced by the competition between supply chains, which is composed of consumers, suppliers, research and development centers, manufacturers, distributors and service providers and other partners to business collaboration and success. Under this condition, supply chain management is becoming increasingly important.

With the globalization of markets, customers' individual demand for products or services has become increasingly evident, and more and more enterprises target customer satisfaction. In this new situation, a company, which relies solely on itself to upgrading competitiveness, can not meet the needs of the market, but must be linked from the upstream supplier to the downstream customers to compete. Companies must turn into the horizontal integration to enhance the entire supply chain competitiveness[1].

In this paper, the model, based on coordinated decision-making centers and workflow technology, is put forward to improve enterprise agility, enabling companies quickly and accurately to meet customer demand.

2 The Agility Analysis of Supply Chain

In the competitive and dynamic market environment, as the core of the whole agile supply chains, the group or enterprise must plan, coordination and control the logistics, information flow and cash flow to form a supply network, which involves all suppliers,

D. Jin and S. Lin (Eds.): Advances in MSEC Vol. 2, AISC 129, pp. 583–588.
springerlink.com   © Springer-Verlag Berlin Heidelberg 2011

vendors until the end-users, can integrate throughout the business process, including outsourcing, production, inventory, sales, transportation and customer service. In order to actively respond quickly to changing market demands, the core group agility of supply chain is mainly reflected in the following three levels, as shown in Figure 1.

(1)The agility of supply chain Establishing. The core group can rapidly reconstruct and adjust the supply chain based on dynamic alliance formation and disintegration.

(2)The establishing and reconstructing of supply chain management information system. With the dynamic alliance formation and dissolution, and how quickly the group complete the reconstruction of the management information system. With the help of information technology, the group and other business partners can use effective methods and technology to integrate and reconstruct the existing management information systems, to ensure their information smooth among the whole supply chain[2].

(3)The agility of logistics, information flow and capital flow among the whole supply chain. The group and its suppliers must keep the internal business processes flexible, logistics, information flow and capital flow pass automatically and intelligently.

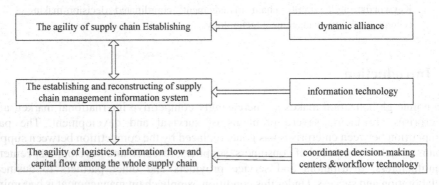

Fig. 1. The agility Hierarchical model

3   The Overall System Design of Agile Supply Chain Management

The core capabilities are to product development, design, sales and assembly, using outsourcing of a large number of parts to other companies. As the core of the entire supply chain, the group is responsible for the production planning, collecting and disseminating information, and control entire supply chain. The overall structure of supply chain management is shown in figure 2.

The production and circulation patterns is driven by demand, can respond quickly to market changes, reduce the product obsolescence and markdowns risk, help reduce inventory and promote corporate cash flow, and increase the profit.

A group should focus on core business, outsourcing non-core businesses; establish core competencies and a clear position in the supply chain[3]. So the group can strengthen the main industry, increase business flexibility through the cooperation among business partners. The group can collaborate closely with suppliers, sharing risks and benefits. From raw material suppliers to end users, all business partners should improve the supply chain together to pursue the overall competitiveness and profitability, reduce transaction costs, improve long-term competitiveness of the supply chain.

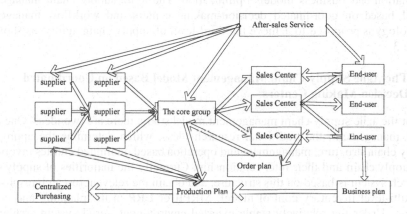

Fig. 2. The entire supply chain

4 Agile Supply Chain Management Model Based on Coordinated Decision-Making Centers and Workflow Technology

All business partners among the supply chain have the same interests to meet customer needs maximally. Because supply chain operations are cross-enterprise collaboration, every member company has the resources and some decision-making power, and the supply chain relationships exist its internal various entities. Internal supply chain system of every member is triggered by the external, need constantly adjust their organizational models, business processes and resource allocation. At present, supply chain management model is lack of flexible mechanisms, can't deal with cross-enterprise business activities, will undermine the good atmosphere for cooperation between enterprises.

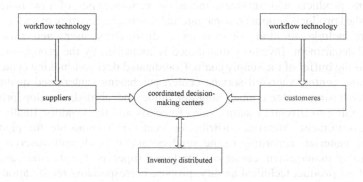

Fig. 3. The agile supply chain management model

Therefore, a coordinated decision-making center should be established to manage the entire supply chain, to promote cooperation. When the customer's needs are broken down into various members of the supply chain enterprise, member company can use workflow technology to achieve process handling standardization, management automation and business models optimization. The agile supply chain management model based on coordinated decision-making centers and workflow management technology is proposed to achieve the third level of supply chain agility, as shown in figure 3.

4.1 The Agile Supply Chain Management Model Based on Coordinated Decision-Making Centers

About the agile supply chain management model study, there are two ideas. One is the serial management, the traditional research ideas, which determine the appropriate supply chain structure, management and operation based on limited entities involved in the supply chain and their inter-relationship. Currently the majorities of supply chain research system is based on this structure, and obtain the relevant research results such as multi-level inventory control model, MRPII or ERP in the planning and control process. Under the relatively stable external environment, serial system architecture and related planning and control process is appropriate, but in competitive and dynamic environment, this is inappropriate, will lag behind the environment change.

Under normal circumstances, when the manufacturers ready to sell product, its price has been determined, at this time vendors can determine the order quantity and selling price, which meet their biggest interest goals, regardless of the manufacturers. For the producers, their attention is focused on production, orders and sales price; they must analyze the biggest sellers to meet their own goals and their vendors. In order to encourage vendors to make decision scientifically, only consider their own interest, the members of the supply chain should provide a coordinated mechanism for sharing information about logistics system, information flow, and capital flow.

The agile supply chain management model, based on coordinated decision-making centers[4], is proposed for effective planning, coordination, scheduling and control, as shown in Figure 4.

The entire supply chain should be optimized based on product time-to-market, quality, cost and service. In the Figure 4, the suppliers will meet the core group demand for various products and services, including various types of raw materials and semi-finished products suppliers, some internal functional department and transporters. Customers include all kinds of customers' distributors, users and some internal functional department. Inventory distributed is controlled by the group, owned by its partners, is the buffer of the supply chain. Coordinated decision-making center consists of marketing center, after-sell service center, purchasing centers and quality control center. Purchasing center contact supplier home and abroad, develop procurement plans according to inventory status, sales forecasts and the vendors, finally carry out unified procurement. Materials distribution center is responsible for planning and scheduling materials according to the vendors' actual needs and material inventory status. Quality management center will guide and supervise the statistics, analysis and control of the product technical quality, propose corresponding rectification program. Pre-and after-sales service centers is responsible for maintaining site information, personnel management, statistical information of billing and costs. And it is responsible for end-user information maintenance, repair and maintenance information query statistics.

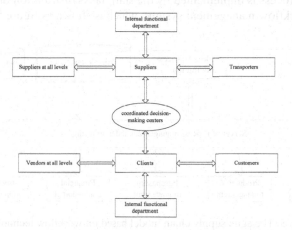

Fig. 4. Coordinated decision-making centers

Compared to the serial supply chain management model, a coordinated decision-making center mainly has the following advantages:

(1)Coordinate decision-making center has relatively complete information, which share among the supply chain, can reduce or eliminate the phenomenon of information distortion and amplification.

(2) It is conducive to separate the logistics from information flow and capital flow, solve Logistics plug and inter-entity chain reaction.

(3) It is useful to bring in more upstream suppliers and downstream users, expand the production scale.

(4) The member company has various channels to get a lot of demand and supply information to enable it to be the most satisfactory and timely supplies.

4.2 The Agile Supply Chain Management Model Based on Workflow Technology

A coordinated decision-making centers can, from the overall situation, develop, monitor, coordinate orders implementation plan to improve the entire supply chain, flexibility and agility.

For a specific member, how does it improve the efficiency to meet customers' needs? Workflow management technology is brought out to carry out the automation of logistics, information flow and capital flow, reduce the order completion time.

Every member entities has its own organizational model, process model, information model is not exactly the same, with some relative independence. To make the logistics, information flow, capital flow deliver fast and efficiently among the entire supply chain, the system must be designed to adapt to these changes.

Workflow management technology can separate system execution model from business process model, organization model, form a hierarchical structure, which is flexible[5].

A business process is implemented by the staff, tasks distribution of business can be done by the workflow management system, which is shown as figure 5.

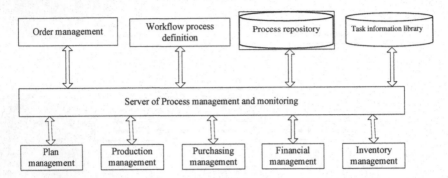

Fig. 5. The agile supply chain model based on workflow technology

The model achieve the separation of logic and processing functions, form independent human-machine interface and the server, The man-machine interface includes plan management, order management, production management, purchasing management, financial management, The server includes process management and monitoring, the process repository, task information library.

5   Conclusion

The theme of this paper is agile supply chain management, the three levels of supply chain agility is analyzed, system structure of agile supply chain management is put forward, the agile supply chain management model is proposed, which improve rapid response to market, expand the competitive advantage.

References

1. Wang, L., Lv, K.: Research Review of Agile Supply Chain. Logistics Technology (2), 169–171 (2010)
2. Jia, G., Zhang, C.: Analysis of the Agility of Agile Supply Chain. Industrial Engineering Journal 7(4), 7–11 (2006)
3. Zhang, L.: Research on the Problem of Information System in Agile Supply Chain. Logistics Sci-Tech (2), 27–29 (2006)
4. Chen, M.: Agile supply chain system based on coordinated planning. Computer Aided Engineering 15(20), 59–64 (2006)
5. Ye, L., Lu, Y., Wang, Z., Tian, J.: Research on Workflow Management System of Manufacturing-Oriented Supply Chain Management Platform. Machinery 42(5), 42–44 (2004)

Using the Knowledge Integration Teaching Method to Improve the Teaching Quality in C Language

ChengLie Du, XueFeng Jiang, and JunRui Liu

Northwestern Polytechnical University Computer College of Xi'an, China

Abstract. After the analysis of the current C language teaching existence, this paper proposes using a knowledge integration teaching method in the teaching process of C language. In this method, the professional knowledge and the application outside the course are integrated in C language teaching, the teaching combines closely with the practical application, C language is used to solve practical problems, according to the application requirements the teachers may modify the teaching direction, and in the teaching process the application value of C language is fully affirmed, so the students' learning interest and their motivation are cultivate, and the teaching achieves good results. The results of the practice show that the knowledge integration teaching method can consolidate learning effect, the students can draw inferences about other cases from one instance, and their ability of solving the problem is enhanced.

Keywords: C language, the knowledge integration teaching method, teaching students according to their aptitude, divergent thinking, innovation practice.

1 Introduction

At present, the C language is a basic computer course of the domestic and foreign science and technology universities, and even the most professional undergraduate liberal arts colleges. However, in the actual teaching process, the teaching effect of C language is not very satisfactory, most students only master the basic knowledge of C language, no the idea of programs and software engineering, their ability of using C language to solve practical application is poor, and they cannot grasp C language practical application value.。 In this paper, through the analysis of the current C language teaching problems, the author puts forward the knowledge integration teaching method in C language teaching, through the cooperation between the teachers and the students, this method integrates the knowledge of other professional and applied field which is outside of the C language course in the teaching process, and it implements the practical implementation in the entire teaching process, so that the students can understand the relationship between the C knowledge and the computer application practice, their interest in learning C language is enhanced, finally the C language teaching can achieve the ideal model which the science ,the technology and the application are with consecutiveness.

D. Jin and S. Lin (Eds.): Advances in MSEC Vol. 2, AISC 129, pp. 589–593.

2   The Existing Problems in the Teaching Process

The reasons of the problems that the teaching effect of C language is poor and the theory and the practice are independent are the shortcoming of the teaching process, mainly in the following two aspects:

(1) The knowledge points are very fragmented, and in the teaching process the overall concepts of programming cannot be carried out.

In most of the domestic institutions, the C language knowledge is various, and the teaching task of C language is heavy. Under the pressure of the huge teaching task, the teachers confined to explain the course content, and in most colleges the course has the assessment, the curriculum evaluation and the related computer examination also requires the students to master a lot of language details, thus the teachers and the students also pay much attention to the language details in the teaching process, while they ignore the whole concept of programming. The result is that although the teachers' teaching content is very substantial while the students study very hard and they have a lot of dazzling "knowledge point pearl", but they could not string these pearls into a colorful "C language programming necklace", and they cannot use the fragmented knowledge to solve practical problems.

(2) The teaching content is limited, and deviates from practical application
The reasons of the deficiency are mainly the following three aspects:

In first, because of the heavy task teaching, the teachers take the matter on its merits and lack the specific application extension of the course knowledge in the teaching.

In second, the teachers who teach C language graduated from the computer science, but the students who study C language study in computer major and non-computer major. The students of different majors will use the C language in different application environment, and use the C language to solve different practical problems. Most of the teachers do not understand the interdisciplinary knowledge and application scenarios, and unable to enumerate and explain the relevant examples to the students according to their actual demand in their professional applications, and because the students are beginners and they lack the ability of using the C language knowledge to solve the professional problems.

In third, most of the teachers who teach C language lack the sufficient research and development experience, their teaching is limited to the theoretical knowledge, and they cannot make the C language teaching close to the practical application through their own development experience.

The problems in the students' learning process are described as following:

(1) The students could not be interested in the curriculum, and they will forget the studied content soon.

Firstly, because in the teaching process the teachers pay too much attention to details, and cannot make many C language knowledge point into effective series, so the students are very difficult in the memory of the fragmented knowledge points; Secondly, due to the engineering characteristics of this course, most of the teachers cannot teach vividly, so the teaching is not attractive and the class effect is poor; Thirdly, at present, the teaching condition is poor in most schools, the C language teaching cannot be put in the laboratory, and the students lack the practice and they cannot enhance the knowledge absorption.

(2) The students' learning is at the shallow level, and they cannot learn knowledge to infer other things from one fact.

In the teaching process, many teachers all have the same experience: they explained an example to the students in detail, the students understand very well, but they cannot search the solving method of the similar problem from the current examples, even it is an example of particularly high topic similarity. The students' Learning stays at a very shallow level and the students are unable to learn knowledge flexible.

(3) The students lack the practical ability, and they cannot use the C language to solve the actual application.

Due to various reasons, the teachers give many mathematical examples in the classroom, and the practical application example is insufficient, they cannot guide successfully the students to put into the C language in the practical application problem solving.

Modern talents should possess the knowledge ability and the capacity of combining theory with practice, otherwise they will not be able to adapt to the knowledge economy social development. Therefore, according to the listed problem that the teaching effect is not ideal, this paper puts forward the knowledge integration method in the C language teaching, its idea and specific implementation methods are described in detail as following.

3   The Knowledge Integration Teaching Method

The knowledge integration teaching refers to fuse the knowledge of other courses and other application field in the C language teaching, and it explains the practical solution to the problems through integrating the knowledge outside of the course. So the teachers and the students broaden one's horizon in the teaching process, breakthrough the course limits, and turn in the direction of practical applications. Therefore, the teaching effect is improved and the students' ability of mastering and using the course knowledge technology is enhanced.

The idea of the knowledge integration teaching method meets the demand of the talents orientation perfectly. Currently on the market the talents is very lack who not only have outstanding experience in professional skills, but also should have high skills, as well as have the knowledge, but also can compose the theory and the practice into an organic combination. The students trained by the knowledge integration teaching method just have such characteristics, they not only master the professional knowledge, but also have strong computer application skills and able to use learned theoretical knowledge to guide the practice application.

The knowledge integration teaching method promotes the professional cross, the knowledge fusion and the technology integration, it not only is a good method used to enhance the effect of C language teaching, but also is consistent with the basic education idea of the various education institutions to cultivate talents.

4   The Implementation Strategy in the Teaching

In the teaching process, the implementation of the knowledge integration method needs the cooperation between the teachers and the students closely. The means are described as following:

(1)The teachers should teach the students in accordance with their aptitude in the setting of syllabus, and the course content should be divided in two parts such as the self-study content and the teaching content according to the knowledge difficulty. The teachers can freed from the heavy teaching task, and have enough energy to full play in the teaching process, they can extend to explain the knowledge points and instill in students the whole concept of programming.

(2) At the situation that the teaching task is reduced, the teachers can change the original teaching modes, use divergent thinking in the teaching process, think the related knowledge when teach a knowledge point, and guide the students to search the all solution of the problem, then analyze and compare with each other, so they can understand the change modes of the problems and the solution, and can search more solving methods. Finally, the students can understand the knowledge and the application examples, and learn to infer other things from one fact and the knowledge integration and flexibility.

(3) According to the different needs of specific professional, the teaching contents should be adjusted appropriately and increase in the curriculum increases the application example for the professional needs. The teachers should do their best to communicate with the teaching supervisor and understand the C program curriculum requirements proposed by the major's follow-up training. If the conditions permit, the teachers should collect the classic specialty applications of the C language, and take these as the class examples to explain. In the classroom teaching, the teachers explain these practical application topics, then the students' interest in learning can be enhanced, the students can feel o the C language practical application value, and their ability of using C language to solve practical application is promoted.

(4) The students are the main roles in the practice. The students are allowed to design innovative exercises, and they change the professional problems and the actual life problems into the C language application problems and successfully solve them, so that their theory in practice ability is increased.

(5) In addition to encourage the students to design their own practice problems and accumulate these valuable teaching resources, the teachers should give the students a supplementary content outside of the textbook, such as the C high-level programming knowledge, the interface programming, the database access, the graphics programming, the multimedia development, and provide the students with adequate technical support for their autonomous practice, so the students can overcome the fear about the technology problems.

5   Conclusions

The knowledge integration teaching method breaks through the original category of the C language course teaching, and in the broad application field it let the students to understand the C language powerful vitality, and let the students recognize fully the practical value of the C language, and thus the students clear out the learning objectives, they have a strong interest in learning and reserves sufficient motivation to learn, therefore the C language teaching effect can be greatly improved. In recent years, the author and his colleagues have implemented the knowledge integration teaching method and the teaching effect is good, we developed a number of excellent software development personnel, and they are involved in a variety of College Students' innovation project, Microsoft and associative organization competition and won the outstanding results. All these confirm the superiority of the knowledge integration teaching method.

The knowledge integration teaching method breaks through course limits in the teaching process and fuses all kinds of knowledge, carries out the guidelines that theory guides practice and the theory should be used in the practice, and it meet with the personnel training strategy. This teaching method is not only suitable for the C language course teaching, also suitable for engineering and liberal arts teaching, and it is an excellent teaching method.

References

1. Ma, T., He, R.: Welcome the challenges of Course cross melts- the reform direction and strategy in China's higher engineering education seen from the request of industrial. Institute of Higher Education Research 26(2), 66–68 (2007)
2. Tang, J., Yun, J., Wang, Z.: The analysis of the curriculum overlapping problem. Journal of Changchun University 16(3), 98–100 (2006)
3. Tian, J., Peng, X.: Some thoughts and suggestions in the fostering of the students of the setting of cross curriculum. Optical Technique 11, 305–306 (2007)
4. Lei, D.: How to improve the teaching quality of "VC + + Programming" course in the applied type universities. Chinese Adult Education in 2009 21 period, 176–177 (2009)
5. Ding, H.: The reform of the arts "Visual FoxPro programming" course teaching. Journal of Yunnan University 31, 544–547 (2009)
6. Liu, Z.: The research and practice of the project teaching method in C language teaching. Chinese Adult Education 4, 139–140 (2010)
7. Guo, H.: The application of the interest teaching method in C language experimental teaching. China Power Education 28, 148–149 (2010)

Collaborative Learning on Multi-agent in M-Learning

Xiaohan Zhang[1], Honghua Xu[1], Lin Hu[2], and Shuying Zhuang[2]

[1] Changchun University of Science and Technology,
Changchun, Jilin Province, China
[2] Jilin University of Finance and Economics,
Changchun, Jilin Province, China
{xiaohanzh,honghuax,huhu315,shuyingz}@yahoo.com

Abstract. Mobile learners can't perceive other collaborative learners' information and communicate with them because of mobile devices' small screen, small storage and data-processing ability. To allow mobile learners to visit the existing collaborative learning platform and making use of multi-agent system's autonomy, coordination and intelligence, this paper puts forward a middleware structure based on multi-agent system. In this paper, middleware structure is given out and it elaborates on the composition and function of each the application feature of the structure.

Keywords: collaborative, agent, m-learning, intelligent.

1 Introduction

In m-learning, collaborative learning is an active research field in which collaborative learners hope to realize collaborative learning by means of the latest result of mobile information technique [1]. During collaborative learning, learners belong to different groups and they consider learning as the sole collaborative result. Sometimes, learners may act as trainer or tutor of teaching activity. Collaborative activity needs to be organized through email, BBS, video conference and other communication tools. Mobile collaborative learning puts emphasis on mobile learning environment. Mobile information technique includes wireless communication technique, embedded wireless communication module etc. Although laptops in some ways have made collaborative learning realized to some extent, the ideal demand of distant mobile environment is far from satisfaction. At present, the ideal tools for collaborative learning under mobile environment are still cell phone and PDA which have the weak abilities of small screen, small storage, weak processing and weak connection. So we need to compensate the hardware deficiency collaborative and satisfied with the clients' personal demands. Taking advantage of the intelligent, collaborative and self-discipline features [2-4], this paper puts forward a collaborative learning solution based on multi-agent system.

D. Jin and S. Lin (Eds.): Advances in MSEC Vol. 2, AISC 129, pp. 595–601.
springerlink.com   © Springer-Verlag Berlin Heidelberg 2011

2 Multi-agent System Structure

Middleware supporting mobile collaborative learning based on multi-agent contains mobile client platform and application program, fixed client, fixed user agent, server's mobile user agent, task agent, group collaborative agent, mode conversion agent, knowledge base, mobile device information database, collaborative learning information base and virtual classroom application program interface. They collaborate with each other to accomplish learners' learning tasks.

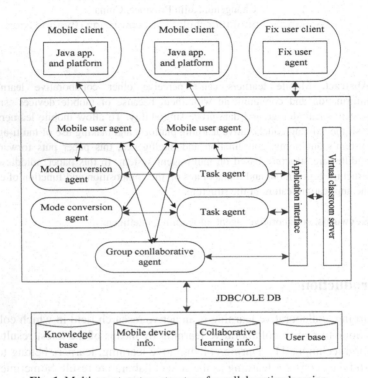

Fig. 1. Multi-agent system structure for collaborative learning

2.1 Client

Mobile client system platform and application program are composed of Java application program and Java platform. Its main functions are as follows:

- Display button style virtual classroom interface such as logging on, course selection and speech interface etc.
- Display subsidiary function interface such as sending and receiving message and inquiring teaching effect etc.
- Receive and transmit demands from mobile clients and output demands.
- Output information from other client agents.

- Tell user how to deal with possible malfunctions such as wrong input password, sending signal failure, busy net etc.

The system establish fixed user agent at the fixed user end. Fixed user agent includes teacher agent and student agent. Fixed user agent has the following main functions:

- Search and collect information such as mobile device model and email address etc to maintain users' information list of mobile devices according to performance parameter list stored in information base of mobile devices.
- Search and analyze course names that students are interested in during learning process. Receive learning materials from students to maintain student knowledge in knowledge base and collaborative learning information in collaborative learning information base.
- Receive and record the information from fixed user which may include user commands. Through its communication module, they can be transmitted to collaborative agent or to the application server of virtual classroom.
- Receive information from collaborative agent and then transmit it to fixed users.
- Send KQML message to group collaborative agent while student's logging out virtual classroom.

2.2 Application Server

Mobile user agent is to add more special functions related with mobile devices on the basis of fixed user agent, i.e., mobile user agent nears the functions of fixed user agent. Besides, mobile user agent has the following functions:

- Receive and record mobile user message which may include user's commands from mobile gateway.
- Analyze information including user's commands. Deal with affairs or submit relative commands to task agent or group collaborative agent.
- Receive relative message from mode conversion agent and then push the message to mobile gateway.
- Receive collaborative perception information sent out from group collaborative agent and then put the information to mobile gateway.

Task agent is mainly responsible for drawing out commands from mobile user, analyzing the meaning, analyzing task and executing. To be specific, task agent has the following main functions:

- Receive KQML [5] information of user command from user agent.
- Extract user commands from the message and then decompose it to a subtask sequence by its logical control module.
- Describe some subtask with KQML and then send out information to relative virtual classroom application adapter.
- Receive KQML message including initial handling result from the virtual classroom application adapter.
- Transmit the initial handling result which meets the requirements to mode conversion agent.

In group cooperation agent, the definition group is similar to the definition class in everyday learning, but the group members only attend one course in virtual classroom, i.e., the relation between a group cooperation agent and the course in virtual classroom is one to one. Group cooperation agent has the following main functions:

- Maintain the lists of registered staff and online staff to consult.
- Receive KQML message including learner's learning situation from user interface agent.
- Collect and sort out learners' learning situation in this group and then analyze the indexes such as learning schedule of the whole class.
- Receive the consultant requirement from teacher's interface agent and then return to the current teaching evaluation result.
- Compare and analyze the place of a certain learner in the whole class.
- Receive the message including collaborative task from agent and other groups' collaborative agents and then organize staff to cooperate to finish task.

Mode conversion agent is mainly responsible for drawing out commands from task agent, analyzing meaning, decomposing mode conversion task and performing. Mode conversion task is mainly related to the conversion from html file to wml file. To be specific, task agent has the following main functions:

- Receive KQML message including initial handling result from task agent.
- Decide the following working sequence according to per formative in KQML message.
- Search for mobile device information list of this group in terms of user ID of KQML message content layer.
- Call up mode conversion algorithm to process the initial handing result in KQML information content layer according to working sequence and mobile user device information.
- Send out the final processed result to mobile user agent.

Virtual classroom [6] application adapter is directly visited by task agent or collaborative agent and is mainly responsible for indirect calling up virtual classroom application adapter by mobile user. To be specific, the interface of virtual classroom application adapter has the following functions:

- Receive subtasks from task agent or collaborative agent.
- Transmit parameters and call up corresponding virtual classroom application adapter.
- Return the performing result of application adapter to task agent or collaborative agent.

2.3 Database End

Knowledge base is used to store knowledge and rules which are related to students' learning, agent reasoning and teacher's teaching. To be specific, the knowledge base in the system stores three aspects of knowledge including user knowledge, rule knowledge and domain knowledge.

- User knowledge includes learning materials shared among learning partners and learners who attend virtual classroom.
- Rule knowledge includes the reasoning rule and collaborative rule of mobile user agent task agent, mode conversion agent and collaborative agent. The above agents are initialized according to their respective rule knowledge whole established.
- Domain knowledge includes common questions and answers in class, teaching evaluation methods, teaching suggestions, teaching plans and reference information.

Mobile device profile database maintains and stores of common mobile devices according to type, brand and model. To be specific, performance parameters of mobile devices include horizontal character number, vertical row number, horizontal pixel number, vertical pixel number and ram volume value. In addition, it can store and maintain mobile device information list to read for mode conversion agent.

Collaborative learning information base is mainly to store perception information of learning in the process of collaborative learning, which language point a certain collaborative learning is learning, how long he spends on it etc.

3 System Structure Feature

3.1 Individualization

While mobile learner visits virtual classroom, system may provide learners for characterized learning service according to the performance of mobile device and learning record. This characterized learning service means the visiting service to the courseware in virtual classroom and producing the list learning partners.

3.2 Intelligentization

Intelligentization refers that the system may automated arrange tasks without user's instruction to reach a goal. It includes the following three aspects:

Firstly user interface agent automated digs learners' information of interest and habit while interacting with learners.

Secondly, collaborative agent positively contacts and organizes user interface agent to finish tasks together according to collaborative task commands.

At last, system refreshes information of knowledge base regularly and automated.

3.3 Collaboration

Many agent cooperates with each other to finish tasks presented by mobile users. A demonstration is given out to show how to satisfy collaborative learning through collaborative work.

Student A would like to review knowledge points related with WSTK. He logs on virtual classroom through cell phone. After he inputs key words "WSTK installation" in search box, the searching command is sent to virtual classroom application server through mobile gateway. The corresponding mobile user agent in mobile user end writes down the key words and then transmits the search command and user ID to task agent and group collaborative agent. Task agent first sends search command to remote application adapter interface of virtual classroom. Then it consults student A historical learning record in collaborative learning information base. Remote application adapter interface calls up courseware search and returns search results to task agent. Task agent decides which is student A required courseware according to student A historical records.

Task agent draws out the courseware from courseware base further analyzes its html file refers to the key words in search command and at last draws out the following html code and then sends back to mode conversion agent together with user ID.

Mode conversion agent consults student A device parameter in mobile device information base according to user ID. If finding out the model of the cell phone, monochrome monitor performance parameter, displaying lines and the number of character handling method to delete redundant spaces and divides overlong text page into several pages. Then with image processing program, it processes color pictures to black-and-white pictures which is suitable for the cell phone to show. Next the left codes will be handled in turn. In this way, mode conversion agent transforms the previous html code into the following wml code. At last it will send it to user interface agent of student A.

Group collaborative agent consults shared learning materials in knowledge base after receiving the search command and user ID from mobile user agent. It finds out student B has electronic books concerning WSTK, but student B hasn't uploaded them. Group collaborative agent first sends message to student A mobile user agent and knows that student A would like to download them to his email. Then group collaborative agent searches for student A email address in collaborative database and sends message to student B fixed user agent to inform student B to send condensed e-book to student A email. At last, it sends message to student A mobile user agent to remind student A of examining his public email in virtual classroom.

4 Conclusion

With agent's intelligentization, collaboration and self-discipline, this paper designs a middleware structure based on multi-agent collaborative learning. This paper macroscopically describes the middleware structure. Then it respectively elaborates on the functions and compositions of each compositions of each component in this middleware structure from three aspects including user end, application server and database end. At last, it analyzes the features of collaborative learning middleware structure based on multi-agent. This structure may compensate the shortcomings of mobile devices from the aspect of distance and it also supports mobile learners to visit the current collaborative learning application platform to interact with other collaborative learners.

References

1. Koeniger-Donolue, R. Handheld computers in nursing education: A PDA pilot project. Journal of Nursing Education, 47(2), 74-77 (2008)
2. Lundin, J., & Magnusson, M. Collaborative learning in mobile work. Journal of Computer Assisted Learning, 19(3), 273-283 (2003)
3. Pinkwart, N., Hoppe, H. U., Milrad, M., & Perez, J. Educational scenarios for cooperative use of Personal Digital Assistants. Journal of Computer Assisted Learning, 19(3), 383-391 (2003)
4. Attewell J, Savill Simith. Learning with mobile devices: research and development. Learning and Skills Development Agency. London, Juyl (2004)
5. M. Cutkosky, E. Engelmore, R. Fikes, T. Gruber, M. Genesereth, and W. Mark. PACT: An experiment in integrating concurrent engineering systems. IEEE Computer, January, 28-38 (1993)
6. Matyska, L., Hladká, E., Holub.Virtual Classroom with a Time Shift. Kumamoto, Japan: Kumamoto university. Proceedings of the Eigth Conference on International Technology Based Higher Education and Training, Kumamoto, Japan, 131-135 (2007)

Wavelet Bases and Decomposition Series in the Digital Image Watermarking

Chen Li, Cheng Yang, and Wei Li

Information Engineering School, Communication University of China,
Dingfu East Road 1, Beijing, China
lichen100808@126.com, cafeeyang@gmail.com,
liwei601@126.com

Abstract. This paper analyzes and compares the performance of different wavelet bases in the digital image watermarking and the effect of different wavelet decomposition series for the digital image watermarking embedding based on the application of wavelet in the digital image watermarking. The experiments proved the digital image watermarking embedding based on biorthogonal wavelet better than others, and summed up that too much or too little decomposition series would not good for conducting digital image watermarking embedding, the suitable decomposition series should be selected which based on cover image and digital watermark image.

Keywords: Digital watermarking, Wavelet bases, Biorthogonal wavelet, Decomposition series.

1 Introduction

With the development of digital technology and network technology, many multimedia digital products have appeared, however, the spread, convenience and insecurity of communication in digital products are exploded. In order to solve the problem, a new effective digital copyright protection and security maintenance technology – digital watermarking – has developed. This technology embedded in many digital media products, effectively protected the digital message's integrity and the digital products' copyright though the watermark detection and analysis. There are many digital watermarking such as image watermarking, video watermarking, audio watermarking and so on. This paper focuses on digital image watermarking.

In the digital image watermarking, not only the robustness of watermarking, but also the human visual system (HVS) should be considered. It asks for proposing a more effective method based on human visual system (HVS). Because the wavelet transform's space-frequency characteristics similarities with the HVS's some characteristics, who is good at the robustness and the invisibility ,the wavelet transform has become widely use in digital watermarking [1].

Wavelet transform is also known as multi-resolution analysis. It is a signal analysis tool based on functional analysis, Fourier analysis, the spline analysis and harmonic analysis, so its application is strong. The basic idea of wavelet transform is unfolding

D. Jin and S. Lin (Eds.): Advances in MSEC Vol. 2, AISC 129, pp. 603–609.

the signal into the weighted sum of a family of wavelet bases function, and using those wavelets bases functions to represent or approximate the signal [2, 3].

This paper analyzes the features of wavelet transform in digital image watermarking pretreatment which based on the robustness and invisibility, and discusses the dependents of selecting wavelet bases which used in cover image pretreatment and determining decomposition series. The experimental proved the selection of wavelet bases and the determination of decomposition series.

2   The Wavelet Bases in Cover Image Processing

Wavelet usually can be divided into continuous wavelet, discrete wavelet, biorthogonal wavelet and multiwavelet etc. Different wavelet bases have different features, which need to use special wavelet bases in special application. The commonly wavelet bases used in digital watermarking is haar wavelet, multiwavelet, biorthogonal wavelet and so on.

Haar wavelet is one of the most common wavelet bases. Compared with db wavelet, sym wavelet, morlet wavelet and so on, haar wavelet has excellent performance. In the compactly supported orthogonal wavelets, only haar wavelet has symmetry. The wavelet of symmetry can reduce the distortion in the signal reconstruction to get the larger compression. But haar wavelet's localization performance is bad, which would get blocking effects [4] in the coding algorithm. It will affect the human's visual perception, so rarely used in practice.

Multiwavelet is a promotion of single-wavelet. It has features such as orthogonal, symmetry, short support set and high vanishing moments etc. Because the characteristics of two-scale equation are impossible to single-wavelet. But when multiwavelet in fact of cutting in large area, the smooth region will be destroyed and the lost message is more than rich-texture region. So watermark image with multiwavelet is weak in cropping, rotating, shifting and so on. The robustness is bad too. At the same time, the algorithm of multiwavelet is difficult and high complexity [5].

Biorthogonal wavelet is good at image processing and liked by researchers because it taking orthogonal, compactness and symmetry into account. Biorthogonal wavelet is including two scaling function and two wavelet function - analysis wavelet and synthesis wavelet. The positions of analysis wavelet and synthesis can be interchanged. The robustness and invisibility of biorthogonal wavelet is good [6, 7]. The calculating algorithm is simple and calculating speed is fast. According to the literature 6: the robustness of due 7/9, 5/7 is far better than the dual 9/3 and 5/3 (the embedding strength of 5/3 wavelet is much larger than 3/5), so the effects of spline wavelet 9/3, 3/9, 3/5, 5/3 are not very good. At the same time, the length of biortholonal wavelets 9/7, 7/9, 5/7, 7/5 are nearly equal, the pp and PSNR values are nearly in the same circumstances, so they are more suitable for application in digital watermarking.

Combined with the robustness and the invisibility in image processing, it is known that haar wavelet's visual perception is bad, multiwavelet's robustness is bad and the computation is large, biorthogonal wavelet's robustness and invisibility are relatively good. So it is more suitable to use biorthogonal wavelet as wavelet base.

3 Selection of Wavelet Decomposition Series in Cover Image Processing

Watermark embedding can be seen as a weak signal (watermark message image) folds on a strong background (cover image). If the over-lay signals strength below the contrast threshold, the visual system couldn't feel the presence of the signal. According to the literature 5: the more of decomposition series, the stronger of low-frequency approximated sub-image to embed watermark message. So the robustness of the watermark is enhanced. It is means that the wavelet decomposition series should be increase based on the number of watermark message embedding which needed in the watermarking.

However, the decomposition series also depend on the complexity of the image and the filter length. Because the wavelet transform of non-separable two-dimensional orthogonal wavelet filters is different from the wavelet transform of separable one-dimensional wavelet filters used tensor products, it divided each band into high-frequency sub-band and low-frequency sub-band [8]. The decomposition series should need that entropy after the decomposition less than entropy before the decomposition.

When selecting the decomposition series, it should not only meet that the number of decomposition series increase as many as possible under the premise of watermark embedding, but also pay attention that entropy after the decomposition less than entropy before the decomposition in order to avoid unnecessary computation.

4 Simulation and Experimental Results

Figure 1-4 represents the cover image and the watermark message image with different wavelet bases and those images with salt-pepper noise. It is can be known from figure 1-4 that when watermark message image embedding with different wavelet bases, the performance of extracting watermark message image in multiwavelet transform with salt-pepper noise is very bad, the performances of cover image and watermark message image in biorthogonal wavelet, especially biorthogonal 7/9 wavelet, are better than the others based on invisibility.

In the table 1, PSNR is peak signal to noise ratio between the original cover image and the cover image with watermark message, SNR is signal to noise ratio between the original cover image and the cover image with watermark message, PSNR1 is peak signal to noise ratio between the original watermark image and the extracted watermark image, SNR1 is signal to noise ratio between the original watermark image and the extracted watermark image, PSNR2 is peak signal to noise ratio between the original cover image and the cover image with salt-pepper noise , SNR2 is signal to noise ratio between t the original cover image and the cover image with salt-pepper noise, MSE is mean square error between the original cover image and the cover image with watermark message, CQ is correlation quality between the original cover image and the cover image with watermark message.

cover image after watermarking in haar wavelet   cover image with salt-pepper noise

the extracted watermark image   watermark image with salt-pepper noise

传媒
大学

传媒
大学

Fig. 1. Images after watermarking in haar wavelet and it with salt-pepper noise

cover image after watermarking in multiwavelet   cover image with salt-pepper noise

the extracted watermark image   watermark image with salt-pepper noise

传媒
大学

传媒
大学

Fig. 2. Images after watermarking in multiwavelet and it with salt-pepper noise

cover image after watermarking in bior3.9 wavelet cover image with salt-pepper noise

the extracted watermark image   watermark image with salt-pepper noise

传媒
大学

传媒
大学

Fig. 3. Images after watermarking in bior3.9 wavelet and it with salt-pepper noise

cover image after watermarking in bior7.9 wavelet cover image with salt-pepper noise

the extracted watermark image   watermark image with salt-pepper noise

传媒
大学

传媒
大学

Fig. 4. Images after watermarking in bior7.9 wavelet and it with salt-pepper noise

Table 1. Performance evaluation in different wavelet bases

Performance evaluation in different wavelet bases	Haar wavelet	multiwavelet	Bior3.9 wavelet	Bior7.9 wavelet
PSNR	41.3255	40.8110	40.5533	41.0075
PSNR1	28.0794	35.5784	32.3333	44.2531
PSNR2	19.0032	17.0032	19.0794	19.0889
SNR	35.9043	36.3121	36.1820	35.2329
SNR1	16.2455	34.0206	31.4738	44.0193
SNR2	13.7236	11.7812	13.7292	13.7892
CQ	143.1412	143.0035	143.1484	143.2950
MSE	5.2076	4.3743	5.4046	6.0018

It is can be known from table 1 that different wavelet bases' PSNR1, SNR1 are different largely, but biorthogonal 7/9 wavelet is the best. It is illustrated that the performance of extracting watermark image is the best. Comprehensive comparison of the PSNR, SNR, PSNR1, SNR1, PSNR2, SNR2, CQ and MSE, the performance of digital watermarking in biorthogonal is better than the others, especially the biorthogonal 7/9 wavelet base.

Table 2. Performance evaluation of different decomposition series

Performance evaluation of different decompositio n series	Wavelet decomposi tion of two	Wavelet decomposi tion of three	Wavelet decomposit ion of four	Wavelet decompositi on of five	Wavelet decomposi tion of six
PSNR	41.2625	40.8110	40.7969	42.2475	43.3551
PSNR1	31.0847	35.5784	34.3108	28.9456	25.7377
SNR	35.9529	35.5014	35.4873	36.9379	38.0454
CQ	142.8054	143.2606	144.1412	143.4127	142.9593
MSE	4.4883	4.9800	4.9962	2.8417	2.020

In the table 2, PSNR is peak signal to noise ratio between the original cover image and the cover image with watermark message, SNR is signal to noise ratio between the original cover image and the cover image with watermark message, PSNR1 is peak signal to noise ratio between the original watermark image and the extracted watermark image, MSE is mean square error between the original cover image and the cover image with watermark message, CQ is correlation quality between the original cover image and the cover image with watermark message.

It is can be known from table 2 that the visual perception of different decomposition series in wavelet transform isn't very distinctive, but the more decomposition series, the better of the visual perception. At the same time, too many decomposition series will make the PSNR1, CQ, MSE declinable, it is mean that the quality of the extracted watermark image becoming deterioration and the robustness of watermark image being worse. So too few or too many decomposition series can affect the digital image watermarking embedding, a suitable decomposition series should be found based on the cover image and the watermark image to improve the robustness and invisibility of digital watermarking.

5   Conclusion

This paper main discusses the features of different wavelet bases and decomposition series in the wavelet transform of digital watermarking embedding based on the robustness and invisibility. The theoretical analysis and experimental results shows that the impact of different wavelet bases is different, and the biorthogonal wavelet in the wavelet bases is good at PSNR, SNR, CQ and MSE based on the robustness and invisibility. At the same time, the theoretical analysis and experimental results also shows that too few decomposition series can affect the invisibility and too many decomposition series can affect the robustness of watermarking embedding. So selecting a suitable decomposition series is better in watermarking embedding.

Acknowledgments. The work on this paper was supported by National Nature Science Foundation of China (60902061), and the 3rd phase of 211 Project of Communication University of China.

References

1. Boggess, A., Narcowich, F.J.: Wavelet and fourier analysis based, pp. 183–239. Electronics Industry Press, Beijing (2002)
2. Geng, X.: The digital image watermarking technology based on wavelet transform. Maser degree's thesis of P.L.A. Information Engineering University (2007)
3. Liu, J.-F., Huang, D.-R., Hu, J.-Q.: Orthogonal wavelet in the digital watermarking. Electronics and Information Technology Journal 04 (2003)
4. Chen, X.-X.: Real wavelet transform method to extract phase information. China Electrical Engineering Journal 27(22), 8–12 (2007)
5. Digital image watermarking technique research based on the characteristics of BOM multiwavelet. P.H.D. thesis of Jilin University (2009)
6. Liu, J.-F., Huang, D.-R., Hu, J.-Q.: Biorthogonal wavelet in the digital watermarking. Sun Yat-sen University Journal 07 (2002)
7. Liu, J.-F.: Wavelet theory and its application in image compression and digital watermarking. P.H.D. thesis of Zhejiang University (2001)
8. Wang, Y.: Orthogonal compactly supported wavelet and its application in digital watermarking. P.H.D. thesis of Dalian Maritime University (2007)

Research of Software Human-Computer Interface Based on Cognitive Psychology

Peng Shi

Research of Software Human-Computer Interface based on Cognitive Psychology Beijing, China
Spanish2001@163.com

Abstract. With the development of society, people's pursuit and attention transferred from the material content to spiritual satisfaction. People dislike the interface of software is lifeless and frosty in every day use, people have emotional needs, and software design also needs to meet people's spiritual needs.in software world, the need of user is not only function of software, more important is enjoyable experience and good feelings when interactive with computer. The perfect interaction relationship between software and human make the software not only become a tool, but also like emotional partner. interface is an interactive media between man and computer, the learning, understanding, mastering of software is experienced a cognitive process from shallow into a deep. The people's cognitive process includes: perception, attention, memory, thinking, associate, emotional experience. Primary cognitive stage lay foundation for the advanced stage of emotional experience. How to design the human–machine interface to make users reaches perfect experience, satisfying the customer emotional demands? This issues is discussed how the interface design will be to satisfy people's emotion need, according to cognitive psychology.

Keywords: cognitive psychology, human-computer interaction, interface design.

1 Introduction

People in human-computer interface field has been doing a lot of research and practical work, also make considerable achievements. But today, with the development of different kinds of software, new user groups appear ceaselessly, software function is becoming more and more powerful and interface conveys greater quantity information, therefore, put forward higher request for the man-machine interface. [1]

How to design the software interface to further improve the level of man-machine interactive, what kind of design for the user more meaningful, what kind of design will produce positive interaction with Users. The problem today is list as follow----1)the function of software become more and more complex, this makes the interface, tend to be complex which confused many users. 2)due to a variety of software be in market, many designer seeking for difference: decoration, style, if interface are distinctive from each others, and makes the design of man-machine interface great arbitrariness, thus deviated from the man-machine interface design fundamental principles. 3)even similar product has the same function, the interface are different in many aspect :icons, color, graphics, symbols,operation unit, ways of information feedback and so on.

D. Jin and S. Lin (Eds.): Advances in MSEC Vol. 2, AISC 129, pp. 611–616.
springerlink.com © Springer-Verlag Berlin Heidelberg 2011

2 Research on Customer's Material Need and Perceptual Need

Maslow's hierarchy theory—According to Maslow's hierarchy theory, human nature needs include: the physiological needs;the security needs;attribution and love needs;respect needs;self realization needs.The Five levels are gradually increased, when junior need obtained and relatively satisfied, higher level need will appear and when satisfied, further level will emerge. [2]

Software and user's relationship are also function and demand, satisfy and being satisfied—So the man-machine interface is a medium, when software function realized, user's needs were met. [3] The function of software is also multiple levels, the realization of lower function is the foundation for the realization of higher function, interface like a belt to connect need and function. what level will be achieve depends largely on the man-machine interface is good or bad, Software function, although very powerful, but not accepted by the user, unable to produce effect, is a defective software.

People's need to man-machine interface—According to Maslow's hierarchy theory, we can deduce the demand for man-machine interface of software can be divided into: 1, Basic function; 2, Functional quality; 3 Aesthetic; 4 Symbolic; 5, Emotional needs. Among them, the latter three belong to high level demand; they realize is based on the fulfillment of low-level needs. [4]

3 Analysis Levels of Needs for Man-Machine Interface

Basic function—Basic function is the basis of man-machine interface, whether design can achieve the basis need of users, is the basic requirements of the man-machine interface.[5]

Interactive quality—The interactive quality of man-machine interface is refers to improve the efficiency of man-machine interaction, how to design software for user to operate the software convenient and completes the work quickly.

Aesthetic demand—Beautiful interface will help relax the mood, concentration and improve working efficiency and make users feel more cheerful. pursuit art and beauty is nature of human in life.

Symbolic demand—Style of Software can affect the emotional and mental state of users, when people interact with computer. People with the nature of searching for self-respect and fulfillment with success. The existing of style and decoration of man-machine interface that can create a specifically environment or atmosphere, thus determined whether users reached an ideal working condition.

Emotional needs—during long state of human-computer interaction, one will have an emotion to software, this kind of emotion is probably love, it may be come from self-confidence to operating software freely and create tacit understanding as partner, also may be dislike to software etc, due to different psychological experience in working process.

4 Use Congnative Psychology Analysis Human-Computer Interface

What conditions User's needs in cognitive process—Cognitive psychology thought, people's cognition must fulfill three conditions: 1) have certain experience.2) thing can provide enough information .3) associate activities can contact people's experience and information. [6] All of these factors are essential and indispensable. When these three factors practice into interface design, we should pay attention to: 1) studies on People's experience and habit, 2) what information will communicated by man-machine interface, 3) person's cognitive activities, there are integrate together, any time interface design should be considering the relationship between those factors.

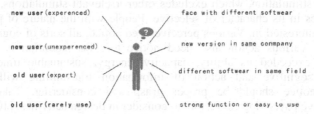

Analysis of what may happen in practice—For Several different user with different knowledge and intellectual backgrounds, so the cognitive process will be fast or slow, cognitive to software will be deep or shallow. Some problem is well understood by old users, but will be difficult to understand by new users. [7] Those problem should be note during software design process. In this way, software can very well accepted by most people.First, we need to study whether target users have the knowledge, ability and experience to achieve well understanding of the information software interface conveyed. Its roughly include two kind of knowledge ---- human's general experience and their special experience in their professions. General experience have been accumulated by people in every day software application, Special experience comes from professional knowledge in one's field and from other software learning and using, It should be pointed out that a lot of professional software in same fields with very different overall layout and detail, insufficient consideration to user's experience, which will cause handicap to user, because their former experience will be useless. Therefore the user may feel depressed and cause confusion and frustration to great extent in study and use of those software.

User's cognitive process theory—The Theory of cognitive psychology thought people experience five stages in cognitive process ---- perception, attention, memory, thinking, associate, feelings, the person cognition to software is through the man-machine interface to realize. The process of human-computer interaction is also experienced such five processes. Based on a study to the characteristics of each stage, will conduct more information for users. [8]

Perception—Perceptions is people who formation of a reflection to the objective things, based on their experience they have, in man-machine interactive activities, objective things stimulate to users largely depend on the person's experience, Potential demand and preferences, So interface design should try best to cater people's appetite and interest, at the same time do every possible to Fill the gaps between people's demanding and providing of Software, what Providing is precisely the things user need. Such as design style, inclination of interface and general color tendency, guidance effect of symbols in Operation panel, clew, warning way for the possible problem may come appear, etc. [9] Such as industrial software user-interface design style has certain industrial characteristic which catering to the aesthetic preferences of engineer very well, but other users will feel these designs style cold and inflexible. This suggests that different types of people can bring about different feeling on the same thing; this depends largely on the environment which user exists in, personal experiences and professional background and so on. Users interface design should catch the first sense of users, in the first contact will generate goodwill, So that users can be delight to use software and produce favorable interaction between user and software.

Attention—Attention in the cognitive psychology is ability to make awareness focus on a certain stimulation, which excludes other irrelevant stimulations, the essence of the notice lies in its character of selective. People with the nature of pay attention to things they interested in. Various perceive need notice, all sorts of cognitive processes need notice, various action also need notice too. But attention is a very limited resource, if exceeded its ability, capacity, energy, sustainable time, users in the perception, cognitive and action of information transmitting will make errors. Therefore, notice should be proper grasp and considering, Take notice as a comprehensive psychological factors to consider in design activities. [10] For example, style of control menu should be let people feel at a glance, unnecessary appear button may be hidden and management through the hierarchy of control mode. Be like again, various ICONS on the control panel, how to arrange the relationship between main menu and sub-menus clearly, what kind of design won't presumptuous guest usurps the host's role.

Memory—The basic process of memory is to remember, maintain, memory or cognitive. That basic process is inseparable. To remember, to keep the memory lay foundation for cognitive, recall and cognitive is the results to remember and keep memories. What kind of interface is convenient for the user to remember, to large extent relevant to user's thinking habits, experience, so the key still depends on the analysis to user groups. For instance, menu, buttons should be arranged according to function classification, convenient for the user to memory, to looking for through logic thinking. Still another example, design of buttons and icons is appropriate to express meaning of function, let the user easy to understand the meaning of ICONS, feel impressive and easy to memory after reading.

Association—Process of associate sensor information though methods of comprehension and abstract generalizations. Take the knowledge in memory as media, reflect the nature of things and relation to each other. This behavior is proceed by the form of concept, judgment and inference, with characteristics of indirect and generalizes. In this phase, the user experience is important, and the objective conditions also play a big role, If a truth more obvious and easy to understand, the more likely to understand by users. [11] in the use of a new software, when software properties, Structure and logical vein can be clearly grasped , then the master of software may more thoroughly from rational aware, and thereby more familiar with connotation and essence of the software.

Emotion—Emotion is essential, constant and variability of subjective experience and attitude depend on whether the information communicated to audience according with one's physiology or psychological need. This kind of emotion stimulate by information but it is different from understanding, not the direct reflection of things characteristics, it is a kind of subjective experience and attitudes which take object as a activate factor and take his own material, spirit need as intermediary. Reflects the effect relations, value relations between needs of user and meet the requirements of user in practical activities.

Cognitive psychology thought things cognition is experienced a process from the shallower to the deeper, in this man-machine interactive process different cognitive stage mobilize the user different psychological experience, Meanwhile bring different psychological experience to user which is corresponding with process gradually satisfying the people's needs .for example Basic function is a foundation for software, no guarantee for basic function ,won't arise high-level significance of distillation.

5 Conclusion

How to design software in accord with the people cognitive psychologies at the same time satisfy all the needs of people, there some rules list as follow: [12]

From the people's demand hierarchy—In software design, we needs to consider the people's demand is progress from low to high, for the satisfaction of demand can't suppressed, for requirements level can't be step over, such design can be vitality.basic function consider as a foundation , First we must consider meeting basic function, secondly to consider how to let users operating software more convenient, If not be able to meet the basic needs of users, or software functions have deficits and can't be acceptedby the user.Appropriate interface optimization and humanized form will improve work efficiency, when users use software with relax and pleasant mood, But we should command the degree, excessive formalized and decorate brings negative influence to basic functions and the operational efficiency.

From the cognitive process—For various users, their preferences to software differ, so user's background is very important, should let users be attracted naturally when first contact to the software. Whether those factors like ICONS, work space planning and, menu decorate and planning etc can be quickly notice, memory, understand and accept by user is problems need researchers to consider.Whether Software interface which involved symbols, semantic expression become linkage to contact users' experience and meaning of function, whether this kind of mode of thinking can easily be learned, mastered, memorized and understood by users.As Users is familiar with each part of software, then further deepen understanding to software system and operating system, finally produced good psychological experience, impression and feelings, consequently improve to emotional experience. [13]

Comprehensive consideration the relationship between man's needs and the people's cognitive process is the key for man-machine interface design—Two analyses have process from low to high, cognitive psychology theory emphases on cognitive stage and process, emphasizes on the grasp of the individual cognitive process sequence and cognitive characteristics of each stage, explains with a scientific way how the people's cognitive did happen. Theory of People's demand is focuses on the research of demand of all levels, fulfillment of low-level requirement is lay foundation for senior demand and the senior demand is the result of low-level requirement. Meanwhile, people's demand is different, only satisfy the needs of the user, design of software interface will meaningful.

level	demand for softwear	Maslow's hierarchy	congnitive procrss	expression
low	basic function	physiological needs	Perception	structure
				easy to understand
	interactive quality	the security needs	attention	
				founction
	Aesthetic demand	attribution and love needs	memory	aesthetic
				humanism
	symbolic demand	respect needs	Association	taste & statue
high				
	emotional needs	self realization needs	Emotion	injoyment
				culture

References

[1] Li, Z.: Design and Implementation of E-Learning System Based on SOSC. A Thesis of Master Degree, Northwest University (2006)

[2] Yan, J., Gou, X.: SOA-based Network Curriculum Sharing Platform Research. Computer Science 36, 161–162 (2009)

[3] Yuan, B., Ruan, Q., Wang, Y., Liu, R., Tang, X.: A Conceptual Model and Features of New Generation (Fourth Generation) Human-Computer Interactive Systems. Acta Electronica Sinica 31, 1945–1954 (2003)

[4] Gong, C.: Human-Computer Interaction: Process and Principles of Human-Computer Interface Design. In: Proc. IEEE Symp. International Conference on Computer and Automation Engineering (ICCAE 2009), pp. 230–233. IEEE Press (2009), doi:10.1109/ICCAE.2009.23

[5] Chun, J., Zhang, N.: The Standard Design Principles of Human-Computer Interface Based on Cognizance Theory. Chinese Journal of Ergonomics 11, 32–34 (2005)

[6] Zhang, Y., Wang, J.: Human-Computer Interaction Design Strategy for Color Identification on Multi-media Visual Interface. In: Proc. IEEE Symp. International Conference on Computer-Aided Industrial Design and Conceptual Design (CAID & CD 2009), pp. 1874–1877. IEEE Press (2009)

[7] Hu, W., Deng, X.: Exploration and Research of the Remote Network Education System Based on SOA. Journal of WUT 31, 736–739 (2009)

[8] Li, R.: The Research of Using Human-Computer Interface Design Method on College Learning Web. A Thesis of Master Degree, Tianjin University (2007)

[9] Shu, B.: Research on Web-Based Personalized Teaching and Learning Management Environment and its Implementation. A Thesis of Master Degree, Shanghai Jiao Tong University (2001)

[10] Zhang, W., Huang, S.: The Research on Personalized Modern Distance Education System. Journal of Guangdong Radio & TV University 15, 10–13 (2006)

[11] Zhang, Z.: The Application of SOA Base on the Sharing of Online Course Resourses. A Thesis of Master Degree, Liaoning Normal University (2009)

[12] Liu, Y.: Study of the Interaction Design of Man-machine Interface Based on Product Usability. Packaging Engineering 29, 81–83 (2008)

[13] Cai, X.: Principles of Human-Computer Interaction in Game Design. In: Proc. IEEE Symp. Second International Symposium on Computational Intelligence and Design, ISCID 2009 (2009)

Structured Evaluation Report of an Online Writing Resource Site

Xing Zou

Department of Foreign Languages
Wuhan Polytechnic University
Wuhan, China
Chriskie2006@yahoo.com.cn

Abstract. Nowadays online writing has become a new way of developing one's writing ability. Through structured evaluation of an online writing resource site, this paper attempts to analyze the writing as a process, as a genre and its practicability in Chinese context. It turns out that writing is a social practice rather than a solitary activity. Learners will gain new insight into writing and it also has pedagogical significance for the teachers.

Keywords: online writing, evaluations, process, genre.

1 Introduction

The name and the URL of the online writing resource are Writing @CSU, Writing Resources (http://writing.colostate.edu/learn.cfm). I will choose the most significant part to analyze.

1.1 One New Insight into Writing

Online writing or writing online is a new way of learning how to write and developing one's writing ability. As for me, the unique sense of online writing is that writing is a social practice rather than a solitary activity---"Many of us think of writing as a solitary activity -- something done when we're alone in a quiet place. Yet most of our writing, like other forms of communication -- telephone conversations, classroom discussions, meetings, and presentations -- is an intensely social activity." (http:// writing.colostate.edu/guides/processes/writingsituatio ns/)

1.2 My Comments

I agree to the above comments in terms of communicative writer-reader relationship. Any writing involves conveying ideas to the audience and the audience will respond to the writing, so writing becomes an effective way of communication like a face-to-face conversation, which let the writers bear the readers in mind, expect the readers' reactions and intend to share by making sense of their own writing so that they can find a sense of community and become more confident and less solitary. Actually, the

D. Jin and S. Lin (Eds.): Advances in MSEC Vol. 2, AISC 129, pp. 617–621.

online-writing websites provide readers and writers with helpful guide, lots of writing materials and chances of exchanging ideas, sharing what one has learned through writing experience and benefit from the wisdom of the others including some experts. They are supported by online-writing Guides; they can make use of Writing Tools; they can join in a Writing Activity, and get feedback on their writing, etc. All these writing activities make writing easier and interesting for the writers to share the pleasure of writing online, more important, they make writing socially significant.

If you have more than one surname, please make sure that the Volume Editor knows how you are to be listed in the author index.

2 Writing as a Process

Process approaches focus primarily on what writers do as they write rather than on textual features. As Figure 2.6 shows, the process approach includes different stages, which can be combined with other aspects of teaching writing (Coffin, C., Curry, M. J., Goodman, S., Hewings, A., Lillis, T. M., and Swann, J., 2003, P33-34). The iterative cycle of process approaches delineated in this Figure showed that this approach involves the linguistic skills of prewriting, planning, drafting, reflection, peer review, revision, editing and proofreading. This approach see writing primarily as the exercise of linguistic skills, and writing development as an unconscious process which happens when teachers facilitate the exercise of writing skills (Richard Badger and Goodith White, 2000, P155). In the On-line lab, I found one part that exhibits writing as a process:

Preparing to Write
Starting to Write
Conducting Research
Reading & Responding
Working with Sources
Planning, Drafting, & Organizing
Designing Documents
Working Together
Revising & Editing
Publishing
(http://writing.colostate.edu/guides/)

Each of the above steps has been illustrated in detail. In fact, the processes listed on the website are slightly different from what we learned from traditional course book (A Handbook of Writing)

1. To fix on a subject or theme

To list an outline
To write the first draft
To revise the first draft: content, structure, sentences, diction, etc.
To compose the final version
To check the final version

Through comparison, one may find that the processes of online writing are of more mutual exchange, which indicates that writing is a social act. Actually, the non-linear process can be got access to in various orders at different points. One needn't follow it strictly and move through the stages one by one. Just like the on-line resource, you can click any button to get a clear idea about the stage and it gives you a hand-on guide for you to practice.

The printing area is 122 mm × 193 mm. The text should be justified to occupy the full line width, so that the right margin is not ragged, with words hyphenated as appropriate. Please fill pages so that the length of the text is no less than 180 mm, if possible.

3 Writing as a Genre

Genre refers to abstract, socially recognized ways of using language. And genre approaches see ways of writing as purposeful, socially situated responses to particular contexts and communities (Ken Hyland, 2003, P25). This approach see writing as essentially concerned with knowledge of language, and as being tied closely to a social purpose, while the development of writing is largely viewed as the analysis and imitation of input in the form of texts provided by the teacher (Richard Badger and Goodith White, 2000, P156). In the on-line lab, according to the social context of creation and use, texts with similar features are grouped together, therefore a linguistic community is established in each type, which involves interaction of the writer, reader and the text. Writing can be guided in different genres:

Composition & Academic Writing
Argument
Fiction
Poetry
Creative Non-Fiction
Writing about Literature
Scholarly Writing
Business Writing
Science Writing
Writing in Engineering
Writing for the Web
Speeches & Presentations
(http://writing.colostate.edu/guides/index.cfm?)

Major text types that student might frequently write or find difficult to write are listed here. Each one provides the learners with an explicit understanding of the text and a metalanguage by which to analyze them, which can help students understand that texts can be explicitly questioned, compared and deconstructed. Writing is not abstract activity, it becomes a social practice, varying from one community context to the next. It can be safely concluded that if a writer wants to create a certain composition, he must get himself familiar with the characteristics of such writing which appears in a certain context and appeals to certain group of readers. A successful writing depends on a desirable interaction within the community of the writer, reader and the text.

4   Overall Suitability as a Self-access Resource for Tertiary EFL Students in China

Students in China can benefit a lot from the writing lab. Since most of them don't form the habit of reading such on-line instructions outside the class, they rely too much on the in-class teaching, which is far from enough for them to improve their writing proficiency. This on-line lab will be of great help for them to raise their interest in and enhance their motivation for writing.

As for non-English majors, their writing competence has been prescribed in the book *College English Curriculum Requirements* (2007) which is the guideline for college English teaching in China. From its basic requirements to advanced requirements, writing ranges from daily message, letters, notes to thesis writing, presentation, paper report, etc. Students may refer to "writing as genres" and practice related activities.

And for English majors, they have more demands for linguistic skills. Sufficient in-class teaching time plus more motivation enable them to learn writing systematically. They can start with constructing a sentence, telling a story or with lots of exposure to reading. And this lab can provide a good platform for them to learn, to practice and to share. So they may combine these two approaches more. From this website, they can see the values of reading and writing and gain a lot of insights into them so that they can make sense of themselves. Apparently this lab will be of great significance to them both academically and socially.

5   Using the Online Resource for Teaching

On-line resources are also of pedagogical significance for the teachers, for it provides a detailed teaching guide for us. Different teachers may choose different way of teaching. During my preparation for the lesson, I'd like to use the Sample Lesson Format (http://writing.colostate.edu/guides/teaching/lesson_plans/pop2f.cfm) in the lab as a guide, considering my own students' needs. In class, this on-line resource will also serve as a reference, and I prefer to combine the process and genre approach, using Write To Learn (WTL) (http://writing.colostate.edu/guides/teaching/planning/wtl.cfm). For example, when it comes to the topic "Pollution", before writing, I'll show them some pictures or short articles on pollution, carrying out a brain-storming activity using the "concept map" or "graphic organizer" to come up with the related category, cause, consequence and possible solutions. Then a model text will be deconstructed to raise students' consciousness of the structure and related social purposes. In the following session students will begin to plan a composition by joint construction with the teacher or partners. When their first drafts come out, they will start the peer review so that they can revise with the help of the teacher. After I give my final evaluation, students are required to write a reflection on what they've learnt from the writing, which I think they will benefit a lot from this experience. After class I will write my own reflection on the goal and result of the lesson so that I can adapt some part of my plan.

This on-line resource provides the teachers with new insight into both writing and teaching writing. Our Chinese teachers and students, in particular will find it a rewarding and meaningful experience in exploring this resource, though it will be a great challenge for us to adjust our traditional way of writing and teaching writing. So even if I don't teach writing in China, I really want to make an attempt to teach after I come back to China.

Acknowledgment. This work is supported by the Teaching and Researching Foundation of my university this year. Thanks to my learning experience in Singapore, I am grateful to Professor Ramona Tang in Nanyang Technological University and teachers of the university, who have helped me a lot in ushering me into the fascinating world of teaching methodology.

I am especially indebted to my husband, my colleagues and family members for their continuous support all the time. Without their encouragement, this paper could not be possible.

References

[1] Badger, R., White, G.: A process genre approach to teaching writing. ELT Journal 54(2), 155–156 (2000)

[2] Tomlinson, B.: Developing Materials for Language Teaching Continuum, illustrated edition. International Publishing Group Ltd. (2003)

[3] Coffin, C., Curry, M.J., Goodman, S., Hewings, A., Lillis, T.M., Swann, J.: Teaching Academic Writing: A Toolkit for Higher Education, pp. 33–34. Routledge, London (2003)

[4] Grabe, W.: Reading in a Second language: Moving from Theory to Practice. Cambridge University Press, New York (2009)

[5] Hyland, K.: Second Language Writing. Cambridge University press, Cambridge (2003)

[6] Zhang, L.: A Study into the Chinese English Learners Psychological Process in L2 Discursive Writing. Foreign Language Learning Theory and Practice 03 (2009)

[7] Liu, X.-Q.: "China English" Study and English Writing. Journal of Nanhua University 04 (2004)

[8] Lu, Z.H.: Discourse Analysis and the Teaching of English Writing. Shangdong Foreign Languages Journal 04 (2003)

[9] Muncie, J.: Finding a place for grammar in the EFL composition class. ELT Journal 56(2), 180–186 (2002)

[10] Wang, W., Wang, L.: L2 Writing Research in China: an Overview. Foreign Language World 03 (2004)

The Discussion of the Designing of Teaching Plan

ShuKun Liu, XiongJun Wen, JinPeng Tang, and Zhen Chen

Department of Computer, Hunan International Economics University, Chang sha, China
{liu_shukun,wenxiongjunce,spiderpong,chenzhenp}@163.com

Abstract. The duty of teacher is making an excellent teaching which can not be absent of the teaching plan. The teaching plan is the pri-condition and base of good teaching. During the process of designing the teaching plan, the teacher must make the clear necessary and the demand of the designing teaching plan and the elements of the teaching. They are the main three factors of the teaching. The technology of teaching is an important part of denoting the level of teaching of a teacher. The teacher must make sure that the character and the principle of the teaching design. This paper mainly describes the achievement of the private college in the idea of teaching, with the emphasis on the discipline, content, patterns and methods of designing.

Keywords: course, teaching plan, discussion.

1 Introduction

The course design of teaching plan before the class is a necessary step. The task of teacher is not only designing the whole teaching plan but also the teaching aim, teaching method, teaching emphasis, and teaching difficulties which is the reference of action of teaching.

The vocational education in the private college is a new education pattern. In order to make sure the quality and characters of teaching, we should make the teaching plan according to the type and the level of teaching object and the difference of major and course and aim of training. The designing of teaching plan is one of the most important contents of organizing of teaching which make the teaching must pay attention to the designing of teaching plan. Especially the young teachers who come from the company must design the proper teaching plan according to the private college.

2 The Goal of Designing Teaching Plan

Teaching plan is a work blueprint with the goal of effective teaching. Nowadays teaching plan is based on the modern teaching theory according to the characters of the teaching objects and the teaching idea of teacher owns and teaching experiences. Teaching plan is the achievement of teaching preparing the teaching content. The proper and scientific designing of teaching plan is the base of the activities of teaching practice and good teaching. In a whole word, there are five important roles of teaching plan for teaching practice.

D. Jin and S. Lin (Eds.): Advances in MSEC Vol. 2, AISC 129, pp. 623–628.

Guidance. Teaching plan is a well-designed blueprint and all the next ideas for teaching for teachers to organize and guide the teaching activities, such as goals to be achieved, tasks to be accomplished by taking various measures have been reflected in teaching programs.

System integrity. Teaching is a complex system which is composed by a variety of teaching elements. Teaching program is a combination of these various elements.

Preview. The essentially of the process of preparation of class and design the writing teaching programs is the every aspect of the actual teaching activities, each step in the minds of teachers in the rehearsal process. It enables teachers to more immersive real teaching situation, teaching activities, careful consideration of every detail, careful planning for the smooth progress of teaching and learning activities provide a reliable guarantee.

The purpose of this study is analyzing and revealing the structural features of good teaching program based on the outstanding teaching in our school program. So the excellent teaching program designed can be sum up for the majority of private college teachers to learn from each other.

3   The Definition of Teaching Plan

In short, the teaching plan that is teaching program. It is a teaching plan which is made according to teaching object, teaching content, and teaching goals. The contents of teaching plan partly is divided into class teaching plan and chapter teaching plan which is composed of teaching content, teaching hours, teaching aims ,teaching forms and teaching method and so on.

As for the teaching techniques and teaching arts which are used in teaching they can not only come from one who has prepared a plan in preparation time but also can come from a long-term teaching practice improvisation. They are one of the necessary contents using which can obtain better teaching effect. This is essential to achieve better teaching results of the content, but do not have to be written in the teaching plan.

Teaching design is a process which relates to an organization of teaching, such as teaching objectives, teaching content, learners, teaching strategies, teaching media, and a model which combined with those basic elements. There are about one hundred instructional design models now which were divided into the ADDIE model of instructional design and the ASSURE model.

The teaching mode of ADDIE includes five elements: analysis of teaching, designing, development, implement and evaluation.

ASSURE instructional design model includes six aspects :analysis of learner characteristics (Analyzelearners), statement of goals (State objectives), select media and materials (Select media and material), the use of media and materials (Utilizmedia and material), requires the learner to participate (Require learneparticipation) and evaluation and modification (Evaluate and revise).

With the development of Today's information communication technology named ICT (Information Communication Technology), more and more difficulties come up which are very big challenge to the designer of the teaching and teachers. The theory of traditional teaching design mainly based on the behavior theory, with the idea that teaching process is a factory production line. In such situation, 80% of the instructional design process will focus attention on the problem 20 % part. This linear instructional design model is clearly not suitable for e-Learning learning environment, especially mobile learning environment. Because in the information technology environment, the knowledge often means hypertext organization and presentation, knowledge can not be divided into small parts learning.

4   The Written Principle and the Designing of Teaching Method

According to our experience and studies, we have found that in order to realize the proper guidance and teaching programs teaching specification role, higher vocational education teaching programs must be designed to follow the "programmatic principles, guiding principles, the purpose of normative principles and service principles".

4.1   Principle of Program

The program principle refers to that whether it is clarifying the content, description of teaching methods, setting priorities, and teaching difficulty, the teaching program should be designed not to be too complicated, only focused on the vital. Teaching program is mainly written for teachers themselves which is a standard to teachers. Using the principle teacher can accurately grasp the whole process of teaching. Therefore, the design principle of teaching plan is made by the teachers themselves. Teaching program is the key link, the script is present.

4.2   Guiding Principles

The guiding of the teaching is limited by teaching program itself because the plan has the function of instructing the whole process of teaching. So in the process of designing teaching plan and writing teaching program must not obtain the rule that "write a thousand words, but the content if empty". In the process of designing teaching plan, the base requirement is the plan must contain true things, and the guide of the teaching plan is the base principle of the teaching. The teaching plan plays an important role in the process of teaching with the function of guide. It is necessary in our teaching. So we may do not write the script, but the teaching plan must be written.

4.3   Normative Principles

The normative principle which is mentioned here refers to that the design principle of teaching program, the content of the teaching plan and the degree of implementation of the teaching should be principle. For example, the five principles mentioned in this article should be followed; The requirements of the teaching itself such as the purpose of teaching, teaching pattern, teaching methods, teaching difficulties, teaching requirements and assessment methods are necessary. Once the teaching plan designed in strict accordance with the teaching program, the implementation of the teaching must obey the teaching plan strictly, otherwise the teaching plan will lose its relevance.

It should be noted that normative principles does not mean absolutely not be changed forever. Its design and writing should be relatively inflexible, that is, teachers can design the teaching plan flexible according to their own advantage, the contents of the special nature of teaching, professors and others. Flexibility and normative is one body which have two characters of contradiction and unity. The teaching plan should show flexibility with the normative guidance, that is seeking change in the unify. Only in this way we can reflect our teaching of artistic charm. There are many difference between our teaching examples in the last of this paper, at the same time each chapter is written teaching program is not entirely consistent.

4.4   The Principle of Service

Teaching programs must serve the purpose of teaching. The design of teaching plan must be consistent to the purpose of teaching. Those four principles are described above also serve for teaching purposes. If we do not adhere to this principle, everything else would be meaningless. Therefore, the fundamental principle of the teaching design and writing is service principles with the aim of service.

5   Examples of Teaching Programs

5.1   Teaching Plan of Software Engineering

Software engineering is an engineering major with which can guide software development and maintenance of software. With engineering concepts, principles, techniques and methods, it can combine management technology which is proved is right with development technology and method which are best technology can be obtained. This course emphasizes software engineering is an engineering major. In future courses, we will introduce the software development process in a standard pattern which has four steps that is "concepts + principle + method + tools". Students should also follow this model to study this course. Studying this course will occupy full 108 hours and four credits will be obtained.

The teaching content includes two parts:

Basics knowledge: software engineering, software engineering principles (72 hours).
Skill practice: design of software engineering course (36 hours).
Teaching Purpose: To master the software, software engineering, software process, the basic concepts of software life cycle; understand the whole process of software development
Teaching pattern: Unit centralized.
Teaching methods: " synchronized between teach and practice"
Teaching ways: teaching the basic theory and essential knowledge with multi-media teaching.

5.2   Suggestions

In a word, in order to improve the quality of teaching programs, design a scientific and rational teaching plan, improve teaching and improve teaching quality, we should know the following questions:

(1). We should learn and maintain the advantages of excellent teaching programs, continue to strength the goals of teaching and the teaching conditions in the process of designing teaching plan. We must pay attention to the teaching difficulties and emphasis in the teaching.

(2). In order to enhance the analyzing of the situation of students, we should continue to studying and exploring the theory and method of teaching and pay more attention to the design of the teaching time and the choice method of teaching method.

6   Conclusions

(1). The design and writing of teaching plan is an essential content of the work which is a base task of each teacher.

(2). In order to improve the quality of teaching and strength the teaching effect, the teaching plan must be prepared carefully. So checking the teaching plan should also be a routine task of the department master.

(3). Adhering to the normative principles of teaching programs benefit to the design and writing of teaching program and the inspection of the administration departments, designed to help teachers in the popularity of writing, is conducive to teaching department inspection, and acceptance. But we should not require the complete reunification, should be promoted under the guidance of a unified, personalized and creative.

Acknowledgments. This work is supported by scientific research fund of Hunan Provincial Education Department (**A Project Supported by Scientific Research Fund of Hunan Provincial Education Department under Grant No.11B073**) and is supported by Scientific Research Fund of Hunan International Economical University (**Project name: The research of program design of contract based on Java modeling language, No.4**) and is supported by Hunan Provincial Natural Science Foundation of China (No.10JJ6092).The work is also Supported by Project Teaching Reform Fund of Hunan Provincial Education Department(**No..Xiang Norimichi [2010] 95) and Xiang Norimichi [2009]321**).

References

1. Jiang, R.: Written of Application Paper. The Press of ShangHai Jiaotong University, ShangHai (2000)
2. Yu, W.: Secretarial Practice. Higher Education Press, Beijing (2003)
3. Wang, J.: The Curriculum Setup and Teaching Methods for Teaching Education. Curriculum Teaching Material and Method 27(1) (January 2007)

628   S. Liu et al.

4. Shi, Y.: The Mode of "the communication of teaching plan and learning plan" in Physics Teaching. Journal of Liaoning Provincial College of Communications 5(3) (September 2003)
5. Lu, F., Zheng, D.: Reasearch to the Contents, Patterns and Management of Educational Technology Training of University Teachers. Modern Educational Technology 17(3) (2007)
6. Zhang, L.: Reflection and Practice on Constructing of Practice-Teaching Pattern. Journal of LianYungang Teachers College (1) (March 2008)
7. Jin, G.: On the Method and Mode of Modern Foreign Language Course Design. Sino-US English Teaching 5(1) (January 2008) (Serial No. 49)
8. Nan, J., Yin, C.: The Characters of Excellent Teaching Plan. The Science of Education (2003)

Study of Agent-Based Simulation of Parking Management System of Intelligent Residence

Qicong Zhang

School of Information Engineering
Shandong Institute of Trade Unions' Administration Cadres
Jinan, 250001, China
zqc198002@163.com

Abstract. Based on the theory of agent and the technology of the internet of things, this paper builds a model of parking management and operation of intelligent residence. Cars and parking spaces are abstracted as different agents, and a parking management agent is also introduced. The operation of residential parking system is simulated through the interaction of different agents. In order to solve the unreasonable phenomenon of the existing residential parking, we propose a classifiable and real-time scheduling algorithm based on the dynamic adjustment. Experiment and analysis show that, the algorithm can not only improve the utilization of parking spaces effectively, but also ease the pressure of traffic and parking around the residence.

Keywords: Intelligent agent, the internet of things, residential parking model.

1 Introduction

It has become a hot problem that the increase of cars leads to the difficulty of parking in the residence. At present, most of residences take the way of parking management of fixed parking spaces. That is to say, parking spaces are sold or rent to owners. The method is simple but the rate of utilization is low. And it often causes the conflict phenomenon that one parking space is free while one car has no space to park. With the development of the Internet Things, the smart concepts such as intelligent residence and intelligent parking appear. Some scholars have achieved many results [1] [2] [3] based on this technology. The character of these researches is that the management of parking spaces is mastered entirely by the central system, and cars have no right to choose a parking space. This method is suitable for the public parking places with simple layout. But the different situation of different areas will lead to difficulties of practical implementation. And it needs specific persons to guide or navigation devices mounted in the car, both of which will make cost increase. If exotic cars have no navigation devices, this method fails. Some researchers point out that, people have individual differences in the parking choice, which will lead to parking uncertainty [4]. Therefore, the residential parking system has the characteristics of complex systems such as open and randomness. Agent-based modeling and simulation is very suitable for the research of this kind of problem. At present, scholars have done some research [5].

D. Jin and S. Lin (Eds.): Advances in MSEC Vol. 2, AISC 129, pp. 629–634.
springerlink.com

Based on the theory of agent and the technology of the internet of things, we build a model of parking management and operation of intelligent residence, composed ʹof the central control part, cars and intelligent parking spaces installed of automatic identification, control and communication device. In the model, cars and parking spaces are abstracted as different agents and a parking management agent is also introduced. The operation of residential parking system is simulated through the interaction of different agents. In order to solve the conflict phenomenon of low utilization and intensive parking spaces in the residential parking and the real problem of difficulties to park in the city, we propose a real-time scheduling algorithm based on the dynamic adjustment. The algorithm overcomes the defects of the current residential parking management methods, but also allows the cars in accordance with conditions to park. The model and algorithm are implemented in netlogo platform. Experiment and analysis show that, the algorithm can not only improve the utilization of parking spaces effectively, but also ease the pressure of traffic and parking around the residence.

2   Agent-Based Model of Parking Management and Operation of Intelligent Residence

Agent-based model of parking management and operation of intelligent residence consists of five types of agents, which are parking space agents, a parking management agent, VIP car agents, common car agents and exotic car agents, respectively. The structure of the system is shown in figure 1.

Parking space agent can receive the messages from the parking management agent and adjacent parking space agents. If the parking space agent receives parking allowance from the parking management agent, it will start the function of detecting automatically car agents. If a car agent that meets the parking condition arrives near the parking space agent and the parking space agent is the first to detect the car agent compared to other parking space agents, the parking space agent will start to unlock and send the message to the adjacent parking space agents to unlock for the parking of the car agent. If the car agent parks at this parking space agent, the parking message is sent to the adjacent parking space agents and the parking management agent while this parking space agent closes the detection. If the car agent parks at other parking space agents, this parking space agent receives the message from the adjacent parking space agents or the parking management agent. And the parking space agent locks again and closes the detection.

The parking management agent achieves the management of parking spaces through sending messages to parking space agents. And the parking management agent updates the historical records through receiving the messages from parking space agents. According to the interaction with car agents, the parking management agent identifies and classifies cars to deal with. By the history of parking spaces, the parking management agent masters the detailed information of current parking spaces. Through the statistic and analysis the history of the parking spaces and time of common car agents, the parking management agent would provide reasonable parking information.

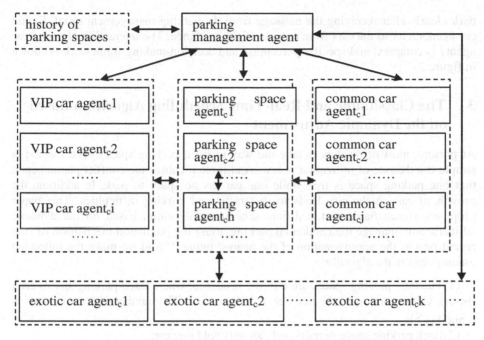

Fig. 1. The structure of the agent-based parking system of intelligent residence

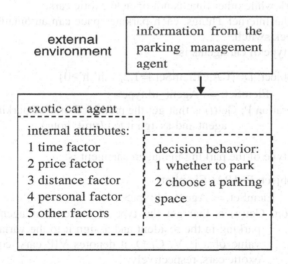

Fig. 2. Perception and decision-making framework of exotic car agent

Car agents are divided into three types, which are VIP car agents, common car agents and exotic car agents, respectively. VIP car agents have fixed parking spaces in the residence. And VIP car agents would drive directly into their own parking spaces after the simple interaction with the parking management agent. Common car agents have parking spaces but not fixed in the residence. And common car agents would park clearly after receiving the message from the parking management agent. Exotic car agents refer to the cars from outside of the residence. The interaction of exotic car agents is complex, and specific perception and decision-making framework is shown in figure 2.

3   The Classifiable and Real-Time Scheduling Algorithm Based on the Dynamic Adjustment

At present, most of residences take the way of fixed parking spaces. The method is simple but the rate of utilization is low. And it often causes the conflict phenomenon that one parking space is free while one car has no space to park. In addition, the growth of cars in the city leads to the traffic and parking difficulties. This paper proposes a classifiable and real-time scheduling algorithm based on the dynamic adjustment to manage the residential parking, using the prediction mechanism of "the recent past as the approximation of the nearest future ". And we make the following assumptions in the algorithm:

(1)The total parking spaces are h in the residence, where fixed parking spaces are i ($i<<h$). Correspondingly, the number of VIP cars is i. The number of common cars is j, and $i+j \leq h$.

(2)Each parking space permits and can only hold one car.

(3)The parking rate is low in 9:00-17:00 per day. And it is the time that allow exotic car sto park while other time is not open to exotic cars.

(4)Based on the Internet Things, each parking space can automatically detect cars and control its lock switch.

Abstract data type of car agents is:

DSP{Agent object: D={$CA_i \in$ElemSet,i=1,2,....h',h'$\geq$0} ;

ElemSet: =< Agent_id,type> ;

Basic operation P: Get(t) is that get the predicted time of parking of current car agent and assign it to t (unit: hour).

}

Abstract data type of the parking management agent is:

DSP{Agent object: P={PA$\in$ElemSet} ;

ElemSet:=< Agent_id,type> ;

Basic operation: Get(a) is that get the type of current car agent that apply for parking in the resident and assign it to the variable a. When the value of a is V, C, O, it denotes VIP cars, common cars and exotic cars, respectively.

Tim(): the time of current bus arrival (unit: hour);

Sum(): the number of current free parking spaces;

}

The core of algorithm is described as follows:

```
Service_Adjust()
{ PA.Get(a);
   if  a=V
```

then permit ingress ;
else { if a=C
then search the history of this car, provide the parking spaces information
to choose, and permit ingress;
else { if PA.Sum()>0
then CA$_i$.Get(t);
{if PA.Tim()+t<17
then choose the suitable parking space according to the specific parking time, due
to consider reserved parking spaces;}}
else provide the shortest time to wait for recent free parking spaces ; }}

4   Model Implementation and Experiment Analysis

Based on netlogo platform [6], a agent-based model of parking management and
operation is implemented in this paper. In the phase of model construction, we first
create a scene of residential parking, including parking space agents and parking
management agent. The attributes and conduct rules of these agents are also
initialized. And then, plot and monitor tools are set to track and record the relevant
data during the operation. The residential parking management system will stop at the
pre-set time.

The parameters of the simulation system are set as follows:

(1)The residence has 200 parking spaces, that is h=200;
(2)The residence has 10 fixed parking spaces, which are one-to-one correspondence
with VIP cars. That is i=10;
(3)The residence has 180 common cars, that is j=180;
(4)The arrival rate of cars is 1 per min during the residential open time to exotic cars;
(5)The residence has no restrictions of parking time. The parking time of exotic
cars is from 0.5 hours to 4 hours.

The simulation system is divided into two groups. One group takes the way of fixed
parking spaces and no open to exotic cars. The other group takes the classifiable and
real-time scheduling algorithm based on the dynamic adjustment. In accordance with
the time ratio of 1:10, we simulate the residential parking system for 24 hours.
Experiment data are recorded as follows:

Table 1. Experiment data

		fixed parking spaces	the classifiable and real-time scheduling algorithm based on the dynamic adjustment
average parking number	9:00-17:00	10	48
	0:00-24:00	21	29
average occupancy rate of parking spaces	9:00-17:00	22%	61%
	0:00-24:00	67%	87%

The experiment data show that the second group is obviously superior to the first group in the average parking number and the average occupancy rate of parking spaces. That's because exotic cars have no parking spaces to stop in 9:00-17:00 and 17: 00 to the next day 9: 00 is the same is the same. With the way of fixed car spaces, the data of whole day are better than that of 9:00-17:00. With the way of the classifiable and real-time scheduling algorithm based on the dynamic adjustment, the average parking number of whole day is decreased due to no egress of cars after 17:00. From our analysis, we can find that the classifiable and real-time scheduling algorithm based on the dynamic adjustment can basically solve the problem that one parking space is free while one car has no space to park. The algorithm can not only improve the utilization of parking spaces effectively, but also ease the pressure of traffic and parking around the residence.

5   Conclusion and Future Work

Based on the theory of agent and the technology of the internet of things, this paper builds a model of parking management and operation of intelligent residence. In order to solve the unreasonable phenomenon of the residential parking spaces management and usage, we propose a classifiable and real-time scheduling algorithm based on the dynamic adjustment. Experiment data demonstrate that the algorithm is feasible and effective. The model could not only realistically simulate the phenomenon of the residential parking system operation, but also make full use of the residential free parking spaces during low peak period under the premise that the residential owners have spaces to park. Thus, our model improves the utilization rate of parking spaces, and eases the pressure of traffic and parking around the residence. With the development of technology, how to use the technology of the Internet Things to realize the information sharing between different areas and make cars obtain the parking information of adjacent residences when the parking spaces of current residence is full, is our further work.

References

1. Oncü, S., Kirci, M.: A solution to the parallel parking problem. In: Proceedings of the 2nd International Workshop on Intelligent Vehicle Control Systems, pp. 93–98 (2008)
2. Khang, S.C., Hong, T.J., Chin, T.S., Wang, S.: Wireless mobile-based shopping mall car parking system (WMCPS). In: Proceedings - 2010 IEEE Asia-Pacific Services Computing Conference, APSCC 2010, pp. 573–577 (2010)
3. Idris, M.Y.I., Leng, Y.Y., Tamil, E.M., Noor, N.M., Razak, Z.: Car park system: A review of smart parking system and its technology. Information Technology Journal 8(2), 101–113 (2009)
4. Vaghela, V.B., Shah, D.J.: Vehicular parking space discovery with agent approach. In: International Conference and Workshop on Emerging Trends in Technology, pp. 613–617 (2011)
5. Fryklund, I.: Human factors in parking enforcement. Ergonomics in Design 8(2), 4–10 (2000)
6. Wilensky, U.: NetLogo. Center for Connected Learning and Computer-Based Modeling, Northwestern University, Evanston, IL (1999),
   http://ccl.northwestern.edu/netlogo

Multi-agent Based Supermarket Queuing Model and Optimization

Qicong Zhang

School of Information Engineering
Shandong Institute of Trade Unions' administration Cadres
Jinan, 250001, China
zqc198002@163.com

Abstract. In order to solve the problem of customers waiting too long in the supermarket, an agent-based simulation model of customer queuing system is built. The customer and the cashier are abstracted as different agents. And the cashier management agent and the guide agent are also introduced. The operation of supermarket queuing system is simulated through the interaction of different agents. A dynamic adjustment algorithm of rapid settlement is proposed in our model. Experiment data and analysis show that, our model can not only simulate the operation of supermarket queuing system actually, but also reduce the average waiting time of customers and the operation cost of supermarkets.

Keywords: Intelligent agent, supermarket queuing model, algorithm optimization.

1   Introduction

With the dramatic increase of customers, queuing phenomenon is more serious in the supermarket. Waiting too long will result in the decline of customer satisfaction. The problem is particularly prominent to the customers that purchase fewer types of shopping and could pay rapidly. If the problem can't be solved, it may cause the loss of customers and the decrease of supermarket benefits. At present, scholars have already started research through mathematical modeling method and achieved some results [1]. But the supermarket queuing problem is open and random, computing experimental method can describe this problem better [2][3]. In our paper, an agent-based simulation model of customer queuing system is built. The customer, the cashier and the guide are abstracted as different agents. The operation of supermarket queuing system is simulated through the interaction of different agents. Literature [4] denotes that nearly half of customers' queue length tolerance is 3 people, and 90% of customers are difficult to tolerate the queue length of more than 6 people. Based on this conclusion, we propose a dynamic adjustment algorithm of rapid settlement, using the prediction mechanism of "the recent past as the approximation of the nearest future ", fractionizing the shopping types and payment ways of customers and enhancing the capability of guiding and shunting customers. Experiment results show that our model can not only simulate the operation of supermarket queuing system

D. Jin and S. Lin (Eds.): Advances in MSEC Vol. 2, AISC 129, pp. 635–640.
springerlink.com   © Springer-Verlag Berlin Heidelberg 2011

actually and control the operation cost effectively, but also reduce the customer average waiting time and improve the customer satisfaction.

2   Agent-Based Model of Supermarket Queuing System

The model of supermarket queuing system firstly creates two types of agents, which are the custom agent and the cashier agent. In order to improve the efficiency of the interaction of customer agents and cashier agents, the cashier management agent and the guide agent are introduced. The structure of the system is shown in figure 1.

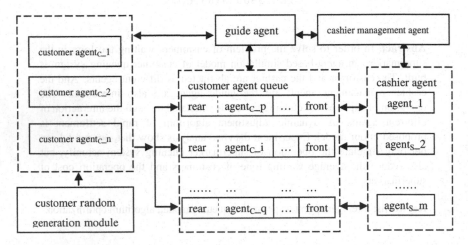

Fig. 1. The structure of agent-based supermarket queuing system

A customer agent is generated at the arrival rate λ by the customer random generation module. Meanwhile, the attributes of the customer agent are initialized, such as the generated location, the amount and types of shopping, payment ways and the personal characters and their priority weights are randomly assigned. Then, the customer agent obtains the queuing information of the queues composed of other existing customer agents, such as the length of queues and the forward speed of queues. Combined with the internal attribute that have been set the priorities, the customer agent makes a decision and implements the behavior of inserting into the rear of the chose queue. During the queuing period, the changes of the external environment information, such as the adjustment of the amount and types of cashier agents, new message provided by guide agents and other unexpected external events, will likely make the attribute values of the customer agent change. If the customer agent perceives these changes, it may adjust decision-making behavior. That is leaving the current queue and inserting into the rear of the new chose queue. The structure of specific perception and decision-making behavior of customer agents is shown in figure 2. After the interaction with a cashier agent, the customer agent will die out.

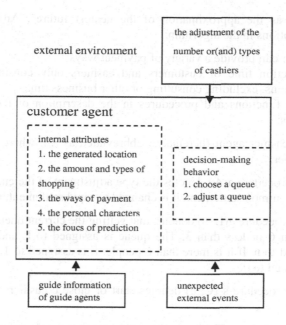

Fig. 2. The framework of customer agents' perception and decision-making behavior

Cashier management agents can obtain the messages about customer agents, which are sent by cashier agents and guide agents. After the comprehension, analysis and forecast of the messages, the cashier management agents will set or adjust the amount and types of cashier agents. And this message will be sent to the cashier agents and the guide agents.

Cashier agents are mainly responsible for the interaction with customer agents and sending this message to cashier management agents. Meanwhile, the cashier agents receive the messages sent by cashier management agents, and set and adjust the amount and types according to the contents of the messages.

Guide agents are primarily responsible to send customer agents the messages about the setting and adjustment of the amount and adjustment of cashier agents. Meanwhile, guide agents obtain the messages about customer agents and send the messages to cashier management agents. And guide agents receive the messages from cashier management agents.

3   The Dynamic Adjustment Algorithm of Rapid Settlement

The large supermarkets provide a plurality of cashiers, and each cashier can satisfy different ways of customers' payment. Most supermarkets will encounter the phenomenon of too long queue and too long waiting when the number of customers increases dramatically. However, it is particularly unreasonable for the customers that purchase fewer types of shopping and could pay rapidly. This paper proposed a dynamic adjustment algorithm of rapid settlement, using the prediction mechanism of "the recent past as the approximation of the nearest future". And we make the following assumptions in the algorithm:

1) Each cashier can provide a variety of payment ways;
2) The interaction time of customers and cashiers only consider the time of payment transactions, excluding consulting or other business time.

The concepts, functions and procedures in the description of the algorithm are defined as follows:

Definition 1: a cashier agent (hereinafter abbreviated as A_i), where i=1,2,...m,m≥0 denotes the number.

Definition 2: a procedure a(A_i) denotes the type adjustment of the current settlement. If the current is common settlement, it is changed to rapid settlement, or conversely.

Definition 3: a procedure g(L, n) denotes that getting the current queue of the waiting people more than 6 or less than 3. The queue is assigned to L and the number of people is assigned to n. If it is more than 6 people, L is assigned to 1. And if it is less than 3, L is assigned to 0.

Definition 4: a procedure s (C, c) denotes start or close c cashiers. If start, C = 1. Otherwise, C = 0.

Definition 5: a procedure t(T, t) increases or decreases t rapid settlement cashiers. If it is increase, T = 1. Otherwise, T = 0.

Definition 6: a function n(a): statistics the number of the customers that meet the condition of rapid settlement in the recent a customers.

Assumed that the system has M cashiers, and each cashier can meet all kinds of payment ways. At present, m cashiers are open and m_q cashiers are rapid settlement cashiers. The initial value of m_q is 0. The algorithm is described as follows:

Service_Adjust()
{g(L, n);
if L=1 and n>0
    then (if m+n $\leq$ M
then s(1, n)
else s(1, M-m))
        else (if L=0 and n>0
then s(0, n)) ;

$$m_a = \left\lceil \frac{n(m \times 10)}{10} \right\rceil - m_q ;$$

if m_a > 0
then (if m>m_a $\geq$ 1
then t(1, m_a)
else if m_a=m
then t(1, m-1))

else(if $m_a < 0$
then $t(0, -m_a))$;
$m_q := m_q + m_a$;
$a(A_i)$}

4   Model Implementation and Experiment Analysis

In this paper, agent-based model of supermarket queuing system is implemented with the netlogo application platform[5]. In the initialization phase of the model, a supermarket queuing scene is simulated, including: firstly it creates cashier agents and guide agents, and initializes attributes and conduction rules of these agents; secondly it creates cashier management agents responsible for receiving the messages sent by cashier agents and guide agents, and makes the decision of adjusting the number and types of cashiers. In addition, plot and monitor tools are set to track and record the relevant data in the operation. The operation of supermarket queuing system ends at the pre-set time.

The parameters of the system are set as follows:

1) the supermarket has 8 cashiers, that is M = 8;
2) the threshold of closing a cashier is less 3 people in a queue, that is $Thr._{min}=3$;
3) the threshold of opening a cashier is more 6 people in a queue, that is $Thr._{max}=6$.

Due to the different supermarket environments, the arrival rates of different types of customers are not same. In our system, the two set of experiments are simulated by setting different arrival rates of customers. Assume that the total number of customers is 200. In the first experiment, the number of customers of less types shopping or rapid payment and the number of other customers are 100 and 100, respectively. In the second experiment, the number of customers of less types shopping or rapid payment and the number of other customers are 140 and 60, respectively. The system runs 2 hours, and the experiment data are recoded in Table 1.

The experiment data show that, using the dynamic adjustment algorithm, the average waiting time of customers of less types of shopping or rapid payment in the

Table 1. Experiment data (unit: second)

		The average waiting time of customers of less types shopping or rapid payment	The average waiting time of other customers	The average waiting time of all customers
first experiment	dynamic adjustment algorithm	72	91	77
	common algorithm	83	89	86
second experiment	dynamic adjustment algorithm	56	75	61
	common algorithm	70	76	72

first experiment can be reduced greatly while the average waiting time of other customers in the second experiment can be remained stable. Although the number of other customers in the second experiment increases, the dynamic adjustment of cashiers can reduce the queuing length of other customers effectively. The analysis concludes that the dynamic adjustment algorithm not only takes care of the customers of less type shopping or rapid payment, but also takes into account the benefits of other customers.

5 Conclusion and Future Work

In this paper, an agent-based simulation model of supermarket queuing system is constructed and implemented. In order to solve the problem that long waiting time causes the decline of customers' satisfaction, we propose a dynamic adjustment algorithm of rapid settlement, which fractionizes the shopping types and payment ways of customers and enhances the capability of guiding and shunting. Experiment results show that, the model can not only simulate the operation of supermarket queuing system actually and control the operation cost effectively, but also reduce the customer average waiting time and improve the customer satisfaction.

The rapid development of electronic technology provides a new concept of operation to the traditional supermarkets, such as the self-help settlement service based on Radio frequency Identification (RFID). Therefore, how to combine new technology with the existing operation mode of supermarket under the condition of reducing the management cost, is our further work.

References

1. Feng, H.: Simulation and Optimization for Checkout Counter of Supermarket. Systems Engineering 19(2), 61–65 (2001)
2. Macal Charles, M., North Michael, J.: Agent-based modeling and simulation. In: Proceedings of the 2009 Winter Simulation Conference, WSC 2009, pp. 86–98 (2009)
3. Wilkerson Michelle, H.: Agents with attitude: Exploring coombs unfolding technique with agent-based models. International Journal of Computers for Mathematical Learning 14(1), 51–60 (2009)
4. Chen, R., Peng, L., Qin, Y.: Supermarket shopping guide system based on Internet of things. In: IET International Conference on Wireless Sensor Network 2010, IET-WSN 2010, vol. 575 CP, pp. 17–20. IET Conference Publications (2010)
5. Wilensky, U.: NetLogo. Center for Connected Learning and Computer-Based Modeling, Northwestern University, Evanston, IL (1999), http://ccl.northwestern.edu/netlogo

Exploration of Computer Specialty Enterprise Work Practices for Ordinary University

Ying Cai and Weizhen Zhou

Dept. of Computer Science and Technology,
Beijing Information Science & Technology University
Chaoyang District, Beijing, 100101, P.R. China
{ycai,zwz}@bistu.edu.cn

Abstract. In this paper, we explored the enterprise work practice including the background, related work and enterprise training program. We present the contents of enterprise work practice including goal and task, content and requirement, teaching methods, requirement time and place, employer feedback, assessment methods, Promote school-enterprise cooperation, strengthening of social practice. We also give the more detail of corporate training program. At the same time we improved concepts, formed a detail enterprise work practice curriculum program, and do it in practice. And we improved concepts, formed a teaching program do it in the 7th semester for the undergraduate in practice.

Keywords: Enterprise work practice, the practice of teaching, University-enterprise cooperation, corporate training program.

1 Introduction

The concept of school-enterprise cooperation was originated in Europe. In order to develop technical skills and improving efficient personnel They have the education system reform, and gradually introduce relevant laws and regulations, make production departments and the school fulfill their respective rights and obligations in the Europe and other developed countries [1]. Currently popular "combining with production and school, school-enterprise cooperation" model, schools and businesses not only play their respective advantages, but also cultivate the talented person that needed by community and market. And it is one of mode in training the talented person which get the benefits both universities and enterprises [2].

With the rapid development of China's economic, high-quality application-oriented talents of growing demand is increasing. The upgrading of the Information Technology(IT) industry as one of the fastest industries, requiring that their talents can absorb and learn advanced work methods and experience from the industry in time, grasp the advanced technology and hardware. It is challenge to the computer professional education for ordinary university. Therefore, as a computer science specialty of ordinary university must use all resources to enhance the overall collaboration between schools and enterprises, improve the general ability to undergraduate students in practice and computer professionals jobs overall competitiveness.

D. Jin and S. Lin (Eds.): Advances in MSEC Vol. 2, AISC 129, pp. 641–647.

2   Related Work

In 2008 we presented the curriculum program which has enterprise work practice including two sections. One is for 6 weeks and another is for 15 weeks. We introduce the contents of enterprise work practice including goal and task, content and requirement, teaching methods, requirement time and place, employer feedback, assessment methods, Promote school-enterprise cooperation, strengthening of social practice.

2.1   Goal and Task

It is an important process of teaching practice in the cultivation of computer science and technology specialty for undergraduate student. Its goal is to let the student get the actual project experience of computer science and technology [3]. At the same time they gain practical work experience by going through to the enterprise to work in practice. It creates favorable conditions for employment. It is great significant for students to broaden their horizons, develop the ability to integrate theory with practice, improve the professional quality, to enhance the vocational skills in the independent practice teaching. Corporate working practices will enable students to learn practical skills in computer application technology and experience to enhance students' perceptions of computer technology, consolidation of theoretical classroom teaching content, in-depth understanding and mastery of basic computer science concepts, principles, and methods. It gives students the real-world experience before the college graduation and lays the foundation in the future to for computer applications.

Each student completes all the required practical work submit a summary report of business according to curriculum program. The students should have engaged in the basic ability of professional work in an enterprise upon completion of this course. Our enterprise work practice is carefully designed and tailored to meet the individual needs of employers and the educational goals of student participants.

During work periods, students apply what they've learned in the classroom. The benefits of enterprise work practice both student and employer--students gain practical experience; employers gain top-quality, young, professional talent.

2.2   Contents and Requeirment

The contents should be closely integrated with computer science and technology professional fields. It can engage in the actual software and hardware product development, such as the requirements for research, requirements analysis, system design, programming, system testing, and system maintenance work. Now we give several examples as following.

(1) The students can engage in the actual product development of software and hardware.

(2) The service outsourcing skills training.

(3) The marketing of software and hardware product and market management in real-world.

(4) Information technology enterprise management.

The students who participate in the enterprises work practice should have teachers including corporation guide teacher and college supervisors. The guide teacher will be responsibility with training and teaching for the students during the practice. And at the same time they do management, check and assessment for the students. Supervising student teachers will be responsible for full supervision of the practice stage and track, urging students to complete practical training tasks, and coordinating the practice phase problems, participate in student performance assessment. Guided by experienced teachers, familiar with internships, strong sense of responsibility, there is a certain organizational and management capabilities of enterprise officer. Supervising teachers will be from the School of Computer Science teachers.

The instructor in the corporation should be combined with the specific conditions of business units, developed in conjunction with practical practice of the staff work plan and distributed to students before the practical work. The course content includes: professional guidance, organization management, scheduling, security and confidentiality, discipline and so on.

(1) Enterprises during practice, students should obey the guidance and supervision of teachers; complete the task seriously according to the practical work plan requirements.

(2) Do good notes, writing practice work diary.

(3) Submit the final report of enterprises work practice.

(4) Not participate in practice which has nothing to do with work tasks.

(5) The students should comply with the disciplinary system and lifestyle.

(6) Comply with the rules and regulations.

(7) Comply with school and enterprise safety requirements.

Your primary requirements as a participating employer are to:

(1) Provide work experiences directly related to the student's field of study.

(2) Provide appropriate supervision and related appraisal of the student.

(3) Provide assistance in helping the student achieve his/her learning objective.

2.3   The Practice Teaching Methods

We take on two ways which specify the enterprise and searched by the students. The students enroll in the former way which is designated by the college and teachers guidance. The last one the students contact the corporation by themselves in the last one. The acceptance letter will be issued by the corporation and reported to college for approval. The college teachers will be arranged to the corporation inspection, guidance, communication, help resolve the matter from time to time.

Enterprises work practice can learn practical ways of teaching specific software and hardware companies' pre-job training or job training for a variety of service outsourcing training mode.

2.4   The Time and Place Requirements

The enterprises practice arranges in the 7th semester. Students to participate in the practice of business units can have long-term partnership with the Academy's training base. Also it can be other corporation to meet the following conditions: well-known software and hardware companies, with the appropriate qualifications of software and hardware companies, technology parks, as well as internships, practical training center. Selected business units should be suitable for professional needs, attention to student internships, give students hands-on opportunity to ensure personal and property safety.

2.5 Employer Feedback

Employer feedback is a critical component of the co-op process. It helps students to grow and develop as a professional. We will provide an different evaluation form for employers to use in their appraisal process. The students' Professional Practice faculty advisor will help students learn from their performance and prepare for the future based upon this critical appraisal. The university can also learn much about its educational processes through this connection with industry and make modifications to enhance its ability to provide you with the best qualified students.

2.6 Assessment Methods

After the students finish the enterprises work practice should immediately submit the following documents to the College:

    (1) Guidance teachers and supervisors sign an opinion of the business practices of teachers work record.

    (2) Provide the enterprise work practices identified table which sign an opinion, signed the official seal. Before students leave the corporation, expert opinion given by the instructor and follow by excellent, good, medium, pass and fail results.

    (3) Provide the enterprises work practice summary report.

The student should get one of the following materials in addition to assessment by the College:

    (1) Intention of the contract or employment agreement.
    (2) Country or industry professional qualification certificate.
    (3) Training certificate issued by the agency training.
    (4) Official seal of the product applications that report or project appraisal.
    (5) Other results can reflect the actual certificate or certificates.

Students must complete all the tasks corporate work experience. Teachers in the Computer school participate in corporate working practices of each student assessment. Final test results will be based on the content of students' practice, performance, guidance and supervision of teachers, teacher observations; comprehensive business appraisal results are given. The 6 credits are available to those who pass the final results. The 6 credits are capable of reaching the "WEB application system design" with 3 credits, "professional training" with 1 credit, professional elective with2 credits.

2.7 Promote School-Enterprise Cooperation, Strengthening of Social Practice

We set up a new cooperation projects with the North China Research Institute for the practice center. We focus on the effectiveness of practice training. We made a detail practice plan about mobilization in place, carefully prepare, and strengthen management, the pursuit of effectiveness. Many students during the practice period were awarded the "black-box testing certificate" by the enterprise. Many students reflect that the practice enhance the practical ability, responsibility sense and teamwork ability and have a great harvest through the actives.

We adjust the curriculum and teaching schedule to comply with the Ministry of Education requirement more than half a year at the practice of enterprises for the engineering students. We develop program of traditional arrangements maintaining the relevance of curriculum. We increase the first six semester class hours and then we arrange students to practice job-related enterprises in order to gain practical work experience at the seventh semester. This will offer opportunities and conditions of their career.

3   Corporate Training Program

3.1   The Training Objectives

The training program's goal is for the company to develop application-oriented software industry professionals, higher education and corporate universities to find the balance between market demand. Qualified students with training will have professional software development, project development capability and professionalism of software engineers and technical level.

3.2   Corporate Training Plan

There are several aspects about corporate training plan as following.

(1) The train direction.

(2) The training period.

(3) The implementation process: In early March each year the project team set up school-enterprise cooperation, according to the company's business situation to determine the list of partner institutions, on the April the project leader from the company negotiates with partner institutions, schools, companies establish joint project team.

(4) Confidentiality and requirement for students taking this program.

(5) Company commitment: After complete business practice plan, and all outstanding assessment results, the company give job offer, directly into the company. If assessment results were good and passing the interview, the students can enter the company after graduation.

3.3   Corporate Training Program

(1) Training time: the annual summer vacation before a month

(2) Training method: basic training (1 month) + Internship (5 months);

(3) The responsible person: Human Resources

(4) Assessment methods: Students need to complete the appropriate assignment jobs which made write-protected PDF documents including review the content learned that day. The teacher will mark it after the students upload the assignments to a specified mailbox. Final score will include the assignment and final test results.

(5) Training program implementation.

3.4   The Conditions and Configuration of Instrutor

There are some hardware and software conditions, characteristics and advantages of the enterprise.

The standard for appointments of teachers:

(1) At least manager with many years experience of interns.

(2) Accomplish something in the professional field.

(3) Strong sense of responsibility, there is a management capacity.

(4) Responsible for full supervision and tracking of students, urging students to complete learning tasks, coordination problems in the teaching process, participate in student performance evaluation.

3.5   An example for the Projects

We give some projects of a corporation that can be chosen. The detail is as shown in the Table 1 including projects name and students numbers.

Table 1. Project Name and Require Student Numbers

No	Project Name	Numbers
1	Chinese word segmentation	1
2	The memory file system on Linux user mode	1
3	Specific components and component development	2
4	iPad application development	1
5	The research, development and implementation of Identity management	1
6	Text mining cloud computing platform	2
7	Internet public opinion Management System development V4	2
8	Entity (person, organization) information tracking system	2
9	Data mining software develop based on the Full-text database	2
10	WEB instant messaging system based on Java	2
11	E-mail data parsing of the and interface show	1
12	Desktop Search	1
13	TCM remote task management and storage system	1
14	Integration and use of Full media resources	2
15	Gallery Templates management	2
16	The citation analysis of medical institute	1

4   Conclusion

In this paper, we explore the enterprise work practice including the background, related work and enterprise training program. At the same time we improved concepts, formed a detail enterprise work practice curriculum program, and do it in practice. We hope that this paper can provide others can learn from the experience and assist.

Acknowledgment. The research is supported by the General program of science and technology development project of Beijing Municipal Education Commission under Grant No.KM201010772012, Funding Project for Academic Human Resources Development in Institutions of Higher Learning under the Jurisdiction of Beijing Municipality Grant No.PHR201007131.

References

1. Yuan, Z.: Universities and enterprises: a win-win strategic management concept. Technology and Management 10(1), 133–134 (2008)
2. Wang, H.J., Lu, Z.: University management staff in the school-enterprise cooperation in the role of bridge. Technology and Innovation Management 27(5), 33–35 (2006)
3. Ying, C., Zhou, W., et al.: Exploration and practice of computer speciality cultivation for ordinary university. In: Proc. Int. Conf. Comput. Sci. Educ., ICCSE 2009, pp. 1497–1501 (2009)

Quality Traceability of MES Based on RFID

Jiwei Hua[1,2], Xiaoting Li[3], Wei Xia[1], and Yang Liu[1]

[1] The College of Computer and Information Engineering,
Tianjin Normal University, 300130 Tianjin, China
[2] TEDA OrKing Hi-tech Co. LTD
[3] School of Control Science and Engineering, Hebei University of Technology
huajiwei@yeah.net

Abstract. The conception of quality traceability and the development process from easy quality sampling to overall standard quality management are introduced. A novel production quality traceability system is proposed. At the same time, the function model of quality traceability management is established by the proposed quality traceability management method based on RFID. Then, the QTIPM (Quality tracing information process model) is analyzed, which includes RFID real-time information and the business process of this system.

Keywords: RFID, quality traceability, MES (manufacturing execution system), QTIPM (Quality tracing information process model).

1 Introduction

With more and more food safety accidents and product recall accidents happened in the market, not only we are stricter and acuity on the quality of product but also do we realize the importance of product quality tracking mechanism. This requires the products to have good quality while it also ensures that when unexpected quality problem happens we could track back in time. So the quality management is one of the cores of MES while quality traceability becomes an important research direction. As the production workshop of a manufacturing enterprise materialized center during the process of production it produce management information and a lot of real time production information. Whether can we use the real time production information efficiently becomes the key of quality traceability. RFID technology is suitable for the data acquisition and process control in the industrial sites. Therefore this paper proposes the quality traceability management method that is based on RFID.

2 The Quality Tracking System of MES Based on RFID

Product quality tracking system is consist of government regulators, the third party audit services, the public quality that is traceable information service, manufacturing enterprise quality that is traceable management system. It makes the local area network that is the enterprise's own and the Internet as its transmission media support. Manufacturing enterprise quality traceable management system is the base of the

D. Jin and S. Lin (Eds.): Advances in MSEC Vol. 2, AISC 129, pp. 649–654.

whole quality traceability system. It provides fundamental quality information for the consumers, sales, upstream and downstream production enterprise. It provides reliable basis for the enterprise's own analysis the potential defects of the quality of products, control and adjusts the technical factors, raw material factors, human factors that produce the defects, achieve the quality traceability. To realize quality traceability, the enterprise must establish perfect and effective quality management mechanism and perfect records product label. Product label is the only mark during the manufacturing process that is identified and recorded. It includes some key information such as product number, batch number ,serial number and so on, and at the same time it builds the accurately and fast retrieval mechanism of these key information. Quality tracking focuses on different points during different products. Some focuses on the process, some on the products, some on the batch while some on the specific components. So the key contents of the product label record are different. Such as flow type industry, its products mainly records raw material batch number, batch number, and the process parameters, etc. When it happens any quality problems, we can trace back the design, production, packaging, transportation and the last sale's information through the product label [1-2].

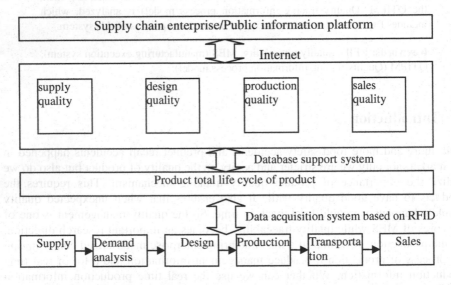

Fig. 1. Product life cycle quality tracking framework based on RFID

3   Quality Tracking Management Structure of Product Life Cycle Based on RFID

Relying on the RFID technology, we can realize the real time data acquisition to every stage of the whole life cycle. Realize overall real time share to the whole supply chain enterprise quality information through collecting, processing and arrangement .And truly to realize the goals of the quality traceability management of the whole life cycle. Because of quality information has a wide coverage, mutually relates, random distribution and complicated, publish the production quality information on time, roundly, accurately is crucial to the entire supply chain enterprises and the whole society consumers. According to the distribution characteristics of the product life cycle we can take the system into four management nodes. They are supply quality tracking management, design quality tracking management, production quality tracking management and sales quality tracking management [3-4]. Product life cycle quality tracking management structure is shown in Fig 1.

4   Quality Tracking Information Process Model Based on RFID

4.1   Product Traceability Organized Method

The production methods of discrete industries and process manufacturing are different, witch determines they are different on the formation of product quality information, flow, traceability processing. Discrete industrial product quality information is consisted of its parts quality information. This includes increase the processing quality information of each procedure. So in the quality traceability we can unfold the quality information of finished products step by step according to the BOM of product. Due to the stronger structural of the products of discrete industry, its quality tracking is simpler. And the products of process industry form the raw material to the final productive process are continuous. So its production quality information is also continuous. And the process parameters of each procedure is a great number, the production environment is complex while the effect factors of the product quality of process industry is so many. At the same time which can't be ignored is that, the process industry products produce many kinds of products after the chemical and physical transformations. So the quality issues appeared in the production process will affect many downstream products. Thus when we do quality traceability to process industrial products we need to reverse back to raw material according to the production process of the product [5-7].

4.2   Quality Tracking Function Model Based on RFID

"Person, machine, material, method, link" are the most important factors that affect the products quality. They are: personnel factor in the whole process; Machine elements of the manufacturing processing equipment; Factors of all kinds of raw materials and spare parts; Design factors in the process and environmental factors of production, storage, transport. Product quality tracking process is the process of recording and restoring all kinds of personal factors and non-human factors that affect product quality in the product manufacturing process. Now commonly used in automatic identification technology the cost of bar code and magnetic card are low, but they are both easy to wear and both have short lifespan; The price of contact IC card is higher but its data storage capacity is bigger, safer, but also easy to wear; RFID radio frequency identification tags have advantages of father read distance, no mechanical wear, long lifespan, able to be encrypted, programmable, great capacity data storage and faster storage and reading speed. At the same time it has characters of waterproof, antimagnetic, thermo stability. So RFID radio frequency identification technology is the most suitable for application in quality tracking technology in manufacturing field. According to the structure of manufacturing quality traceability and the character of RFID radio frequency identification technology, on the basis of functions we can divide the quality traceability system into five subsystems: raw materials quality management subsystem, process quality management subsystem, quality tracking subsystem, quality analysis subsystem, quality documents management subsystem [8-9]. Its function structure is shown in Fig 2.

4.3   The Quality Tracing Process Analysis Based on RFID

At present the whole society pay more and more attention to product safety problems, so it is necessary to improve enterprise quality management level and achieve the product quality traceability. Therefore, combined with the present situation of Chinese manufacturing's information integration, an integrated enterprise quality tracing system is urgent. Facing the information integration, quality tracing system need to cover the whole life cycle of products, including raw material quality information, design quality information, production process information, production, storage environment, transportation, packaging information, and the final sale and consumption user information. Faced with such large and complex information, enterprise should not only make their integration of information, but also cooperate with the upstream and downstream cooperative enterprise to make the information connection seamless [10].

The classification of quality tracing process
Quality tracing system can use RFID automatic identification technology, database technology and network technology. Pasting label on the packing of product, each product is equivalent to having the identity certificate.

When the product quality problems appear, based on stored in the electronic tags quality tracing system can tracking products' vendors, manufacturers, raw material purchase and other information, and according to all kinds of information's associated content track of quality information chain transversely, from the sale information of the product directly back to raw material information, and quickly and exactly find out the specific link of quality problems.

When quality problem appear, and through the transverse tracing find out the problems, then need to longitudinal trace with this link .In another word, specific problem specific analysis. First of all, according to the reason causing quality defect to confirm the product batch, track products inventory, and receiving department information of the same batch in real time, ensures the range of problems product. And then analyze the specific reasons causing quality defects majorly and improve quality control method, so as to improve the quality of products.

Analysis of quality tracing system information process
Due to the difference of the enterprise production process, quality tracing modes are also different. We extract the common ground of manufacturing enterprises' production process, and then constructed quality tracing system of the information process based on RFID technology. Based on RFID technology quality tracing system by information classification and flow can be divided into four layers. They are decision layer, analysis layer, execution layer and detection layer. This system's information is supported by relation database and real-time database, and through establishing the special data interface and send these messages to other information system of the enterprise, to integrate the information effectively.

Decision layer is located in the top of the system, and it is the command center of the enterprise operation, makes the enterprise information policy, examines and approves major quality problem and coordinates quality problem between the departments and conflict of interest. The main information includes document information and comprehensive information, quality decision information.

Analysis layer collections the bottom quality information together, through the special mathematics model analysis, using FMEA (Potential Failure Mode and Effects Analysis), SPC (Statistical Process Control), MSA (Measurement System Analysis) and other methods, then pass the trimmed and analyzed quality information to decision layer, to support the leadership layer for making management decisions ,controlling and prevent the production process quality. Analysis layer's information mainly includes design analysis information, process analysis information and quality cost analysis information.

Execution layer is the core of the quality tracing system. Its specific function is responsible for quality diagnosis, identify quality defects, trace quality, confirm problems range, and then go through the problem analysis, send the quality defects information, quality diagnosis, traceability information and the final analysis results to Decision layer, in order to guide quality management improvement to raise the level of

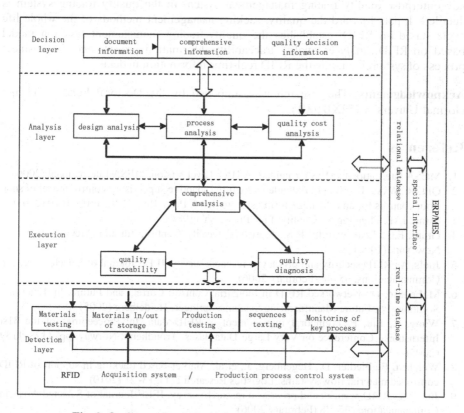

Fig. 2. Quality tracing information process model based on RFID

product quality. Through the set up quality tracing system, enterprise can solidify the key quality management process and the idea of quality control in the system, establish the enterprise-level quality management model. Execution layer includes the quality diagnosis information, quality traceability information and quality comprehensive analysis information.

Detection layer is in the bottom of the system, and it is the information basis of quality traceability system. Detection layer in the process of production take RFID technology as the core technology.

5   Summary

This paper introduces the quality tracing's development from simple quality to comprehensive standard quality management, of recent years, then illustrates the significance of quality traceability with the quality and safety frequent events around the world, and raises four aspects of the product quality tracking system which are based on the government supervision service, the third party audit services, public traceability information service quality and manufacturing enterprise quality traceability management system and introduces their respective functions. Finally the key enterprise quality tracing management system in the quality tracing system is detailed. It puts forward the quality tracking management methods of the whole life cycle based on RFID, establishes the quality tracing management function model based on RFID, analyses quality traceability information process and the business process of system including the RFID real-time information in detail.

Acknowledgments. The research was supported by the Doctoral Fund of Tianjin Normal University (52XB1001).

References

1. MESA International. MES Explained: A High Level Vision. MESA International (2006)
2. Qiu, R., Wysk, R., Xu, Q.: Extended structured adaptive supervisory control model of shop floor controls for an e-Manufacturing system. Int. J. Prod. Res. 41(8), 1605–1620 (2003)
3. Juran, J.M.: Planning for Quality. Free Press, NY (1988)
4. Grant, E.L., Leavenworth, R.S.: Statistical Quality Control, 7th edn. McGraw-Hill Inc., New York (1996)
5. Juels, A.: RFID security and privacy: a research survey. IEEE Journal of Selected Areas in Communications, 35–48 (February 2006)
6. McFarlane, D.: Networked RFID in Industrial Control: Current and Future. In: Emerging Solutions For Future Manufacturing Systems. Springer, Heidelberg (2005)
7. Wang, F., Liu, P.: Temporal management of RFID data. In: Proceedings of the 31st International Conference on Very Large Data Bases, Trondheim, Norway, pp. 1128–1139 (2005)
8. Weichert, F., Fiedler, D., Hegenberg, J., et al.: Marker-based tracking in support of RFID controlled material flow systems. Logistics Research 2(1), 13–21 (2010)
9. Juels, A.: RFID security and privacy: a research survey. IEEE Journal of Selected Areas in Communications, 35–48 (February 2006)
10. Riekki, J., Salminen, T., Alakarppa, I.: Requesting Pervasive Services by Touching RFID Tags. IEEE Pervasive Computing 5(1), 40–46 (2006)

The Study of RFID Oriented Complex-Event Processing

Jiwei Hua[1,2], Chuan Wang[3], Wei Xia[1], and Jin Zhang[4]

[1] The College of Computer and Information Engineering,
Tianjin Normal University, 300130 Tianjin, China
[2] TEDA OrKing Hi-tech Co. LTD
[3] School of Control Science and Engineering, Hebei University of Technology
[4] China Mobile Group Tianjin Co., Ltd.
huajiwei@yeah.net

Abstract. The theory of complex event processing is applied to RFID events processing. The RFID primitive event and RFID complex event are described by the proposed RFID data processing model. In order to simplify the processing, a novel RFID event processing language ROCL (RFID Oriented Complex-Event Language) is presented. At the same time, the filtration of RFID data and the matching of RFID complex event are analyzed detailed.

Keywords: RFID, complex event, MES (manufacturing execution system), ROCL(RFID Oriented Complex-Event Language).

1   Introduction

With the rapid development of RFID technology, the application range of RFID is more and more widely, the application degree is more and more deeply. Through the analysis of RFID data, the advanced RFID complex event can be extracted from simply RFID data, thus the high-level management application can be supported more efficiently. The traditional data processing method can not be applied on RFID data processing, because the features of RFID data are instability, dynamic, massive and spatial-temporal correlation. The complex event processing is an emerging technology field, which can be applied to process massive simply data, therefore the valuable events can be extracted. Therefore the theory of complex event processing is applied to process RFID events by this paper. The RFID data processing model is definitude by this paper, the RFID primitive event and RFID complex event are described by that RFID data processing model. A novel RFID event processing language ROCP is presented by this paper. At the same time, the filtration of RFID data, the matching of RFID complex event is analyzed detailed.

2   Problem Facing in RFID Data Processing

On the practical application environment of RFID, the massive dynamic data about state change of RFID tags will generated through the real-time information interaction between readers and RFID tags. Comparison with traditional data, the RFID data is

D. Jin and S. Lin (Eds.): Advances in MSEC Vol. 2, AISC 129, pp. 655–661.
springerlink.com   © Springer-Verlag Berlin Heidelberg 2011

instability, dynamic, massive and spatial-temporal correlation. The characteristics of RFID data bring great challenges to RFID practical application. Because both false negative reading and false positive reading is the characteristics of RFID data, the data cleaning is the key of pretreatment work. It is difficult to apply traditional data cleaning method on RFID data processing directly, because the RFID data is the real-time dynastic data stream. With the real-time dynastic characteristic of RFID data, the time factors and spatial-temporal characteristic should be considered fully in the RFID data cleaning to ensure the reliability and real-time RFID data processing.

After RFID data cleaning, the reliability of the primitive RFID data will be ensured basically. There is much implicit context state and related background information in the RFID data stream, which is closely related to advanced application logic. It is finally purpose that the implicit complex business application information should be extracted from the massive and simply primitive RFID data [1-2].

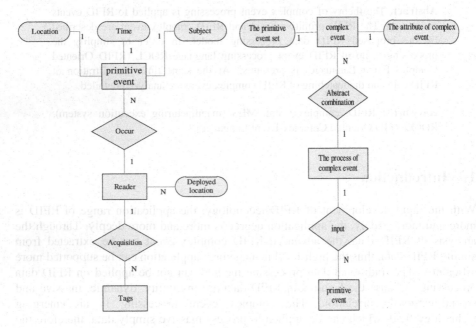

Fig. 1. The primitive event E-R and primitive event E-R of RFID event

3 RFID Oriented Complex-Event Processing System

The RFID data is collected by RFID reader from RFID application system, which is considered as RFID primitive event. The RFID primitive event is the actual event in the physical layer. Based on the event model defined by RFID application system, the RFID complex event can be derived from RFID primitive event. From the event processing perspective, the massive RFID data is firstly cleaned. After pattern matching, the RFID primitive event is converted to RFID complex event. As the output of event processing, the RFID complex event can be applied to the management system [3].

3.1   The Mechanism of Complex-Event Processing

The complex-event processing is new information search and finish technique for distributed message-based systems, it can hit and abstract various levels of advanced event according to the authorities of users. The advanced event could be the bottom processing data, or the top business decision-making information. The complex-event processing builds high-level complex events by filtration and abstract combination, processes these generated complex events as other events. Filtration, event-poset as the input, is effectively reducing the amount of events by saving meaningful events which is conformed to defined model. Abstract combination, also takes event-poset as the input, is matching them to defined model, and then output valid defined complex events. Filtration and abstract combination are working on transmission layer of the whole complex-event processing system. The system primitive event inputs from transmission layer, and then abstract combination in event processing mode which is filtrated and combined, afterwards repeats the steps of filtration and abstract combination by next layer, the output events in each layer can be display and process by analysis tools [4-5].

3.2   RFID Oriented Complex-Event Model

RFID application oriented event processing system is a processing platform which can realize the definition of filtration rules as well as pattern matching, and efficiently execute the rules and pattern; it includes systematic entirety structure and formalized processing language. The working steps of RFID application oriented event processing system are divided into three parts. First of all, abstract and finish the process logic of RFID application, it means build the event process model of RFID application using formalized representation method; Secondly, through the running of the entire system, efficiently execute the filtration rules and pattern matching; At last, digging out the implicit advanced application events from massive RFID event stream in input system. RFID oriented event processing model has an abstract description on RFID event processing action object, it extends the data model of RFID, increases the abstract description on relationship among the RFID event substances [6-7].

In RFID event model, events are divided into two shapes: primitive event and complex event, their definitions are as follows:

Definition 3.1. RFID primitive event
RFID primitive event—Event is a triple, Event=(RID, TID, Timestamp), RID refers to the mark of readers, it shows the location where event occurs; TID refers to the only mark of tags, it shows the subject of the occurring event; Timestamp refers to the time when event occurs.

RFID primitive event is the event that readers effectively get the acquisition of the information of the tags, it is the basic information source of RFID application system, and definitely reflects the time, location, subject of each primitive event, thus provide the key to basic information for composing complex event. Fig 1 shows the relationship between primitive event and substances [8-9].

Definition 3.2. RFID complex event
Complex event is a combination of primitive events; these primitive events are correlated by the operation through some specific events. RFID complex event

Composite Events = (RID, TID, Timestamp, EA, EC), RID refers to the mark of readers in each primitive event that constitute complex event, it shows the location where event occurs; TID refers to the only mark of tags in each primitive event that constitute complex event , it shows the subject of the occurring event; Timestamp refers to the time when each primitive event that constitute complex event occurs; EA refers to the attribute in each primitive event that constitute complex event, it's a quadruple, EA= (EID, T,C, V), while EID refers to events identifier which is the only identifier separate the event from others in the event set, T is the class name of the event, it's used in separating major categories by the information showed in the event, C refers to event classification, it can distinguish primitive event from complex event, V refers to the weight of the event; EC refers to the relationship among primitive events that constitute complex event [10].

The object of RFID complex event is abstract constituted by the primitive events that constitute the complex event and the events composing by the attribute of the complex event. Fig 1 describes the relationship between complex event and its substances. The primitive event set is a set of primitive events that constitute the RFID complex event, it includes the attributes of each primitive event, and the attribute of complex event is a new attribute where many primitive events compose to complex event.

Definition 3.3. Event operators

Event operators are the operators that form more complex event by related events; the operation of events is defined as primitive events use event operators repeatedly. Event operators include AND, OR, NOT, PRE-ORDER, their priority level from higher to lower is: NOT, PRE-ORDER, AND, OR.

3.3   RFID Oriented Complex-Event Description Language

Structure of RFID Oriented Complex-Event Language

ROCL (RFID Oriented Complex-Event Language) can accurately make filtration and abstract combination on RFID event stream, it can match various complex events which are formed by conditional factors such as time-based, location-based, process logic-based with primitive event stream, and give definition to actions and events triggered by successfully matching. The grammatical structure is:

```
CEP < Define CEP Name>
EVENT <event_stream expression> ::=
   { operator(event category event alias)} [ ,...n]
     [FIELTER< filter expression> ::=
         {time_filter(expression)}[,...n]
         {position_filter(expression)}[,...n ]
         {business_logic_filter(expression)}[,...n]]
      [WHERE<search_condition>]
TODO<todo_expression>
```

In the grammatical structure of complex-event description language, EVENT assigned input event stream that extract complex event from specific primitive event stream, FIELTER clean the input event stream according to the given rules, as to remove redundant data; WHERE describes the logic relationship among primitive events that constitute complex event mode, complex-event description language give birth to complex event after the input event stream mode matching followed by these rules.

RFID Event Cleaning Strategy

In RFID oriented complex-event processing system, the cleaning work of primitive event is divided into two groups. First, clean the edge layer, filtrate data delivered by each reader, and then clean the data that is after the first cleaning followed by integral process logic rules. These two steps are illustrates in details next.

Edge layer cleaning:

RFID primitive event delivered directly through the readers is a triple, E= (ReaderID, TagID, Timestamp), ReaderID refers to the message of readers, that is the location message of tags, Timestamp refers to the time message of events, for these two parts ,the system formulate the constraint rules to filtrate and clean data. There may have lots of redundant data in RFID primitive event, the reasons might be as follows:

In the short time, the same reader read the data of the same tags many times. For example, when a tag stays longer within the scope of the reader, the reader will read the information of the tag repeatedly.

More than one reader read data of the same tag. As RFID systems rely on radio signals to transmit information, therefore it's highly vulnerable to external environmental interference, such as the placement of reader antenna, tag distance from reader, In order to minimize the occurrence of leakage reading, usually set more than one reader in an important position, this will result in more than one reader simultaneously read the same tag, so the duplication of information acquisition occurs.

In regard to the problem that the same reader read the same tag, system uses Time Slide Window to set the time constraints. Suppose there are event $e_1 = (r_1, o_1, t_1)$ and $e_2 = (r_2, o_2, t_2)$, if $r_1 = r_2, o_1 = o_2$, and $|t_1 - t_2| < \delta$ (δ refers to the given time window), then event e_1 and event e_2 are each other redundant events, one of them can be clean.

Process logic cleaning:

As the RFID event have the characteristics of Space-time correlation and semantic richness, therefore, the RFID data stream implicit contains numerous information closely related to the upper application logic, after i cleaning the edge layer, summary of RFID event in the process logic develop a more accurate cleaning rules based on the actual process logic, it can be more effectively remove redundant data in order to ensure the accuracy and timing of data.

The Analysis of RFID Complex Event Mode

Complex event is a composite event that composes by operators with a group of related events based on time correlation, position correlation and logic correlation.

Complex event possesses a higher level of abstraction; it can express richer semantic information.

4   Summary

This paper analyzes the characteristics of RFID data at first, for it has the characteristics of instability, dynamic, huge amount, space-time correlation and so on, the traditional data processing methods are difficult to apply to RFID data processing in manufacturing execution systems. Secondly, this paper analyzes the problems RFID data processing are facing, and the RFID data processing method and specification at present, thus proposed the use of the theory of complex event processing to process RFID event, gives the interpretation and the definition of related models of RFID data processing, using the model, primitive event and complex event are described. And then, this paper presents a new RFID Oriented Complex-Event Language (ROCL), ROCL fully use of the space-time correlation of RFID data, accurately make filtration and abstract combination on RFID event stream at edge layer and process logic layer, it can match various complex events which are formed by conditional factors such as time-based, location-based, process logic-based with primitive event stream, and trigger the well-defined senior events after successfully matching. This paper focus on the characteristics of RFID data, according to the definition of filtration rules in Complex-Event Language and complex event mode filtrate and match the input event stream, solve the problem of further combination of complex event according to RFID complex event, it proved that the method can greatly reduce the complexity of complex event detection model.

Acknowledgments. The research was supported by the Doctoral Fund of Tianjin Normal University (52XB1001).

References

1. Palmer, M.: An Overview and History of Complex Event Processing. The Event Processing Blog (April 2007)
2. Michelson, B.M.: Event-Driven Architecture Overview. Patricia Seybold Group (February 2006)
3. Wang, F., Liu, P.: Temporal Management of RFID Data. In: Proceedings of the 31st VLDB Conference, Trondheim, Norway, pp. 1128–1139 (2005)
4. Wang, F.-s., Liu, S., Liu, P., Bai, Y.: Bridging Physical and Virtual Worlds: Complex Event Processing for RFID Data Streams. In: Ioannidis, Y., Scholl, M.H., Schmidt, J.W., Matthes, F., Hatzopoulos, M., Böhm, K., Kemper, A., Grust, T., Böhm, C. (eds.) EDBT 2006. LNCS, vol. 3896, pp. 588–607. Springer, Heidelberg (2006)
5. Wu, E., Diao, Y., Rizvi, S.: High-Performance Complex Event Processing Over Streams. In: Proceedings of the ACM SIGMOD 2006, Chicago, Illinois, pp. 407–418 (2006)
6. Wu, E., Diao, Y., Rizvi, S.: High-Performance Complex Event Processing over Streams. In: ACM SIGMOD International Conference on Management of Data, Chicago, Illinois, USA, pp. 27–29 (June 2006)

7. Gyllstrom, D., Wu, E., Chae, H.-J., Diao, Y., Stahlberg, P., Anderson, G.: SASE: Complex Event Processing over Streams. In: Biennial Conference on Innovative Data Systems Research (CIDR), Asilomar, California, USA, pp. 7–10 (January 2007)
8. Palmer, M.: Principles of efferctive RFID data management. Enterprise Systems (March 2004)
9. Wang, F.S., Liu, S.R., Liu, P.Y.: Complex RFID event processing. The International Journal on Very Large Data Bases 18(4), 913–931 (2009)
10. Zang, C.Z., Fan, Y.S., Liu, R.J.: Architecture, implementation and application of complex event processing in enterprise information systems based on RFID. Information Systems Frontiers 10(5), 543–553 (2008)

The Study of RFID-Based Manufacturing Execution System

Jiwei Hua[1,2], Yao Lu[3], Wei Xia[1], and Xiaolei Hua[4]

[1] The College of Computer and Information Engineering,
Tianjin Normal University, 300130 Tianjin, China
[2] TEDA OrKing Hi-tech Co., Ltd
[3] School of Control Science and Engineering, Hebei University of Technology
[4] Emerson Process Management (Tianjin) Valves Co., Ltd
huajiwei@yeah.net

Abstract. With adopting RFID, the MES (manufacturing execution system) gets some advantages, and brings a lot of benefits to manufacturing industry. The textile industry is taken as the study object, and it's production processing is analyzed. The MES model based on RFID and the seamless integration scheme between the RFID and MES are proposed, which provide a perfection system solution for realizing enterprise real-time informatization.

Keywords: RFID, MES (manufacturing execution system), textile enterprise, RFID middleware.

1 Introduction

Nowadays, enterprise has not satisfied to the accuracy of the information, but in-time and integration of it which is important to further promotion of enterprise management. Therefore, enterprise must integrate production information and logistic information during the production into the integral management system platform real-timely, so as to realize in-time sharing with real-time information, provide decision basis for the management, eliminate the delay of service process, response to market changes rapidly, shows the best configuration resources, make production plan, and maximize the value of various information produced from informatization construction the enterprise invest. Manufacturing execution system(MES) is production workshop oriented, workshop, as the materialized center of manufacturing enterprise, it's not only the specific executor of production plan, but also the feedback individuals of production information and the distributing centre of large amount of real-time production information. Therefore, resource management, logistic control and information integration of workshop is vital to enterprise information integration system, it is the center of enterprise informatization. RFID is a new automatic identification technology, it has the characteristics of long-distance reading, non-mechanical wear, passable of secret code, programmable, large capacity of data storage, fast speed of storage and read, and it is waterproof, antimagnetic, resistant high temper, so RFID is suitable to data acquisition and process control in industrial production scene. Rely on these characteristics, manufacturing enterprise could obtained and integrated the real-time information, connected real-time information to upper management system seamlessly, meanwhile transfer the upper management decision to the production. The paper makes a detailed analysis about technology advantage of RFID-used manufacturing execution system, take textile enterprise for example, put forward a manufacturing execution system model which is based on RFID, it provides a perfect system solution for real-time informatization [1-3].

2   The Advantage of RFID Using in Real-Time Manufacturing Execution System

During the process of production manufacturing, rely on RFID, enterprise could obtained and integrated the real-time information, connected real-time information to upper management system seamlessly, meanwhile transfer the upper management decision to the production. Using RFID, write process data, materials data, quality data and some other production data of each process during production process into the product RFID label in real-time, so enterprise could real-time master production schedule, the quantity of raw materials between processes, WIP and some other workshop production information, lean production can truly be realized in workshop by analysis these real-time information. Meanwhile, real-time integrated these dynamic production information into enterprise management system, so the management could dynamic analysis the execution situation of production plan, response to market changes rapidly, adjust production plan and purchase plan in time. And it also can realize the dynamic management of quality, provide basis for the future quality tracing. Because RFID are used in these links, the enterprise obtains the real-time information during the process of production, increases production efficiency, decreases production cost, makes production management improved, also ,these information can be transferred from upper enterprise to bottom enterprise using RFID label, sharing and using the information improves the value of real-time information. Because of RFID, the enterprise management system is more and more real time transformation, intelligent and automatic [4-5].

3   The Design of System

3.1   The Analysis of System Background

In the textile industry, especially in cotton textile industry, the production process are more like the composite of process and discrete, the production process of cotton approach to process industry, The process are showed in Fig.1.Because of different equipment ,size, fabric texture, process route are made based on fabric variety and production situation.

From this product process, it shows the production of cotton is semi-continuity processed, it can be considered as a process production in general, but there is a lot of parallel production equipment with same or different sizes. And the features are: 1)multi-computer serial; 2)multi-host parallel; 3)some products have the characteristic of reverse process and leap process; 4)varied specification in production line at the same time; 5)many production constraint (equipment capacity, exchange station time, production process etc.) ; 6)complicate production path[6-8].

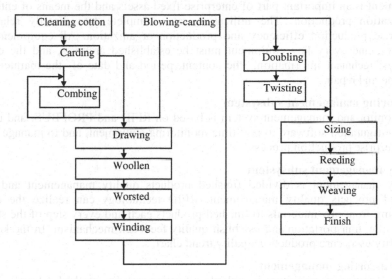

Fig. 1. The production process of cotton spinning industry

3.2 System General Structure

Manufacturing execution system are divided into three layers: production control layer, process monitor layer, field control layer. In field control layer, using PROFIBUS field bus to real-time monitor electrical equipment in production field, using RFID to real-time acquaint key data to production process, material inventory between processes and production personnel attendance. Process monitor layer mainly monitor real-time data produced in production field, real-time scheduling and optimized production [9-10]. Production control layer integrate the connect of production and management, so the manager can master the running situation in production field in time and accurately; it also integrate the seamlessly connect of control system and manage system, realize data sharing, comprehensive analyze and unified plan multiple control system, undergo deeper data digging to resource control system.

3.3 System Function Design

After the investigation of the current situation of the enterprise, with the feasibility analysis, we divide manufacture execution system into six subsystems, each subsystem specific functions as follows:

workshop management subsystem
The workshop management of the production management plays a part in the acceptance production instructions, organization production, recording the production

data and controlling the production status, providing feedback information to superior departments, ensuring the production steady and timely finishing the task role.

equipment management subsystem
Equipment is an important part of enterprise fixed assets and the means of enterprise organization production. The utilization rate, complete rate directly related to enterprise production efficiency and production organization. All equipment must register , and every key equipment must be established archives, and the content includes: technical information, the content, period and date of the maintenance , examine and repair .

monitoring management subsystem
The monitor and management system is based on RFID and PROFBUS, and use the FIX configuration software to real-time monitoring equipment, and to manage the all the enterprise production process.

quality management subsystem
Quality management is divided finished products quality management and semi-finished products quality management. RFID technology can realize the quality following from raw materials to finished products each and every step of the storage, processing, transportation,and establish quality feedback mechanism. In the key link of quality users care production quality trend chart.

cotton assorting management
Cotton assorting management is with classification queuing method. Classification is suitable spinning a product or especial yarn into a class. Lining up is making the congener of raw cotton yarn which has the basic adjacent region, nature, type and different batches arranged in a team, and using the raw cotton team by team.

planning and production scheduling
Enterprise's actual production needs to ensure the full load operation, and after the new production orders received the enterprise should stop the other varieties of products production, and arrange equipment and special equipment used to make new contract's production.

4   Integration Applications of RFID and Manufacture Execution System

RFID middleware extends enterprise middleware technology to RFID field, shield the diversity and complexity of the RFID equipment, and provide powerful support for background business system, so as to drive RFID applications more extensively and more abundantly. RFID middleware's structure of the system is shown in figure 2.The reader and writer interface process the RFID data from different types of reader and writer, get uniform pattern of the data, examine the data validation, according to user defined agreement package RFID data. At first news management put the news has been encapsulation in the message server cache, then based on the message content make these news classification and make similar news integration in the same message queue. Filtering data module will filter out repeat RFID, and filtered RFID data will be transplanted to center database. At the same time, the access data module

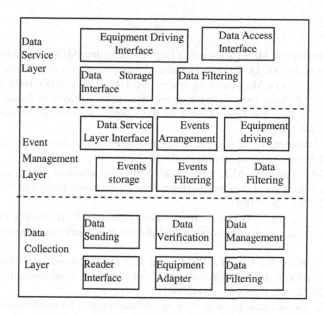

| Data Service Layer | Equipment Driving Interface | Data Access Interface | |
| | Data Storage Interface | Data Filtering | |

Fig. 2. The function model of RFID middleware

of data interface provides a interface accessing to the central database, and provides services for enterprise applications and remote applications.

5   Summary

The RFID, integrating with existing manufacturing information system, can build a more powerful integrated information chain, as well as transmit the accurate data timely , thereby to increase productivity, improve asset utilization ,improve the levels of quality control and various online measurements, provide information base for production process monitoring management and production plan optimization. Thus enterprise can achieve differential production and lean production, according to the enterprise production ability and market change, formulate the production operation plan, as well as raw materials and spare parts of auxiliary work plan. At the same time the enterprise must concretely analysis own actual situation, give consideration to both the new and the original system, avoid repeated waste, make reasonable design scheme of the system. It is more important to constantly optimize the management process, through the implementation of manufacture executive information integration system, so that from the management personnel to engineering technical personnel of the enterprise have the fundamental change about the production and management theory , improve overall quality of employees and the enterprise strength.

Acknowledgments. The research was supported by the Doctoral Fund of Tianjin Normal University (52XB1001).

References

1. MESA International. MES Explained: A High Level Vision. MESA International (2006)
2. Qiu, R., Wysk, R., Xu, Q.: Extended structured adaptive supervisory control model of shop floor controls for an e-Manufacturing system. Int. J. Prod. Res. 41(8), 1605–1620 (2003)
3. Tsai, Huang: A Real-Time Scheduling and Rescheduling System Based on RFID for Semiconductor Foundry Fabs. Journal of the Chinese Institute of Industrial Engineers 24(6), 437–445 (2007)
4. Riekki, J., Salminen, T., Alakarppa, I.: Requesting Pervasive Services by Touching RFID Tags. IEEE Pervasive Computing 5(1), 40–46 (2006)
5. Jeffery, S.R., Franklin, M.J., Garofalakis, M.: An adaptive RFID middleware for supporting metaphysical data independence. The International Journal on Very Large Data Bases 17(2), 265–289 (2008)
6. Leong, K.S., Ng, M.L., Grasso, A.R.: Synchronization of RFID readers for dense RFID reader environments. In: Applications and the Internet Workshops, January 23-27, p. 4 (2006)
7. Cheong, T., Kim, Y., Lee, Y.: REMS and RBPTS:ALE-compliant RFID Middleware Software Platform. In: 8th International Conference on Communication Technology, February 20-22, vol. I, pp. 699–704 (2006)
8. Chen, R.-S., Tu, M.: Development of an agent-based system for manufacturing control and coordination with ontology and RFID technology. Expert Systems with Applications 36(4), 7581–7593 (2009)
9. Chen, R.S., Tu, M.A., Jwo, J.S.: An RFID-based enterprise application integration framework for real-time management of dynamic manufacturing processes. The International Journal of Advanced Manufacturing Technology (March 18, 2010)
10. McFarlane, D.: Networked RFID in Industrial Control: Current and Future. In: Emerging Solutions For Future Manufacturing Systems. Springer, Heidelberg (2005)

Visual Research on Virtual Display System in Museum

Peng Shi

School of Mechanical and Vehicular Engineering
Beijing Institute of Technology
Beijing, China
Spanish2001@163.com

Abstract. With the progress of society and development of science and technology, People's standard of living rise extraordinary, make people constantly seeking for more high-grade leisure and enjoyment, also pay more attention to improvement of their own quality and literacy, which made all sorts of entertainment and leisure project feel sprit willing but flesh is weak. Through this new interpretation and design for concept museum which can take as an ideal spot, a new economic growth point, combining leisure and learning, bring merry experience for consumers. Not only bring knowledge, but release people's physical and mental pressure. This paper discusses the new concept of virtual interactive display system in the museum which appears inevitable, it integrated with multiple advantages, and is benefit to sustainable development of economy. Combining with museum own advantages, very suitable for requirement of consumer. Based on related fields of scientific and technological achievements, demonstrates the development of the virtual interactive in Museum is feasibility and practicality.

Keywords: Virtual Display system, human-computer interaction, vitual technology.

1 Introduction

Meaning of development of virtual display system in Museum—with the progress of society, the development of science and technology, people's standard of living rise extraordinary, make people constantly seeking for more high-grade leisure and enjoyment, also pay more attention to their own improvement of capability and literacy, More consumers tend to combine consume experience like leisure, entertainment activities with learning and studying, make their body and mind rest and pressure released, it can improve their own quality, ability, learn more knowledge at the same time the body get trained.

The museum is not only a show window of region characteristic and culture, but also a new economic growth point which attract tourist and bring economic improvement, no economic support, a region's cultural and characteristic hard to maintain and develop. Ordinary museum exhibition methods are mostly by text or picture, cooperate with some real items and model has been difficult to meet consumer's interest, which also has brought much inconvenience for consumers accept information. The machine-made exhibition methods will lead to consumer's antipathy. [1]

D. Jin and S. Lin (Eds.): Advances in MSEC Vol. 2, AISC 129, pp. 669–674.

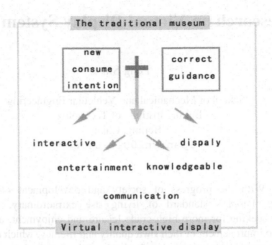

Virtual interactive display is inevitable stage in museum development—The development of science and technology brings designers infinite imaginary space, introduction of multimedia virtual interactive system in museum will keep pace with pulse of Times and steps of social development and fulfill demand of consumers. [2]Many cell phone game development like iphone's balance ball game is originated from the man-machine interactive concept, application of digital three-dimensional film, image recognition technology provide reliable technical which guarantee to various virtual interactive and multidimensional exhibition technology.

Museum exhibit methods and characteristics—The traditional museum transmit cultural knowledge, reveals certain scientific principle, technology and natural phenomenon to people through the use of words, images, object. It's have following features:

Uniqueness—the exhibits not batch production, the design activities is combined with local humanism, history and culture to undertake, choice of exhibits will usually also combined with local technology advantage, so even show the same scientific principle, its way of display will also be different.

Participation—exhibition methods have some interactivity and participation through certain media which can interactive with viewers in enjoyable experience and achieve better effect to spread knowledge and culture.

Understandable—the interactive manner in show generally simple to operate, not need to undertake special training.

Durable—hardware and interface general durable, is not easy to damage. [3]

2 Analysis of Virtual Interactive Way in Museum

Nowadays virtual interactions are widely used in all aspects of the social life, affecting people's thinking, invention, life, communication. [4] The unprecedented direct communication between People and computer is the most important features of virtual interaction. Virtual interactions are applied to more and more fields, these areas also changed under its influence and meanwhile promoting the development of virtual interaction.

Motion capture—The most famous man-machine interactive games, is the Nintendo game machine and Nintendo's Wii which invested massive manpower and material resources in visual interactive study, not only the Nintendo's Wii games were enjoyed by young and old people, but also become a kind of communication tool between parents and children. It can encourage people to conduct more physically exercise, it not only promote the progress of visual interactive technology but also bring about significant profits and influence.Based on visual interaction technology the efficiency of interactive is much higher than the traditional input methods. Also add a lot of fun and exercise experience to people through a more direct way to control your own virtual role. [5] We need a more natural way to interact with computer, virtual interactions with bring about very wide application field, just like applied to the digital entertainment field r, will bring more natural and immersive gaming experience for the users.

Technology of three-dimensional model and three-dimensional imagine scanning—fantastic scene and character in Hollywood movies not only leave deep impression to Audiences but also embodies the rapid development of computer simulation and image generation technology, technology can make the impossible possible and make does not exist become existence and has very big development space, furthermore the hardware invest is one-off, which can better save resources, space and manpower.

Technology of three-dimensional model and three-dimensional imagine scanning—fantastic scene and character in Hollywood movies not only leave deep impression to Audiences but also embodies the rapid development of computer simulation and image generation technology, technology can make the impossible possible and make does not exist become existence and has very big development space, furthermore the hardware invest is one-off, which can better save resources, space and manpower. [6]

3   Taditional Information Commuicating in Museum

Static information—the way of Static information display including: picture, Instructions brand , Propaganda album, This way is the earliest and most common form appears in museum display that can introduce basic information, the characteristic of communicate is similar to the way of books, visitors can independently obtain information by themselves

Models—models, physical displayed is more visualized compared with text, this kind of way with the characteristic of more tangible and visualize. Real objects and model can be better interest the viewer and the visual experience more deeply, the shortcoming is tourists will slightly regret, because to prevent damage most model can not be touched. [7]

Voice Introduction

record—In Voice introduction, visitors will trigger the introduction record in each fixed display position, its characteristic is machine-made, and information cannot be choice and lack of vivid.

Hierophant—Way of Commentator with the characteristics of real-time information transmission, at the same time visitors can perform simple communication and ask questions with Commentator, defect is under the influence of quality of Commentator, the information may also exist errors.

4 The Concept of Museum Virtual Display System

Virtual man-machine interactive display
Compared with the traditional museum exhibition methods, the features of virtual interactive exhibition emphasize on idea of interactive design which original from information and emotional communication between people and products. The introduction of human-computer interaction design concept to display will expand the scope of human-computer interaction and have more advantage compared with traditional museum information display. [8]

- Virtual display provides new way of display
- Virtual display provides new way of information communication
- Virtual display provides new way of interactive communication between people

Compared with the traditional exhibition methods, the biggest characteristic of virtual display is vivid and interactivity, which can make the viewer better receive information, even produces experiences and feeling of at your fingertips and Immersive. [9] You can also enhance the interaction between the human and virtual objects or Virtual Commentary, this interaction reduces the possible errors, and improve the timely of information communication ,enhance the accuracy of the information transmission, also allows the user to choice of information very well, virtual display satisfy the emotional needs of visitors.

Practice of new technology on Virtual interactive system

3-Dmodel—Through three-dimensional virtual technology, there can be a Three-dimensional Exhibits which also can be observe with various angles, the details of three-dimensional virtual Exhibits can be infinitely enlarged as long as the model allowed, place a product in the appropriate virtual Environment will increase the Realistic and exhibit effect, those effect is difficult to achieve in traditional exhibition.

Stereoscopic film technology—Stereo images camera can capture stereo images, the viewer wearing stereo glasses, can see the three-dimensional images. create a virtual scene Through computer, the viewer feel like being among in the real scene, The principle of three-dimensional film technology is to simulate different images the viewer's eyes see, through wearing stereo glasses, viewers can see images overlying, resulting in a real sense of distance, as if the object at your fingertip.

Solid User Interface—Solid user interface is an interactive interface can be perceived both visually and tangible, solid interactive facility consists of touch pad, keyboard, buttons, operating handle, with good prospects for development. Drawback is that the maintenance and investment of hardware is costly and not easy to ensure, which increasing the operating costs and implementation difficulties for Museum

Virtual User Interface—Virtual user interface is a combination of a variety of the latest technology to create the real, direct user interactive experience. The research and development of virtual user interface is based on Solid User Interface, and is more suitable for museums and other public places compared to Solid User Interface. Its technology is based on image recognition technology, combined with motion capture technology, through computer programming, can identify the location of body parts, facial expression, various human movements etc in three-dimensional space, utilize this technique to design the virtual experience for viewer, so that the viewer will feel their virtual interactive tours more realistic, nature and interesting.

Voice User Interface—Appearance of Voice user interface is introduction voice into interactive system during the man-machine interface process. Voice user interface is a useful complement to visual user interface and physical user interface, three of them constitute to the trinity of human-computer interaction. And its main form includes interactive voice, speech recognition, voice prompts, background music, the feedback sound to operation, etc. The introduction of voice interface will strengthen people's perception of the visual image which can affect people's mood, stimulate association and promote the function effectively.

5 Basic Principle of Man-Machine Interface Design in Museum

Ease to use—Principle of easy to use is let the user convenient to use, through a variety of man-machine interactive way, the method of man-machine interactive is not limited to a single operation mode, users can through various means to achieve purpose and get desired effect. Easy to use, accept and comprehend is core of human-computer interaction. Simplified human-computer interaction is easier to use, understanding, and can reduce the possibilities of wrong choice. Human short-term memory have characteristics of unreliable and limited, so in the design process we should consider limits of human brain. To the viewers, browse is more easily than operation, ideal design is understand the interactive way and correct operate the system without explanation or consideration.

Regulation and aesthetic—Excellent human-machine interface, its relationship between content and structure should clear and keep consistent with each other, the utilization of visual elements and form not only follow the necessary rules of aesthetics but should comply with the requirements of the relevant standards and regulations. The Style and decoration of Interface should be match with the content and the exhibition environment characteristics, when shaping the virtual image and interactive manner. we should pay attention to integral style of interface in display and also notice with the consistency and continuity of cultural characteristic.

Reasonableness—Design of human-computer interaction according with people's thinking habit will provide users to easily achieve ideal interactive experience. In Generally condition, the users always operate according to their own understanding and logical thinking, and relying on their daily life experience. Therefore, we should consider from the point of view of user's and their knowledge level. By comparing with the distinction between the real world and virtual world, through appropriate guidance, create effective mode of virtual human-machine interactive display.

Safety—The design of human-computer virtual interaction should be consideration to all possible problems as thorough as possible, minimize the possibility of mistakes or misunderstandings will be happen when used by users. For example, the user can make a free choice among variety of options; give some hints or Warnings, when users make irreversible or dangerous choices. ; To avoid invalid operate and the operation should be according with users' intention as much as possible, in order to enhance stability. Public interactive facilities should consider with the different ability of people, because all sorts of people will participate into the interactive activities, at the same time the overall physical condition should be grasped for the users.

6 Forecast

Practicability and realizable will determine market prospect of virtual interactive display and development of Man-machine interactive experience technology. Take Philosophy of return to nature and people-oriented design as a guide, innovate the way of interactive exhibit in Museum. Through the technology of high-quality virtual Man-machine interactive experience that could lead consumers to focus on culture, knowledge, meet the internal needs of the consumer, meanwhile, to Guide social culture to a more ideal direction.

References

[1] Bowman, D.A., Kruijff, E., Laviola, J.J., Ivan Poupyrev, L.: 3D User Interfaces Theory and Practice (2006)
[2] Apple Compute Inc. Human Interface Guidelines (1987)
[3] Levin: Computer Vision for Artists and Designers : Pedagogic Tools and Techniques for Novice Programmers. Journal of Artificial InteIllgence and Society (2005)
[4] Bolullo, R.: Interaction design: More than interface (2002)
[5] Preece, J., Rogers, Y., Sharp: Interactive design (2003)
[6] Sato, K.: Multimodal Human Computer Interaction for Crisis Management Systems. In: IEEE Workshop on Applications of Computer Vision, pp. 203–207
[7] Laurel, B.: Computers as Theatre. Addison-Wesley Publishing Co. (1993)
[8] Zhang, W., Huang, S.: The Research on Personalized Modern Distance Education System. Journal of Guangdong Radio & TV University 15, 10–13 (2006)
[9] Zhang, Z.: The Application of SOA Base on the Sharing of Online Course Resourses. A Thesis of Master Degree, Liaoning Normal University (2009)

Dynamic Public Transport Passenger Flow Forecast Based on IMM Method

Zhenliang Ma[1], Jianping Xing[1,*], Liang Gao[1], Junchen Sha[1], Yong Wu[2], and Yubing Wu[2]

[1] School of Information Science and Engineering, Shandong University, 250100 Jinan, China
[2] Center of Bus Transmit Information, Public Traffic General Company, 250100 Jinan, China
sdumzhl@hotmail.com, xingjp@sdu.edu.cn

Abstract. In this paper, an dynamic urban public transport passenger flow forecasting approach is proposed based on interact multiple model (IMM) method. The dynamic approach (DA) maximizes useful information content by assembling knowledge from correlate time sequences, and making full use of historical and real-time passenger flow data. The dynamic approach is accomplished as follows: By analyzing the source data, three correlate times sequences are constructed. The auto-regression (AR), autoregressive integrated moving average (ARIMA), seasonal ARIMA (SARIMA) models are selected to give predictions of the three correlate time sequence. The output of the dynamic IMM serves as the final prediction using the results from the three models. To assess the performance of different approaches, moving average, exponential smoothing, artificial neural network, ARIMA and the proposed dynamic approach are applied to the real passenger flow prediction. The results suggest that the DA can obtain a more accurate prediction than the other approaches.

Keywords: Urban public transport passenger flow forecast, IMM, DA.

1 Introduction

The capacity to accurately forecast passenger flow in real time is expected to facilitate the provision of public transport services suitable for the needs of bus travelers. With the development of the Intelligent Public Transport System, the dispatcher can obtain timely passenger flow information. However, it is difficult to predict passenger flow exactly and reliably applying a usual forecasting method indicating that the information of passenger flow cannot be effectively applied to real-time scheduling.

At present, there is little insight into public transportation passenger flow prediction. However, there exist many prediction methods used in the areas such as traffic flow prediction, railway passenger flow prediction etc., including regression forecasting method [1],[2] time series prediction method [3],[4], Bayesian network approach [5],

---

* Corresponding author.
Research Interests: ITS, Data Mining, WSN.

D. Jin and S. Lin (Eds.): Advances in MSEC Vol. 2, AISC 129, pp. 675–683.
springerlink.com

Jenkins forecasting method [6], neural networks approach [7],[8], Grey forecasting method [9], intelligence computation method [10],[11], SVM prediction approach [12]-[14], and Markov forecasting method [15].

However, the critical limitation of these methods is ineffective use of real-time passenger flow information. For example, in Ref [8], the output of the trained NN serves as the final prediction, though the real-time information factor is considered in the NN model input, the trained NN model is based on the historical data and it cannot change the sub-model output weights dynamically with real-time passenger flow information acquired by the dispatcher. In this case, if an unexpected event happens, the repeatable pattern of passenger flow is lost and the prediction accuracy will be affected. In addition, most of these techniques specialize in specific application scenarios, and the traditional forecasting methods always lead to inaccurate results due to the regularity and randomness of the public transportation passenger flow rate. A new method is, therefore, very much needed.

Aiming at improving the prediction accuracy, a dynamic approach for public transportation passenger flow forecasting is presented that is based on the IMM method. The dynamic approach is accomplished as follows: By analyzing the source data, three correlate times sequences are constructed. The AR, ARIMA and SARIMA models are selected to give predictions of the three correlate time sequence. The output of the dynamic IMM serves as the final prediction using the results from the three models.

2   Data Structure and Forecast Framework

In this section, we describe the dynamic approach for urban public transport passenger flow forecasting. The passenger flow data were provided by the Jinan Public Transport Corporation and were aggregated in 30-minute periods from 5:00 AM to 22:30 PM per day. The data were recorded 365 days from January 1, 2010 to December 31, 2010 and the bus route 1 of JN transit operating in three districts in Jinan is chosen as the concerned route of this study.

2.1   Correlate Time Sequence Construction

Let $p(t)$ be the 30-minute passenger flow that is recorded within the time interval $(t-1,t]$ or t for short, where $p(t)$ is the source time sequence and t is an integer. By analyzing the observed passenger flow data, it can be found that the passenger flow pattern is almost cyclical every week and that it is similar every weekday (Monday to Friday) and similar every weekend (Saturday and Sunday). Thus, three relevant time series are constructed for the dynamic approach. They are the weekly similarity time sequence $p_w(t)$, the daily similarity time sequence $p_d(t)$, and the hourly time sequence $p_h(t)$

1) $p_w(t)$ is a set consisting of passenger flow record in sequential k_1 weeks before $p(t)$ and the data will be on the same weekday or weekend

$$p_w(t) = \left\{ p(t-7\times24\times k_1), p(t-7\times24\times(k_1-1)), \cdots p(t-7\times24) \right\} \quad (1)$$

For example, to forecast the passenger flow at time interval t on a certain Tuesday, data at time interval t on the previous k_1 Tuesday are selected.

2) $p_d(t)$ is a set consisting of previous passenger flow record within the same time interval on k_2 days before $p(t)$

$$p_d(t) = \left\{ p(t-24\times k_2), p(t-24\times(k_2-1)), \cdots p(t-24) \right\} \quad (2)$$

For example, to forecast the passenger flow at time interval t on a certain day, data at time interval t on the previous k_2 days are selected.

3) $p_h(t)$ is a set consisting of previous k_3 passenger flow record before $p(t)$

$$p_h(t) = \left\{ p(t-k_3), p(t-(k_3-1)), \cdots p(t-1) \right\} \quad (3)$$

For example, to forecast the passenger flow at time interval t on a certain day, data on the previous k_3 intervals on the same day are selected.

2.2 Forecast Framework

Let $\hat{p}_w$, $\hat{p}_d$ and $\hat{p}_h$ be the forecast value that results from weekly model, daily model and hourly model for time sequence p_w, p_w and p_w, respectively. In the dynamic stage, an IMM method is used to produce the final predictions

$$\hat{p}_{HA}(t) = f_i(\hat{p}_w(t), \hat{p}_d(t), \hat{p}_h(t)) \quad (4)$$

Where $f_i(\bullet)$ is the dynamic function that is determined by the IMM method, $i = 1,2,3,\cdots N$. The function adjusts itself dynamicly accoding to the observed real-time passenger flow data.

Finally, the process to produce the forecast value $\hat{p}_{HA}(t)$ by the dynamic approach is implemented in the following manner (see Fig. 1). Using the historical data to construct the time sequence $p_w(t)$, $p_d(t)$ and $p_h(t)$, and then selecting the AR, SARIMA and ARIMA models to produce the forecast value $\hat{p}_w(t)$, $\hat{p}_d(t)$ and $\hat{p}_h(t)$, respectively. In the dynamic stage, an IMM method is used to produce the final forecast value.

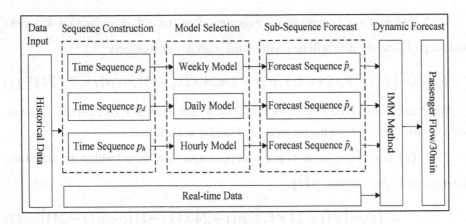

Fig. 1. Framework of the dynamic approach

3   Formulation of DA for Time Sequence Forecast

\In this section, we introduce the model for each time sequence forecast and formulate IMM method for data hybrid strategy. x_t denotes the actual value at period t and $\hat{x}_{t+1}$ is the forecast value for the next period.

3.1   Individual Sub-models

1)   *AR Model:* An AR model of order k is computed by

$$\hat{x}_{t+1} = \varphi_{1t}x_t + \varphi_{2t}x_{t-1} + \cdots + \varphi_{kt}x_{t-k+1} + \varepsilon_{t+1} \tag{5}$$

Where k is the mumber of terms in the AR and ε_t is the white noise sequence which represents the random error at time t.

2)   *ARIMA Model:* A general ARIMA model of order (r, d, s) can be written as

$$\phi(B)\nabla^d x_t = \theta(B)\varepsilon_t \tag{6}$$

Where B is a backward-shift operator defined by $Bx_t = x_{t-1}$ and $\nabla^d = (1-B)^d$, and d is the order of diffencing. $\phi(B)$ and $\theta(B)$ are the AR and MA operators of order r and s, respectively, which are defined as

$$\phi(B) = 1 - \phi_1 B - \phi_2 B^2 - \cdots - \phi_r B^r, \theta(B) = 1 - \theta_1 B - \theta_2 B^2 - \cdots - \theta_s B^s$$

Where $\phi_i(i = 1, 2, \cdots, r)$ are the AR coefficients and $\phi_j(j = 1, 2, \cdots, s)$ are the MA coefficients.

3) *SARIMA Model:* A general SARIMA model of order $(r,d,s)(R,D,S)_w$ can be written as

$$\phi(B)\Phi(B^w)\nabla^d\nabla^D_w x_t = \theta(B)\Theta(B^w)\varepsilon_t \tag{7}$$

Where $\nabla^D_w = (1-B^w)^D$, D is the order of seasonal diffencing and w is the time sequence period. $\Phi(B^w)$ and $\Theta(B^w)$ are the seasonal AR and seasonal MA operators of order R and S, respectively, which are defined as

$$\Phi(B) = 1-\Phi_1 B-\Phi_2 B^2 -\cdots-\Phi_R B^R, \Theta(B) = 1-\Theta_1 B-\Theta_2 B^2 -\cdots-\Theta_S B^S$$

Where $\Phi_I(I=1,2,\cdots,R)$ are the seasonal AR coefficients and $\Theta_J(J=1,2,\cdots,S)$ are the seasonal MA coefficients.

3.2 DA Based on IMM

Consider the system with a certain model set $M = \{m_i \mid i = 1,2,\cdots,m\}$

$$x(k+1) = F_i(k)x(k)+G_i(k)w_i(k) \tag{8}$$

$$z(k) = H_i(k)x(k)+v_i(k) \tag{9}$$

where $x(k)$ is the state vector and $z(k)$ is the observation process. The subscript i denotes the model. The matrix functions $F_i(\bullet)$, $G_i(\bullet)$ and $H_i(\bullet)$ are known. $w_i(k)$ and $v_i(k)$ are model-dependent process noise and model-dependent measurement noise with covariance $Q_i(k)$ and $R_i(k)$, respectively.

Let M_j^k denote the model j at time k and π_{ij} denote the transition probability from model i to model j. μ_i is the probability of model i.

Step 1: Calculating mixed initial probability for filter matched to model M_j^k

$$\mu_{i|j}(k \mid k) = \frac{\pi_{ij}\mu_i^{k-1}}{\overline{c}_j} \tag{10}$$

where $\overline{c}_j = \sum_{i=1}^{m}\pi_{ij}\mu_i(k-1)$ is the normalization constant.

Step 2: calculating the mixed initial state and corresponding covariance for the filter matched to model M_j^k

$$\hat{x}_{0j}(k \mid k) = \sum_{i=1}^{m}\mu_{i|j}(k \mid k)\hat{x}_i^{k-1} \tag{11}$$

$$P_{0j}(k \mid k) = \sum_{i=1}^{m} \mu_{i|j}(k \mid k) \left\{ P_i^{k-1} + \left[\hat{x}_i^{k-1} - \hat{x}_{0j}(k \mid k) \right] \times \left[\hat{x}_i^{k-1} - \hat{x}_{0j}(k \mid k) \right]^T \right\} \quad (12)$$

where $\hat{x}_i^{k-1}$ is the estimation of state based on the i th Kalman filter at time $k-1$, and the corresponding covariance is P_i^{k-1}.

Step 3: Filtering .

The Kalman filter produce outputs $\hat{x}_j^k$ and P_j^k

Step 4: Combining the state estimations and corresponding covariance according to the updated weights

$$\hat{x}_I(k) = \sum_{j=1}^{m} \mu_j^k \hat{x}_j^k \quad (13)$$

$$P_I(k) = \sum_{j=1}^{m} \mu_j^k \left\{ P_j^k + \left[\hat{x}_j^k - \hat{x}_I(k) \right] \times \left[\hat{x}_j^k - \hat{x}_I(k) \right]^T \right\} \quad (14)$$

The update weight of model M_j^k is $\quad \mu_j^k \triangleq P\{M_j^k \mid Z^k\} = \dfrac{1}{c} \Lambda_j(k) \overline{c}_j \quad (15)$

$$\Lambda_j(k) = N\left(\left(z_j(k) - \hat{z}_j(k \mid k-1) \right) \mid 0, S^j(k) \right), c = \sum_{j=1}^{m} \Lambda_j(k) \overline{c}_j \quad (16)$$

4 Experiments and Analysis

Several single-source models, including the ES, ARIMA and NN models are applied to time sequence $p_h(t)$. The ARIMA model is the same as that in the DA approach. We compare their performance on the same testing sets together with the proposed DA model.

Table 1. Comparison of MAE and RMSE among the forecast methods

Forecast Method	Testing set 1		Testing set 2	
	MAE	RMSE	MAE	RMSE
ES	182.04	252.40	113.43	147.20
ARIMA	127.39	198.63	95.80	127.39
NN	95.35	125.97	99.40	123.41
DA	**34.08**	**45.13**	**40.06**	**53.97**

Note that all of the single-source models used for comparison are employed to produce prediction for hourly time sequence $p_h(t)$. We choose the coefficients by observing the best fitting or forecasting and evaluate their prediction using the same test set. As the hourly patterns of weekday and weekend are different, there are two testing sets. Testing set 1 is the passenger flow on Wednesday, November 10, 2010. Testing set 2 is the passenger flow on Sunday, November 14, 2010. The training sets for the above models cover the data from January 1, 2010 to December 31, 2010. The validation set for the NN model is the observed passenger flow record on November 3 and 7, 2010.

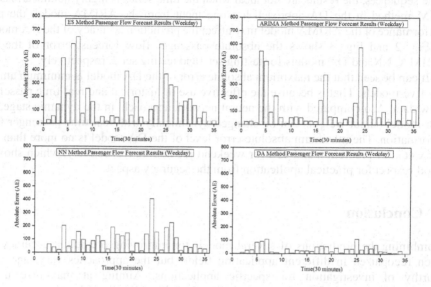

Fig. 2. Absolute passenger flow forecast error for testing set 1

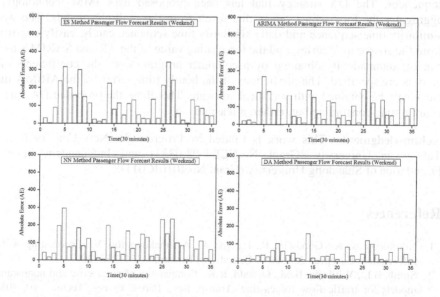

Fig. 3. Absolute passenger flow forecast error for testing set 2

Table 1 shows the MAE and RMSR of the forecast methods (ES, ARIMA, NN and DA) using two testing sets.

From the table, it can be seen that the predictions that result from the DA model are better than the predictions that results from the ES, ARIMA and NN models. The predictions that results from the NN model are better than the predictions that result from the ES and ARIMA models. Although the ARIMA models are quite flexible in many time sequences, the results are not ideal when the time series is highly nonlinear. As the IMM method in the DA approach has one input from the ARIMA model, the poor performance of the ARIMA model may affect the prediction accuracy of the DA model.

Fig. 2 and Fig. 3 shows the absolute passenger flow forecast error of the ES, ARIMA, NN and DA models for testing set 1and testing set 2, respectively.

It can be seen that the maximum absolute error of the DA model is minimum among the five models. That is because the effective use of historical and real-time passenger flow data. And compared with the neural network approach, in the dynamic stage, the sub-model output weights can be changed dynamically with real-time passenger flow information. The maximum absolute error level of the DA model is no more than 100 for weekday prediction and 150 for weekend prediction, respectively, which shows a good respect for practical application from the accuracy aspect.

5   Conclusion

Combining the predictions of several models for improving prediction accuracy has been recognized in different application fields, but the approaches vary, and thus, worthy of investigation for specific applications. Aiming at maximize useful information content of the source passenger flow data, three correlate time sequences have been constructed, and three models have been selected to forecast these sequences. The DA strategy that has been proposed uses IMM technology and aggregates the forecasting values that result from the multiple models. The weekly similarity time sequence and daily similarity time sequence can be easily constructed from the source time series, and the forecasting value of the AR and SARIMA models can be automatically obtained by a computer program once the coefficients of the models are specified. The predictions of the hourly time series by the ARIMA model can easily be obtained with statistical software. Therefore, the proposed DA approach is not a time-consuming but, rather, a feasible job.

Acknowledgments. This work is funded by Program for New Century Excellent Talents in University (Grant No.NCET-08-0333) and Independent Innovation Foundation of Shandong University (Grant No.2010JC011).

References

1. Nikolopoulos, K., Goodwin, P., Patelis, A., Assimakopoulos, V.: Forecasting with cue information. European Journal of Operational Research, 354–368 (2007)
2. Smith, B.L., Williams, B.M., Oswald, R.K.: Comparison of parametric and nonparametric models for traffic flow forecasting. Transp. Res., Part C-Emerg. Technol. 10, 303–321 (2002)

3. Williams, B.M.: Multivariate vehicular traffic flow prediction: Evaluation of ARIMAX modeling. Transp. Res. Rec. 1776, 194–200 (2001)
4. Williams, B.M., Hoel, L.A.: Modeling and forecasting vehicular traffic flow as a seasonal ARIMA process: Theoretical basis and empirical results. J. Transp. Eng. 129, 664–672 (2003)
5. Sun, S., Zhang, C., Yu, G.: A Bayesian network approach to traffic flow forecasting. IEEE Trans. Intell. Transp. Syst. 7, 124–131 (2006)
6. Lu, Y., AbouRizk, S.M.: Automated Box–Jenkins forecasting modeling. Automation in Construction, 547–558 (2009)
7. Dia, H.: An object-oriented neural network approach to short-term traffic forecasting. Eur. J. Oper. Res. 131, 253–261 (2001)
8. Yin, H.B., Wong, S.C., Xu, J.M., Wong, C.K.: Urban traffic flow prediction using a fuzzy-neural approach. Transp. Res., Part C–Emerg. Technol. 10, 85–98 (2002)
9. Rafiul Hassan, M.: A combination of hidden Markov model and fuzzy model for stock market forecasting. Neurocomputing 72, 3439–3446 (2009)
10. Chen, H., Grant-Muller, S.: Use of sequential learning for short-term traffic flow forecasting. Transp. Res., Part C–Emerg. Technol. 9, 319–336 (2001)
11. Huisken, G.: Soft-computing techniques applied to short-term traffic flow forecasting. Syst. Anal. Model. Simul. 43, 165–173 (2003)
12. Castro-Neto, M., Jeong, Y.-S., Jeong, M.-K., Han, L.D.: Online-SVR for short-term traffic flow prediction under typical and atypical traffic conditions. Expert Systems with Applications 36, 6164–6173 (2009)
13. Chen, Q., Li, W., Zhao, J.: The use of LS-SVM for short-term passenger flow prediction. Transport 26(1), 5–10 (2011)
14. Chen, X.: Railway Passenger Volume Forecasting Based on Support Vector Machine and Genetic Algorithm. In: Qi, L. (ed.) FCC 2009. CCIS, vol. 34, pp. 282–284. Springer, Heidelberg (2009)
15. Rafiul Hassan, M.: A combination of hidden Markov model and fuzzy model for stock market forecasting. Neurocomputing 72, 3439–3446 (2009)

Teaching Reformation on Course of Information Theory and Source Coding

Xin-qiang Wang and Guo-xiong Xu

Zhongyuan University of Technology
450000 Zhengzhou, China
dqxwxq@sina.com

Abstract. With the rapid development of science and technology, tele-communications technology in particular, more and more universities attach importance to information theory in all communications or telecommunications majors, usually as a professional theory course. Since the course has more abstract concepts, strong theories, complex mathematical calculations, it is often difficult for teachers and students to express and understand in the teaching process, so motivation to learn is often not high. In response to this widespread phenomenon, the article sums up some common problems easily arised in teaching process with own reality, and proposes reformation on teaching content, teaching methods and other aspects, the results show that the good effect of teaching can be achieved.

Keywords: information theory, teaching content, teaching methods, teaching reformation.

1 Introduction

With the rapid development of information technology and digital technology, how to achieve information transmission efficiency, reliability and security in the communication process becomes the focus of our research, in particular the encoding and decoding technology is applied widely in communications, network and multimedia environment, so just information theory of which these technologies is as the main content becomes more and more important. Meanwhile, the domestic colleges and universities used to offer this course only at the graduate level before and now the information theory and coding is offered at the undergraduate. As a very complex course, it involves extensive knowledge and is widely applied in various disciplines in modern society. The course is based on mathematics, probability theory and stochastic process, with mathematical knowledge principles of encoding and decoding technology is as the main object of study on digital communication systems, a lot of its basic theories are used in communications, computer science, broadcasting, information security and other fields, and even physics, biochemistry, literature and art. Course elements include the Shannon information theory on information entropy, channel capacity, information rate distortion function, the Shannon three coding theorems and some specific sources, channel coding methods. At the undergraduate

D. Jin and S. Lin (Eds.): Advances in MSEC Vol. 2, AISC 129, pp. 685–689.
springerlink.com   © Springer-Verlag Berlin Heidelberg 2011

this course is usually arranged 48 lessons, the theories are based, supplemented by computer simulation, the students generally feel that the concepts are abstract, difficult to master the basic theory, mathematical derivation is too complicated, these have serious impact on student's interest in learning, thus to affect the teaching effect of course. In this paper, the problems of "Information Theory and Source Coding" course in undergraduate teaching process is studied, with the actual situation of the unit and the author's own teaching experience, teaching content, teaching methods and teaching focus are discussed[1].

2   Main Issues and Analysis

2.1   Characteristics of Intense Theory and More Mathematics

"Information Theory and Source Coding" has characteristics of intense theory and more mathematics. The discipline of information theory is established as a symbol of the "Mathematical Theory of Communication" published by Shannon, so it can be said to be from the disciplines of mathematics, its theorems and conclusions are based on rigorous mathematical derivation, in the teaching process many mathematical proofs and derivations are also inevitable, it is very difficult to grasp for the most of undergraduate students. And these mathematics include the probability theory, advanced mathematics, linear algebra or modern algebra, etc, but many students have more or less difficulty in learning these courses, such as learning content is not complete, some simply have no studied, and some even learned mathematical formulas are now blurred and so on. These problems have a direct impact on the study of information theory for the fear to mathematical[2].

2.2   Concepts and Theorems Are Too Abstract and Hard to Understand.

"Information Theory and Source Coding" of concepts and theorems are too abstract and difficult to understand. In this discipline, many of the basic concepts and theories have been applied in daily life, but they become a section of arcane concepts and theorems when they are described in word into the textbook, such as the definition on the amount of information. The common understanding is that the events which happen frequently or normally should have a large amount of information carried as the message, and those events which infrequently or impossibly happen do not make sense to us and should not have any information of course. This understanding is contrary to the calculation of information defined by Shannon .Moreover there are the degree of channel doubt, the degree of channel distribution, three coding theorem and so on. They have students find it difficult to understand.

2.3   Symbolic Terms and Parameters Are Not Easy to Distinguish

"Information Theory and Source Coding" has many new concepts and physical quantities, symbolic terms and parameters are not easy to distinguish. Such as the source entropy $H(X)$ has unit of bits per source symbol, and the discrete extended sources have an average symbol entropy $H_N(X)$, the encoded code has a corresponding entropy has unit of bits per code symbol in the Shannon coding theorem and so on.

These definitions themselves as well as their physical meaning, physical units are not well understood, particularly confusing, The beginners are hard to master difference of these definitions[3].

2.4 Practical Is Less, Means of Experiment Is Single

In the teaching process of information theory, practical is less, means of experiment is single. Currently, most schools have theory lectures as main, very few experimental lectures, almost no special laboratory equipment. And compared with other specialized courses with many experiments, information theory itself is very abstract and more intuitive, more students are not interested[4].

2.5 Coding Techniques Develop Relatively Quickly

Coding techniques in information theory develop relatively quickly. A lot of coding techniques which are efficient, reliable and suitable for different environment and requirements of the distortion are constantly updated, how teachers deal with updated knowledge is to solve the issue in the teaching process.

3   Reformation of Teaching Content

3.1 The Allocation of Teaching Cycles

According to requirements of different major, a reasonable choice of teaching content and organization of teaching cycles is important. As the author is mainly faced by the communication engineering undergraduate classes, the teaching guide is to culture communications engineer who is suitable for the development of new communication technologies and focus on the practical ability, so encoding and decoding techniques have practical significance for most of the engineering undergraduates. In addition, information theory coding is divided into two pieces: source coding, channel coding. Channel coding techniques require the basis of modern algebra, undergraduate students have not learned it, which need to specially arrange cycles to learn in advance. Therefore, when we develop a specific teaching program, the information theory and source coding allocate 30 cycles as a part, channel coding allocate separately 30-44 cycles as an additional part . Here we investigate the information theory and source coding.

3.2 The Choice of Teaching Content

Because content of information theory can be expanded and should pay more attention to the basics at the undergraduate stage when choose the content, so emphasis are placed on basic concepts, principles, methods, highlight the basic, typical knowledge points. Information theory includes mainly: source and amount of information, channel and channel capacity, information rate-distortion function as well as three Shannon coding theorem. In this process, given the current trends of technology and actual situation of students, we highlight the information entropy of discrete memoryless source and discrete stationary source, discrete memoryless channels capacity, rate-distortion function of discrete memoryless source. For nondiscrete source, we emphasize how to discretize nondiscrete source for processing. Source coding is based on encoder and the related concept , students should master mainly the simple variable length coding --- Huffman, Shannon, fano coding, and several improved practical coding --- run-length, arithmetic, predictive and transform coding[5].

4 Reformation of Teaching Methods

- In introduction of this course we emphasize students should pay attention to characteristics and focus of the course, application and development trend of information theory and coding currently, it should enable students to learn the significance of this course. Such as information theory, of application in the rapid communication, data compression, information security, automatic control and so on. Through brief introduction to CELP audio coding technology in 2G, SMV encoding in 3G and computer video encoding technology ---JPEG, MPEG and so on, students can have a more intuitive acquaintance with information theory and coding techniques initially, so as to greatly increase their interests in learning.

- According to the situation of relevant mathematical knowledge grasped by students, we arrange lessons that help students review relevant tmathematics and enable students to be proficient in the relevant content

- To combine the course content with communication theory, computer networks, mobile communications and other courses, as in the teaching of limited-distortion coding combine with 13 line and 15 line of PCM coding, in communication principle channel capacity can combine with calculations on mobile channel and fiber channel capacity. So that allow students to link the professional knowledge in context, to be more conducive to understand theory and application of information theory further[6].

- Due to the existence of a lot of mathematical derivations in this course, the complexity derivation often becomes the focus of students, and that the conclusions, understanding and application of the theorem should be the focus is often overlooked. Therefore, we should note that in addition to explaining in detail the necessary, representative mathematical derivation in teaching, the physical meaning and practical applications of derived conclusions should be emphasized. Such as the rate-distortion function itself is quite abstract, we teach its physical meaning through combining with common examples in daily life. The same music stored in MP3, the WAV format and MP3 format, the different capacity and sound quality, the reason lies in different distortion D of two formats, that output rate R(D) are different under different distortion limits. In this example, students can be allowed to clearly understand that the seemingly boring, useless mathematical definitions and formulas learned are not useless, just the opposite the latest, most practical technologies are based on these formulas, no theory can not grasp these technologies. Thus student 's interest in learning is greatly improved. In addition, when teach the amount of information, it can help increase the impressive and effective interest in the definition through making students themselves determine the size of the amount of information on two sentences of "Yangtze River is freshwater "and" Yangtze River is saltwater".

- Take full advantage of multimedia teaching, adopt teaching methods of multimedia teaching complementary to writing on the blackboard. It's easy to use ppt to show some textual, graphic, icons and animation, while writing on the blackboard is better to the derivation of some formulas and conclusions, it's interactive and students can easily understand.
- To increase designing experiments to improve students' interest in learning, develop practical ability and creative cogitation by using matlab or systeview simulation.
- In teaching process, we used the more common symbols in information theory materials currently, and keep the continuity to facilitate student understanding. At the same time, select some representative, comprehensive exercises to enable students to solve by using learned knowledge synthetically, so as to consolidate knowledge and deepen understanding.

5 Conclusion

According to characteristics of information theory and source coding in the undergraduate, analyzes the main problems on courses, propose measures for improving teaching effectiveness from allocation of cycles, teaching content, teaching methods and organization, etc. Through years of teaching practice shows that these measures play a significant role to stimulate the enthusiasm of students, improve teaching effectiveness.

Acknowledgment. Thanks for being supported by the fund of teaching reformation project of Zhongyuan University of Technology.

References

1. Deng, J.-X.: Teaching Reformation on the Course of Information Theory and Coding. Journal of EEE 29(2), 51–54 (2007)
2. McEliece, R.J.: Information Theory and Coding. Publishing House of Electronics Industry, BeiJing (2007)
3. Zhang, Z.-Y., Zhang, B.: Application of MATLAB in Experiment Teaching of the Theory of Information and Coding. Laboratory Science 19(03), 109–112 (2010)
4. Feng, G., Lin, Q.-W.: Information Theory and Coding Technology. TsingHua University Press, BeiJing (2007)
5. Li, Z.-Q., Pan, L.-B.: The Discussion of Teaching Study for Information Theory and Coding. Journal of Inner Mongolia University for Nationalities, Natural Sciences 10(04), 460–461 (2008)
6. Liu, Y.-R., Li, X.: The Teaching Theory and Experiment of Information Theory and Coding. China Computer & Communication 20(09), 130–131 (2008)

A Distributed Multicast Protocol with Location-Aware for Mobile Ad-Hoc Networks

Tzu-Chiang Chiang[1,*], Jia-Lin Chang[2], and Shih-Wei Lin[1]

[1] Depart. of Information Management, Tunghai University, Taiwan, R.O.C.
[2] Department of Transportation Technology and Management, Kainan University, Taiwan
steve312kimo@thu.edu.tw

Abstract. Mobile ad-hoc networks mean a group of mobile devices which communicate to each other without infrastructure assistant in wireless networks. Since the nodes are mobile, the network topology may change rapidly and unpredictably; therefore, broadcast storm problem becomes a critical issue for stable and energy-saving routings. We propose a Location-based Distributed Clustering Algorithm (LDCA), which utilizes the advantages of region partition with clustering multicast method. LDCA separates several partitions in area and searches the cluster header for each partition which utilizes the routing protocol of mobile devices and transports packets by each cluster header.

Keywords: Mobile ad-hoc networks, MANET, DCA, MAODV, Multicast.

1 Introduction

Mobile ad hoc networks can connect nodes and transmit messages immediately, so that it can save a lot of construction time and financial costs. MANET routing architecture generally can be divided into flat structure and the hierarchical structure [1-3]. In the flat architecture shown in Figure 1, all nodes within the region are playing an identical role which requires the functions of routing and discovers other nodes with maintenance.

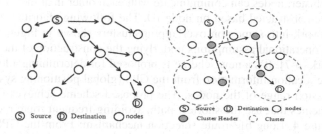

Fig. 1. A Flat Architecture for MANETs   **Fig. 2.** Routing of Hierarchical Architecture

---

* Corresponding author.

D. Jin and S. Lin (Eds.): Advances in MSEC Vol. 2, AISC 129, pp. 691–697.
springerlink.com   © Springer-Verlag Berlin Heidelberg 2011

Flat architecture is suitable for simple network topology, for it can construct in a short time and cost fewer resources. However, if used in a sophisticated network topology, it would cause broadcast flooding easily and paralyze the entire network in the worst case. In the hierarchical architecture shown in Figure 2, all nodes will be divided into groups of cluster according to the difference of the algorithm. In each cluster, it will select a node as the cluster-leader, before different hierarchical cluster nodes transmit packets, they must communicate with their own cluster-header first then the cluster-header will transmit the packet into the destination node. In most common situation, each node in the hierarchical architecture just keeps up its cluster-header with the survival of the nodes. Without broadcasting into all nodes and waste network resources, there will be a better performance in a sophisticated network topology. On the contrary, hierarchical architecture cost great amount of time and resources than flat architecture in the beginning of the construction. In multicast, many scholars had proposed the routing algorithm, for example, Multicast Ad Hoc On-demand Distance Vector Routing (MAODV), MZRP, ODMRP, DCMP and Distributed clustering algorithm (DCA), etc., all of them are the refinement of wireless multicast routing protocol. The present study uses the existing DCA and MAODV as the basis to reinforce the original routing protocol algorithm, hoping to increase the performance of multicast and broadcast in MANET.

2   Related Works

The traditional wireless network users access the Internet service through mobile device such as mobile phone, PDA and laptop to connect the base stations constructed by the Internet service provider. However, under the construction of MANET and the conception of ubiquitous computing, the movement of mobile node changes the network topology. In an MANET, the routing and resource management are used a distributed manner so that all nodes are able to coordinate and communicate with each other. The routing protocol of MANET has significant difference with wired network protocol; under this circumstance, the past studies had proposed many solutions [4]. In the end, we will consider the broadcast flooding problem which bring out by the packets, such as redundant rebroadcast, contention and collision. Many localization schemes have been proposed in the past few years. For the cluster algorithm, we have so far assumed that transmission power is fixed and is uniform across the network. Within each cluster, nodes can communicate with each other in at most two hops. The clusters can be constructed based on node ID. The following algorithm partitions the multi hop network into some non-overlapping clusters shown in Figure 3. We make the following operational assumptions underlying the construction of the algorithm in a radio network [3-7]. It's a new scheme is proposed for determining a long-life route based on node location information from the GPS (global positioning system) and the radio transmission range of the nodes. The proposed scheme achieves a more stable route than schemes that use the shortest path, avoiding frequent route reconstructions shown in Figure 4. Long-life route selection mechanism: From the GPS, each node obtains its own location information (latitude (s) and longitude (y)), but not that of other nodes. In addition, it is assumed that each node knows its current radio transmission range. Using this information, each node estimates how long the route will last until route failure. A source generates a route discovery packet that contains source position information and its current transmission range [8-15]. This packet is propagated to all neighboring nodes as in other source-initiated on-demand routing protocols.

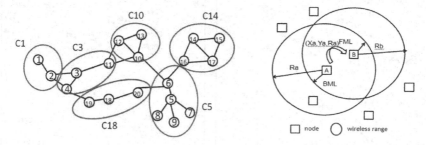

Fig. 3. LDCA Algorithm   Fig. 4. FML and BML of Long-Life Scheme

3 Location-Based Distributed Clustering Algorithm

The present study uses location distributed and multicast cluster algorithm as the base to establish Location-based Distributed clustering Algorithm (LDCA). We use LDCA to distribute different amounts of partition which measured of area in a rectangle area and search for the cluster-header in each partition. In the end, we use the way of MAODV to search, establish, and maintain the route. First, we set a rectangle area with all nodes in the range and all the nodes know their own coordinates. The start coordinate of the rectangle is (0, 0) and the end is (x, y). So the central point of this rectangle area is (x/2, y/2). Secondly, we distributed the rectangle into several partitions, each partition is 100m*100m and it has their own id. We are trying to find out the node which has the shortest distance from the central point in the each partition. The node then will be set as the cluster-header.

LDCA Algorithm in Practice - In fig 5, it is a 400 square meter open area distributed with 80 wireless nodes. Each node in this area has their unique id and knows their coordinate by the GPS.

Location Distribute - First, we marked the left-top corner as the starting point, the coordinate is marked (0,0). The right-bottom corner is the ending point, marked as (x , y), therefore we can know the central point of the partition is (0+x/2 , 0+y/2). In this case, we define the length of the side in each partition is 100 meters, so we will get a (x/100*y/100) distributed sub-partition. Each node in the sub-partition will use the GPS to compute and get its own coordinate. We used (xi/100)+(yi/100)(x/100) to compute the belonging sub-partition of each node. The (xi , yi) is the coordinate of the node and x is the length of the x axle in the partition. Fig 5 is a 400 square meter partition which divided with 100 meters. This partition will distributed into 16 sub-partitions, each sub-partition represent a cluster. All the nodes in the cluster will have their own coordinate location; nodes can know their partition through their coordinate. For example, if the coordinate of the node is (132, 257), the coordinate will be located at 1+2*4=9, which is the 9th partition.

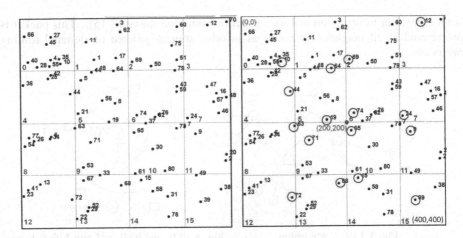

Fig. 5. Distributed 4*4 partitions   **Fig. 6.** Cluster Header Selection in each partition

Establish Cluster Headers - For pages other than the first page, start at the top of the page, and continue in double-column format. The two columns on the last page should be as close to equal length as possible. Secondly, we will compare the distance between each node and the central point. We will measure the distance between two nodes first and then broadcast out the distance with the central point to other nodes around itself. While other nodes receive the packets, they will compare the partition location of the two nodes. If the nodes were not in the same partition, the packet will be abandoned; on the other hand, if they were in the same partition, the nodes will compare which one was closer to the central point and find out the closest node to the center point through the above procedure. The closest node then will be the cluster-header in the sub-partition, it will announce message to other nodes in that partition. After other nodes receive the message, they will join the group. As shown in fig 6, node 25 in partition 4 will deliver its message (id, (x,y,), partition_id) to other nodes, after other neighboring node 42, 44, 36, 28, 55, 10, 4, 35 and 40 received, they will see whether their partition_id is coordinate with node 25. Node 42, 44 and 36 are in the same partition as node 25, so they will keep this packet, other nodes which are not in the same partition will abandon this packet. After that, node 42, 44, and 36 will compare with node 25 the distance to the central point (200,200). Node 44 will inform other nodes that it has the closest distance to the central point. Eventually, node 44 will repeat this procedure until the cluster-header in that partition is selected. The other nodes in that partition will join in and become a group.

Path Establish and Packet Delivery - Footnotes should be Times New Roman 9-point, and justified to the full width of the column. After the cluster-header was selected in each partition, if transmit packet to the node in other partition was need, the node have to transmit the packet to the cluster-header in its partition. As the cluster-header received the source node packets, it will transmit to the destination node. Before the route was established, we will search the destination node through the following procedure. When the source node needs to deliver packet to the destination node, it will broadcast the packet of RREQ to its cluster-header, the cluster-header will broadcast out to look for the destination node. After RREQ reach the destination node, the destination node will deliver RREP packet to reply the source node. Both RREQ and RREP packets will be delivered by the cluster-header. Each cluster-header on the routing table will note down the transmitting routing information. The information gathered by RREQ and RREP and other routing information will be preserved in routing table. When the source node wants to transmit a packet to a certain destination node, it will ask its cluster-header whether there is a routing information first. The cluster-header will check its routing table to determine whether there is a route to reach the destination node exists. If the path is existed, it will relay transmit the packet from next cluster-header to the destination node. If there is no path in the routing table, the source node has to broadcast RREQ packet to the nearby cluster-header. The RREQ packet includes source IP address, destination node IP address, and the unique id of the node.

4 Simulations

The present study simulated an experiment to discuss whether the size of the region partition and the number of cluster-header will affect the performance of Internet transition in region partition. We separated the partition size into un-partition, 2*2 partition and 4*4 partition types. Also, the protocol of wireless network was using MAODV, the area was 800m*800m and the transition range of wireless nodes was 150m, whereas the numbers of nodes were 50. In addition, the present study used network simulator– NS2 to simulate the packet delay, packet jitter and the throughput in different partition size during packets transition. Figure 7 is the result of packet delay. At first, under the unpartition situation, the more partition is, the higher packet delay will be. However, after the cluster header was confirmed, the transfer of the nodes become faster, the packet delay will gradually become lower and the packet delay area of the unpartion area will become higher instead. The figure 8 is the packet jitter. The packet jitter was lower than the other two situations in the 4*4 location distributed. But under the un-partition and the 2*2 partition situation, which the partition is too small, there is no significant difference in the figure. Table 1 is the throughput of the partition in different sizes and there is no significant difference. In the 4*4 partition, the average speed is slightly higher, while in the peak speed, the un-partition becomes higher. Overall, throughput has little effect to partition.

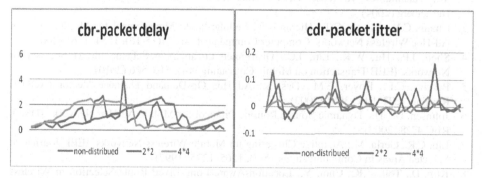

Fig. 7. The comparisons of packet delay in different partitions

Fig. 8. The comparisons of packet jitter in different partitions

Table 1. The comparisons of the throughput in different partition

Partition / Throughput	Un-distributed	2*2	4*4
Average rate	14.68KB	13.17KB	14.83KB
Peak rate	20.26KB	19.22KB	18.7KB

5 Discussions and Conclusions

The study implies a way of how to distribute each sub-partition efficiently in a huge range of partition environment and uses the sub-partition as cluster to find out the cluster header in each cluster. Each partition node will use the cluster header to transfer packet in order to transfer and receive the packet more efficiently. So that the transfer efficiency in the huge range partition of MANET will be improved and therefore reduce the waste of network bandwidth and network risk caused by unnecessary broadcast. From the experimental data in the study, 4*4 partition has a more effective outcome than un-partition in aspect of more nodes and longer transfers time. However, when it comes to 2*2 partition, there is no significant improved. The further research can consider the difference of transfer range size between nodes and the condition of mobile nodes.

Acknowledgments. This work was supported by the National Science Council of Republic of China under grant NSC 100-2221-E-029 -016.

References

1. Royer, E.M., Toh, C.K.: A review of current routing protocols for ad hoc mobile wireless networks. IEEE Personal Communications 6(2), 46–55 (1999)
2. Papadopoulos, A.A., McCann, J.A.: Towards the Design of an Energy-Efficient, Location-Aware Routing Protocol for Mobile, Ad-hoc Sensor Networks. In: Galindo, F., Takizawa, M., Traunmüller, R. (eds.) DEXA 2004. LNCS, vol. 3180, pp. 705–709. Springer, Heidelberg (2004)
3. Chang, C.Y., Chang, C.T.: Hierarchical Cellular-based Management for Mobile Hosts in Ad-Hoc Wireless Networks. Computer Communications 24(15-16), 1554–1567 (2001)
4. Shen, J.P., Hu, W.K., Lin, J.C.: Distributed Localization Scheme for Mobile Sensor Networks. IEEE Transaction on Mobile Computing 9(4), 516–526 (2010)
5. Perkins, C.E., Royer, E.M., Das, S.: Ad Hoc On-Demand Distance Vector (AODV) Routing, IETF RFC 3561 (2003)
6. Johnson, D.: The Dynamic Source Routing Protocol for Mobile Ad Hoc Networks (DSR), RFC 4728 (2003)
7. Lin, C.R., Gerla, M.: Adaptive Clustering for Mobile Wireless Networks. IEEE Journal on Selected Areas in Communications 15(7), 1265–1275 (1997)
8. Kim, D., Toh, C.K., Choi, Y.: Location-Aware Long-Lived Route Selection in Wireless Ad Hoc Network. In: IEEE Vehicular Technology Conference, vol. 4, pp. 1914–1919 (2000)

9. Khan, F.A., Song, W.C.: A Location-Aware Zone-Based Routing Protocol for Mobile Ad hoc Networks. In: The Joint International Conference on Optical Internet and Next Generation Network, COIN-NGNCON (2006)

10. Royer, E.M., Perkins, C.E.: Multicast Operation of the Ad Hoc On-Demand Distance Vector Routing Protocol. In: Proceedings of the 5th ACM/IEEE International Conference on Mobile Computing and Networking, pp. 207–218 (1999)

11. Xiang, X., Wang, X., Yang, Y.: Stateless Multicasting in Mobile Ad Hoc Networks. IEEE Transactions on Computers 59(8) (2010)

12. Royer, E.M., Perkins, C.E.: Ad Hoc On-Demand Distance Vector Routing. In: IEEE Workshop on Mobile Computing System, pp. 90–100 (1999)

13. Cheng, C.T., Tse, C.K., Lau, C.M.: A Bio-Inspired Coverage-Aware Scheduling Scheme for Wireless Sensor Networks. In: Vehicular Technology Conference, pp. 223–227 (2008)

14. Liao, W.H., Sheu, J.P., Tseng, Y.C.: GRID: A Fully Location-Aware Routing Protocol for Mobile Ad Hoc Networks. Telecommunication System 18(1-3), 37–60 (2001)

15. Rohini, S., Indumathi, K.: Consistent Cluster Maintenance Using Probability Based Adaptive Invoked Weighted Clustering Algorithm in MANETs. In: National Conference on Innovations in Emerging Technology, pp. 37–42 (2011)

Exploring of New Path for Job Opportunities with New Model in IT Education

Hongbo Zhao, Chen Wang, and Dejun Tang

Dalian Neusoft Institute of Information, Dalian 116023, China
{zhaohongbo,wangchen,tangdejun}@neusoft.edu.cn

Abstract. SOVO which is the abbreviation of Student Office & Venture Office is a brand-new educational theory and process introduced by Dalian Neusoft Institute of Information. With eight years' experiences in both theory research and practice, an outstanding methodology has been created for the purpose of educating students in software industry. Based on the revolution of education in TOPCARES-CDIO in our school, a series of changes have been made to the original SOVO management and process. It helps to emphasis the idea of combining IT trainings with practical activities in real life, and finally figure out an interactive education method between IT education and IT industry.

Keywords: SOVO, Entrepreneur education, TOPCARES-CDIO.

1 Introduction

With the development of higher education, university education in this country has turned from the phase of "elitist education" into "public education". Low training quality in professionalism field for students will drag their career training period and increase working pressure. In these cases, self-employment seems to be a "better" option for student after graduation. Believing "education brings the best out of students", Students Enterprise Center (SOVO) was founded in 2002 by Dalian Neusoft Institute of Information. By applying the "Virtual company" model, center helps students not only consolidate their knowledge from books, but also enable them to practice in reality to entirely understand the working process in the modern enterprises. The aims are cultivating students in their creativities, practical skills, team-work spirits and self-management skills, which will make students very competitive in the current job market.

The revolution of TOPCARES-CDIO is actually the international project industrial education revolution plan developed on the TOPCARES-CDIO theory. It includes the TOPCARES-CDIO theory itself and relevant research objectives, realizing and assessing standards and a series of methodologies in planning, designing and implementing and reviewing, as well as utilizing resources. As part of our education processes, it involves more students get the chance to work with the "virtual companies" and practice in the real company environment. In order to achieve SOVO targets, SOVO applies the revolutionary theories of TOPCARES-CDIO industry in converting the original management structures and process system, and adopts some

D. Jin and S. Lin (Eds.): Advances in MSEC Vol. 2, AISC 129, pp. 699–704.
springerlink.com © Springer-Verlag Berlin Heidelberg 2011

beneficial trials and explorations into education in Dalian Neusoft Institute of Information.

2   SOVO Guidance in Operation Process– Adopting TOPCARES-CDIO Industry Education Model and Methodology

The concept of TOPCARES-CDIO education model means, working through the whole process of product manufacturing and system in planning, designing, implementing and going production, students are able to combine theory with practice voluntarily and improve their study and research skills. Their science and IT knowledge, self-learning ability, communication and team-working skills as well as abilities in developing and inventing of products in corporations in reality can be improved significantly.

In SOVO, it is emphasized to improve the comprehensive sides of students, including opening minds, creativities, professionalism, communications and team-works skills, great attitudes and habits as well as capable of being responsible. SOVO innovated the original management model and process system. Through the corporation of the virtual companies, it benefits students by making them even more competitive and helping students to make good preparations of their careers.

3   SOVO Management Model – Building Professional Virtual Companies, Simulating Work Process in Companies, Managing by Departments and Institution, Emphasizing the Comprehensive Quality Improvement.

"V-Company method" in SOVO allow students from SOVO to build virtual companies in different domains. Students manage the companies and corporate the business internally in their spare time. Except for the registered legal body corporation part, all management works are processed following real companies' rules. To expand the range of students joining the SOVO virtual companies and effectively utilize the resource of our school, SOVO is now operated by second-level department and institution in Dalian Neusoft Institute of Information.

There is a SOVO management platform in institution, and all the virtual company management department, system management department and comprehensive operation department are set up upon this platform. SOVO management platform provides the strategy management and services in perspective of the whole institution. The departments manage the virtual companies in every department. Based on the goals of their own department education, they can arrange students to build virtual companies in different domains while providing operational and technical guidance and supports.

From involving in the works in virtual companies, students can turn their knowledge in theory to the practice skills in reality effectively. In Virtual companies, they also nominate a general manager, who can establish various departments and arrange positions according to the requirements and peculiarities of demand.

In different companies, students can join in and work for virtual companies to greatly improve their teamwork and skills and develop their professional skills. There are total 34 virtual companies at the moment, which ranges from website design, embedded system design, graphic design, business, English and Japanese. And there are 1000 official employees in virtual companies.

To simulate a more realistic corporation and management environment and let students feel the fierce competitions in reality, SOVO calculate the costs of virtual companies regularly and review the performance of the companies. It helps students to gain experience in management and become an actual platform in cultivation and practice of students' capabilities.

The comprehensive system can ensure the operation of SOVO effectively. Since founded, a serial of standard documents have been released including "SOVO employee management guidelines", "SOVO mentor guidance quality review system", "SOVO performance review system", and many other documents, such as "SOVO mentor standard guidance".

4   SOVO Students Interview Model – Adopting Corporation Process, Combination Study and Practice, and Improving the Responsibility and Career Attitude of Students

To improve the responsibility, career attitude and professional habits for students, SOVO has been operated strictly under standard operation of real companies. Students need to be interviewed and employed to join in SOVO. The hiring process is strictly followed with the process in real companies. Every employee should sign a confidential contract and will be reviewed for the performance every half of a year. Therefore any employee who cannot pass the performance review will be sacked from the virtual companies. Every official SOVO employee should guide two interns working in the virtual company. And all the interns must pass the training phase to become official SOVO employees. In each virtual company, automatic attendance system is widely adopted. The weekly attendance reports will be published weekly. Meanwhile, the employee files are also archived. Employee who wishes to alter his job position should fill in form to request for the change.

Every employee working in the virtual companies should strictly follow the rules in "SOVO Employee Management System" and "SOVO Employee working behaviors policy". The complete information system is set up in SOVO. All the reservations for meeting rooms, training rooms are required to be made from the system. In conclusion, all the essentials are provided in the SOVO process.

The strict promotion and management systems help to ensure the SOVO employees work hard, help them understand the performance management and enterprise culture better in real companies and cultivate their responsibilities and career attitude in career.

5 SOVO Cultivation Model – Cultivating Comprehensive IT Students, Improve the Expertise Skill and Comprehensive Abilities for Students

To truly improve students' technical skills, comprehensive abilities and competitions in career, SOVO arranges a serial of trainings in IT technology, project management, enterprise management, sales and financing, and invite famous entrepreneurs all over the world to give presentations and sessions to students regularly. Many successful entrepreneurs have been invited by SOVO to present for our SOVO employees. According to the requirements of actual projects, specific trainings divided by topics are also arranged for students. Individual's formal survey will be collected to insure the expected results achieved afterwards.

It is estimated that almost 300 training sessions have been arranged and totally 20000 employees have attended during the last 8 years, on the average of once a week. Furthermore, considering the requirements of projects and companies, hundreds of various videos have been made available for students.

6 SOVO Project Development Model – PM Quality Management Process, Which Cultivates Students the Theory of Planning, Designing, and Implementation Skills and Quality Ideas

To greatly cultivate the abilities in planning, designing and implementation for students, SOVO encourages students to get involved with projects based on ISO9001, 2000 quality ensuring system and CMM related specifications. All the projects follow the PM standards strictly. The scientific and standard processes of project management help students to understand the latest IT specifications in project managements through involving projects in SOVO. Through working in the projects, students accumulate large amount of experience in development and management of the project, and acquire the latest technical skills required by IT corporations. And they can improve significantly in software development skills and management skills, become mature in career. In the meanwhile, it can also help students to find their weaknesses in their theory studies, thus inspire them concentrate on their studies in class, and work harder to pass examinations and achieve good marks. It also helps students contacting with society, facing companies and making exact assessment and selecting suitable projects for themselves.

In additional, from the detailed work in real projects, it accumulates students to enrich their experience in career, and shorten the periods of training in entry of enterprises. Students need to write complete project documents, understand the entire working process of project development and implementation, which helps students to cumulate their professional abilities and skills. Through working in real project, students can convert their knowledge in theory to practice skills in reality.

SOVO has now completed the website development for Dalian small-medium size companies' service system. And total number of websites in development and maintenance has reached over 100.

7 SOVO Mentor Mechanism – Recording in Project, Encouraging by Mentors and Improving the Open Minded and Creation Abilities for Students

To improve the open minded and creativity for students and avoid students make mistakes during projects, every virtual company is assigned a professional mentor and entrepreneur mentor to guide students through the whole process of project by SOVO. Each mentor is asked to attend scheduled meeting; the creation mentor is required to report project status every month so that mentors can fully grasp the whole status for each team members and can do their guidance works better. All the mentors are selected from the senior technical members, who are very experienced in development and research. Mentors always keep in touch with team members, and support their difficulties in techniques with meeting scheduled every month. They provide guidance in both technical design and management skills. The SOVO mentor system supplies strong assurance in guidance, so that to keep the success of operation and the improvement of their practice skills.

8 SOVO Generator Machine – Build the Create Atmospheres, Supply Service for Students

Graduation generator is another important characteristic in SOVO model. SOVO Generator Company can supply a serial of supports to the real companies build by graduates. It encourages graduates to create companies growing up quickly and better. The SOVO model helps students make good preparations to create their enterprises during the period of study, and accumulate strong functions for their future creation. And in SOVO, a serial of generator company mechanism, can strongly support graduates the create companies in both hardware and software. It can reduce the chance of failures in the beginning of their company's creation, and help to build their companies up in the real markets.

From the graduates, there are totally six real companies generated by SOVO. The founding of students creation club, make an essential of build a platform for project in ordering, supporting and creation experience. On the other hand, SOVO also contact with the creation and working departments in province and civic, to win more chances and supports from government. To largely improve the competition of occupation for students, SOVO engages in helping students to obtain employment in high quality and creation companies with full preparation. With the effort from many sides, in the April of 2011, SOVO delegating with Dalian Neusoft Institute of Information, signed the contact and join in the "cultivation plan" with the students creation center in Sha He Kou District in Dalian. The contact aims to build a platform for graduates who desire to create enterprise in ordering projects, supporting, building up inspiration mechanism and creation experience. We can believe that with the sign of the contract, it will definitely make a prompt for the work of obtaining employment and creation corporations for Dalian Neusoft Institute of Information.

9 Conclusion

SOVO cultivation model has been around for about eight years. And it has been widely acknowledged by entrepreneurs and students. From the beginning of SOVO, there are totally more than 4000 students joined SOVO. With the higher pressure in job market, the occupation rate of graduations from SOVO reached to 100% till now. And with the high competition, many students joined in top 100 worldwide companies, such as GE, Citibank, Accenture.

The purpose of the issue aims not only to emphasize the importance of education in enterprising, but also to arise the enthusiasm from entrepreneur, and explore the improvement of education in job opportunities with new model and path, so that to promote the education level in enterprising and keep the improvement continually.

Reference

1. Crawley, E., Malmqvist, J., Ostlund, S., Brodeur, D.: Rethinking Engineering Education, the CDIO Approach. Springer, Heidelberg (2007)

Author Index